Biotechnology and Safety Assessment

THIRD EDITION

Biotechnology and Safety Assessment

THIRD EDITION

Edited by

John A. Thomas

University of Texas Health Science Center
San Antonio, Texas

Roy L. Fuchs

Monsanto Company
St. Louis, Missouri

Amsterdam Boston London New York Oxford Paris
San Diego San Francisco Singapore Sydney Tokyo

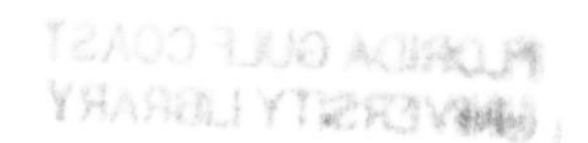

This book is printed on acid-free paper.

Academic Press
An imprint of Elsevier Science.
525 B Street, Suite 1900, San Diego, California 92101–4495, USA
http://www.academicpress.com

Academic Press
84 Theobalds Road, London, WC 1 8RR, UK
http://www.academicpress.com

Library of Congress Catalog Card Number: 2002101526

International Standard Book Number: 0–12–688721–7

PRINTED IN THE UNITED STATES OF AMERICA
02 03 04 05 06 07 MM 9 8 7 6 5 4 3 2 1

Contents

Chapter 5
Plant Biotechnology Products with Direct Consumer Benefits

Maureen A. Mackey and Roy L. Fuchs

Chapter 6
Animal Feeds from Crops Derived through Biotechnology: Farm Animal Performance and Safety

Marjorie A. Faust and Barbara P. Glenn

Chapter 7
Preclinical Immunotoxicology Assessment of Cytokine Therapeutics

Robert V. House

Chapter 10
Biotherapeutics: Current Status and Future Directions

James E. Talmadge

Chapter 11
Food Allergy Assessment for Products Derived through Plant Biotechnology

Steve L. Taylor and Susan L. Hefle

Chapter 12
Biotechnology: Safety Evaluation of Biotherapeutics and Agribiotechnology Products

John A. Thomas

Contributors

Numbers in parenthesis indicate page numbers on which authors contributions begin.

Gary A. Bannon (p. 1) Department of Biogeochemistry & Molecular Biology, University of Arkansas for Medical Sciences, Little Rock, Arkansas 72205. E-mail: bannongarya@uams.edu

Detlef Bartsch (p. 13) Department of Biology, Aachen University of Technology, Aachen, Germany. E-mail: bartsch@rwth-aachen.de

George A. Burdock (p. 39) Burdock Group, Vero Beach, Florida 32962. E-mail: gburdock@burdockgroup.com.

Bruce M. Chassy (p. 87) College of Agricultural, Consumer, and Environmental Sciences, University of Illinois at Urbana-Champaign, Urbana, Illinois 61801. E-mail: b-chassy@uiuc.edu

Marjorie A. Faust (p. 143) Iowa State University, Ames, Iowa 50011. E-mail: mafaust@iastate.edu

Roy L. Fuchs (pps. 117, 435) Monsanto Company, St. Louis, Missouri 63198. E-mail: roy.l.fuchs@monsanto.com

Barbara P. Glenn (p. 143) Federation of Animal Science Societies, Bethesda, Maryland 20814. E-mail: bglenn@faseb.org

Richard E. Goodman (p. 435) Monsanto Company, St. Louis, Missouri 63198. E-mail: richard.e.goodman@monsanto.com

Kathryn Hamilton (p. 435) Monsanto Company, St. Louis, Missouri 63198. E-mail: kathryn.a.hamilton@monsanto.com

Susan L. Hefle (p. 325) University of Nebraska-Lincoln, Food Allergy Research and Resource Program, Lincoln, Nebraska 68583. E-mail: sheflel @unl.edu

Robert V. House (p. 191) Covance Laboratories Inc., Madison, Wisconsin 53704. E-mail: robert.house@covance.com

Yan Lavrovsky (p. 253) Serona Pharmaceutical Research Institute, Geneva, Switzerland. E-mail: yan.lavrovsky@serona.com

Maureen A. Mackey (p. 117) Monsanto Company, St. Louis, Missouri 63198. E-mail: maureen.a.mackey@monsnato.com

Michael J. McKee (p. 233) Monsanto Company, St. Louis, Missouri 63198. Email: michael.j.mckee@monsanto.com

Thomas E. Nickson (p. 233) Monsanto Company, St. Louis, Missouri 63198. E-mail: thomas.nickson@monsanto.com

Arun K. Roy (p. 253) University of Texas Health Science Center at San Antonio, San Antonio, Texas 78229. E-mail: roy@uthscsa.edu

Gregor Schmitz (p. 13) Botanical Garden, University Konstanz, Konstanz, Germany. E-mail: gregor.schmitz@uni-konstanz.de

James E. Talmadge (p. 281) University of Nebraska Medical Center, Omaha, Nebraska 68198. E-mail: jtalmadg@unmc.edu

Steve L. Taylor (p. 325) University of Nebraska-Lincoln, Food Allergy Research and Resource Program, Lincoln, Nebraska 68583. E-mail: sltaylor@unlnotes.unl.edu

John A. Thomas (p. 347) University of Texas Health Science Center, San Antonio, Texas 78216. E-mail: jat-tox@swbell.net

Jennifer A. Thomson (p. 385) Department of Molecular and Cell Biology, University of Capetown, Cape Town, South Africa. E-mail: JAT@molbiol.uct.ac.za

François Verdier (p. 397) Aventis Pasteur, Marcy L'Etoile, France. E-mail: francois.verdier@aventis.com

Mike Wilkinson (p. 413) Department of Agricultural Botany Plant Sciences Laboratories, Reading, United Kingdom. E-mail: m.j.wilkinson@reading.ac.uk

Preface

The first edition of Biotechnology and Safety Assessment, edited by John A. Thomas and Laurie A. Myers was published in 1993 with its major emphasis on emerging molecular biology techniques used in the production of recombinant DNA-derived drugs as well as describing early protocols designed to ensure their pre-clinical safety and efficacy. Advances in transgenic animal models and safety evaluation approaches to genetically modified (GM) foods were also described.

The Second edition of Biotechnology and Safety Assessment, edited by John A. Thomas in 1999 heralded an expansion of topics in the fields of biotherapeutics and agribiotechnology. It encompassed the latest advances in antisense therapeutics, molecular modification of cytokines, the clinical toxicity of interferons, and the pharmacology of recombinant proteins. This Edition was greatly expanded into areas of agribiotechnology including risk/benefit issues, environmental considerations, food and feed safety assessment and allergens in GM and non-GM foods.

The Third Edition continues to highlight major advances in areas of biotherapeutics and agribiotechnology. The Third Edition is more comprehensive than previous editions and provides important global perspectives on the safety and commercialization of GM crops and newer, more potent therapeutics agents. Biotechnology and Safety Assessment, 3rd edition is edited by John A. Thomas and Roy L. Fuchs and contains chapters written by internationally recognized experts in the fields of molecular genetics, nutrition, food science and safety/risk assessment. It contains a wide spectrum of topics yet integrates them into an overall approach involving safety testing, regulatory oversight and post-marketing surveillance. Many topics are especially important to the toxicologist, the pharmacologist, the nutritionist and those responsible for assessing risk/benefit and environmental impacts and the safety of GM pharmaceuticals, microbial products and plant products.

Through recent advances in agribiotechnology there is a transition from crop genetically modified for improved insect, weed and disease control to crops with enhanced nutritional properties such as vitamin and other micronutrients or safer foods with decreases allergenic concerns.

There is truly a revolution in food technology and one that will lead to helping feed the burgeoning world population in the 21st century. Finally, chapters are specifically devoted to the pre-clinical safety of GM microorganisms used in food processing and fermentation, and to immunotoxicological testing protocols for cytokines and other therapeutic proteins. The environment, non-target species, and risk/benefit topics are covered in significant depth to make this Third Edition a valuable resource for the corporate technical library and for the medical center library. It will also be very beneficial to the biomedical scientist's book shelf whether their field is molecular genetics, agronomy, microbiology, nutrition or a healthcare provider seeking to better understand the rapid progress being made in biotechnology.

The Editors

Chapter 1

Using Plant Biotechnology to Reduce Allergens in Food: Status and Future Potential

Gary A. Bannon
Department of Biochemistry & Molecular Biology
University of Arkansas for Medical Sciences
Little Rock, Arkansas

Food allergic reactions affect 6–8% of children and 1–2% of the adult population. The incidence of IgE-mediated reactions to specific food crops is increasing, particularly in developed countries, likely owing to increased levels of protein consumption. Many allergic reactions are to foods of plant origin, including peanuts, soy, wheat, and tree nuts. Allergic reactions are typically elicited by a defined subset of proteins that are found in abundance in the food. The increased prevalence of allergic reactions coupled with the sometime severe clinical symptoms has led many scientists to explore methods of reducing the allergenicity of some crops. This chapter explores the potential to reduce allergenicity of plants used as food crops by both traditional breeding practices and genetic engineering methods.

Biotechnology and Safety Assessment, 3rd edition

INTRODUCTION

A mere 20 years ago the improvement of crop productivity and heartiness was a trial-and-error process; sometimes it took years to determine whether a desired trait was stable in a new hybrid. This process depended on the existence of natural variation in the plants of interest or on our ability to create variability by chemical or irradiation mutagenesis coupled with our ability to identify specific phenotypic characteristics that might improve a plant's production potential. Once desirable phenotypic qualities had been identified, the laborious task of crossing and back crossing plants was started in the hope of moving whatever genetic material was responsible for this phenotype into the new hybrid line, without introducing any undesirable traits. There are obvious limitations to this approach, primarily the requirement that there be a naturally occurring variant with the desired phenotypic trait or the ability to create such variation via mutagenesis or other methods, and the time-consuming and labor-intensive process of hybrid production. Even with these limitations, crop scientists and geneticists were able to improve most crop yields severalfold to feed an ever-expanding world population.

With the advent of molecular biology and biotechnology it became possible not only to identify a desirable phenotypic trait but also to identify the precise genetic material responsible for that genetic trait. Recombinant DNA and plant transformation techniques have made it possible to alter the composition of individual plant components (lipids, carbohydrates, proteins) beyond what is possible through traditional breeding practices. The thrust of most plant biotechnology programs has been to enhance or reduce the level of specific components naturally found in the plant or to introduce a component not naturally found in the plant. One example of a naturally occurring component in a plant that has been increased in a biotechnology-engineered crop is the starch content in potatoes. Starch consists of three components in varying amounts depending on the plant source: the large linear molecules of amylose, complex branched amylopectin, and a smaller size amylose (Baba and Arai, 1984). As might be expected, the starch biosynthetic pathway is complicated, with many enzymes involved in producing the final product. However, the product of the ADPG pyrophosphorylase gene (reviewed by Smith *et al.*, 1995) appears to control the overall flux through the starch biosynthetic pathway. In this example, Stark *et al.* (1992) utilized the non-feedback-inhibited ADPG pyrophosphorylase gene from *E. coli* to increase the starch content of potatoes.

One example of a biotechnology-engineered crop distinguished by a component that is not naturally found in a plant is "golden rice." This biotechnology-derived rice line was developed to combat vitamin A deficiency, the leading cause of severe visual impairment and blindness among children in developing countries. In this rice line, genes encoding proteins necessary for the production of β-carotene, the precursor of vitamin A, were introduced into the genome. Successful integration and functioning of the genes resulted in rice plants that produced yellow-tinted kernels with the intensity of color indicating the amount of β-carotene present (Friedrich, 1999). Genes were also introduced to increase the level and bioavailability of iron, another important nutrient.

In addition to investigations aimed at improving the nutritional quality of food crops, a large body of work has been targeted at improving resistance to insect predation. Gene transfer work utilizing the bacterial (*Bacillus thuringiensis*) crystal protein (Bt-Cry) produced genes with resistance to a range of lepidopteran insects. Cauliflower, corn, and tomato varieties have been successfully transformed with vectors expressing insecticidal Bt-Cry proteins with no significant changes in key nutrients, overall composition of unknown metabolites, or *N*-glycans.

It is important to recognize that modifying plant genomes introduces the possibility of altering the allergenic potential of foods whether that change is brought about by classic plant breeding practices or by directed gene approaches. This can happen by increasing the allergenic potential of resident allergenic proteins or by introducing completely novel proteins that have characteristics of food allergens. Methods and safety assessment approaches have been developed and applied to address these concerns (Taylor and Hefle, Chapter 11, this volume). However, biotechnology can also be used to directly decrease the levels of known allergens or their allergenicity. With this in mind, this chapter focuses on some of the approaches being taken to reduce the allergenic potential of foods derived from major crops.

CHARACTERISTICS OF FOOD ALLERGENS

Before reviewing the different approaches to reducing the allergenic potential of foods, it is important to mention the components of foods that are classified as allergens. There are about 26 major allergens identified for about 17 different food items (IUIS Allergen Nomenclature Subcommittee, 1997). From biochemical analysis of this limited number of allergens, certain characteristics shared by most but not necessarily all can be identified. For example, food allergens are typically low molecular weight glycosylated proteins that are relatively abundant in a food source. In addition, they have acidic isoelectric points, as well as multiple, linear IgE binding epitopes, and are resistant to denaturation and digestion (Stanley and Bannon, 1999). These characteristics are purported to be important to the allergenicity of a protein for various reasons. The low molecular weight and glycosylation pattern of food allergens were believed to be responsible for facilitating their ability to move across the gut mucosa and gain access to the immune system to stimulate a Th_2-type (IgE-producing) response. To account for the observation that most food allergens are relatively abundant in the food source, it was suggested that the immune system was more likely to encounter these proteins than any allergens present as a small percentage of the total protein ingested. The acidic isoelectric point of some food allergens, which precipitate at the low pH encountered in the stomach, may lead to longer transit times in the gastrointestinal (GI) tract. Resistance to denaturation and digestion of an allergen is thought to be an important characteristic because the longer the significant portion of the protein remains intact, the more likely it is to trigger an immune response. Finally, most food allergens have multiple, linear binding epitopes so that even when they are partially digested or denatured, they are still capable of interacting with IgE and causing an allergic reaction (Maleki *et al.*, 2000).

An example of a food allergen that has these characteristics is the peanut allergen *Ara h* 2. Only about 17 kDa in size, this allergen represents up to 4% of the total protein in peanuts and is recognized by IgE from more than 95% of peanut-allergic patients (Burks *et al.*, 1992). In addition, *Ara h* 2 is glycosylated containing a high mannose carbohydrate side chain with a $\beta_{1,3}$-linked xylose residue. This allergen has an acidic isoelectric point (pI 4.5), is resistant to denaturation and digestion by enzymes commonly encountered in the GI tract, and contains at least 10 linear IgE binding epitopes (Astwood *et al.*, 1996; Stanley *et al.*, 1997).

While most food allergens exhibit these characteristics, there are notable exceptions such as the major potato allergen, patatin (*Sol t* I). This is the major allergen identified by IgE from potato-sensitive patients. Sensitivity to potatoes is quite rare, and the reactions in those who are sensitive is typically not severe. Patatin represents 40% of the total protein in potato and contains multiple linear IgE binding sites. It is partially glycosylated, but the type of carbohydrate side chain has not yet been reported. Even though patatin has many of the characteristics of an allergen, one notable difference is its instability to enzymes commonly encountered in the human GI tract (Seppala *et al.*, 1999; Astwood *et al.*, 2000a).

Another allergen that exhibits characteristics not commonly found in allergens is the wheat allergen γ-thionin. In this case, the allergen is resistant to the action of GI tract proteases but represents only 0.02% of the total protein in wheat (Astwood *et al.*, 2000b). These examples illustrate the difficulty in predicting whether a protein has the potential to be allergenic simply based on our current knowledge of protein characteristics.

TRADITIONAL PLANT BREEDING METHODS FOR REDUCING ALLERGENICITY

Traditional plant breeding methods can be used to address the problem of allergenicity in a variety of crops by taking advantage of the inherent variability in protein expression levels that occur in most plant varieties. This approach requires a method for detecting allergens contained in the food source and germ plasm from the different varieties of the crop. Typically, methods for detecting the allergens involve using either serum IgE from patients who have documented hypersensitivity to the food in question or monoclonal and polyclonal antibodies that have been developed to the specific allergens. Whether serum IgE or allergen-specific antibodies are used, the goal is to develop a semiquantitative assay to estimate the amount of allergen present in a protein extract from different varieties of a food crop. One or more crop varieties currently used for food production is the standard to which other varieties are compared in search of potential varieties that exhibit lower amounts of either IgE binding proteins or specific allergens. Once identified, the varieties exhibiting lower amounts of allergenic proteins can be used in traditional breeding schemes to produce a hypoallergenic crop variety. Alternatively, chemical or irradiation mutagenesis can be used to try to produce a plant with reduced levels of one or more of these allergens.

This strategy is being applied in work with peanut cultivars. Peanut allergy affects only about 0.6% of the population and is considered to be one of the most severe of the food allergic reactions (Sicherer *et al.*, 1998). The major peanut allergens have been identified, and numerous allergen-specific antibodies are available for quantitative assessment of total allergen content (Burks *et al.*, 1991, 1992; Eigenmann *et al.*, 1996). A consortium of investigators from the University of Georgia, the U.S. Department of Agriculture, Alabama A&M University, and Texas A&M University are beginning to identify peanut varieties that show amounts of IgE binding proteins quantitatively lower than those found in the current production varieties (Allergy Research Forum, Fairfax, VA, August 2001). These results, which confirm that different varieties can express different amounts of allergenic proteins, demonstrate the potential for success of this approach in producing a peanut variety with reduced levels of allergens. However, there are still several hurdles to overcome before this approach can be fully realized. Most food crops of agronomic importance have been selected over many generations for a variety of characteristics including, growth rate, yield at harvest, and heartiness. Any crop variety that is eventually identified would have to meet agronomic and food processing criteria deemed important for that crop.

USE OF GENETIC ENGINEERING TO REDUCE ALLERGENIC POTENTIAL

Genetic engineering can be used to reduce the levels of known allergens by post-transcriptional gene silencing or to reduce the critical disulfide bonds by means of thioredoxin or by directly modifying the genes encoding the allergen(s). Examples of these are discussed now.

POST-TRANSCRIPTIONAL GENE SILENCING

Post-transcriptional gene silencing is a targeted mechanism that has been successfully used to reduce the levels in food crops of specific plant proteins, including proteins with allergenic potential. In general, this mechanism suppresses the accumulation of a gene-specific mRNA so that it cannot be translated into protein, thereby reducing the quantity of that protein (Baulcombe, 1996). Regulation of gene expression by this mechanism is a naturally occurring phenomenon in bacteria (Mizuno *et al.*, 1984; Simons, 1988) and some eukaryotic cells (Dolnick, 1997). While there is controversy about the mechanism by which this approach works, all models require the production of RNA in the reverse orientation of that required to produce the targeted protein, that is, an antisense RNA. Such an antisense ribonucleic acid could be produced by direct transcription, in response to overexpression of the transgene, or in response to the production of an aberrant sense RNA product of the transgene (Baulcombe, 1996).

The RNA antisense approach has been successfully used to reduce the allergenic potential of rice. Most rice allergens have been found in the globulin fraction of rice seed (Shibasaki *et al.*, 1979). The globulins and albumins have been estimated to comprise about 80–90% of the total protein in rice seeds (Cagampang *et al.*, 1966). From this fraction a 16-kDa α-amylase/trypsin inhibitor–like protein was identified as the major allergen involved in hypersensitivity reactions to rice (Matsuda *et al.*, 1988, 1991; Nakase *et al.*, 1996). Using this antisense RNA approach, Nakamura and Matsuda (1996) generated several rice lines that contained transgenes producing antisense RNA for the 16-kDa rice allergen. These authors successfully lowered the allergen content in rice by as much as 80% without a concomitant change in the amount of other major seed storage proteins.

There are several advantages to using a posttranscriptional gene silencing approach to reducing allergenicity. These include the targeted nature of the method and the relatively small amount of information about the allergen that is required to apply this technique. However, as Nakamura and Matsuda (1996) observed, there were wide fluctuations in the amounts of allergen reduction between seeds, even within a single transformant, indicating that the method can give variable results. Furthermore, it is difficult to totally eliminate the allergen, which may be required to make the food safe for highly sensitive individuals, especially if the protein is a potent allergen like *Ara h* 2 from peanut. In addition, it would be difficult to apply this method to a food that has a number of major allergens.

Reduction of Disulfide Bonds by Thioredoxin

Thioredoxins represent a family of 12-kDa proteins that undergo reversible redox change through a catalytically active disulfide site (Buchanan, 1991; Buchanan *et al.*, 1994; Williams, 1995; Holmgren 2000a,b). Thioredoxins have been shown to reduce intramolecular disulfide bonds from a wide variety of proteins, many of which are considered to be allergens (Gomez *et al.*, 1990; Ogawa *et al.*, 1991; Teuber *et al.*, 1998). The biological activity of this ubiquitous protein, coupled with the observation that many food allergens are proteins containing intramolecular disulfide bonds that may be important to their allergenicity (Lehrer *et al.*, 1996), raises the possibility of using thioredoxin to reduce the allergenic potential of some foods.

This concept was tested on allergens in wheat and milk by Buchanan and colleagues, who showed a significant reduction in the allergic symptoms elicited from sensitized dogs (Buchanan *et al.*, 1997; del Val *et al.*, 1999). Briefly, the authors exposed either the purified allergens or an extract from the food source containing the allergens to thioredoxin purified from *E. coli* and then performed skin tests and monitored gastrointestinal symptoms in a sensitized dog model. Allergens that had their disulfide bonds reduced by thioredoxin elicited greatly reduced skin reactions and gastrointestinal symptoms. These results provide a critical proof of concept for this approach, one that had to be obtained before the work of constructing transgenic wheat lines that over produce thioredoxin could begin.

The advantage of using thioredoxin is that it is a general approach that will be useful for reducing the allergenicity of any food crop whose allergens depend on disulfide bonds for their activity. However, the approach may be somewhat limited, especially for food allergens whose IgE binding epitopes can elicit an allergic response in the absence of intact disulfide bonds.

Modification of Genes Encoding Allergens

One of the more ambitious approaches to reducing allergenicity of food crops is by modification of the genes encoding the allergens so that they produce hypoallergenic forms of these proteins. This approach is based on the observation that most food allergens have linear IgE binding epitopes that can be readily defined by using overlapping peptides representing the entire amino acid sequence of the allergen and serum IgE from a population of individuals with hypersensitivity reactions to the food in question. Once the IgE binding epitopes have been determined, critical amino acids can be identified that, when changed to another amino acid, result in loss of IgE binding to that epitope without modification of the function of that protein. Any changes that result in loss of IgE binding can then be introduced into the gene by site-directed mutagenesis.

Serum IgE from patients with documented peanut hypersensitivity and overlapping peptides were used to identify the IgE binding epitopes of the major peanut allergens *Ara h* 1, *Ara h* 2, and *Ara h* 3. At least 23 different linear IgE binding epitopes located throughout the length of the *Ara h* 1 molecule were identified (Burks *et al.*, 1997). In a similar fashion, 10 IgE binding epitopes and 4 IgE binding epitopes were identified in *Ara h* 2 and *Ara h* 3, respectively (Stanley *et al.*, 1997; Rabjohn *et al.*, 1998). Mutational analysis of each of the IgE binding epitopes revealed that single amino acid changes within these peptides had dramatic effects on IgE binding characteristics. Substitution of a single amino acid led to loss of IgE binding (Stanley *et al.*, 1997; Shin *et al.*, 1998; Rabjohn *et al.*, 1999). Analysis of the type and position of amino acids within the IgE binding epitopes that had this effect indicated that substitution of hydrophobic residues in the center of the epitopes was more likely to lead to loss of IgE binding (Shin *et al.*, 1998). Site-directed mutagenesis of the cDNA encoding each of these allergens was then used to change a single amino acid within each IgE binding epitope. The hypoallergenic versions of these allergens were produced in *E. coli* and tested for their ability to bind IgE from peanut-sensitive patients. The modified allergens demonstrated a greatly reduced IgE binding capacity when individual patient serum IgE was compared with the binding capacity of the wild-type allergens (Bannon *et al.*, 2001).

Even though the results just described are encouraging, there are still significant hurdles to overcome before this approach can be fully explored. For example, while it is extremely easy with today's technology to reintroduce the modified gene into the plant genome, it is not possible to completely inactivate the resident wild-type gene. Yet such inactivation must be achieved before this approach can be fully exploited. Furthermore, the reduction in binding would have to be shown to be sufficient to reduce reactions in patients who are sensitive to that allergen or food in clinical trials. Thus human studies will be required before any claims of hypoallergenicity are made for a plant variety or food product.

CONCLUDING REMARKS

World population is expected to increase by 2.5 billion people in the next 25 years. Concomitantly, the food requirements for this growing population are expected to double by the year 2025. In contrast, there has been a decline in the annual rate of increase in cereal yield such that the annual rate of yield increase is below the rate of population increase (Somerville and Briscoe, 2001). To feed this growing population, crop yield will have to be increased, and some of the increase in yield will be due to genetic engineering of foods. In addition, the incidence of food allergies appears to be on the rise, particularly in developed countries (Taylor *et al.*, 1987; Sicherer *et al.*, 1998). The convergence of these two phenomena will require that we continue to explore novel ways of lowering the potential allergenicity of food crops through biotechnology. Any modification of the major food proteins in grain designed to reduce allergenicity will have to undergo testing to determine whether allergic reactions are reduced in affected individuals and new consumers are not sensitized. In addition, the resulting grain would have to undergo processing and functionality studies to determine whether the modifications altered any food characteristics of the product.

REFERENCES

Astwood, J. D, Leach, J. N., and Fuchs, R. L. (1996). Stability of food allergens to digestion in vitro. *Nat. Biotechnol.*, **14**, 1269–1274.

Astwood, J. D., Alibhai, M., Lee, T., Fuchs, R., and Sampson, H. (2000a). Identification and characterization of IgE binding epitopes of patatin, a major food allergen of potato. *J. Allergy Clini. Immunol*, **105**, 555a.

Astwood, J. D., Tran, K., Liang, J., Goodman, R., and Sampson, H. (2000b). Digestibility and allergenicity of gamma-thionin from wheat flour. *J. Allergy Clin. Immunol*, **105**, 419a.

Baba, T., and Arai, Y. (1984). Structural characterization of amylopectin and intermediate material in amylomaize starch granules. *Agric. Biol. Chem.*, **48**, 1763–1775.

Bannon, G. A., Cockrell, G., Connaughton, C., West, C. M., Helm, R., Stanley, J. S., King, N., Rabjohn, P., Sampson, H. A., and Burks, A. W. (2001). Engineering, characterization and in vitro efficacy of the major peanut allergens for use in immunotherapy. *Int. Arch. Allergy Immunol.*, **124**, 70–72.

Baulcombe, D. C. (1996). RNA as a target and an initiator of post-transcriptional gene silencing in transgenic plants. *Plant Mol. Biol.*, **32**, 79–88.

Buchanan, B. B. (1991). Regulation of CO_2 assimilation in oxygenic photosynthesis: The ferredoxin/thioredoxin system. Perspective on its discovery, present status, and future development. *Arch. Biochem. Biophy.*, **288**, 1–9.

Buchanan, B. B., Schurmann, P., Decottignies, P., and Lozano, R. M. (1994). Thioredoxin: A multifunctional regulatory protein with a bright future in technology and medicine. *Arch. Biochem. Biophys.*, **314**, 257–260.

Buchanan, B., Adamidi, C., Lozano, R. M., Yee, B. C., Momma, M., Kobrehel, K., Ermel, R., and Frick, O. L. (1997). Thioredoxin-linked mitigation of allergic responses to wheat. *Proc. Natl. Acad. Sci., USA*, **94**, 5372–5377.

Burks, A. W., Williams, L. W., Helm, R. M., Connaughton, C., Cockrell, G., and O'Brien, T. (1991). Identification of a major peanut allergen, *Ara h* I, in patients with atopic dermatitis and positive peanut challenges. *J. Allergy Clin. Immunol.*, **88**, 172–179.

Burks, A. W., Williams, L. W., Connaughton, C., Cockrell, G., O'Brien, T. J., and Helm, R. M. (1992). Identification and characterization of a second major peanut allergen, *Ara h* II, with use of the sera of patients with atopic dermatitis and positive peanut challenge. *J. Allergy Clin. Immunol.*, **90**, 962–969.

Burks, A. W., Shin, D., Cockrell, G., Stanley, J. S., Helm, R. M., and Bannon, G. A. (1997). Mapping and mutational analysis of the IgE-binding epitopes on *Ara h* 1, a legume vicilin protein and a major allergen in peanut hypersensitivity. *Eur. J. Biochem.*, **245**, 334–339.

Cagampang, G. B., Cruz, L. J., Espiritu, G., Santiago, R. G., and Juliano, B. O. (1966). *Cereal Chem.*, **43**, 145–155.

de Val, G., Yee, B. C., Lozano, R. M., Buchanan, B., Ermel, R. W., Lee, Y.-M., and Frick, O. L. (1999). Thioredoxin treatment increases digestibility and lowers allergenicity of milk. *J. Allergy Clin. Immunol.*, **103**, 690–697.

Dolnick, B. J. (1997). Naturally occurring antisense RNA. *Pharmacol. Ther.*, **75**, 179–184.

Eigenmann, P. A., Burks, A. W., Bannon, G. A., and Sampson, H. A. (1996). Identification of unique peanut and soy allergens in sera adsorbed with cross-reacting antibodies. *J. Allergy Clin. Immunol.*, **98**, 969–978.

Friedrich, M. J. (1999). Genetically enhanced rice to help fight malnutrition. *JAMA*, **282**, 1508–1509.

Gomez, L., Martin, E., Hernandez, D., Sanchez-Monge, R., Barber, D., del Pozo, V., de Andres, B., Armentia, A., Lahoz, C., and Salcedo, G. (1990). Members of the alpha-amylase inhibitor family from wheat endosperm are major allergens associated with baker's asthma. *FEBS Lett.*, **261**, 85–88.

Holmgren, A. (2000a). Antioxidant function of thioredoxin and glutaredoxin systems. *Antioxidants Redox Signaling*, **2**, 811–820.

Holmgren, A. (2000b). Redox regulation by thioredoxin and thioredoxin reductase. *Biofactors*, **11**, 63–64.

IUIS Allergen Nomenclature Subcommittee. *Official List of Allergens*. (1997). International Union of Immunological Societies, San Francisco.

Lehrer, S. B., Horner, W. E., and Reese, G. (1996). Why are some proteins allergenic? Implications for biotechnology. *Crit. Rev. Food Sci. Nutr.*, **36**, 553–564.

Maleki, S. J., Kopper, R. A., Shin, D. S., Park, C. W., Compadre, C. M., Sampson, H., Burks, A. W., and Bannon, G. A. (2000). Structure of the major peanut allergen *Ara h* 1 may protect IgE-binding epitopes from degradation. *J. Immunol.*, **164**, 5844–5849.

Matsuda, T., Sugiyama, M., Nakamura, R., and Torii, S. (1988). Purification and properties of an allergenic protein in rice grains. *Agric. Biol. Chem.*, **52**, 1465–1470.

Matsuda, T., Nomura, R., Sugiyama, M., and Nakamura, R. (1991). Immunochemical studies on rice allergenic proteins. *Agric. Biol. Chem.*, **55**, 509–513.

Mizuno, T., Chou, M., and Inoue, M. (1984). A unique mechanism regulating gene expression: Translational inhibition by a complementary RNA transcript (micRNA). *Proc. Natl. Acad. Sci., USA*, **81**, 1966–1970.

Nakamura, R., and Matsuda, T. (1996). Rice allergenic protein and molecular–genetic approach for hypoallergenic rice. *Biosci. Biotechnol. Biochem.*, **60**, 1215–1221.

Nakase, M., Alvarez, A. M., Adachi, T., Aoki, N., Nakamura, R., and Matsuda, T. (1996). Immunochemical and biochemical identification of the rice seed protein encoded by cDNA clone A3-12. *Biosci. Biotechnol. Biochem.*, **60**, 1031–1042.

Ogawa, T., Bando, N., Tsuji, H., Okajima, H., Nishikawa, K., and Sasaoka K. (1991). Investigation of the IgE-binding proteins in soybeans by immunoblotting with the sera of the soybean-sensitive patients with atopic dermatitis. *J. Nutr. Sci. Vitaminol.*, **37**, 555–565.

Rabjohn, P., Helm, E. M., Stanley, J. S., West, C. M., Sampson, H. A., Burks, A. W., and Bannon, G. A. (1999). Molecular cloning and epitope analysis of the peanut allergen *Ara h* 3. *J. Clin. Invest.*, **103**, 535–542.

Seppala, U., Alenius, H., Turjanmaa, K., Reunala, T., Palosuo, T., and Kalkkinen, N. (1999). Identification of patatin as a novel allergen for children with positive skin prick test responses to raw potato. *J. Allergy Clin. Immunol.*, **103**, 165–171.

Shibasaki, M., Suzuki, S., Nemoto, H., and Kuroume, T. (1979). *J. Allergy Clin. Immunol.*, **64**, 259–265.

Shin, D. S., Compadre, C. M., Maleki, S. J., Kopper, R. A., Sampson, H., Huang, S. K., Burks, A. W. and Bannon, G. A. (1998). Biochemical and structural analysis of the IgE binding sites on *Ara h*, an abundant and highly allergenic peanut protein. *J. Biol. Chem.*, **273**, 13753–13759.

Sicherer, S. H., Burks, A. W., and Sampson, H. A. (1998). Clinical features of acute allergic reactions to peanut and tree nuts in children. *Pediatrics*, **102**(1), e6.

Simons, R. W. (1988). Naturally occurring antisense RNA control—A brief review. *Gene*, **72**, 35–44.

Smith, A. M., Denyer, K., and Martin, C. R. (1995). What controls the amount and structure of starch in storage organs? *Plant Physiol.*, **107**, 673–677.

Somerville, C., and Briscoe, J. (2001). Genetic engineering and water. *Science*, **292**, 2217.

Stanley, J. S., and Bannon, G. A. (1999). Biochemistry of food allergens. *Clin. Rev. Allergy Immunol.*, **17**, 279–291.

Stanley, J. S., King, N., Burks, A. W., Huang, S. K., Sampson, H., Cockrell, G., Helm, R. M., West, C. M., and Bannon, G. A. (1997). Identification and mutational analysis of the immunodominant IgE binding epitopes of the major peanut allergen *Ara h* 2. *Arch. Biochem. Biophys.*, **342**, 244–253.

Stark, D. M., Timmerman, K. P., Barry, G. F., Preiss, J., and Kishore, G. M. (1992). Regulation of the amount of starch in plant tissues by ADP glucose pyrophosphorylase. *Science*, **258**, 287–292.

Taylor, S. L., Lemanske, R. F., Jr., Bush, R. K., and Busse, W. W. (1987). Chemistry of food allergens, in *Food Allergy*, R. K. Chandra, ed, St John's, Newfoundland, Nutrition Research Education Foundation, pp. 21–44.

Teuber, S. S., Dandekar, A. M., Peterson, W. R., and Sellers, C. L. (1998). Cloning and sequencing of a gene encoding a 2*S* albumin seed storage protein precursor from English walnut (*Juglans regia*), a major food allergen. *J. Allergy Clin. Immunol.*, **101**, 807–814.

Williams, C. H., Jr. (1995). Mechanism and structure of thioredoxin reductase from *Escherichia coli. FASEB J.*, **9**, 1267–1276.

Chapter 2

Recent Experience with Biosafety Research and Post-Market Environmental Monitoring in Risk Management of Plant Biotechnology Derived Crops

Detlef Bartsch
Department of Biology
Aachen University of Technology,
Aachen, Germany

Gregor Schmitz
Botanical Garden, University Konstanz
Konstanz, Germany

Scientific research directed toward gaining a better understanding of the risks associated with genetically modified plants (GMPs) can be broadly characterized by noting when the research is conducted in the life span of the GMP. We use the term "biosafety research" to refer to studies conducted prior to commercialization. Similarly, "monitoring" refers to studies performed after the product has been placed on the market. In this instance, monitoring is equivalent to postregistration monitoring. This chapter describes our recent work in both areas. We describe biosafety research designed to assess the risk associated with gene flow from a sugar beet (Beta vulgaris *ssp.* vulgaris *var.* altissima) *modified to resist viral infection. Since risk is the product of the frequency of gene flow and the probability that some harm (e.g., enhanced weediness) may result, our work focused on assessing the fitness of the outcrossing event (i.e., the hazard or consequence of gene flow). We used this approach because it was known that gene flow between cultivated sugar beet and a wild relative* (B. vulgaris *ssp.* maritima) *was likely to occur. This biosafety research demonstrated that there is no increase in fitness provided by the virus-resistant trait derived from biotechnology that would alter the weediness of the wild relative. Our second major effort was to define appropriate monitoring of insect-protected maize in Germany. More specifically, efforts to design appropriate methods to assess potential nontarget insect effects from* Bt *maize are described. Monitoring can serve several useful purposes, one of which is to confirm the results of biosafety research. Because biosafety research and monitoring are costly in terms of time, money, and resources, we recommend concentrating on developing thorough, science-based experiments for all GMPs, while using a cautious, tiered approach with organisms defined as ecologically "riskier."*

INTRODUCTION

Risk assessment is the scientific standard used to assess the risk of any new technology that might affect human and animal health or environmental safety. Since risk is a product of both exposure and hazard, it is scientific consensus that risk research must target both the exposure (frequency) and the possible consequences (potential hazards) associated with a new technology. Our research with genetically modified plants (GMPs) has been focused in two areas: biosafety research (precommercial) and monitoring (postcommercial). Furthermore, our biosafety research has been concentrated on understanding the potential negative consequences of gene flow rather than the frequency (exposure). More recently, we have initiated monitoring research designed to confirm the biosafety research conducted on products that have been approved for commercialization through the process set forth in European Union Directive 90/220 (now 2001/18). Within the traditional scientific risk assessment paradigm, biosafety research should complement monitoring.

A thorough review of the available literature on the subjects of "biosafety research" and "monitoring of GMPs" reveals a substantial body of existing knowledge. The International Centre for Genetic, Engineering and Biotechnology (ICGEB) maintains one of the most comprehensive databases available, and according to its website http://www.icgeb.trieste.it/~bsafesrv/ the number of biosafety publications related to transgenic organisms has increased within the decade 1990–2000 to more than 2600 records. A large number of these reports concern gene flow and impacts to nontarget organisms. Extensive field studies support the view that, so far, no negative effects to nontarget organisms detected in laboratory studies have been reported in the field. In addition, biosafety studies have typically demonstrated that there is no difference in hybridization frequencies between GMPs, or non-GMPs, and crossable wild populations. Since the *phenomenon of gene flow* by itself is not an adverse effect, the focus of our research with GMPs has been on the novelty of the introduced trait, regardless of whether this trait provides a selective advantage and whether any other adverse effects may be present.

Published studies have typically demonstrated that the current commercial GMPs behave ecologically in a manner similar to non-GMPs when environment and experimental conditions offer no advantage for the modified trait. However, GMPs may perform better than their traditional counterparts if the new phenotype experiences ecological conditions favoring the modified trait. Indeed, this scenario is identical to that for traditionally bred traits, and thus there should be little difference in the assessments performed with plants developed through classical breeding. For example, virus-resistant crops developed through traditional breeding or by biotechnology could generate essentially identical phenotypes and genotypes. This means that the environmental impact does not depend on the process used to produce the variant, but on the end product. This chapter describes two examples, biosafety research with virus-resistant sugar beet and monitoring of *Bt* corn. Also, a discussion is presented depicting how case-specific biosafety research results and monitoring can be connected. We conclude with the recommendation that the combination of an intelligent selection of relevant indicator species and appropriate baseline data are needed to increase the effectiveness of monitoring.

END POINTS AND DEFINITIONS

At the outset of risk assessment work, before one can characterize results as hazardous, it is necessary to harmonize definitions and clearly link end points to adverse effects. In the debate over GMPs, there is much confusion about how to use the term "risk." Many biosafety research studies (cf. Hoffman, 1990; Rissler and Mellon, 1996) tend to regard "risk" as equivalent to "exposure" or to assume that "gene flow" equals "hazard." If this were the case, evolution would be classified as a harmful process since hybridization, introgression, and gene flow are essential to speciation, especially in plants. Figure 1 gives one definition relating risk and environmental concerns. The establishment of broad monitoring plans, including potential cumulative long-term effects associated with human activities such as agriculture, may give useful information that could be applied as a baseline for future environmental risk assessment work. Currently, baseline data and specific protocols for studying long-term effects are lacking, thus limiting the possibility of assessing and accurately characterizing environmental effects such as the evolution of aggressive weeds from crop plants. Absent this information, some parties are resorting to classifying natural phenomena of biological systems as inherently risky when viewed in the context of GMPs.

One place to start this analysis is to consider "*weediness*" as associated with hazard identification for plants. Although there are various ways of defining weeds, there is no generally accepted classical approach (Amman *et al.*, 2001). Popular as well as subjective concepts define weeds as plants of any kind growing in the wrong place, causing damage, conferring no benefit, and suppressing cultivated plant species. *Economical concepts* reflect the view of agronomists, who concentrate on the reduction in yield, thereby stressing the aspect of damage. A weed problem is solved as soon as the plant no longer creates damage in the field, a state that is reached by means of weed control (crop rotation, tillage, herbicide application). In contrast, *ecological concepts* of a weed consider habitats that lie outside agrosystems and are colonized by a natural (unmanaged) community of plants. An aggressive weed can cause damage not only in agrosystems (cultivated fields), gardens, roadsides and embankments, but also in (semi)natural plant communities where they outcompete desirable species and effectively reduce the overall diversity of the local plant community (Fig. 2).

We use the term *biosafety research* for the examination of environmental effects of GMPs before commercialization. Biosafety research is used to make the initial conclusion concerning any risk of transgenic organisms. One example we shall give is from our work with sugar beets and virus resistance in the Po Valley of Italy. In biosafety research we need to understand the basic changes (or lack of changes) to a plant's fitness and the impact on gene flow prior to commercialization. Biosafety research workers should start by developing clear and interpretable end points, since research is costly in terms of money, time, and resources. Much inefficiency can be avoided if clear

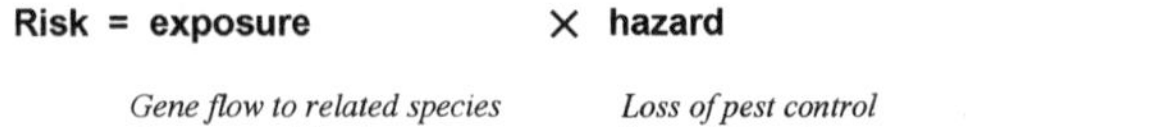

Figure 1 Definition of risk indicating environmental concerns targeted by the two multiplicands.

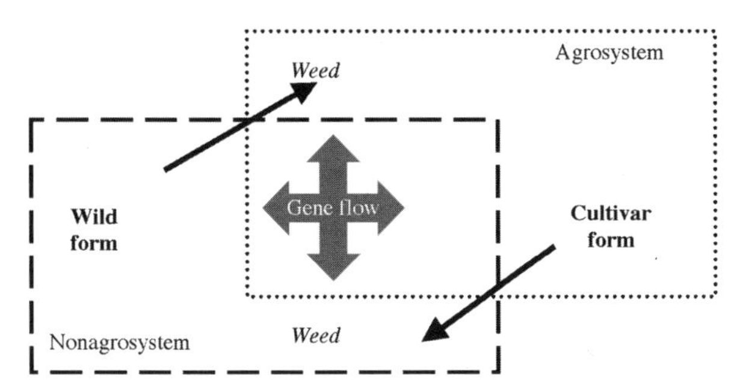

Figure 2 Intraspecific gene flow in a wild weed–cultivar system. Weeds are plants in the wrong place, here, either as cultivars outside agricultural areas or as wild plants in cultivated ground. Weeds may also evolve as a result of gene flow. Wild and cultivar forms of a single species can be protected as a genetic resource in one place and eradicated as weed in another.

assessment end points and definitions of unwanted environmental effects are established at an early stage of the risk assessment process.

Two general conclusions can be drawn from past biosafety research on the invasiveness of transgenic plants. First, research has concentrated more on gene flow probability than on the ecological consequences resulting from the escape process itself. Since the majority of crops have sexually compatible wild relatives growing sympatrically somewhere in the world (Table 1), it is becoming increasingly clear that gene flow from cultivated plants to wild relatives is inevitable with *Brassica* (Mikkelsen *et al.*, 1996) or *Beta* (Bartsch and Pohl-Orf, 1996) species, and other examples have been examined for this phenomenon. Empirical work has largely focused on the gene flow process (i.e., on experiments addressing whether a crop and a wild relatives are able to hybridize under field conditions) and on descriptive studies addressing whether introgression from crops has occurred in populations of adjacent wild relatives. As stressed by Bartsch *et al.* (1999), potential effects on the genetic diversity of natural populations and the potential fitness associated with transgenes need to be analyzed in environmental risk assessment.

The second major conclusion evident from biosafety research completed to date concerns ecological tests of invasiveness of genetically modified plants (GMPs). Earlier publications have involved traits with little ecological relevance such as herbicide tolerance (Kareiva *et al.*, 1997; Crawley *et al.*, 2000). Consequently, the failure of these studies to reveal any increased risk of invasion was not surprising. Transgenic plants are considered to be a potential risk if they contain a trait conferring a large fitness advantage in natural situations. A trait that could alter the competitive ability and survival rate might be considered to be ecologically "riskier." One way in which a plant may gain a fitness advantage is by escaping its natural enemies. This effect of

Table 1

Important Crops That Hybridize with Wild Relatives Somewhere in Their Cultivation Area[a]

Crop	Relative(s)
1. Wheat	Wild *T. turgidum* ssp., some *Aegilops* species
2. Rice	Wild *Oryza* species
3. Maize	Wild *Zea mays* ssp.
4. Soybean	*Glycine gracilis, G. soya*
5. Barley	*Hordeum spontaneum*
6. Cotton	Wild *Gossypium* species
7. Sorghum	Wild *Sorghum* species
8. Millet	*Eleusine coracana* ssp. *africana*, wild *Pennisetum* species
9. Beans	Wild *Phaseolus* species
10. Oilseed rape	Some wild Brassicaea species
11. Peanuts	No report
12. Sunflower	Wild *Helianthus annuus*
13. Sugarcane	Wild *Saccharum* species
14. Sugar beet	Some wild *Beta* species

[a]Sugar beet is added to the list of the 13 most important crops worldwide (Ellstrand *et al.*, 1999) because of the importance of the *Beta* species in Europe.

"ecological release" has long been postulated as one of the major causes of successful invasion by exotic species (Mooney and Drake, 1986).

The term *monitoring* is used for any postcommercialization measure that provides data on the fate or effects of GMPs in the environment. Monitoring should be based in part on baseline data on the evolution of a given (eco)system structure and system process (Fig. 3). Both indirect and direct methods are helpful for detecting the possible impact of GMPs or their products. Environmental monitoring of agricultural crops and crop production practices is generally needed, not because of any specific, identified risk, but to enhance our ability to develop more sustainable food production practices. We present an example of monitoring transgenic *Bt* maize toward the end of this chapter.

Monitoring is a well-accepted tool closely associated with risk assessment and decision making (Nickson and Head, 1999). Monitoring of genetically modified organisms (GMOs) is conducted to achieve any of four specific objectives: to confirm compliance with regulatory requirements, to collect information necessary for controlling and managing potentially adverse

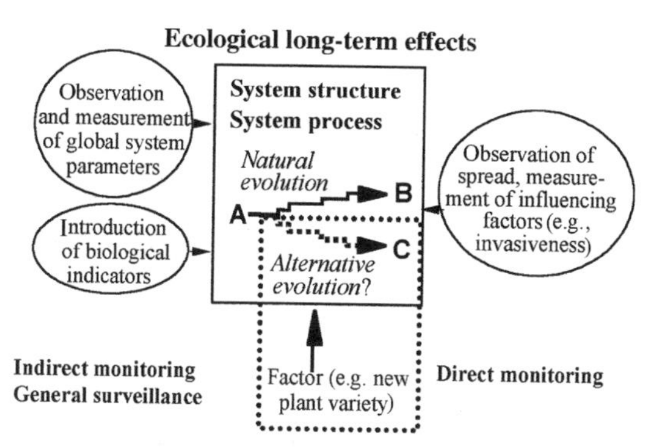

Figure 3 Ecological long-term effects and the role of monitoring. The remaining question of whether system structure or process C or B is unwanted must be assessed on the basis of scientific knowledge.

environmental situations or systems, to assess environmental quality, and to detect "unexpected" and potentially damaging effects (Suter, 1993). As such, monitoring may be recommended to reduce uncertainty remaining from risk assessment, to confirm conclusions with additional data, or to provide informational feedback on system status or condition. Monitoring is not a substitute for biosafety research or risk assessment. Rather, it is integrated with research and risk assessment to ensure that ecological systems and processes of value are being protected. Ideally, a decision to require monitoring is based on the scientific information provided in the risk assessment or some other scientific rationale that a risk is possible. Where a conclusion of minimal risk is made based on scientific data, no monitoring should be required to concentrate limited resources on more significant areas.

Nickson and Head (1999) have divided monitoring of GMPs into two basic approaches: general and specific. General monitoring, which is also referred to a surveillance, is not necessarily based on any specific hypothesis of risk. It could be accomplished by using expertise and infrastructure already present in agricultural systems and within conservation efforts. By gaining familiarity and experience with GMPs through general monitoring, one can conduct "range finding" and possibly better define the nature of a perceived risk and benefit. Specific monitoring, however, must be based on a scientific hypothesis. It is science-based monitoring that relies on a protocol with specific interpretable end points. Eventually, information from general monitoring could be refined through the development of specific monitoring protocols designed to determine what, if any, correlations existed between practices, technologies, activities, and so on used in agriculture and the overall condition of the system (Nickson and Head, 1999).

For monitoring to be more than a "good idea," a well-defined purpose must be formulated. Suter (1993) states that "[m]onitoring fails for lack of well-defined purpose." Furthermore, the purpose must be grounded in clearly delineated assessment end points. Assessment end points are defined by Suter (1993) as "a quantitative or quantifiable expression of the environmental value considered to be at risk." For practical purposes, assessment end points are operationally defined by an ecological entity and its attributes. For example, a sugar beet or field is operationally defined by its ability first to produce sugar for food use and second to perform the services of sequestering carbon and purifying water and habitat for organisms that do not have an unacceptable impact on the primary purpose of producing sugar beets. There is a third assessment end point as well: sugar beet fields in agriculture are part of the landscape, in terms of both aesthetics and biodiversity preservation. Many Europeans regard this point differently from people in other parts of the world. Obviously, appropriate assessment end points for sugar beet fields will differ depending on the ecological systems and specific country. However, the interface between adjacent systems (e.g., farm fields and unmanaged areas) must be considered in choosing appropriate assessment end points for monitoring. Finally, assessment measurements must be selected based on their relationship and interpretability to the assessment end point. Because of the numerous challenges and complications in selecting operationally definable assessment end points, the difficulties associated with designing relevant measurements, and the history of failed monitoring programs, some have questioned the value of monitoring overall (Thorton and Paulsen, 1998). Clearly, monitoring can be problematic if care is not taken to develop a program that is scientifically sound and relevant to the concerns of stakeholders.

REGULATORY ASPECTS

Some proposals for systematic and structured approaches to monitoring have been presented, but methodologies still need to be implemented worldwide. The new European Union Directive 2001/18/EC, however, specifically emphasizes harmonization of procedures and methodologies, as well as additional initiatives that will be taken to establish a common methodology, including monitoring, for risk assessment. Monitoring has several functions = to confirm risk, to confirm that risk mitigation practices are working as they should, to ensure that environmental goals are being achieved, to assess environmental conditions, and so on. A company registering a GMP has to ensure that monitoring is carried out according to specified conditions, such as being traceable at all stages of the market. Although constituting the fundamental principle of precaution for risk assessment, and acknowledging case-by-case treatment and step-by-step introduction of a GMO into the environment, the directive does not provide information on the methods to be used or applied for the risk assessment and monitoring.

BIOSAFETY RESEARCH ON VIRUS-RESISTANT SUGAR BEET

We here present results from biosafety research focusing on an ecologically relevant transgenic trait with the potential to increase invasiveness in wild plant hybrids. Beet (*B. vulgaris* L.) is an important subject for invasiveness studies for two reasons. First, gene flow has been demonstrated between cultivated sugar beets (*B. vulgaris* ssp. *vulgaris* provar. *altissima* Döll) and wild beets (*B. vulgaris* ssp. *maritima* Arcangelí), as evidenced by the introgression of the annual habit into cultivated beets (Bartsch *et al.*, 1999) and the introgression of (conventional) genes of sugar beet from areas of seed production to wild beet populations (Desplanque *et al.*, 1999, Mücher *et al.* 2000). Second, hybridization between wild beets and sugar beet breeding plants leads to a hybrid form ("weed beet") that is able to bolt and flower among the biennial sugar beet varieties during the cultivation period. Annual weed beets are a serious problem in parts of Europe. Wild and cultivated *Beta* species have also invaded the New World (California) (Bartsch and Ellstrand, 1999). This is important baseline and background information for integrating into a risk assessment of potential GMP effect patterns.

FIELD EXPERIMENTS

Transgenic traits are genetically dominant in heterozygotes and inherited like conventional genes in wild beet populations (Dietz-Pfeilstetter and Kirchner, 1998). In our study, the transgenic beets expressed tolerance to rhizomania caused by the virus BNYVV, a disease that has spread through the sugar beet fields of Europe, California, Japan, and China. Rhizomania is transmitted via the fungus *Polymyxa betae* Keskin (Cooper and Asher, 1988), and the disease leads to decreased sugar beet yields and losses of up to 30% sugar content. The advantages of tolerance to BNYVV may appear in ecological performance parameters at different life stages of beet, including first-year vegetative growth, overwintering, or second-year bolting and seed formation. We wanted to know whether transgene-mediated virus tolerance conferred fitness over and above naturally virus-tolerant beet genotypes, especially wild beet hybrids.

In our field experiments, we compared the ecological performance of three beet genotypes. Two were hybrids between wild beet and sugar beet (F1 and F2); the third was a pure sugar beet variety. All three exhibited a naturally selected virus tolerance originating from wild beet by hybridization or conventional breeding, but one of the wild beet hybrids carried an additional transgene for virus tolerance. In all other respects the two wild beet/sugar beet hybrids were genetically equivalent. The methods chosen for competition and overwintering experiments, including determination of winter cold sum, as well as geographical description of the field site with virus infestation and the virus-free control site, are described elsewhere (Bartsch *et al.*, 1996, Pohl-Orf *et al.*, 2000). For phytosanitary reasons, inoculation of virus-free sites was not possible, and thus the comparability of both sites, with their different soil physiochemical and climatic conditions, was limited. Seed biomass production was measured from flowering plants in pollen-impermeable field chambers. The plants were grown with or without virus infection, and with different levels of competition by a common weed (*Chenopodium album*).

Independent of virus presence or weed competition level, we found no significant differences in first-year biomass production among the three genotypes (Fig. 4). In the competition experiments beet biomass production was decreased at the highest weed pressure. The plants at the virus infestation site tended to have lower biomass production in comparison to virus-free conditions. Enzyme-linked immunosorbent assay (ELISA) data demonstrated a low level of virus infection in all plants grown under virus infestation, although earlier studies had revealed high virus infection of susceptible genotypes (Bartsch *et al.*, 1996). At a given field site, no significant differences among the three plant genotypes were found in overwintering capacity, biomass production of bolters resulting from the surviving beets, or seed production. The only exception was the sugar beet cultivar, which showed a significantly weaker overwintering rate at our virus-free site but produced better bolter biomass under virus infestation (Fig. 5). However, the field location had a clear effect on the overwintering rate and bolter biomass in general, although not on seed biomass production: the plant survival rate was significantly lower at the virus-free site that at the infestation site, most likely because of the colder winter at the virus-free site (winter cold sum: – virus / + virus: −27 °C / −17 °C). Indeed, the survival rates fit well with the known correlation between survival and winter cold sum (Pohl-Orf *et al.*, 1999).

Owing to the potential for hybridization between cultivated and wild beets, it is important to know whether transgenic virus tolerance could also increase the fitness of wild beet populations. To date, no field release study has assessed the performance a wild plant of an ecologically relevant transgenic trait – such as virus tolerance – in comparison to an isogenic conventional genotype or a conventional newer virus-tolerant cultivar. Our experiments

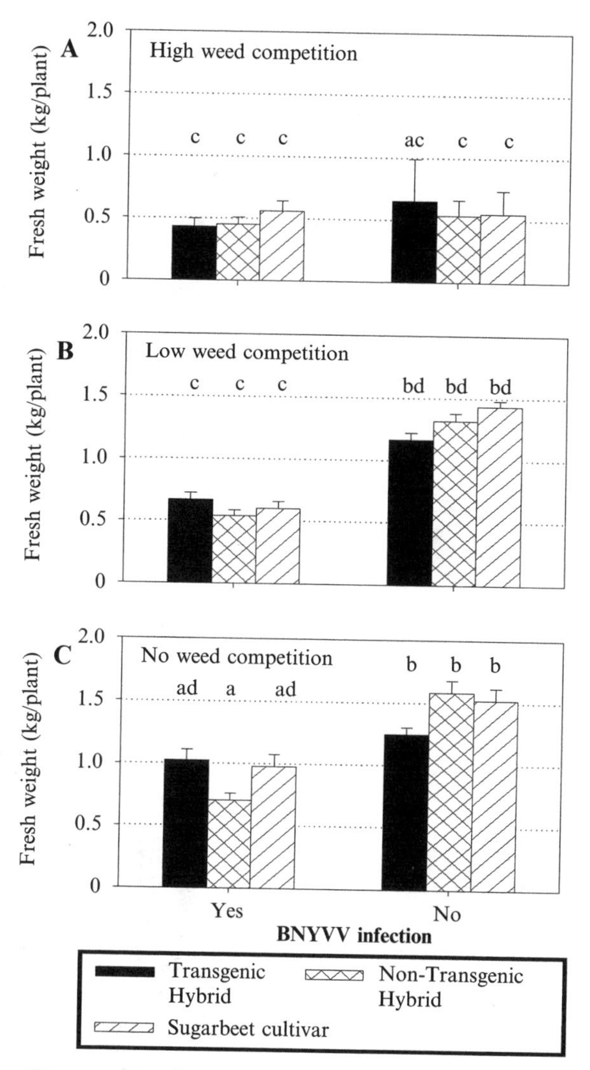

Figure 4 Competitiveness of beet (measured as kilograms of biomass production per individual) grown under different weed competition treatments (A-C) with and without infection by virus BNYVV. Means of biomass production with the same indicator letter are not significantly different by three–way ANOVA – Tukey test.

could be extrapolated to address such questions. Indeed, it was most likely due to the genetic wild beet background of the cultivars that we found no difference between transgenic and nontransgenic hybrids. Natural virus tolerance

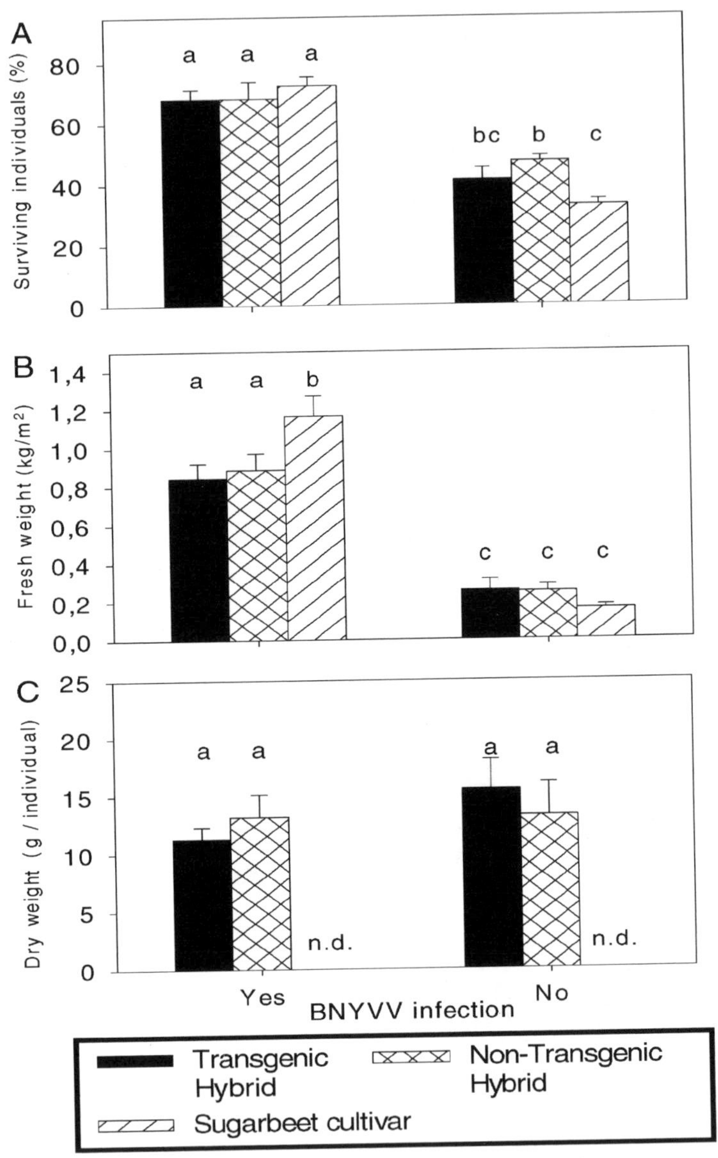

Figure 5 Ecological parameters of beet grown with and without virus infection. (A) Hibernation, (B) bolter biomass, and (C) seed production. Mean levels of characters with the same indicator letter are not significantly different by two-way ANOVA – Tukey test (SEM, standard error of mean; n.d., not detected).

was inherited from wild beet in all three genotypes tested: in the transgenic and isogenic control genotype by our hybridization with wild beet, and in the virus-tolerant cultivar by recent hybridization with wild beet and subsequent backcross-breeding to a high performance cultivar. Our results allowed us to conclude that addition of a transgenic virus tolerance trait will not significantly increase the ecological dominance of naturally tolerant genotypes.

Considering that gene exchange between wild and cultivated beet has been observed, we would predict a risk of transgenic tolerance to rhizomania in wild beet hybrids similar to that of naturally tolerant beets. These results may alter the important debate about genetic traits incorporated using traditional breeding methods and traits incorporated by using recombinant DNA technology (Tiedje *et al.*, 1989). One concern in this debate is that the process of mutation and selection can be accelerated by genetic engineering and that gene technology may break the "rules" of evolution. An escape of transgenes into wild beet populations will surely occur, although strategies are being developed to minimize gene flow (Gray and Raybould, 1998; Saeglitz *et al.*, 2000). Our studies with transgenic and nontransgenic plants show that no ecological advantage will result, and wild beets can and will develop resistance to rhizomania owing to natural mechanisms of disease tolerance. That the transgenic wild hybrids were not more fit under conditions of virus infestation suggests that the risk of this particular transgenic trait is comparable to the risk present with conventionally bred virus-tolerant beets. In addition, the virus seems to be excluded from coastal wild beet habitats, where unfavorable salt conditions in the soil depress virus transmission of the fungus vector (Bartsch and Brand, 1998). Finally, in this specific case, there is no science-based, risk-based reason to minimize or mitigate gene flow beyond the precautions already being used in sugar beet seed production.

BASELINE DATA: EXAMPLE OF USING BIOGEOGRAPHICAL DATA FOR BIOSAFETY RESEARCH

The genus *Beta* descended from the Old World. Cultivated beets have been known for more than 2000 years in the eastern Mediterranean region. In Europe, wild *Beta vulgaris* ssp. *maritima* is largely a coastal taxon, widely distributed from the Cape Verde and Canary Islands in the west, northward along Europe's Atlantic coast to the North and Baltic Seas. It extends eastward through the Mediterranean region into Asia where it occurs in Asia Minor, in the central and outer Asiatic steppes, and in desert areas as far as western India (Letschert, 1993). There is no barrier to crossing between wild and cultivated forms of *B. vulgaris* (Bartsch and Pohl-Orf, 1996). Thus we have proposed that wild populations be recorded as baseline data on existing genetic diversity and abundance of potential recipients of gene flow. Under conditions where gene flow is likely, wild populations should be collected prior to large-scale release of genetically modified organisms.

Our general survey focused on two regions: one in the United States and one in Europe. One of the most sensitive areas for invasive species is California, since the state's enormous species diversity is threatened by nonnative plants that already contribute to 40% of the total species number. Introduced wild beets in California come either: from European *B. vulgaris* or from *B. macrocarpa* (Fig. 6). Substantial evidence was found for hybridization and introgression of *B. vulgaris* alleles into *B. macrocarpa* populations of the Imperial Valley. Therefore, monitoring the potential spread of weed beet populations into protected areas should be concentrated in two local areas: in the Anza Borrego Desert State Park, which is close to the western edge of the Imperial Valley; although no beet has been found there so far (http://theabf.org/statepark.html) and in the Channel Islands National Park, where *B. macrocarpa* is already reported. The latter region has the advantage of extensive, established floristic monitoring programs (Johnson, 1998) (http://www.nps.gov/chis/rm/Index.htm).

The second region is one of Europe's most important sugar beet seed production districts, mentioned earlier (Fig. 7). In northeastern Italy, domesticated beet seed production has been carried out for more than 100 years, with intensification since the 1950s. Commercial sugar beet seeds are produced on 4500 ha, and each hectare contains approximately 50,000 flowering plants. Furthermore, small farmers in the region grow red beet (*B. vulgaris* ssp. *vulgaris* var. *conditiva*) and Swiss chard (*B. vulgaris* ssp. *vulgaris* var. *vulgaris*) for private seed production, which may be an additional source of gene flow. Wild sea beet populations occur on the nearby coastal plain, sometimes within a kilometer of the cultivated fields. Crop-to-wild-gene flow is common but has not led to any adverse effects on the population size or genetic diversity decline (Bartsch *et al.*, 1999). In this region, gene flow from 2.25×10^8 flowering sugar-beets into populations of approximately 4×10^4 flowering wild beets occurs over an area of $4000 \, \mathrm{km}^2$ (Bartsch and Schmidt, 1997; Bartsch and Brand, 1998). In this area the general monitoring needed to study long-term effects of gene flow between cultivated and wild beets is best conducted by local research stations (e.g., the Research Institute for Industrial Crops: http://www.isci.it/).

MONITORING OF INSECT-RESISTANT MAIZE

In the last decade, "*Bt*" genes of *Bacillus thuringiensis* var. *kurstaki* (Berliner) that encode lepidopteran-specific toxins (cry1Ab, cry1Ac, cry9) were

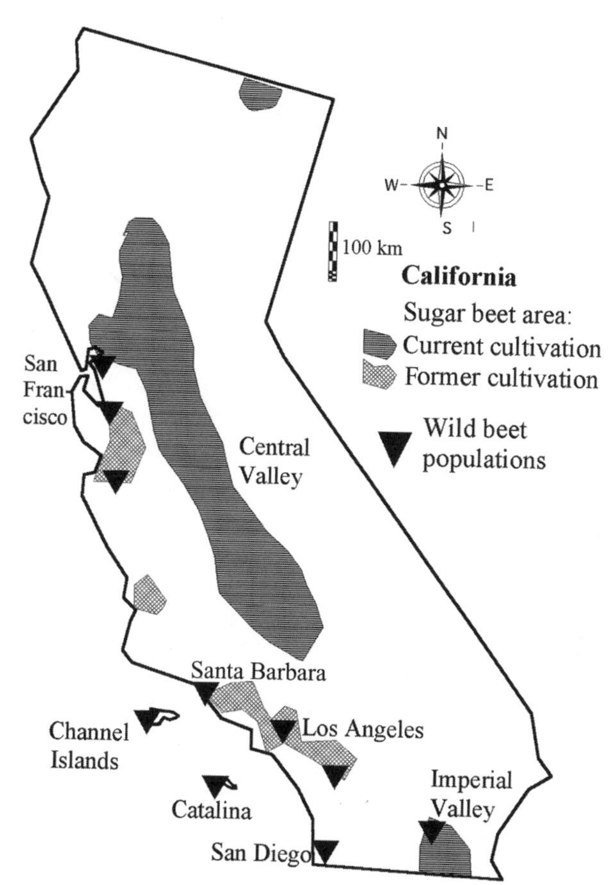

Figure 6 *Beta vulgaris* ssp. *vulgaris* (sugar beet, Swiss chard, table beet) cultivation areas and wild beet (*B. vulgaris* ssp. *maritima* and *B. macrocarpa*) in California (Bartsch and Ellstrand, 1999). The Channel Islands and Catalina, with their intensive floristic monitoring programs, should be designated areas for monitoring in nonagrosystems. The Imperial Valley is important for monitoring owing to potential gene flow in a wild weed–cultivar system.

engineered into maize for protection against the European corn borer [ECB, *Ostrinia nubilalis* (Hbn.)]. Arguments in favor of the introduction of *Bt* maize claim that the ECB and other harmful lepidopterans can be controlled effectively, selectively and in an environmentally friendly manner. However, laboratory studies revealed evidence of potentially adverse effects on non-target organisms. For example, Hilbeck *et al.* (1998a,b) raised concerns about lacewings negatively affected by lepidopteran prey previously fed with *Bt* maize; Losey *et al.* (1999) demonstrated with monarch caterpillars

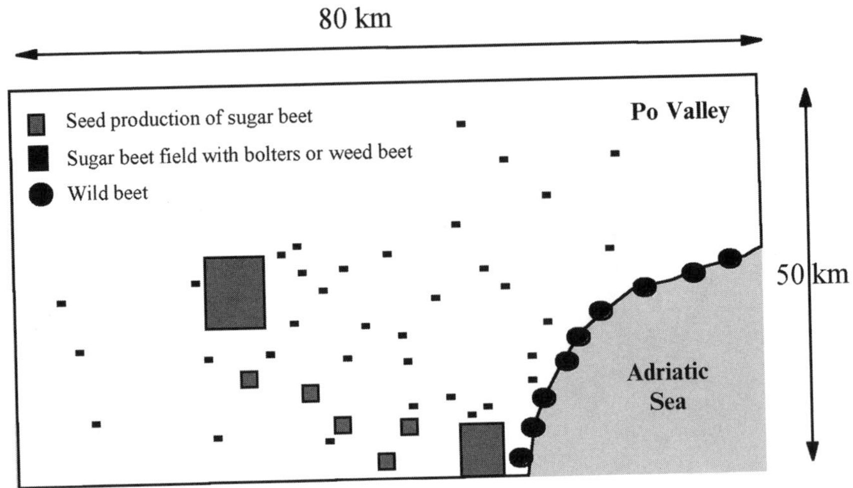

Figure 7 Necessary monitoring area in northeastern Italy with sympatric growth in a wild weed–cultivar system (symbols of population size are proportional to number of generative plants).

(*Danaus pleixppus*) that deposition of *Bt* maize pollen on plants can reduce vitality of phytophagous insects feeding on them. The potential relevance of this interaction stimulted extensive discussions on the environmental risks of genetically modified organisms (Kleiner, 1999) and intensified research into the effect of *Bt* maize pollen deposition on the monarch butterfly. The results of medium- and large-scale field studies in comparison to the toxicity data collected to date demonstrate minimal risk to butterflies (Hellmich *et al.*, 2001; Oberhauser *et al.*, 2001; Pleasants *et al.*, 2001; Sears *et al.*, 2001; Stanley-Horn *et al.*, 2001; Zangerl *et al.*, 2001)

On the other hand, concern was raised that season-long, constitutive expression of the *Bt* toxin in plants may select for rapid resistance development in pest populations (e.g., ECB). Applying a precautionary approach, much effort has been spent on establishing national insect resistance management (IRM) strategies to delay or avoid this effect (Alstad and Andow, 1996; Siegfried, 2000).

General Approaches to Monitoring

Two-tiered approaches to monitoring can be defined based on the starting point of a cause–effect hypothesis. First, the "bottom-up approach" is based on the lowest level of biological organization and tries to focus on resulting effects at a higher organization level: gene → organism → population → community → ecosystem. This first approach is based on building from lower to higher complexity. It is equivalent to the "tiered" approach that is used in biosafety research in a step-by-step, case-specific manner prior to commercialization. The "top-down approach," on the other hand, focuses primarily on nature conservation end points at a high level of biological organization: ecosystem → community → population → organism → gene and tracks observations back to causes from higher to lower complexity. The latter approach may not have a specific hypothesis about cause and effects. The "top-down" approach relies on the development of appropriate indicator species and parameters to detect effects. There is much work needed to reduce theory to practice, since current knowledge and scientifically established correlations between indicator species and ecosystem conditions are limited. The "top-down" approach, based on general surveillence, can be used to identify effects that may not be readily associated with a known stressor. In this case, the source of the hazard would have to be defined through specific experiments.

V.B. Field-Based Monitoring: Agricultural and Nature Conservation Areas

A general approach for the selection of specific, relevant nontarget species for large-scale field studies (> 10 ha) is lacking. Monitoring should concentrate the limited resources on the protection of environmental end points. Since laboratory and field studies can target only a limited number of species (Wagner *et al.*, 1996), broad and long-term research programs must be designed to detect a wide range of potential effects that are considered to be adverse. Those monitoring for adverse effects through the use of specific laboratory toxicology tests therefore confront the problem of having to choose potential nontarget species from the numerous species in the field. The straightforward collection of arthropods from fields and census evaluation may not detect subtle, but important effects. The adequate conservation of herbivores, especially lepidopterans, in the agricultural landscape is important for general environmental protection efforts (Declaration of Rio: New *et al.*, 1995; integrative concept of nature conservation: Plachter, 1991). In addition, integrated pest management (IPM) strategies rely on sufficient nontarget caterpillars that serve as alternative hosts for parasitoids of economic relevance (Franz and Krieg, 1982). Monitoring must also be adapted to European agricultural structures with relatively small fields often close to or even in nature conservation areas, meaning that in contrast to the United States, Red List species are widely distributed and protected in agrosystems.

A monitoring approach calls knowledge of indicator species representing local ecological structure and function. We concentrate here on potential effects of maize pollen with significant contents of *Bt* toxin. To develop and apply a preselection tool for relevant species in Germany for monitoring programs, the following steps can be used:

- Establishment of an index of *Bt* pollen relevance ($I_{Bt\,p}$), which preselects herbivore species based on known data of the circumstances of pollen exposure and on the general susceptibility of certain systematic groups
- Screening of a database of Macrolepidoptera (LEPIDAT) from the German Federal Environmental Protection Agency for a larger scale evaluation of relevant and protected species (Pretscher, 1998).

Selecting Relevant Species

Characterization of potential harmful effects is regarded as one of the first steps in assessing ecological risks potentially caused by genetically modified plants. While a system for assessing the likelihood of outcrossing into native relatives is available (Ammann *et al.*, 1996), the potential risk by exposure to *Bt* maize pollen for nontarget herbivores has not yet been evaluated. The Ammann/Dutch system will not work for exposure to pollen because it measures an end point that does not explicitly factor pollen volume. Exposure to pollen poses a different route of exposure to *Bt* toxin that is relevant for insects other than herbivores that feed on corn tissue (Fig. 8). This latter group of insects appears to be well studied; indeed, a number of publications are available that find no effect of cry1Ab toxins on these nontarget species (Glare and O'Callaghan, 2000).

Our "index of *Bt* pollen relevance" ($I_{Bt\,p}$) is the first attempt to formulate and order criteria and thus standardize the selection of species potentially affected by *Bt* maize pollen (Schmitz et al, submitted, Fig. 9). This index combines both factors for determining the probability of toxin ingestion (exposure assessment) and factors describing the susceptibility of herbivore populations and their conservation status (hazard assessment). It clearly corresponds to various models of chemical risk assessment (cf. review of Strauss, 1990). The decision tree (Fig. 9) and especially the Lepidoptera lists do not directly indicate the risk of cultivation of *Bt* maize per se but could be used to select species for more detailed field and lab studies. This decision tree lacks a key element: hazard assessment. The range of $LC_{50}s$ to specific *Bt* proteins across the order Lepidoptera is highly variable: *Spodoptera frugiperda* is approximately 25,000 times less sensitive to Cry1Ab than monarch larvae. Since large amounts of data are missing for most of the countries' species, the decision tree must be open at this stage. In this context one should clearly distinguish between the vast group of species that theoretically (i.e., under certain local conditions, e.g., Fig. 10, and many species listed in Villiger, 1999) could be affected by deposition of *Bt* maize pollen and the (smaller)

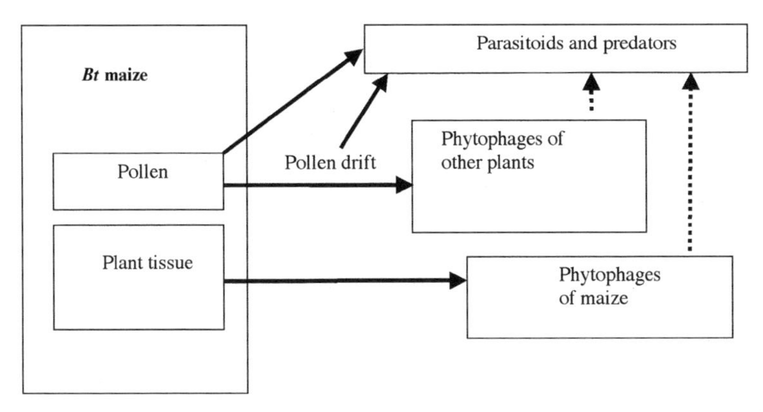

Figure 8 Exposure to *Bacillus thuringiensis* toxin and potential tritrophic effects in maize fields or adjusted field borders.

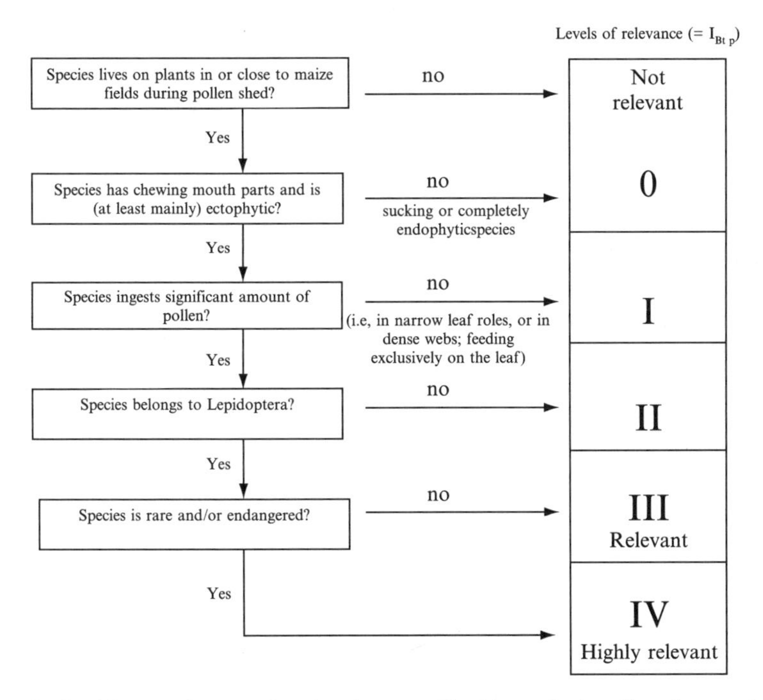

Figure 9 Decision tree for assessing the relevance of herbivores in the adjusted vegetation to investigate the effect of *Bt* maize pollen exposure (for more details, see text).

group of species living primarily in the weedy vegetation of fields and field margins of agricultural landscapes where maize can be cultivated. At least five different factors are considered when relevant species are selected for more detailed risk assessment. These are discussed in the subsections that follow.

Pollen Exposure

The decision steps leading to $I_{Bt\,p}$ relevance levels include criteria of pollen exposure, *Bt* toxin susceptibility, and frequency. The amount of pollen ingestion depends on the feeding mode, which can vary significantly within the life span of the herbivorous stage. While, for instance, the Agrotinae (Noctuidae) as young larvae feed externally on leaves but as mature larvae live hidden at the roots and the stem base, the life of pierid butterflies is first hidden (leaf mines) and later exposed. Species that mainly feed endophytically would ingest pollen only when boring into a new feeding place (*Hadena* spp., Noctuidae). On a microspatial scale, the characteristics of both plant architecture and surface probably influence the amount and duration of pollen load. While on rough, hairy, or glandulous leaves (e.g., species of Boraginaceae, Urticaceae, Lamiaceae) pollen would accumulate, smooth and waxy leaves (e.g., *Brassica* spp.) would minimize pollen deposition owing to self-cleaning effects. In addition, pollen could accumulate on leaves with a honeydew layer.

Susceptibility to Bt Toxin

The susceptibility to *Bt* toxins is first and foremost species specific and is thus influenced by natural physiology (Glare and O'Callaghan, 2000). Studies on chemical pesticide effects show that polyphagous species are in general less susceptible to *Bt* toxins because they have inherent detoxification ability (e.g., Gordon, 1961). In addition, one should consider that young larvae are in general more susceptible than older ones (Glare and O'Callaghan, 2000), meaning that relevance is highest when larvae of a given species hatch from eggs in July. Moreover, it is difficult to assess the effect of a given *Bt* maize

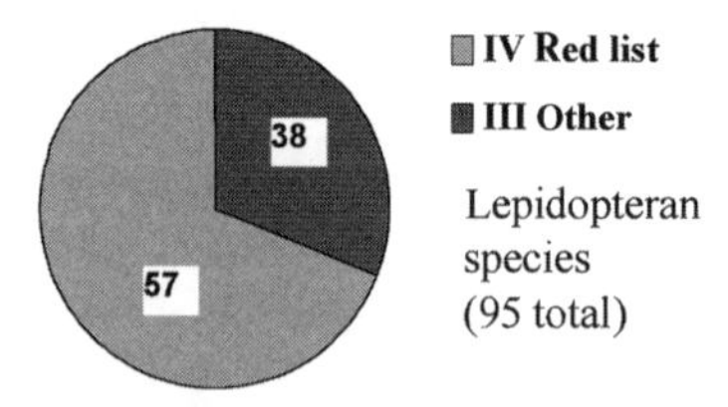

Figure 10 Macrolepidoptera selected from the LEPIDAT database living primarily in or near the border of maize fields in Germany. Red List species are endangered (Binot *et al.*, 1998).

strain on a specific herbivore species because various factors influence these interactions. For instance, the specific amount of *Bt* toxin in the pollen differs significantly between the various strains of *Bt* maize. The cry1Ab protein content (micrograms per gram, fresh wt] varies from 0.09 to 7.1 (transgenic events Mon 810 and Nov 176, respectively) (Fearing *et al.*, 1996). Until now, negative effects have been found only with the pollen of high-expression Bt176. Within other transgenic events (Mon 810) it seems much more unlikely that direct lethal effects on nontarget herbivores will be detected. However, it should be considered that even small amounts of *Bt* toxin may alter physiological and ethological features of the herbivores (pesticides: Theiling and Croft, 1988: *Bt* oilseed rape: Schuler *et al.*, 2001). For instance, an elongated time of larval stage or a change in defense or escape behavior could diminish the effectiveness of attack avoidance by parasitoids and predators. In addition, sublethal effects may also include a reduced number of matings or eggs, or less successful overwintering.

Frequency

The frequency of – and respective threat to – a given species varies significantly between the different natural areas within a country. However, to select species in a solid and thus practicable manner, such detailed differentiation was not recognized. This is because the aspect of practicability plays an important role in monitoring program selection from a list of preselected species.

Macrolepidoptera List

The wealth of knowledge pertaining to the "Macrolepidoptera" (including the Zygaenidae) suggested the practicality of applying the index criteria – as a first step – to this herbivore group and on the electronic LEPIDAT database of the status and ecology of the Macrolepidoptera in Germany. To focus on the species that are potentially highly affected by *Bt* maize pollen, species inhabiting primarily meadows, pastures, hedges, or natural biotopes were excluded. Since species inhabiting a wide range of open-area biotopes would not be likely to be endangered by the *Bt* pollen effects, these were also excluded from the list even though they might colonize agricultural biotopes in high densities. It should be considered that – as with other herbivores – lepidopteran species might differ in their habitat choice from region to region. Species that hardly occur in or at the margin of fields in the northern lowlands of Central Europe may be abundant in the southern hilly parts. This could also be combined with differences in host plant preference, again possibly altering the susceptibility to *Bt* toxins (Glare and O'Callaghan, 2000). As a result, the selection of a species as a candidate for biosafety research, especially long-term monitoring, needs detailed evaluation of the biological key parameters.

Comparison between Local Data and the list extracted from LEPIDAT

A simple listing of diurnal butterflies whose larvae live during pollen seasons in the altitudes of the maize cultivating zones appears to be inadequate. For instance, many of the species listed by Villiger (1999) do not live in close spatial contact to maize fields and would therefore hardly ingest corn pollen in amounts constituting a risk. While the absolute amount of pollen deposition in relation to the distance from the field margin varies between different studies, a sharp exponential decrease in the first few meters was always reported.

In Germany, the order Macrolepidoptera consists of approximately 1400 species (Gaedicke and Heinicke, 1999). Considering habitat preference, we found 95 lepidopteran species reported to live in or at the field margin of maize fields (see criteria above), which represents 7% of the total species list (Fig. 10). Thirty-eight of these are rare or endangered (5.3% of the Red List species). Relative to the total number of species in Germany, the Pieridae are the most highly represented family.

Assuming that species are unlikely to be affected at distances of more than 10 m from maize field margins, the number of species potentially effected is very small. As an example, we found on a local scale only 14 lepidopteran species, which are also not subject in the Red List (Schmitz *et al.*, submitted).

CONCLUSIONS: LINKING BIOSAFETY RESEARCH AND MONITORING

It is a legal and practical requirement that biosafety research be conducted prior to commercialization. Biosafety research must provide sufficient information to permit a science-based decision concerning risk to be made. Practical temporal and spatial limitations, however, compel the investigation of some relevant hypotheses regarding cause and effect on a regional scale. When these studies have been completed, monitoring can be designed to confirm the results of the biosafety research, support risk management decisions, or better define risk. Separate from the decision of what constitutes acceptable risk is the need for monitoring to more broadly define environmental quality and direct the future development of food production practices. A monitoring program could be developed to also address risk uncertainty, which in principle is not different for GMPs compared with traditionally bred plants. Selection of monitoring parameters based on conservation goals is a prerequisite for its overall effectiveness (Fig. 11). Method

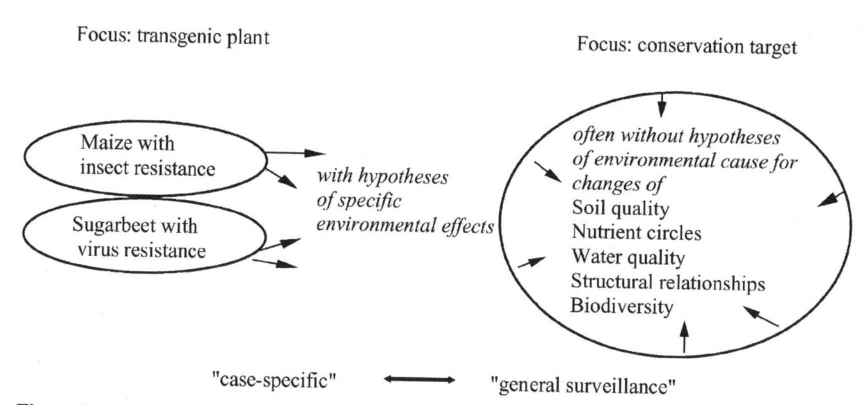

Figure 11 Future task of monitoring: linking the case-specific approach (left) with general surveillance (right).

compendiums for field experiments and monitoring are available, but scale, time period, and number of replicates need to be defined. General studies of nontarget effects in and outside the field should be linked to general environmental monitoring in these habitats. "Monitoring and surveillance of genetically modified higher plants," by Kjellsson and Strandberg (2001) provides measures for analyzing possible environmental effects when GMPs are cultivated. These anothers offer suggestions, including objectives, procedures, and methods for detecting GM plant dispersal and hybridization, effects on ecosystems, vegetation, and organism groups, sampling design, data analysis, and relevant statistical methods. Because detailed monitoring can be costly in terms of time and money, any program worth the expense should be sensitive to detecting potential cumulative long-term effects.

ACKNOWLEDGMENTS

The authors thank Tom Nickson for helpful comments, Peter Pretscher for analyzing the LEPIDAT database, and Avril Arthur-Goettig for reviewing the English manuscript. Part of the work was funded by the German Ministry of Education and Research (BMBF grant, 0310532 and 0310785).

REFERENCES

Alstad, D. N., and Andow, D. A. (1996). Implementing management of insect resistance to transgenic crops. *AgBiotechno. News Inf.* **8**, 177–181.

Ammann, K., Jacot, Y., and Rufener Al Mazyad, P. (1996). Field release of transgenic crop in Switzerland, an ecological risk assessment, in *Gentechnisch veränderte krankheits- und schä-*

dlingsresistente Nutzpflanzen – Eine Option für die Landwirtschaft, E. Schulte and O. Käppeli, eds., pp. 101–158. Schwerpunktprogramm Biotechnologie des Schweizerischen Nationalfonds zur Förderung der wissenschaftlichen Forschung, Bern.

Ammann, K., Jacot, Y., and Rufener Al Mazyad, P. (2001). Safety of genetically engineered plants: an ecological risk assessment of vertical gene flow, in *Safety of Genetically Engineered Crops*, (R. Custers, ed.), pp. 60–87. Flanders Interuniversity Institute for Biotechnology VIB publication, Zwijnaarde, Belgium.

Bartsch, D., and Brand, U. (1998). Saline soil condition decreases rhizomania infection of *Beta vulgaris*. *J. Plant Pathol.* (Pisa), **80**, 219–223.

Bartsch, D., and Ellstrand, N.C. (1999). Genetic evidence for the origin of Californian wild beets (genus *Beta*). *Theor. Appl. Genet)*., **99**, 1120–1130.

Bantsch, D., and Schmidt, M. (1997) Influence of sugar beet breeding on populations of *Beta vulgaris* ssp. *maritima* in Italy. *Journal of Vegetation Science* (Uppsala) **8**, 81–84.

Bartsch, D., and Pohl-Orf, M. (1996). Ecological aspects of transgenic sugar beet – Transfer and expression of herbicide resistance in hybrids with wild beets. *Euphytica*, **91**, 55–58.

Bartsch, D., Schmidt, M., Pohl-Orf, M. Haag, C., and Schuphan, I. (1996). Competitiveness of transgenic sugarbeet resistant to beet necrotic yellow vein virus and potential impact on wild beet populations. *Mol. Ecol.*, **5**, 199–205.

Bartsch, D., Lehnen, M., Clegg, J., Pohl-Orf, M., Schuphan, I., and Ellstrand, N. C. (1999). Impact of gene flow from cultivated beet on genetic diversity of wild sea beet populations. *Mol. Ecol.*, **8**, 1733–1741.

Bartsch, D., Brand, U., Morak, C., Pohl-Orf, M., Schuphan, I., and Ellstrand, N. C. (2001). Biosafety research on genetically engineered virus resistant hybrids between transgenic sugarbeet and Swiss chard. *Ecol. Appl.*, **11**, 142–147.

Binot, M. Bless, R., Boye, P., Gruttke, H., and Pretscher, P. (1998) Rote Liste gefährdeter Tiere Deutschlands.; Schriftenreihe für *Landschaftspflege und Naturschutz* (Bonn-Bad Godesberg), **55**, 1–434.

Cooper, J. I., and Asher, M. J. C. (1988). *Viruses with Fungal Vectors*. Association of Applied Biologists, Wellesbourne, United Kingdom.

Crawley, M. J., Brown, S. L., Hails, R. S., Kohn, D., and Rees, M. (2001). Transgenic crops in natural habitats. *Nature*, **408**, 682–683.

Desplanque, B., Boudry, P., Broomberg, K., Saumitou-Laprade, P., Cuguen, J., and Van Dijk, H. (1999). Genetic diversity and gene flow between wild, cultivated and weedy forms of *Beta vulgaris* L. (Chenopodiaceae), assessed by RFLP and microsatellite markers. *Theor. Appl. Genet.*, **98**, 1194–1201.

Dietz-Pfeilstetter, A., and Kirchner, M. (1998). Analysis of gene inheritance and expression in hybrids between transgenic sugarbeet and wild beets. *Mol. Ecol.*, **7**, 1693–1700.

Ellstrand, N. C., Prentice, H. C., and Hancock, J. F. (1999). Gene flow and introgression from domesticated plants into their wild relatives. *Annu. Rev. Ecol. Syste.*, **30**, 539–563.

Fearing, P. L., Brown, D., Vlachos, D., Meghji, M., and Privalle, L. (1996). Quantitative analysis of CryIA(b) expression in *Bt* maize plants, tissues, and silage and stability of expression over successive generations. *Mol. Breed.*, **3**, 169–176.

Franz, J. M., and Krieg, A. (1982). *Biologische Schädlingsbekämpfung*, Verlag Paul Parey, Hamburg and Berlin.

Gaedike, R. Heinicke W. (1999) Verzeichnis der Schmetterlinge Deutschlands. Fauna Germanica Vol.3. *Entomologische Nachrichten und Berichte (Dresden)* **5**, 1-216

Glare, T. R., and O'Callaghan, M. (2000). "*Bacillus thuringiensis: Biology, Ecology and Safety*". Wiley, Chichester.

Gordon, H. T. (1961). Nutritional factors in insect resistance to chemicals. *Ann. Rev. Entomol.*, **6**, 27–54

Gray, A. J., and Raybould, A. F. (1998). Reducing transgene escape routes. *Nature*, **392**, 653–654.

Hellmich, R. L., Siegfried, B. D., Sears, M. K., Stanley-Horn, D. E., Daniels, M. J., Mattila, H. R., Spencer, T., Bidne, K. G., and Lewis, L. C. (2001). Monarch larvae sensitivity to *Bacillus thuringiensis* purified proteins and pollen. *Proc. Natl. Acad. Sci, USA*, **98**, 11925–11930.

Hilbeck, A., Baumgartner, M., Fried, P.M., Bigler, F., and (1998a). Effects of transgenic *Bacillus thuringiensis* maize-fed prey on mortality and development time of immature *Chrysoperla carnea* (Neuroptera, Chrysopidae). *Environ. Entomol.*, **27**, 480–487.

Hilbeck, A., Moar, W. J., Pusztai-Carey, M., Filippini, A., and Bigler, F. (1998b). Toxicity of *Bacillus thuringiensis* CryIAb to the predator *Chrysoperla carnea* (Neuroptera, Chrysopidae). *Environ. Entomol.*, **27**, 1–9.

Hoffman, C. (1990). Ecological risks of genetic engineering of crop plants. *Bioscience*, 40, 434–437.

Johnson, L. W. (1998). *Terrestrial Vegetation Monitoring 1984–1995*. National Park Service – Channel Islands National Park Technical Report 98–08, 39 pp. http://www.nature.nps.gov/im/units/chis/pdfreports/terrestrial/84–95veg.pdf

Kareiva, P., Parker, I. M., and Pascual, M. (1997). Can we use experiments and models in predicting the invasiveness of genetically engineered organisms? *Ecology*, **77**, 1670–1675.

Kjellsson, G., and Strandberg, M. (2001). *Monitoring and Surveillance of Genetically Modified Higher plants. Guidelines for Procedures and Analysis of Environmental effects*, 119 pp, Birkhäuser Verlag, Basel.

Kleiner, K. (1999). Monarchs under siege. *New Scie.* **162**, 2187.

Lambelet-Haueter C. (1991). Mauvaises herbes et flore anthropogène: II. Classifications et catégories. *Saussurea*, **22**, 49–81.

Letschort J. P. W. (1993) *Beta* section *Beta*: biogeographical patterns of variation and taxonomy. *Wageningen Agricultual University Papers* (The Netherlands) **93**, 153pp.

Losey, J. E., Rayor, L. S., and Carter, M. E. (1999). Transgenic pollen harms monarch larvae. *Nature*, **39**, 214.

Mikkelsen, T. R., Andersen, B., and Jorgensen, R. B. (1996). The risk of crop transgene spread. *Nature*, **380**, 31.

Mooney, H. A., and Drake, J. A. (1986). *Ecology of Biological Invasions of North America and Hawai*, Springer-Verlag, New York.

Mücher, T., Hesse, P., Pohl-Orf, M., Ellstrand, N. C., and Bartsch, D. (2000). Characterization of weed beets in Germany and Italy. *J. Sugar Beet Res.*, **37**, 19–38.

New, T, R., Pyle, R. M., Thomas, J. A., and Hammond, P. C. (1995). Butterfly conservation management. *Annu. Rev. Entomol.*, **40**, 57–83.

Nickson, T. E., and Head, G. P. (1999). Environmental monitoring of genetically modified crops. *J. Environ. Monit.*, **1**, 101N-105N.

Oberhauser, K. S., Prysby, M. D., Mattila, H. R., Stanley-Horn, D. E., Sears, M. K., Dively, G., Olson, E., Pleasants, J. M., Lam, W. K. F., and Hellmich, R. L. (2001). Temporal and spatial overlap between monarch larvae and corn pollen. *Proc. Natl. Acad. Sci., USA*, **98**, 11913–11918.

Plachter, H. (1991). *Naturschutz*, 463 pp. Fischer, Heidelberg.

Pleasants, J. M., Hellmich, R. L., Dively, G. P., Sears, M. K., Stanley-Horn, D. E., Mattila, H. R., Foster, J. E., Clark, T. L., and Jones, G. D. (2001). Corn pollen distribution on milkweeds in and near cornfields. *Proc. Natl. Acad. Sci., USA*, **98**, 11919–11924.

Pohl-Orf, M., Brand, U., Driessen, S., Hesse, P., Lehnen, M., Morak, C., Mücher,T., Saeglitz, C., von Soosten, C., and Bartsch, D. (1999). Overwintering of genetically modified sugarbeet,

Beta vulgaris var. *altissima* Döll, as a source for dispersial of transgenic pollen. *Euphytica*, **108**, 181–186.

Pohl-Orf, M., Morak, C., Wehres, U., Saeglitz, C., Driessen, S., Lehnen, M., Hesse, P., Mücher, T., von Soosten, C., Schuphan, I., and Bartsch, D. (2000). The environmental impact of gene flow from sugarbeet to wild beet – An ecological comparison of transgenic and natural virus tolerance genes, in *Proceedings of the sixth International Symposium on the Biosafety of Genetically Modified Organisms*, July 2000, Saskatoon, Canada, (C. Fairbairn, G. Scoles, and A. McHughen, eds., pp. 51–55.

Pretscher, P. (1998). Rote Liste der Grossschmetterlinge (Macrolepidoptera), *Schriftrenr. Landschaftspflege Naturschutz* (Bonn), **55**, 87–111.

Rissler, J., and Mellon, M. (1996). *The ecological risks of engineered crops*, MIT Press, Cambridge, MA, and London, United Kingdom.

Saeglitz, C., Pohl, M., and Bartsch, D. (2000). Monitoring gene escape from transgenic sugarbeet using cytoplasmic male sterile bait plants, *Mol. Ecol.*, **9**, 2035–2040.

Schmitz, G., Pretscher, P., and Bartsch, D. (submitted). Monitoring of environmental effects of maize pollen containing *Bt* toxins: Tools for selection of relevant non-target herbivores.

Schuler, T. H., Denholm, I., Jouanin, L., Clark, S. J., Clark, A. J., and Poppy. G. M. (2001). Population-scale laboratory studies of the effect of transgenic plants on non-target insects. *Mol. Ecol.*, **10**, 1845–1853.

Sears, M. K., Hellmich, R. L., Stanley-Horn, D. E., Oberhauser, K. S., Pleasants, J. M., Mattila, H. R., Siegfried, B. D., and Dively, G. P. (2001). Impact of Bt corn pollen on monarch butterfly populations: A risk assessment. *Proc. Natl Acad. Sci., USA*, **98**, 11937–11942.

Siegfried, B. D. (2000). *Bt* transgenic plants for pest management – Challenges and opportunities, "*Proceedings of the sixth International Symposium on the Biosafety of Genetically Modified Organisms*, July 2000, Saskatoon, Canada," C. Fairbairn, G. Scoles, and A. McHughen, eds., p. 113–119.

Stanley-Horn, D. E., Dively, G .P., Hellmich, R. L., Mattila, H. R., Sears, M. K., Rose, R., Jesse, L. C .H., Losey, J. E., Obrycki, J. J., and Lewis, L. (2001). Assessing the impact of Cry1Ab-expressing corn pollen on monarch butterfly larvae in field studies. *Proc. Natl, Acad. Sci., USA*, **98**: 11931–11936.

Strauss, H. E. (1990): Lessons from chemical risk assessments, in: *Risk Assessment in Genetic Engineering*, M. A. Levin, and H. S. Strauss, eds, pp. 297–318. McGraw-Hill, New York.

Suter, G. W. (1993). Environmental surveillance, in *Ecological Risk Assessment*, G. W. Suter, ed., pp. 377–383 Lewis Publishers, Chelsea, MI.

Theiling, K. M., and Croft, B. A. (1988). Pesticide side-effects on arthropod natural enemies: A database summary. *Agric., Ecosyst. Environ.*, **21**, 191–218.

Thorton, K. W., and Paulsen, S. G. (1998). Can anything significant come out of monitoring? *Hum. Ecol. Risk Assess.*, **4**, 797–805.

Tiedje, J. M. R., Colwell, R. L., Grossman, Y. I., Hodson, R. E., Lenski, R. E., Mack, R. N., and Regal, P. J. (1989). The planned introduction of genetically engineered organisms: Ecological considerations and recommendations. *Ecology*, **70**, 298–315.

Villiger, M. (1999) *Effekte transgener insektenresistenter Bt-Kulturpflanzen auf Nichtzielorganismen am Beispiel der Schmetterlinge*. WWF Schweiz Eigenverl, Zurich.

Wagner, D. L., Peacock J. W., Carter H. L., and Talley S. E. (1996): Field assessment of *Bacillus thuringiensis* on nontarget Lepidoptera. *Envir. Entomol.*, **25**, 1444–1454.

Zangerl, A. R., McKenna, D., Wraight, C. L., Carroll, M., Ficarello, P., Warner, R., and Berenbaum, M. R. (2001). Effects of exposure to event 176 *Bacillus thuringiensis* corn pollen on monarch and black swallowtail caterpillars under field conditions. *Proc. Natl. Acad. Sci., USA*, **98**, 11908–11912.

Chapter 3

Status and Safety Assessment of Foods and Food Ingredients Produced by Genetically Modified Microorganisms

George A. Burdock
Burdock Group
Vero Beach, Florida

The "learning curve" for biotechnology has gotten ahead of regulations and threatens to erode public confidence in biotechnologically produced foods. This chapter discusses the four key issues: characterization of the organism; pathogenicity, toxigenicity, and antinutritive effects; substantial equivalence; and gene transfer, marker genes, and antibiotic resistance. These issues have shaped the guidelines, also discussed in this chapter. Case studies are provided to demonstrate how these issues are resolved by a petitioner according to the appropriate guidelines. For all these issues, the key ingredient in maintaining public confidence is the judgment of experienced scientists, whether as part of a GRAS (generally recognized as safe) process or a regulatory review process.

Biotechnology and Safety Assessment, 3rd edition

INTRODUCTION

Biotechnology represents a potential quantum leap forward to provide new foods that are better, safer, nutritionally enhanced, and more convenient, while also more abundant and available at less cost. The greatest beneficiaries will be the world's less developed regions, such as parts of Africa, where food production must increase by 300% by the middle of the twenty-first century if the continent is to keep pace with population growth (Mackey and Santerre, 2000). These twin drivers of benefit and necessity ensure that biotechnology will remain a mainstream issue and will not be side-tracked as did food irradiation, and receiving belated acceptance only after deaths occurred from hemorrhagic *Escherichia coli* O157:H7 contamination.

Scope of This Chapter

While most public attention has been focused on biotechnologically produced novel food plants (crops such as soy, corn, and rice) or edible animals (*e.g.*, salmon), this chapter addresses safety assessment of novel food ingredients[1] (including processing aids) and novel foods (including fermented foods: Table 1), produced by microorganisms as the result of altered phenotypic expression following genetic change. These genetic changes are induced through recombinant DNA (rDNA) technologies or are developed by deletion, rearrangement, or suppressed expression of native DNA with exogenous treatment by agents such as chemical mutagens. Theoretically, microorganisms could include members of any of the following groups: bacteria, mycoplasma, chlamydia, rickettsia, protozoa, fungi, algae, and virus, parts of these microorganisms, and any combination thereof. However, for practical reasons, this discussion is limited to genetically modified bacteria and fungi (including yeasts). The European Community specifically designates "microorganism" as encompassing bacteria and fungi (including yeasts); microalgae (viruses and plasmids are outside the scope of the guidelines) (EC, 1997a).

Food ingredients are substances not generally eaten as food as such. For example, the consumption of a food emulsifier, preservative, or texturizer is not generally contemplated. At times, however, food ingredients are consumed for their intrinsic value; for example, some people consume citric acid (as vitamin C), brewer's yeast, or individual amino acids as dietary or nutritional supplements. For the purposes of this discussion, food ingredients

[1]The term "food ingredients" is a category that may be subdivided into two principal categories: generally recognized as safe (GRAS) substances and "food additives," the latter term being a regulatory distinction.

Table 1
Fermented Foods

Substrate	Enzyme/source	Food product
Soy	Koji cultures	Soy sauce, miso
Milk	*Streptococcus thermophilus*	Yogurt
	Saccharomyces cerevisiae	Brewer's yeast
Flour	*S. cerevisiae*	Bread
Sausage	*Lactobacillus curvatus*	Sausage

can be subdivided into those with functionality in the final food (*e.g.*, gums to provide thickness or fortificants such as vitamins) and those without functionality in the final food, such as processing agents (substances used during food processing and remaining in the finished food product, albeit with no functionality in the finished product, such as enzymes in cheese).

In general, fermented foods may be thought of as foods to which fermentive organisms have been added, although at the time of consumption, the organisms are inactive or killed. The obvious exception is probiotic food (*e.g.*, rDNA bacteria fed to ruminant animals to aid fermentation in the stomach or human foods such as yogurt).

Theoretically, conclusion of fermentation or the conclusion of change brought about by the activity of these organisms or their extracts indicates conclusion of the process and readiness for consumption. There are exceptions, however: for example, yogurt and (some types of) beer must be consumed before too much of the substrate has been converted to lactic or acetic acid, respectively, making the product unpalatable. For most substances, such as bread and most commercially produced beer, the conversion process is halted through further processing. The organisms involved in the fermentation of commercial beer and bread exemplify microorganisms as processing agents. An example of a food ingredient with functionality in the final product could be a gum used as an emulsifying agent in ice cream. The action of the gum is static and integral to the fitness of the food product; therefore, its "activity" does not cease. In the case of yogurt, *Lactobacillus* acts as a processing aid in the production of the yogurt, and these bacteria have functionality in the final food as a probiotic.

Fermentation of foods is likely the oldest form of food processing, predating cooking. Deliberate fermentation of food probably came somewhat later, with the production of cheeses and yogurt, the fermentation of dough for bread, and the brewing of beer. Later societies found the fermentation of yams and cassava useful in the production of more nutritious starches, and still later, people discovered the benefits of improved taste, extractability, nutrition, preservation, and other effects following fermentation of tea, coffee, cocoa, soy, meats, and other foods. Just as earlier societies improved the end product with selective use of the organisms that produced the best results, our society can enhance the final result with the use of genetically modified organisms. These benefits include the use of modified strains of baker's yeast to make dough rise faster and modified strains of *Saccharomyces cerevisiae* to increase the number of candidate substrates; improved *Lactobacillus* cultures that preserve freshness and flavor and produces substances (*e.g.*, nisin and pediocin) that inhibit the growth of common pathogens found in dairy products, including *Listeria monocytogenes* and *Salmonella* spp. (Smith, 1994). Over 3500 fermented foods are known (Madden, 1995). Examples of whole foods produced by organisms and consumed with minimal processing include soy sauce, miso, and yogurt.

Whole foods produced microbiologically is a decades-old goal of the food industry. Popular example is single-cell protein (SCP), a novel whole food entirely produced by an organism. At one time (*i.e.*, the 1970s) SCP was thought to be the most likely candidate to fulfill a perceived protein shortage for both animal and human food. However, because of the increase in oil prices, SCP was shelved and largely supplanted with soy products. Fulfilling the concept of a microbiologically produced whole food is a commercially available mycoprotein (Quorn). There are a number of advantages to fungal protein over soy products, in particular the favorable texture resulting from a linear arrangement of the fungal hyphae, similar to that of muscle fibers in meat (Madden, 1995; Rodger, 2001). Like irradiation of food, mycoprotein is an idea whose time has come, at least in part because of problems such as bovine spongiform encephalitis and foot and mouth disease in European livestock.

In terms of *tonnage* consumed, production of whole foods by the fermentive activity of rDNA organisms probably is exceeded only by crop production (*e.g.*, soy, corn, wheat, rice). However, the *number* of food ingredients produced by rDNA technology likely exceeds the combined number of economically viable crops produced by rDNA technology.[2] At the top of this list is the number of enzymes approved for use in food on the Center for Food Safety and Applied Nutrition (CFSAN) website for GRAS notifications alone (FDA, 2001b) (Table 2).

Other novel food ingredients include amino acids, citric acid, vitamins, gums, and sweeteners (aspartame and rThaumatin). Monosodium glutamate, lysine, cysteine, methionine, phenylalanine, and tryptophan are a few of the

[2]Foods and food ingredients obtained from microorganisms are third on a list of priorities provided by FDA U.S. Codex Manager to the director of the International Food Safety Program of Codex Alimentarius Commission (Scarbrough, 2000).

Table 2
Commercially Available rDNA Enzymes

Enzyme	Host	Donor
Lipase	*Aspergillus oryzae*	*Fusarium oxysporum*
Pullulanase	*Bacillus licheniformis*	*B. deramificans*
Xylanase	*Fusarium venatum*	*Thermomyces lanuginosus*
Pectin esterase	*A. oryzae*	*A. aculeatus*
α-Amylase	*B. licheniformis*	*B. lichenformis* and *B. naganoensis*
Pectin lyase	*Trichoderma reesei*	*Aspergillus niger*
Lipase	*A. oryzae*	*Thermomyces lanuginosus*
Aspartic proteinse	*A. oryzae*	*Rhizomucor miehei*

Source: U.S. FDA (2001b).

rDNA-derived amino acids. Recombinant DNA technology has permitted conversion of guar gum into a more valuable commodity, exhibiting favorable attributes resembling the technical advantages conferred by the more expensive locust bean gum and improved methods for production of xanthan gum, eliminating a costly cleanup process. Thus, rDNA technology has already significantly improved food production.

Concepts in Safety Assessment

Because the facts supporting the benefits of biotechnology are indisputable and its continued use inevitable, concern is focused on modalities for acceptance, for safety, and, most specifically, for the detection of unintended effects. The term *unintended effects*, has special meaning in connection with biotechnology. Such effects include but are not limited to pleiotropic effects, gene activation, and silent gene activation (including new metabolic pathway activation) that may give rise to a toxic or allergic effect. The Codex Alimentarius (2001) divides unintended effects as the result of rDNA into two groups: "predictable" (based on metabolic connections to the intended effect or knowledge of the site of insertion) and unintended effects that are "unexpected."

The FDA has clearly assumed a leadership role in the regulation and safety of rDNA products and specifically states "[B]ioengineered foods and food ingredients (including food additives) must comply with the same standards of safety under the act that apply to other food products" (U.S. FDA, 1996). The agency has a mandate to do so embodied in the Federal Food, Drug, and Cosmetic Act (FFDCA) and intends to regulate rDNA products based on two sections of the act (U.S. FDA, 1992, 1996, 2001a). Section §402 of the FFDCA is as follows:

> SEC. 402. [342] A food shall be deemed to be adulterated – (a) (1) If it bears or contains any poisonous or deleterious substance which may render it injurious to health; but in case the substance is not an added substance such food shall not be considered adulterated under this clause if the quantity of such substance in such food does not ordinarily render it injurious to health; . . .

This section provides FDA with the authority to declare a food adulterated and can thus be removed from the market as provided for in a later section of the act, as follows:

> SEC. 409. [348] (a) A food additive shall, with respect to any particular use or intended use of such additives, be deemed to be unsafe for the purposes of the application of clause (2) (C) of section 402 (a) . . . .

Sections 402 and 409 do not apply if the substance is generally recognized as safe (GRAS) or otherwise exempt.

These sections of the act provide the agency with two important and powerful tools, but equally powerful in this theory of regulatory authority are two important concepts to which FDA has adhered: (1) a belief that the product (*i.e.*, the food or food additive), not the process, should be the subject of inquiry, and (2) more subtle, but still evident, the concept of substantial equivalence. The wisdom of the first concept was the knowledge that since the public is exposed to the product, not the process, only the product should be subject to scrutiny. Also, because processes can be modified and improved, one avoids the complication of regulating against a process only to be forced to "unregulate" or approve, a modified, safer version of the process. Prohibition of a process would also tend to stifle innovation needed to improve the process. The regulation still affords the consumer protection however, because if a process produces a contaminated product, FDA may simply place restrictions on the level of contaminants. A good example of a process vs product issue was the contemplation by some European authorities of restricting the temperatures used to produce "reaction flavors," whereas all that was truly needed was a limitation on the harmful contaminants produced by the high temperatures. Much of the same process regulation argument is promoted today: that is, biotechnologically produced substances, simply as the result of the production methodology, and may result in unpredictable consequences for which testing regimens do not exist. The concept of inquiry into the product and not the process is gaining ground and was recently reiterated by the Organisation for Economic Co-operation and Development (OECD):

> There is a widespread scientific consensus that in assessing risks, it is not the process applied in breeding but the genetic outcome and the trait it confers to the plant that matters (OECD, 2000).

The groundwork is laid for the second concept, substantial equivalence, in the "may render" and "added substance" phrases in Section 402 of the FFDCA. This concept is iterated several places, including the following and will be discussed in the next section:

> Based on our present knowledge of developments in agricultural research, we believe that most of the substances that are being introduced into food by genetic modification have been safely consumed as food or are substantially similar to such substances (U.S. FDA, 1996).

The concept of substantial equivalence is a rationale safe harbor for wizened regulators who refused to boxed in via "argument by dilemma," the contention that testing of all new ingredients, chemicals, or their variants must be subjected to the same rote testing methods, without regard for purpose or exposure (*i.e.*, intended use). The deductive argument demands that the listener agree with the rationale presented that all substances must be tested by the same methods. Then the analogous argument for equally thorough testing of an rDNA product is presented, but shown to be impossible to carry out because of the very nature of food, thereby creating the dilemma. The argument usually begins with a comparison of the properties of a single chemical entity and a food (Table 3). While these properties are comparable, the obvious differences between them do not allow for the same methods of safety testing. Because single chemical entities and foods are so different, new methods of safety evaluation must be established. This point was made by Kotsonis *et al.* (2001) in their opening paragraphs on food toxicology, in which they stress the importance of addressing practical and workable approaches to the assessment of food safety: "[f]ood toxicology is different from other subspecialties in toxicology, largely because of the nature and chemical complexity of food."

Undeterred, the dilemma argument is posited by Millstone *et al.* (1999), who also account for those not in agreement with the dilemma as follows:

> [because] new foods or food ingredients must be tested using conventional methods (and an Acceptable Daily Intake (ADI) conferred using at least a 100-fold safety factor) and [because] of the difficulties in testing, no "new" food could be tested at levels to allow for a 100-fold safety factor, [therefore] no substance could be used in food at a level greater than 1% and, [because] industry needs use levels of 10% or more of the diet, [therefore] regulators are cowed from enforcing conventional safety assessment.

The fallacy of the argument is its circular nature: "conventional testing paradigms must be used to prove safety," but because "conventional testing is not applicable," "safety can never be proven." Far from this accusation of being cowed, regulators simply refuse to be hamstrung by dogma and look to new concepts for safety evaluation. Nobody said this was going to be easy.

Table 3
Testing Attributes of Single Chemical Entity and Food

Chemical	Food
Material usually simple, chemically precise substance	Complex mixtures of many compounds
Highest does level should produce an effect	Effects improbable at the maximum dose level that can be incorporated in the diet for the test species
Small dose (usually $< 1\%$ of the diet)	High intake (usually $> 10\%$)
Easy to give excessive dose	Intakes above those normally present in the diet difficult
Acute effects obvious	Acute effects difficult to produce (usually absent)
Generally independent of nutrition	Nutrition dependent
Specific route of metabolism simple to follow	Complex metabolism
Cause/effect relatively clear	Cause/effect, if observed at all, may be confused

Source: MAFF (1999).

ISSUES IN FOOD AND FOOD INGREDIENTS PRODUCED FROM rDNA

There are any number of issues associated with rDNA-produced food, but those of primary relevance to this chapter are as follows:[3]

- Characterization of organism and phenotypic characteristics
- Pathogenicity, toxigenicity, and antinutritive effects of the organism
- Substantial equivalence
- Gene transfer, marker genes, and antibiotic resistance

As described later, only the methods used to produce biotechnological foods and food ingredients are "new"; the rationale by which the safety of these products may be determined is not. As any "FDAer" will say, "safety is the primary mission of the FDA," and the method CFSAN has used for determination of safety has a long record of success. Biotechnological products can be subjected to the same method, but the specific questions need to be

[3]Issues dealing with nutrient content of rDNA foods are discussed at length elsewhere and are not covered here. Other issues that tend to work their way into discussions of biotechnology include food traceability and labeling, but these are only marginally relevant to this discussion and are not discussed in detail. Also, the safety or regulation of biotechnologically produced plants or new plant varieties, discussed in depth elsewhere, is not treated in this chapter.

modified somewhat. Recognizing this long history of success, the framework of this chapter is built on the provision in both law[4] and regulation,[5] which describe the GRAS, or generally recognized as safe concept.[6] FDA has also provided some guidance on the concept of GRAS (*FDA Guidance on GRAS Determination, Fed. Regist.* **62**, 18937, 1997) and guidance on biotechnologically produced substances [*e.g., Statement of policy: Foods derived from new plant varieties, Fed. Regist.*, **57**(104), 22984–23005, 1992)[7] and *Premarket notice concerning bioengineered foods, Fed. Regist.*, **66**(12), 4706–4738, 2001)].

Characterization of Organisms and Phenotypes (U.S. FDA, 1992; WHO, 1996; Pedersen, 2000)

Characterization of the host, donor, and transfer process is integral both to the success of the product and to success in obtaining regulatory approval. Many of these points (see Table 4) were elaborated by Pariza and Foster (1983) some years ago and are true today. For example, no one disputes that characteristics not present in the host and donor will not be expressed in the production organism.

Pathogenicity, Toxigenicity, and Antinutritive Effects of the Organism

The success of any pathogenic microorganism is the result of a combination of traits that enhance its possibility of survival, including but not limited to adhesion, invasion, and toxigenicity. To state the obvious, the microorganism intended for use in food processing should be derived from organisms that are known, or have been shown by appropriate testing, to be free of traits that confer pathogenicity (WHO, 1991). Likewise, if the genetic material transferred to the host organism does not contain traits conferring pathogenicity, it is clear that the modified organism will not be pathogenic.

[4]Federal Food, Drug and Cosmetic Act (FFDCA), Section 201(s).

[5]21 CFR § 170.30.

[6]See also, G.A. Burdock, (2000) Dietary supplements and lessons to be learned from GRAS. *Regul. Toxicol. Pharmacol.*, **31**, 68–76.

[7]The agency makes it clear the notice [Foods Derived from New Plant Varieties, *Fed. Regist.*, **57** (104), 22984-23005, 1992] addresses only foods derived from new plant varieties, not foods or food ingredients derived from algae, microorganisms, or other nonplant organisms, including (1) foods produced by fermentations, where microorganisms are essential components of the food (*e.g.*, yogurt and single-cell protein); (2) food ingredients produced by fermentation, such as enzymes, flavors, amino acids, sweeteners, thickeners, antioxidants, preservatives, colors, and other substances; (3) substances produced by new plant varieties whose purpose is to color food; and (4) foods derived from animals that are subject to FDA's authority, including seafood.

Table 4

Characterization of Host, Donor, and Construct

Characterization of host

- Origin
- Taxonomic classification
- Scientific name
- Relationship to other organisms
- History of use as a food or food source
- History of production of toxins
- Infectivity
- Significant nutrients associated with host species
- Presence of antinutritional factors and physiologically active substances in the host and closely-related species

Characterization of genetic modification and inserted DNA

- Vector/gene construct
- Description of DNA components
- Transformation method used
- Promoter activity

Characterization of modified organism: Genotypic characteristics

- Selection methods
- Phenotypic characteristics compared to host
- Regulation, level and stability of expression of introduced gene(s)
- Copy number of new gene(s)
- Potential for mobility of introduced gene(s)
- Functionality of introduced gene(s)
- Characterization of inserts

Characterization of modified organism: Phenotypic characteristics

- Taxonomic characterization (*e.g.*, traditional culture methods, physiology)
- Colonization potential
- Infectivity
- Ability to colonize the gut[a]
- Host range
- Presence of plasmids
- Antibiotic resistance
- Toxigenicity, antinutrients produced

[a]Applicable for organisms occurring as living cells in the final food product, such as yogurt.

The same is true for toxigenicity, the production of toxins. There are a number of different types of bacterial toxin: an *enterotoxin* is a toxin having action on the *enteric* cells of the intestine; an *endotoxin* is generally a lipopolysaccharide membrane constituent released from dead Gram-negative bacteria (these toxins are nonspecific and stimulate inflammatory responses from macrophages); *an exotoxin* is synthesized and released (usually by Gram-positive bacteria) and is not an integral part of the organism but may enhance its virulence. Some bacteria, such as *Shigella, Staphylococcus aureus*, or *Escherichia coli* (which releases the shiga-like vero toxin), can elaborate both endotoxin and exotoxin (Kotsonis *et al.*, 2001). It is necessary to ensure that the genetic sequence encoding for these traits is not present in the production organism.

Nearly all the fungi used in the production of food ingredients are capable of producing one or more types of mycotoxin. However, mycotoxins are normally produced only under stressful growth conditions, not under the optimal conditions required for maximum growth; that is, "these substances [mycotoxins] are not known to be produced under conditions of current good manufacturing practice" (JECFA, 1990). In fact, screening of the production organism yield for mycotoxins acts as a check on the successful operation of the production process.

Other detrimental characteristics may be more subtle, such as the antinutritive effects (e.g., goiterogens in *Brassica*, trypsin and/or chymotrypsin inhibitors in soybeans, phytates in soybeans) that may bind minerals (zinc, magnesium, phosphate), avidin binding the vitamin biotin, and antithiamines found in fish and plants (Kotsonis *et al.*, 2001).

Also to be considered are the origins of food intolerance or idiosyncratic reactions to food not associated with an immunologic reaction or food toxin. These would include the following (Kotsonis *et al.*, 2001):

- *Anaphylactoid reaction*: mimics an allergic response but is not immune mediated. Includes scombroid poisoning, sulfite poisoning, and red wine sensitivity.
- *Pharmacological food reaction*: non–immune related, pharmacological (but often exaggerated) response to a food component such as tyramine in patients treated with monoamine oxidase inhibitors or endocrine effects from isoflavones.
- *Metabolic food reaction*: inherent toxic property of the food manifested through excessive consumption or improper preparation (*e.g.*, cycasin, vitamin A toxicity, goiterogens).

The origin of these concerns is the knowledge of specific subgroups with particular sensitivities that are nonallergic. Moreover, the advantages conferred by the novel food may result in a significantly increased consumption by some groups (*e.g.*, children), unmasking a vulnerability hitherto existing at subthreshold levels.

Proteins may also have allergic potential, especially those from species most often associated with allergies.[8] Thus unless the peptide epitope can be

[8]FDA believes there is consensus the following substances account for 90% of food allergies: peanuts, soybeans, milk, eggs, fish, crustacea, tree nuts and wheat.
http://www.fda.gov/ora/inspect ref/igs/Allergy Inspection guide.htm (visited September 2, 2001: site

eliminated from the gene product, it is best to avoid donor sequences from these species.[9]

A related concern, but one approaching the limits of the "precautionary principle," is the expression of a transgenic protein that could result in the production of a toxicant not observed in the parent species. This can happen however and an example is the hybrid product of *Solanum brevidans* and *S. tuberosum* (potato) that has been observed to produce the toxicant demissine, not found in either parental line (OECD, 2000). An analogous situation has not been reported in microorganisms used in foods, although a well-designed toxicity screen should allay any fears of the appearance of a hitherto unknown toxin, regardless of the high improbability of one occurring. The testing regimen should also take into consideration the digestibility of the proteins (or other substances) produced and the effects of processing (if any) on the product.

Many of these characteristics of pathogenicity, toxigenicity, antinutritive effects or sources of idiopathic reactions to food can be eliminated early in the process by a thorough characterization of the host, donor, and transformation process.

SUBSTANTIAL EQUIVALENCE

Because rDNA technology held the promise of vastly different foods and both micro- and macroingredients, it was obvious very early that safety assessment methodologies based on traditional toxicology testing methods may not be applicable. A joint FAO/WHO session was convened in 1990 to address this problem and determined that "safety assessment strategies should be based on the molecular, biological and chemical characteristics of the food to be assessed and that these considerations determine the need for, and the scope of traditional toxicological testing" (WHO, 1991).

Thus, the need for a more mechanistic assessment was promoted, and the stage was set for the concept now generally referred to as "substantial equivalence." Substantial equivalence embodies the concept that if the rDNA product is *substantially equivalent* to an existing product, safety assessment of the rDNA product should be made in the context of the existing product. That is, a novel food is essentially the same as that found in nature except for the novel trait. Regulators have used the concept of substantial equivalence for at least 10 years (Tomlinson, 2000). The concept is not new. Indeed, the same principle was used in assessing the safety of new hybrid varieties produced by nonbiotechnological means. For example, potatoes are known to contain solanines, a toxic alkaloid native to potatoes. New potato

[9]A thorough description of allergy is presented in Chapter 11.

varieties are examined for solanines, and if the solanines are within acceptable levels, the potato is used (Love, 2000). The object is to produce a potato for a specific need (*e.g.*, better processing qualities or disease resistance), not necessarily a potato with lower levels of solanines. The rationale of substantial equivalence eliminates the testing for toxins that would not logically be present (tetrodotoxin is unlikely to be produced by *Lactobacillus*) and to avoid the folly of testing animals with whole foods (*e.g.*, onions, potatoes; see review by Hammond *et al.*, 1996) that would only demonstrate nutritional deficiencies, at the highest dose levels, that result from food displacement or other effects unrelated to the change rendered by rDNA.

Admittedly, the concept of substantial equivalence does not address all "what if" scenarios such as uncovering silent genes or inducing new pathways producing toxic products. It decreases the probability of their occurrence, however, because if the host organism possessed the inherent capacity to produce these toxins, this would likely have been demonstrated in a *wild* type at some time in the past.

Substantial equivalence is not a substitute for safety assessment, but a step in an overall assessment of safety. That is, it is necessary to determine the specific context of the comparison, the degree of relevance of that context (*i.e.*, how similar or how different), and finally, the appropriate testing regimen. First, however, it is necessary to characterize the modified (final) product.

The specific context of comparison is critical to ensure compliance with existing regulation that is, the substance should be tested in the form presented to the public.[10] In the case of a food ingredient such as xanthan gum produced by an rDNA organism, it would be necessary to compare the new gum with the existing gum; for a new chymosin, it would be necessary to examine the enzyme, not the cheese it produces. With respect to transferred genetic material (nucleic acids), generally FDA does not indicate that this would be worthy of a petition because nucleic acids are present in the cells of every living organism and do not represent a concern for safety; that is, nucleic acids are GRAS (U.S. FDA, 1992). For example, the introduction of an antisense RNA would not raise concern in and of itself, only the intended effects of ant antisense RNA. Also, if the purpose of the antisense RNA were to suppress an enzyme (as might be the purpose of a deletion or nonsense mutation), the effect on the host organism would have to be considered. Such an effect may be a protein fragment formed as the result of the mutation (U.S. FDA, 1992). The same sentiment is expressed by the World Health Organisation (WHO, 1996).

In the circumstance of the gum and the enzyme, it is not the organism that is consumed, only the *products* of the organism. What happens when the entire organism is consumed? For example, in the case of a yogurt or fermented

[10]This is consistent with FDA policy to assess product safety, not process safety.

sausage produced by an rDNA *Lactobacillus*, would the new sausage or yogurt be tested or just the *Lactobacillus*? In all cases, testing should be conducted on the final product as conceivably consumed, the gum and the enzyme, because this is the form in which they are added to food (i.e., a cell-free extract); the yogurt and sausage as the organisms or their remnants are still present. The WHO expert consultation specifically addresses this latter case:

> Where genetically modified organisms have been determined to be acceptable as a result of the safety assessment, these further strains/varieties should be assessed on their own merits according to practices applied for the assessment of conventionally derived organisms (WHO, 1996).

Further, the levels and variation for characteristics in the genetically modified organism must be within the natural range of variation for those characteristics considered in the comparator and must be based on the appropriate analysis of data (see earlier discussion on phenotypic expression). In the case of the gum and the enzyme, because remnants of the new proteins may be present in the cell-free extract, these should be analyzed separately (more on this later).

Because the degree of equivalence to a previously consumed food is a continuum of "substantial" to unique and not existing in nature, three tiers of degree of equivalence have been proposed (OECD, 1993, 2000; WHO, 1996).

> (a) When substantial equivalence has been established for an organism or food product, it is considered to be as safe as its conventional counterpart and no further safety evaluation is needed.
>
> (b) When substantial equivalence has been established apart from certain defined differences, further safety assessment should focus on these differences: a sequential approach should focus on the new gene product(s) and the(ir) structure, function, specificity and history of use. If a potential safety concern is indicated for the new gene product(s), further *in vivo* and/or *in vitro* studies may be appropriate.
>
> (c) When substantial equivalence cannot be established, this does not necessarily mean that the food product is unsafe. Not all such products will require extensive safety testing. The design of any testing program should be established on a case-by-case basis . . . [with the implication that] further studies, including animal feed trials, may be required, especially when the new food is intended to replace a significant part of the diet (OECD, 2000).

A Substantially Equivalent Food or Food Ingredient

Once the phenotypic characteristics of the rDNA organism have been determined to be substantially equivalent to those of its predecessor, a compositional analysis should be performed on the product for comparison to the appropriate comparator. Analyses for key nutrients and possible toxicants are adequate and analysis of a broader spectrum of components is, in general, unnecessary (WHO, 1996). It is appropriate, however, to review consumption patterns to determine whether particular societies or cultures will make increased use of the product (as the result of its increased value and/or greater availability) and what impact the increased use of the key nutrients or possible (inherent) toxicants may have. That is, while the product has not now been changed such that under Section 402 of the FFDCA, "...the quantity of such substance in such food does not ordinarily render it injurious to health," the increase in consumption may result in untoward effects.

As a result of the genetic modification, there has been an insertion, deletion, or rearrangement of nucleic acid and the possible generation of a new type of RNA and resulting nonfunctional protein. As noted earlier, while regulatory authorities are not generally concerned about nucleic acids, there is concern directed at the safety of expressed protein.[11] While tens of thousands of proteins may be expressed by a cell, only very few are toxic or may provoke an allergic reaction. Therefore, sequence homology of these newly expressed proteins should be compared with those of toxic or allergic proteins, especially if the host or donor organism or both had been known to produce allergic or toxic substances.

Products Substantially Equivalent to Their Comparators Except for Defined Differences

Defined differences, in this context, are the qualities that make the novel food different from the comparator. These defined differences are thus the basis for segregation from the comparator and the entities subjected to safety assessment (Tomlinson, 2000). In this case, the expressed protein is functional itself (*e.g.*, an enzyme) or may cooperate in the production of fats, carbohydrates, or other biological components. Again, proteins not functionally similar to normally expressed protein should be examined for potential toxicity and allergenicity. Thus, sequence homology comparison to known allergens and toxins is mandated. Key also in this assessment is the behavior of the new protein when subjected to proteolytic digestion under both gastric and intestinal conditions (WHO, 1996). Special subpopulations (*e.g.*, achlorhydrics with impaired digestion) should be considered in the study design.

Carbohydrates or fats produced by rDNA organisms are a less serious concern for allergic or toxic effects, especially if well characterized, and the new or enhanced levels of constituents are reviewed by the toxicologist for potential untoward effects. The final question might be, Is this product *substantially equivalent* to the "parent" product such that "common and usual" name of the product could be used on the label?

As noted earlier, consumption of the new product should be determined, especially if the new product may present limitations in tolerance, as might a new source of dietary fiber (Carabin and Flamm, 1999).

[11]Although the protein is ultimately of concern, nucleic acid may, in some cases, be used as a marker to predict the presence of the protein it encodes.

Products Not Substantially Equivalent to Existing Food or Food Components

Occasionally, FDA is critical of a food or food ingredient offered for sale to the U.S. public that has no history of use in this country. However, a shrinking world has led to the consumption of foods once thought exotic, but have now become simply ethnic or even mainstream. As a result, kiwi, gurana, maté, tomatillos, wasabi, or chermoyia hardly rate a second glance (Burdock, 2000). Because the number of naturally occurring foods hitherto unconsumed by substantial portions of the population is becoming vanishingly small, it is reasonable to expect that a "new" food, unlike any existing food, will be the product of rDNA technology.

It should be noted that while substantial equivalence is not a substitute for safety assessment, lack of substantial equivalence does not indicate lack of safety *per se*. Not all such products will necessarily require extensive testing (WHO, 1996), but a determination of effects on bioavailability and the presence of antinutrients may be critical. It should also be noted that foods by their very nature are bulky and complex mixtures of substances. As discussed earlier, testing single chemical entities is not at all comparable to testing foods. Also noted earlier, animal feeding studies with whole foods at high dietary levels have not been successful historically because nutritional problems and diet balancing induced adverse effects not related directly to the material itself (Tomlinson, 2000). Human clinical studies might prove to be of great value with any substance taken in significant quantities. It is also important to test the material as it would be consumed; for example, testing a raw rDNA potato when the potato was intended for consumption following cooking (Walker, 2000).

Gene Transfer, Marker Genes, and Antibiotic Resistance

Although horizontal gene transfer is less a question for cell-free extracts with small amounts of DNA fragments remaining, the relative possibility of gene transfer in the gut from viable rDNA organisms to gut microflora is significantly greater. This transfer of genetic information may take place by one or a combination of two or more of the following mechanisms: (1) transformation of released genetic material (naked DNA), (2) transduction (viral transfer of DNA into bacteria, although these phages have a very narrow host range of infection and this method is not considered a significant method of genetic information transfer between species), or (3) conjugation (requires cell-to-cell contact exchange of plasmid or transposon DNA between compatible bacteria), albeit the possibilities are low (Morrison, 1996; Dröge *et al.*, 1998).[12]

[12]For reviews on this subject see Lorenz and Wackernagel (1994), Dröge *et al.* (1998), and Nielsen *et al.* (1998).

The critical question affecting regulatory acceptance is as follows: If transfer of genetic material is accomplished, does the transfer confer some selective advantage to the new host? Factors amounting to a selective advantage include but are not limited to phage resistance, virulence, adherence, and substrate utilization or production of bacterial antibiotics (WHO, 1996).

One of the most often repeated recommendations for decreasing the possibility of advantageous gene transfer (therefore conferring selective advantage) includes not using selectable marker genes that encode resistance to clinically important oral antibiotics (*e.g.*, kanamycin/neomycin, with the *npt*II gene or the *aad* marker gene conferring resistance to spectinomycin/streptomycin). Further, cell-free extracts do not contain any DNA, thereby thwarting natural transformation.

Just how (relatively) important is the much discussed issue of antibiotic resistance via rDNA? Bacterial acquisition and rearrangement of genetic material is commonplace in nature, and the best evidence is bacterial acquisition of multiple drug resistance (Gasson, 2000). Thompson (2000) indicates that bacterial strains resistant to antibiotics are common in the environment and cites data that ampicillin resistance has been shown in as much as 70% of *E. coli* isolated from diseased calves and 20% from diseased cattle, as well as 69% of isolates from Tennessee groundwater (Thompson, 2000). In a survey of children in Mexico, nearly all the strains of *E. coli* isolated were ampicillin resistant (Saylers, 2001). Resistance by *Salmonella typhimurium* to ampicillin has been demonstrated in up to 60% of diseased cattle assayed. Interestingly, a recent study of unpasteurized cheeses and other foods found that many of the bacteria were resistant to one or more antibiotics (Saylers, 2001). Thompson (2000) and Saylers (2001) conclude that the more rational concern about antibiotic resistance should be focused on overuse in veterinary and human clinical practice and use in animal feed, for these uses *do* confer a selective advantage to gut organisms. Conversely, antibiotics used in genetically modified strains pose a much less serious threat, as the antibiotics used in rDNA applications are old, narrow-spectrum, have less mobilizable resistance genes, and are not involved in clinically important treatment regimens. Thompson (2000) also comments about studies in which animals were orally dosed with organisms containing antibiotic resistance markers that were recovered from the animal tissue. However, Thompson (2000) points out that the tissues in which the markers were found consisted of inflammatory cells, which are expected to contain foreign DNA as part of the foreign body response and elimination process of the reticuloendothelial system.

In all likelihood, the best strategy is removal of the marker gene, since theoretically, the marker gene would have no functionality in the final product, or to find an alternative to antibiotic resistance genes as markers. Marker gene removal could be accomplished by several methods including, but not limited to recombination-based deletion, transposon-mediated relocation, and cotransfer with marker gene on a separate vector (Ow, 2000). Alternatives to antibiotic resistance genes include insertion of trytophan decarboxylase gene (eliminates a toxic tryptophan analogue), activation of cytokinin in adequate quantities for cell division, and incorporation of genes for utilization of unique substrates such as xylose or mannose (Ow, 2000).

Probably of greatest concern is the acquisition of resistance by probiotic organisms, and resistant strains of *Lactobacillus* have been isolated. Although there is no evidence of transmission of resistance to date, the conferred resistance to human pathogens could well become problematic (Saylers, 2001). The answer may, in fact, be a future requirement that all probiotic organisms be screened for resistance to antibiotics as part of good manufacturing practice.

The WHO (1996) concluded that if rDNA does not confer a selective advantage, no further testing of the organism is necessary. If, however, the rDNA can confer a selective advantage," (1) . . . vectors should be modified so as to minimize the likelihood of transfer to other microbes; and (2) selectable marker genes that encode resistance to clinically useful antibiotics should not be used in microbes intended to be present as living organisms in food". That is, information should be developed either demonstrating that the genes do not confer a selective advantage in the gut or influence the existing microflora "under typical or extreme conditions," or, if a selective advantage is conferred, the consequences to the consumer must be defined (OECD, 2000).

In draft guidance for industry, FDA (U.S. FDA, 1998) stated that the use of antibiotic resistance markers should be evaluated on a case-by-case basis, taking into account the following: whether the antibiotic is an important medication, whether it is frequently used, whether it is orally administered, whether it is unique, whether there would be selective pressure for transformation to take place, and the level of resistance to the antibiotic present in bacterial populations. Further, FDA states, "if a careful evaluation of the data and information suggests that the presence of the marker gene or gene product in food or feed could compromise the use of the relevant antibiotic(s), the marker gene or gene product should not be present in the finished food.

CONCEPTS IN SAFETY TESTING

It is fair to say that many issues have come to the forefront in the safety assessment of novel foods. These issues are the product of the last 30 to 40 years of testing single chemical species and the long-term effects of these chemical species, some of which were unpredictable at the time of testing, and some effects still would not be predictable even with the knowledge base we have today. Any effort to test a single chemical species for all possibilities is not practical and even less feasible with such complex mixtures as are represented by novel foods.

To some, testing procedures will always be inadequate, but there will always be legitimate debate calling for expanded effort. For example, in the case of macronutrients (as many novel foods may be), additional issues come into question, including possible drug interactions (*e.g.*, especially with fat substitutes or fiber-based materials), nutrient interactions (*e.g.*, changes in bioavailability as the result of changes in gut flora or changes in gastrointestinal tract pH or water content), or other effects such as those on satiety (Borzelleca, 1992). The need for additional testing is especially acute when particularly susceptible subpopulations are identified. For example, should the ease of rDNA degradation be tested because achlorhydrics or very young children who consume novel foods and will be unable to thoroughly digest rDNA, leading to the possibility of transformation in the gut? How extensive a protocol should be developed for endocrine disruption? Where is the point of diminishing returns, and how can it be identified?

This is not to say that there are no inadequacies in the criteria for screening novel foods. The animal models now in use may not represent the best models for man. For example, there are no good models for predicting allergenicity, there are few animals with exactly the same dietary requirements of man, and some animals are wholly unsuitable for some test substances (*e.g.*, the results of complex carbohydrate testing in dogs cannot be used to predict results in humans). The lessons learned from whole-food testing of irradiated foods should not be forgotten, or like all who forget the lessons of history, we will be condemned to repeat the mistakes of our forebears. The bottom line is that unless appropriate control treatments are included in the study design, or a clearly different toxicological end point for the added trait is known beforehand, complications in interpretation may occur (OECD, 2000).

A frequently mentioned method for checking on safety testing and assessment is postmarket surveillance. Postmarket surveillance requires two groups (populations that consume and do not consume the novel food), knowledge of the consumption patterns of both groups and of the range of consumption of the consuming group, and the health status of both groups on both acute and long-term bases. Variables include temporal or geographic changes or both. Bias certainly will come into play, and measuring acute vs long-term effects is exemplified by a story of public concern about *Bt* spraying in British Columbia. To establish a baseline in the Canadian provinces, aerial spraying of water was carried out soon after the announcement of forthcoming *Bt* spraying. This water spraying was followed by a significant increase in the number of reported health effects (OECD, 2000). Epidemiological studies are difficult, especially long-term cohort studies. A more applicable method of postmarket surveillance might be case–control studies in which a particular outcome (in relation to a novel food) is suspected. Such a study would be "reactive" rather than constituting true "surveillance" (OECD, 2000).

REGULATORY REQUIREMENTS

There is no "one size fits all" regulatory requirement for novel foods and novel food ingredients. As for other approval regimens, differences between regulatory agencies exist in terms of basic philosophy and approach, but also with respect to one current state of the regulations, because this is an evolving science. Because of these differences, we summarize several regulatory approaches to give the reader an understanding of the lack of perfect congruency of regulations, the areas of similarity between regulatory bodies, and the areas different regulators may see as viable issues. Thus readers can design individual approaches to risk assessment to allow registration in more than one jurisdiction without unnecessary duplication of effort.

U.S. Food and Drug Administration

Premarket Notice Concerning Bioengineered Foods: Required Parts [*Fed, Regist.*, 66(12), 4706–4738, 2001]; Advance Notice of Proposed Regulation 21 CFR 192.25[13–15]

Part I: Letter

The first part of the letter attests to the balanced nature of the notice and to compliance of all parts of the regulation. Further, the food (or ingredient) is compliant with the standard "as safe as"—a comparative standard that takes into account circumstances such as the existence of naturally occurring toxicants. Of course, the notifier must justify the comparison food.

The agency proposes that data and information in support of the submission be available for copying and the data and information also be submitted directly to the agency. The notifier must indicate which data or information within the Premarket Biotechnology Notice (PBN) is exempt from disclosure

[13]FDA is also seeking to codify the definition of "bioengineered food": *Fed. Regis.*, **66** 4706, January 18, 2001 http://www.cfsan.fda.gov/~lrd/fr010118.html (Site visited September 3, 2001.)

[14]See also: How to submit a GRAS notice (excerpted from *Fed. Regist.*, **62** 18937, April 17, 1997) <http://www.cfsan.fda.gov/~dms/opa-frgr.html> (Site visited October 3, 2001.)

[15]Presently there are no guidelines or proposed guidelines for bioengineered foods or food ingredients produced by microorganisms. Since guidelines will pertain to bioengineered plants but are expected to embody the general framework upon which guidelines for food or food ingredients produced by microorganisms. Therefore where the word "plant" appears in square brackets [ ], substitute "microorganism."

under the Freedom of Information Act (FOIA) and explain the basis for that claim.

Part II: Synopsis

The synopsis must include the identity of the notifier, the name of the bioengineered food, the distinctive designations(s) that the notifier uses to identity the applicable transformation events, a list of the identity(ies) and source(s) of introduced genetic material, the purpose or intended effect of transformation event, a description of the application or uses of the new food, and a description of any applications or uses that are not suitable.

Part III: Status at Other Federal Agencies and Foreign Governments

The notifier must inform FDA of the status of any prior or ongoing evaluation of the bioengineered food by another agency of the U.S. government (e.g., the U.S. Department of Agriculture, its Animal and plant Health Inspection Service, the EPA), or a foreign government.

Part IV: Method of Development

The FDA proposes that the notifier provide the method of development including (1) characterization of the parent [plant] including scientific name, taxonomic classification, mode of reproduction, and pertinent history of development; (2) construction of the vector used in the transformation of the parent [plant], with a thorough characterization of the genetic material intended for introduction into the parent [plant] and a discussion of the transformation method, open reading frames, and regulatory sequences; (3) characterization of the introduced genetic material, including the number of insertion sites, the number of gene copies inserted at each site, and information on DNA organization with the inserts; as well as information on potential reading frames that could express unintended proteins in the transformed [plant]; and (4) data or information related to the inheritance and genetic stability of the introduced genetic material.

Part V: Antibiotic Resistance

FDA is proposing to require that a PBN include a discussion about any newly inserted genes that encode resistance to an antibiotic. Because scientific methods are evolving, the notifier should contact FDA about the agency's current thinking on this topic.

Part VI: Substances in the Food

FDA is proposing that a PBN include data or information about substances introduced into, or modified in, the food. The data or information would include data or information about the identity and function of these substances, the level of these substances in the bioengineered food, dietary exposure to these substances, the level of these substances in the bioengineered food, dietary exposure to these substances, the potential that a protein introduced into the food will be an allergen, and discussion of other safety issues that may be associated with these substances.

FDA is proposing that a notifier provide data or information about substances introduced into, or modified in, the food. Under the proposed regulation, a "modified substance" would include a (nonpesticidal) substance that is present in the bioengineered food at an increased level relative to that in comparable food. However, FDA has made it clear that there is unlikely to be a safety question sufficient to warrant challenging the presumed GRAS status of the expression products of the transferred genetic material when the expression products do not differ significantly from other substances commonly found in food and are already present at generally comparable or greater levels in currently consumed foods.

Since the 1992 guidelines were drafted, rDNA technology has allowed the introduction of multiple genes to generate new metabolic pathways. However, it is the FDA's view that the substance produced by the new pathways would be presumed to be GRAS if it does not differ significantly from other substances that are currently present at generally comparable or greater levels in food and, as such, are safely consumed.

FDA is proposing that a PBN include data or information about the identity and function of substances introduced into, or modified in, the food and the level in the bioengineered food of these substances. That is, the notifier should include either an estimate of dietary exposure to substances introduced into, or modified in, the food; or a statement that explains the basis for the notifier's conclusion that an estimate of dietary exposure to these substances is not needed to support safety. For example, the amount of an enzyme used in food is very small and not relevant to the safety of the food. Additional discussion is requested on allergenicity (for which FDA is now preparing guidelines) and any other relevant safety data not covered by the regulation.

Part VII: Data and Information about the Food

This part requires identification of a comparable food, its history of use, and a comparison of the composition and characteristics of the bioengineered food to the comparator.

Guidelines for Genetically Modified Microorganisms and Their Products (HPB, 1994)

The Food Directorate of the Health Protection Branch of Health Canada defined "novel food" in a regulatory amendment published in 1999.[16] These definitions may be abstracted as follows:

- A substance, including a microorganism, that does not have a history of safe use as a food
- A food that has been manufactured, prepared, preserved, or packaged by a process that has not been previously applied to that food and causes the food to undergo a major change
- A food that is derived from a plant, animal or microorganism that has been genetically modified such that (a) the plant, animal, or microorganism exhibits characteristics that were not previously observed in that plant, animal or microorganism, (b) the plant, animal, or microorganism no longer exhibits characteristics that were previously observed in that plant, animal, or microorganism, or (c) one or more characteristics of the plant, animal, or microorganism no longer fall within the anticipated range for that plant, animal, or microorganism.

The Health Protection Branch promulgated guidelines for novel foods in 1994. The guidelines (HPB, 1994) generally parallel the concepts voiced by WHO and OECD both before and after publication of the HPB guidelines. The guidelines are summarized in the subsections that follow.

Development and Production of the Modified Organism

The requirement for information about the development and production of the modified organism refers to organisms that have undergone recombinant technology or were developed by deletion, rearrangement, or suppression of native DNA and those that have undergone genetic modification by intentional mutagenesis (*e.g.*, chemical treatment or ultraviolet irradiation) that has resulted in alteration of phenotypic expression. In general, this section outlines requirements for characterization of the organism that will permit correlations to be made to its unmodified counterpart, including degrees of difference between the two.

For the host and donor organism, a natural history should be compiled including information on toxin production (in this and related genera), pathogenicity, and previous food and/or medical use.

[16]*Food and Drugs Regulations*, Amendment (Schedule no.948), as published in the *Canada Gazette*, Part II, October 27, 1999.

For the introduced or modified DNA, information should include function of the DNA, location and extent of any deletion, and location and orientation of any rearrangements. For all introduced DNA, there should be information about sequence, restriction map, vector (if used), lack of sequences known to code for toxic substances, limitation of insert for the essential sequence only, limitation of the effect of the DNA for the intended purpose only, absence or inactivation of potentially harmful markers, and absence of unnecessary intermediate host DNA. For modifications not involving the introduction of foreign DNA, a description of the modification, evidence that the modification is limited to its intended function, and identification of affected genes are required.

A description is required of how the inserted gene or genes are regulated in the modified host. If the nature of the inductive mechanism and the constancy of regulation and expression can be induced, this information should be included, as well.

The modified host description should include a detailed description of the method of construction, purpose, metabolic profile, taxonomic designation, biological growth (including physiology), potential pathogenicity and toxigenicity, and organism maintenance. Documentation must also be submitted describing the potential for secondary effects, stability of the construct under process conditions, and mobile stability of the introduced DNA.

The expressed substance should be characterized. If the novel product is a protein, its identity, functionality, and, if appropriate, similarity to products from similar sources of the material should be characterized. If the expressed substance can alter the expression of traditional constituents or metabolites of the organism, information about these secondary effects should be provided.

Product Information

For microorganisms used in or as food, information needed in addition to the foregoing includes a description of the product, its proposed use, process flow diagrams, quality control and standard operating procedures, and other programs to ensure conformity to good manufacturing practices. Concerning the organism, the petitioner must provide growth characteristics and metabolic profile. Product information should include composition (including data on nutrients, antinutrients, nonnutrients, and amounts of any toxin present), safety data, and characterization of the novel constituent.

For microbial products used in food, data already required for food additives, including enzymes, must be supplied, although additional data may be requested depending on the nature and degree of genetic modification from the comparator.

For products identical to food additives, evidence to support the sameness must be presented. Additional information should include the range of variability in composition of the final product in comparison to the approved additive. If HPB determines that the novel substance is not identical to the approved substance, additional data may be required.

For products that represent new food additives, the submission must meet section B.16.002 of the Canadian Food and Drug Regulations and must include (1) description, area of use, and proposed level of use, (2) efficacy data to justify functionality and level of use, (3) safety data, and (4) if the additive is removed, destroyed, or reacted, residue data.

For microbial products, such as those produced *in situ*, the purified product will be subjected to assessment required for substances produced by traditional processes. Other data required include any change that may be imparted to the cellular constituents or by-products of the food.

Dietary Exposure

Dietary exposure is an integral part of the petitioner's submission because the extent of exposure, along with degree of similarity to the traditional counterpart, plays a significant role in determining the type and amount of toxicity testing. Obviously less concern is expressed for substances that are only the products of microorganisms and are used at relatively low levels in food, as well as for substances identical to a food substance that already has approval. In any event, some assessment should be made of the possibility of the change in patterns of use of the substance as the result of the appearance on the market of the novel food.

Nutritional Data

HPB advises a consideration of the nutritional consequences for the population as a whole and particularly for certain subgroups, such as children. Concern here is based on the possible substitution of the novel food for the traditional counterpart if the former has a lower nutritional value in some respects and on any distortion of nutrient intakes as the result of the presence of an unusual level of a nutrient in the novel food (thus exceeding the upper limit) and/or the presence of antinutritional factors.

Nutritional data should also include (1) proximate composition, (2) content of true protein, nonprotein material (including nucleic acids and aminoglycosides), and usual amino acids (such as D-amino acids from bacterial proteins), (3) quantitative and qualitative composition of the lipid fraction, including saponifiable and nonsaponifiable fractions, fatty acid profile (and possible cyclic fatty acids or toxic fatty acids), phospholipids, and sterols, (4) composition of the carbohydrate fraction, including sugars, chitin, tannins, lignins, and nonstarch polysaccharides, (5) vitamins, (6) antinutritional factors, and (7) storage stability with regard to nutrient degradation. Nutritional bioavailability should be determined and, although nutritional "fingerprinting" to its comparator may be adequate, animal studies may be necessary.

Toxicology Data

The HPB divides the section on toxicology data into two parts: laboratory animal studies and allergenicity considerations. The guidelines recognize the potential impracticality of testing whole foods or macronutrients and recommend a carefully planned approach. For substances more amenable to conventional testing and setting an ADI, standardized protocols such as those promulgated by the OECD are acceptable. For allergenicity testing, the petitioner is asked to consider the known allergic potential of the host and donor organisms. In any event, consulation with the HPB is urged.

European Community Regulation 258/97 Concerning Novel Foods and Novel Food Ingredients

Definition and Exempted Items

The EC defines novel foods and novel food ingredients as those not hitherto used to a significant degree for human consumption the European Community and includes the following:

- Those foods containing or consisting of genetically modified organisms
- Those foods produced from, but not containing, genetically modified organisms
- Those foods with a new or intentionally modified primary molecular structure
- Those foods consisting of or isolated from microorganisms, fungi, or algae
- Those foods consisting of or isolated from plants and food ingredients isolated from animals, except for foods and food ingredients obtained by traditional propagating or breeding practices and having a history of safe food use
- Those foods to which has been applied a production process not currently used, where that process gives rise to changes in the composition or structure of the foods or food ingredients significant enough to affect their nutritional value, metabolism, or level of undesirable substances

Exemptions include food additives, flavorings, and extraction solvents, which are regulated within the scope of other council directives. Foods and food ingredients falling within the scope of Regulation 258/97 must not present a danger to the consumer, mislead the consumer, or differ from foods or food ingredients they are intended to replace to such an extent that their normal consumption would be nutritionally disadvantageous for the consumer (EC, 1997a).

Provision for Containment

Council Directive 98/81/EC provides guidelines on the contained use[17] of genetically modified microorganisms (GMMs), noting that the organisms should first be classified in relation to the risks the organisms present to human health and the environment. To this end, Article 2 indicates that microorganisms should be divided into four classes indicating level of risk. Article 5 provides the classes from 1 (activities of no or negligible risk for which level 1 containment is sufficient) to 4 ("activities of high risk"). Techniques not considered to result in genetic modification are exempted; these include *in vitro* fertilization, natural processes (conjugation, transduction, and transformation), and polyploidy induction. Details on containment levels and specific direction are provided in the directive.

Guidelines (EC, 1997b)

The Commission Recommendation of July 29, 1997, describes the information necessary to support an application for a novel food or novel food additive as provided for by Regulation 258/97. These guidelines are thorough and describe what is needed and the reasoning behind the request. However, the document entitled Guidance on Submissions for Food Additive Evaluations by the Scientific Committee on Food (EC, 2001) specifically stated that new guidance on novel foods and novel food ingredients was in preparation and expected to be finalized during 2001 (EC, 2001). As opposed to FDA guidelines, the EC guidelines emphasize the mechanics of a process and any changes or items derived from the use of a process.

The guideline is divided into eight sections, and following the introduction, the categories of novel foods and novel food ingredients named in Regulation 258/97 are identified. These categories are summerized in the list in the section entitled "Definition and Exempted Items."

Key Issues in Assessment of Novel Foods and Novel Food Ingredients

- *Substantial equivalence (SE)*: acknowledges the definitions set forth by OECD, but adds that the technical approach to establishing SE will differ for

[17]Article 2 of this directive states: "...contained use shall mean any activity in which microorganisms are genetically modified or in which such GMMs are cultured, stored, transported, destroyed, disposed of or used in any other way..."

whole animals, plants, microorganisms, chemical food ingredients, and novel processes.

• *Compositional analysis*: recognized to be of crucial importance not only for substantial equivalence, but also as a prerequisite for nutritional and toxicological assessments. The analyses should include macro- and micronutrients, toxicants, and antinutrients, which might be either inherently present or process derived.

• *Intake*: because consumption patterns may be changed as the result of introduction of the novel food, the guidelines request the establishment of a surveillance program and warn that if surveillance reveals changes in factors that raise concerns regarding wholesomeness, the acceptability of the novel food may be reappraised.

• *Nutritional considerations affecting toxicological testing in animals*: the guidelines recognize the dilemma posed with conventional feeding studies and urge the petitioner to determine the nutritional values of the food prior to embarking on long-term studies. The guidelines also recognize the limitations of these studies and reflect awareness that a safety factor smaller than that traditionally used in safety assessment may be required.

• *Toxicological requirements*: the toxicological requirements should be considered on a case-by-case basis, giving consideration to the principle of substantial equivalence. If SE to a traditional food cannot be established, additional considerations, including chemical structure and physicochemical properties of the novel food, may be explored. Also, if a comparator cannot be identified, dietary exposure becomes critical and the testing program must be made more extensive.

• *Implications of the novel food to human nutrition*: If the novel food will have a substantial impact on human consumption, the effect on human nutrition must be studied. Particular attention should be paid to susceptible subgroups including infants, children, pregnant and lactating women, the elderly, and those with chronic disease (*e.g.*, diabetes mellitus, malabsorption).

• *Novel organisms used in food*: by definition, microorganisms with no traditional use in food production in Europe cannot have a comparator, and the following criteria need to be assessed: containment (*e.g.*, limited to fermentor, remaining alive in food, killed during processing), potential for colonization of the mammalian gut, potential for toxigenicity as well as pathogenicity in mammals, whether genetic engineering was applied.

• The safety assessment of a genetically modified organism should consider the origin of the introduced material (*e.g.*, vectors, regulatory elements, foreign genes including target and marker genes) and whether the system was homologous (self-cloning, where all elements came from the same taxonomic species) or heterologous (donor and host are from different taxonomic species). The implication of horizontal gene transfer in the gut should be evaluated as well.

- *Allergic potential*: as a general principle of assessment, sera of people allergic to the traditional counterpart should be tested for activity against the novel food. Other factors to be taken into consideration are sequence epitope homology of novel proteins with known allergens, heat stability, sensitivity to pH, digestibility by gastrointestinal proteases, presence of detectable amounts in plasma, and molecular weight.
- *Marker genes*: considerations for microorganisms, especially those for antibiotic resistance, must be assessed in relation to the host organism, the biological containment established by the genetic construct, the possibility of colonization of the human gut, and the relationship between the efficacy of antimicrobials and acquired resistance.

Scientific Classification of Novel Foods for the Assessment of Wholesomeness

Novel foods are diverse and often very complex. To facilitate safety and nutritional evaluation, six classes of novel foods have been identified. The following listing of the classes elaborates on those relevant to this chapter.

- *Class 1*: pure chemicals or simple mixtures from non-GM sources. This class comprises foods and food components that are single-chemical defined substances or mixtures of these not obtained from genetically modified plants, animals, or microorganisms. There are two subclasses: novel foods whose source has a history of food use in the EC and sources without a history of use in the EC.
- *Class 2*: complex novel foods from non-GM sources. This class comprises non-GM complex novel foods that are neither sources nor derived from sources. Intact plants, animals, and microorganisms used as foods as well as food components are included. The same subclasses identified for Class 1 apply.
- *Class 3*: genetically modified plants and their products.
- *Class 4*: genetically modified animals and their products.
- *Class 5*: genetically modified microorganisms and their products. Living, genetically modified organisms may be used in the production of food or food ingredients. This class includes all novel foods regardless of the viability of the organisms in the final food. There are two subclasses: those for which there is a history of consumption and those for which there is not.
- *Class 6*: foods produced by means of a novel process. This class comprises foods and food ingredients that have been subjected to a process not currently used in food production, including new types of heat processing,

nonthermal preservation methods, new processes for chilling, freezing, or dehydrating products, and the application of new processes catalyzed by enzymes. However, the resulting product is considered to be a novel food only if the process results in changes in the chemical composition or structure of the food or food ingredient that affect its nutritional value, metabolism, or level of undesirable substances.

Identification of Essential Information for Assessment of Wholesomeness

• *Specification of the novel food*: parameters most relevant to characterizing the product from safety and nutritional points of view should be included. Parameters should include species and taxon, chemical composition (especially as the components relate to nutrition) and antinutritional/toxicological concerns.

• *Effect on production process applied to the novel food*: details must include technical data to distinguish between novel and existing process and to predict whether the process has the potential to induce changes that may impact safety or nutrition.

• *History of the organism used as the source of the novel food*: if there is no history of use in food in the European Community, the species/taxon is considered to be new, and a full description is needed to assess its future role in the European food supply. Information should include past and present methods of obtaining raw materials and food, procedures for fermentation and preparation, description of transport and storage, and the traditional role in the diet at locations outside the EC.

• *Effect of the genetic modification on the properties of the host microorganism*: the parent to the host must be recognized as a microorganism with a tradition in food fermentation in the EC, as a nonpathogenic, biologically advantageous human intestinal commensal, or as a traditionally used production organism for foods, including food additives and technical aids.

• *The genetic stability of the introduced genetic material and expression of the gene.*

• *Specificity of expression of novel genetic material*: factors involved in regulation of gene expression.

• *Transfer of genetic material from genetically modified organism*: animal or *in vitro* gut models may be considered. The food safety consequences of gene transfer need to be considered, including the nature of the gene and its product, the frequency of the transfer, and the level of expression in transformed gut microorganisms.

• *Ability of the genetically modified organism to survive in the gut*: for living genetically modified organisms in the food, consideration should be given to ability to colonize the human gastrointestinal tract and maintain genomic stability. Pathogenicity and gastrointestinal immunity should be considered.

- *Anticipated intake and extent of use of the novel food.*
- *Information from earlier human exposure to the novel food or its source.*
- *Nutritional information*: including nutritional consequences at normal and maximum levels of consumption and the effect of any antinutritional factors.
- *Microbiological information on the novel food*: wholesomeness of a novel food embraces microbiological safety. Information on the source organism should include documentation about its nonpathogenicity and nontoxigenicity.
- *Toxicological information on the novel food*: the extent of information is related to the degree of substantial equivalence. If substantial equivalence to a traditional counterpart cannot be established, the safety assessment must include consideration of the toxicity of individually identified components, toxicity studies *in vitro* and *in vivo*, including mutagenicity studies, reproduction and teratogenicity studies, and long term feeding studies, following a tiered approach, and studies on potential allergenicity.
- In the case of novel microconstituents and isolated novel food components, which differ by identifiable characteristics from traditional foods, or in the case of defined novel products derived from genetically modified organisms, it may be possible to test only the characteristics that differ from those of traditional foods. Conventional methods of safety evaluation may be used and are described in Scientific Committee for Food (SCF) Report 10. Substances in this category may require only a 90-day feeding study in a rodent species.

CASE STUDIES IN SAFETY ASSESSMENT

GRN 000072: Pullulanase Enzyme from *Bacillus Licheniformis* (U.S. FDA, 2001c)

A General Rulemaking Notice 000072 roughly conforms to the Advanced Notice of Public Rulemaking described in 2001 [*Fed. Regist.* **66**, 4706–4738, 2001 the Premarket Notice (PMN)] even though it was originally submitted some years before as a request for GRAS affirmation. The submitter had asked the FDA to convert this request to a GRAS notification and it was awarded GRN 000072. The "no objection" notice is posted on the CFSAN website (FDA, 2001b).

Identity

The petitioned substance is pullulanase enzyme preparation obtained from a strain of *Bacillus licheniformis* that contains a gene encoding pullulanase derived from *B. deramificans*.

- Pullulanase (EC 3.2.1.41) is pullulan 6-glycanohydrolase, which can hydrolyze the $\alpha_{1,6}$-glycosidic linkages of amylopectin and pullulan for the saccharification of starches.
- The host organism is *B. licheniformis* strain SE2 delap1,[18] used for construction of the production strain (*B. licheniformis* strain SE2-pul-int211).
  - *B. licheniformis* strains are commonly found in most soils and are listed in the *Food Chemicals Codex* (FCC, 1996) as a source of enzymes used in food processing.
  - FDA has approved carbohydrases and proteases derived from strains derived from *B. licheniformis* for use in food processing as described in the Code of Federal Regulations (21 CFR §184.1027). The petitioner cites published reports on the cloning and expression of proteins in *B. licheniformis* for use in food products.[19]
- The donor organism is *B. deramificans*; an earlier submission describes the cloning process (which did not raise any safety concerns).
- Production organism is *B. licheniformis* strain SE2-pul-int211.
  - DNA sequencing demonstrated the DNA sequence encoding pullulanase is the same as that derived from *B. deramificans*.
  - Pullanase from the production organism is characterized in comparison to the donor strain. Analyses included molecular weight, optimal pH and optimal temperature for pullulan hydrolysis, ion exchange chromatography profile, pattern of hydrolysis products, and amino-terminal sequencing.
  - Stability of production organism was assessed by means of Southern hybridization (to demonstrate stability of the sequence encoding for the enzyme) and transformation experiments with competent *E. coli* (which indicated no transformation potential from the integrated vector).

[18]FDA notes the host organism derives from the *B. licheniformis* strain SE2. Although *B. licheniformis* strain SE2 contains a gene encoding alkaline protease, this alkaline protease gene has been deleted from *B. licheniformis* strain SE2 delap 1.

[19]Much of this information goes toward the claim of a history of use. Although not the basis for a claim of GRAS with a history of use prior to January 1, 1958, this statement and the one before it indicate an established history of exposure to consumers. In addition, these statements indicate that because products from a similar species are already listed in the *Food Chemicals Codex*, approval of (or no objection to) one more products from this species is easily justifiable.

Manufacturing Information

- Pullulanase enzyme preparation is produced by a submerged, aerobic and pure culture fermentation of the production strain in accordance with current Good Manufacturing Practices (cGMP). The manufacturing process consists of fermentation, recovery and formulation.
  - During the fermentation process, propagation is carried out on a small scale to ensure that the bacilli are viable and are a pure culture. The culture is transferred to a primary fermentor for seed fermentation, which generates an initial biomass. The biomass from the seed fermentation is then used in the main fermentation to generate large quantities of biomass.
  - During the recovery process, the enzyme is separated from the fermentation debris. Then it is purified and concentrated by ultrafiltration, to ensure that the concentrated enzyme solution is free of the production strain and also contains no insoluble components from the fermentation medium.
  - During the formulation process, the enzyme concentrate is stabilized with potassium sorbate and sodium benzoate and standardized at 40% solids with corn syrup.
- Pullulanase enzyme preparation complies with the general and additional requirements for enzymes preparations set forth in the *Food Chemicals Codex* (FCC, 1996).

Grounds for GRAS Determination

Scientific procedures assess pathogenicity and toxicity.

Pathogenicity

Groups of five male and five female rats were given intraperitoneal injections of either live or killed cells in doses of 10^6, 10^9, and 10^{11} bacteria per kilogram of body weight. No animals demonstrated signs of pathogenicity at any does with either live or killed cells.

Toxicity

- *Acute inhalation toxicity study*: no signs of toxic effects.
- *Primary dermal irritation study*: no signs of toxic effects.
- *Feeding studies in rats*: 14- and 28-day feeding studies were conducted in rats with a maximum dose of 5% of the diet; no signs of toxic effects were reported.

- *Genetic toxicity studies*: bacterial reverse mutation assay in *Salmonella typhimurium* (Ames test), an *in vitro* histidine forward mutation assay in mouse lymphoma cells, and *in vivo* mouse bone marrow micronucleus and chromosomal aberration assays were performal. There was no evidence of mutagenic or genotoxic activity.

Technical Effect

Processing aids in the manufacturing of starch hydrolysates were maltodextrins, maltose, glucose, and high-fructose corn sweeteners (HFCS). Pullulanase is used in the wet milling of cornstarch in concert with glucoamylase (hydrolyzes $\alpha_{1,4}$-linkages). This use of pullulanase reduces the level of glucoamylase used (glucoamylase is slower at hydrolyzing $\alpha_{1,6}$-linkages), increases glucose yield, allows the saccharification process to be carried out at higher levels of dissolved solids, and shortens saccharification time.

Exposure

Although 2.3 mg of enzyme protein will be used for each kilogram of starch (*i.e.*, 2.3 ppm), as a result of additional purification steps of the hydrolysate, the amount of enzyme in the final product will be less than 2.3 ppm (and possibly as low as 0.23 ppb in HCFS).

GRN 000043: Lipase Enzyme from *Aspergillus oryzae* (US, FDA, 2000b)

As in the preceding example, GRN 000043 roughly conforms to the 2001 ANPR cited earlier *Fed. Regist.*, **66**, 4706–4738), even though the PMN was originally submitted in April 2000. The "no objection" notice is posted on the CFSAN website (U.S. FDA, 2000b).

Identity

- The petitioned substance is a lipase enzyme preparation derived from *Aspergillus oryzae* carrying a gene encoding lipase from *Thermomyces lanuginosus*.
- Triacylglycerol lipase (generic name: lipase) (EC 3.1.1.3; CAS 9001–62–1) is specific for the 1,3-position ester bonds in triglycerides with broad fatty acid specificity. Molecular weight of 35 kDa; isoelectric point is 4.4, and the total nucleotide and amino acid sequences have been determined.

- The host strain is *A. oryzae* strain IFO 4177 (synonym A1560), a commonly used industrial strain obtained from the Institute for Fermentation in Osaka, Japan (IFO). The genetically modified *A. oryzae* strain has been in use by the petitioner for over 10 years in the production of a commercial lipase enzyme for technical applications.
- Two plasmids were used in constructing the strain, one an expression plasmid and the other a selectable marker plasmid. These plasmids contain strictly defined fungal chromosomal DNA fragments and DNA from well-characterized *E. coli* vectors. The specific DNA sequences include a gene encoding a *T. lanuginosus* lipase enzyme, an *A. nidulans* selectable marker gene, *amdS* (acetamidase), and a well-characterized noncoding regulatory sequence from *Aspergillus niger* and *Aspergillus oryzae*, as well as known sequences from *E. coli* plasmids pUC19 and pBR322.
- A production organism, designated H-1-52/c (synonym AI-11) was constructed by plasmid transformation and classical mutagenesis. The organism complies with OECD criteria for Good Industrial Large-Scale Practice (GILSP) for microorganisms, and the criteria for a safe production microorganism as described by Pariza and Foster (1983) and other experts. The recipient microorganism used in the construction of the production strain is designated strain IFO 4177 (synonym A1560). This classification of A1560 as *A. oryzae* has been confirmed by the Centraalbureau voor Schimmelcultures, Baarn, Holland.
- *Lipase expression plasmid*: The 5.5-kb lipase expression plasmid pBoe 1960, used in the construction of the production strains, contains the following genetic material:
- 1.41-kb DNA from the *A. oryzae* TAKA amylase gene promoter
- 0.92-kb DNA from the *T. lanuginosus* lipase gene
- 0.75-kb DNA from the *A. niger* glucoamylase gene terminator sequence
- 2.69-kb DNA from the *E. coli* plasmid pUC19
- *Selectable marker plasmid*: The 8.96-kb selectable marker plasmid p3SR2, used in the construction of the production strain, contains the following genetic material:
- 5.25-kb DNA from *A. nidulans* encoding the *amdS* gene
- 3.71-kb DNA from the *E. coli* plasmid pBR322
- *Stability of production organism*: this property is Southern hybridization assessed by means of (to demonstrate stability and potential for transfer of the DNA sequences introduced); the transforming plasmid DNA is stably integrated into the *A. oryzae* chromosome, is poorly mobilizable for genetic transfer to other organisms, and is mitotically stable.
- *Antibiotic resistance gene*: Both pBoe 1960 and p3SR2 contain the β-lactamase gene *bla* (from *E. coli* plasmid pBR322), encoding resistance to

ampicillin. These genes are prokaryotic in origin and lack the appropriate sites and signals to be functionally expressed when integrated in a eukaryotic chromosome. In addition, the prokaryotic *bla* gene lacks appropriate eukaryotic signal sequences and processing signals necessary for secretion and export. Therefore, any β-lactamase potentially produced by expression of the *bla* gene would be localized intracellularly and not present in the final product, the lipase enzyme preparation. Tests of *A. oryzae* production strains containing the *bla* gene integrated as part of an expression vector have not shown any evidence of β-lactamase expression.

Manufacturing Information

Raw Materials

All materials used in the fermentation process and recovery are standard in the industry and comply either with current *Food Chemicals Codex* requirements or, for substances not given in the *FCC*, have specifications that are congruent with *FCC* specifications.

Fermentation Process

The lipase is manufactured by means of submerged fed-batch pure culture fermentation. The equipment used is designed, constructed, cleaned, and operated to prevent contamination by foreign microorganisms.

- Each batch of the fermentation process is initiated with lyophilized stock culture of the production organism, which undergoes quality control assays before use.
- Samples should be taken from the seed fermentor and the main fermentor at various intervals to ensure proper performance. Criteria for rejection due to contamination are established.

Recovery Process

Recovery, which consists of purification and formulation, starts immediately after fermentation.

Purification Process The enzyme is recovered by pretreatment (pH adjustment), primary separation (vacuum drum filtration), concentration (ultrafiltration and evaporation), prefiltration and germ filtration (removal of production organisms and as protection from any possible microbial degradation), preservation and stabilization (addition of sodium chloride), and final concentration (evaporation if enzyme concentration is too low to reach target yield).

Formulation and Standardization Processes Depending on use as a lipase formula for fats and oils applications or a lipase formula for baking applications, the liquid enzyme is sprayed onto a mixture of silicon dioxide, cellulose, and dextrin, and a granulate is formed, or mixed with granulation aids (stabilizers or binders) and dried, respectively.

Grounds for GRAS Determination

Scientific procedures are summarized as follows.

- The petitioner indicates that *A. oryzae* is nonpathogenic and nontoxigenic by general agreement within the scientific community and is GRAS as the result of a history of use prior to 1958:
  - It meets the criteria for nontoxigenicity and nonpathogenicity as described by Pariza and Foster (1983) and it is not considered to be pathogenic by JECFA, the Joint WHO/FAO Expert Committee on Food Additives.
  - *A. oryzae* has been used as a method for producing soy sauce since before 1958, and a petition (3G0016) proposing affirmation of enzyme preparations from *A. oryzae* be accepted as GRAS was filed with the agency.
- Safety of the lipase enzyme
  - Lipases, from one source or another, have been in general food use since 1952, are GRAS, and have food additive status.
  - This lipase enzyme is substantially equivalent to other lipases.
    - *Thermomyces lanuginosus*, a ubiquitous, thermophilic fungus, is not known to be pathogenic or toxin producing. The lipase coding sequence from this organism was not changed upon insertion into the host, also a nonpathogenic, nontoxigenic organism.
    - The petitioner's lipase and those lipases in commerce are similar in amino acid sequences and in number of amino acid residues, and there are structural similarities among them.
- Safety studies performed on the lipase included acute oral toxicity in rats, acute inhalation toxicity in rats, subacute oral toxicity in rats, skin irritation in rabbits, eye irritation in rabbits, skin sensitization (delayed contact hypersensitivity in guinea pigs), gene mutation (Ames test with *S. typhimurium/E. coli*); chromosome aberrations (*in vitro* cytogenetics with human lymphocytes), aquatic organism toxicity (*Daphnia* and carp), algal growth inhibition test; and biodegradability and pathogenicity of *A.*

oryzae (spores of host organism and genetically modified recombinant strain) (Greenough *et al.*, 1996).
- The safety of the manufacturing process meets the general and additional requirements for enzymes described in the *Food Chemicals Codex* and is compliant with cGMP.

Technical Effect

This lipase is used in dough, and baked goods, as well as in the fats and oil industry at minimum levels necessary to achieve the desired effect. The lipase preparation would be used as a catalyst in the interesterification of glycerides and acidolysis between glycerides and fatty acids in fats and oils at a maximum level of one kilogram of lipase per ton of triglycerides. The lipase would be used in the hydrolysis of primary ester bonds in triglycerides in dough and baked goods for the purpose of modifying lipid–gluten interactions at a maximum level of 1 to 5 g per 100 kg of flour.

Exposure

Fats and Oils Application

Triglycerides would contain 60 mg of TOS*/per kilogram of triglyceride and based on the average consumption of 69 g of vegetable oil per person per day, total daily intake of TOS would be 4.1 mg/day or 0.058 mg/kg/day for a 70-kg person. The no observed adverse effects level (NOAEL) in the 13-week rat study was 1350 mg/kg, therefore the margin of safety (NOAEL in rats/TOS consumed) is 2.3×10^4.

Baking Application

The maximum TOS in bread is estimated to be 1.88×10^{-6} g TOS/g bread. The average bread consumption of bread is 160 g/person/day, or 2.67 g bread/kg for a 60-kg person. Therefore, the maximum estimated daily intake (EDI) of TOS via bread is 0.005 mg/kg/day. The NOAEL in the 13-week rat study was 1350 mg/kg; therefore the margin of safety (NOAEL in rats/TOS consumed) is 2.7×10^5.

Illustration of the Concept of Substantial Equivalence in Lactic Acid Bacteria in Dairy Products (Bergmans and Knudsen, 1993)

This is an example of the decision process and treatment of an organism (lactic acid bacteria) according to the criteria for examination as they existed for

*Total Organic Solids

OECD in 1993 (OECD, 1993). The authors note that the use of modified lactic acid bacteria[20] has precedent in terms of other modified organism-generated substances such as chymosin and egg white lysozyme. The authors further distinguish traditional products of lactic acid bacteria from novel products.

Traditional versus Novel Products: Two Classes

Organisms are considered to be "traditional" (*i.e.*, not novel) if the levels of expression are comparable to those of traditional organisms. Included in this definition are lactic acid bacteria with cloned homologous genes and cloned genes derived from other lactic acid bacteria such as those encoding proteolytic enzymes and enzymes involved in sugar metabolism, nisin production, or resistance to bacteriocins and bacteriophages.

Novel organisms are those whose gene products have not been actively synthesized in dairy products, although they may have been added (such as egg white lysozyme or chymosin). Included here are lactic acid bacteria carrying heterologous genes: for example, genes encoding fimbriae from other prokaryotic sources (bacteria), genes encoding egg white lysozyme or prochymosin, and genes encoding proteolytic enzymes from eukaryotic plant sources.

Evaluation Procedures

Traditional organisms should demonstrate nonpathogenicity and nontoxigenicity but are otherwise considered to be substantially equivalent unless the gene product is of a type or degree outside the range of normal gene products of the lactic acid designation.

Novel organisms need to be evaluated on a case-by-case basis on two fronts: regarding the gene product, as a substance, in relation to traditional dairy products, and regarding the effect of the new triat in relation to the function of the organism in its traditional habitat, as it relates to food safety issues.

The authors leave the door open to consideration of substantial equivalence if the foregoing evaluations allow. In addition, horizontal gene transfer (see earlier section) should be considered.

Mycoprotein from RNA-Reduced Cells of *Fusarium venenatum*[21]

Information on a mycoprotein from an RNA-reduced mass of fungal hyphae was published in the public literature (Miller and Dwyer, 2001;

[20]The authors note the term "lactic acid bacteria" is a generic term that includes other genera: *Lactobacillus, Leuconostoc, Lactococcus, Pediococcus*, and *Streptococcus*.

[21]While mycoprotein is not an rDNA product, it represents a novel source of food, unlike anything found in nature, and therefore is representative of the third tier of substantial equivalence: "products not substantially equivalent to existing food or food components."

Rodger, 2001) presumably in anticipation of fulfillment of the requirement for public disclosure of data pursuant to a GRAS determination. On the basis of the data presented, the authors reported an Expert Panel concluded that, "to a reasonable certainty, mycoprotein is safe and suitable for incorporation into a wide variety of foods for humans" (Miller and Dwyer, 2001).

Identity

The mycoprotein (Quorn®) is produced from RNA-reduced hyphae of *Fusarium venenatum* (PTA 2684), a native-growing mushroom like plant found in Buckinghamshire in Britain. On the basis of dry weight, it consists of about 50% protein, 13% lipid, and 25% fiber. Ergosterol is present, but not cholesterol. The cell wall constitutes approximately one-third of the dry cell weight and is composed of chitin [poly *N*-acetyl glucosamine and β-glucans ($\beta_{1,3}$ and $\beta_{1,6}$-glucosidic linkages)].

Manufacturing Information

- Mycoprotein is produced by a submerged, aerobic, and pure culture fermentation of the production strain with a continuous feed of nutrients and simultaneous removal of broth. Temperature and pH are controlled, as are the flows of nutrients and other medium components, to maintain optimum growth conditions. Following RNA reduction, the suspension is recovered and dewatered to form a paste with approximately 75% water content.
- Because ingestion by humans of an excess of RNA may cause an increase in serum uric acid, a heat treatment process is used to reduce RNA content in the mycoprotein from a native 10% RNA to less than 2% (dry weight).

Safety Determination

The authors note that safety determinations were carried out in accordance with the Food and Drug Administration's *Toxicological Principles for Evaluating the Safety of Food Ingredients* (U.S. FDA, 2000a).

- Pathogenicity: the authors provided no information.
- Toxicity:
  - *In vitro testing*: because of the high protein content, the authors anticipated a high incidence of false positives when the *Salmonella* histidine reversion test was used. Therefore, the tests were carried out in a suspension, using chicken meat as a control. No mutagenic changes were found with or without S9 metabolic activation.
  - *Primary dermal irritation study*: no signs of toxic effects with intact or abraded skin.

- *Intradermal injections*: in guinea pigs and rabbits, a granulomatous response was induced, but the authors concluded that it was no different from that provoked by mushrooms or other food grade microorganisms.
- *Estrogenic potential*: tested in pigs and mice and found negative.
- Subchronic feeding studies in rats:
  - Rats were fed mycoprotein at dietary levels ranging from 26 to 52% for 22 weeks. Cecal enlargement was noted, presumably due to the high fiber content of the mycoprotein.
  - Rats were fed dried cooked mycoprotein at 13 and 35% in the diet for 13 weeks, using a casein control. Cecal enlargement was noted in the absence of significant changes in growth, blood parameters, organ weights, and histopathology.
  - Rats were fed an undried form of mycoprotein at 13 and 35% in the diet for 13 weeks, using a casein control. No significant differences were found in the categories tested, which included availability and balance of calcium, phosphorus, magnesium, iron, copper, and zinc.
  - Rats were fed mycoprotein at 20 and 40% of the diet for 90 days and an *in utero* exposure group was included with an additional 90-day study in the offspring. At weaning, both sexes at the high dose had lower body weights than the casein controls, and although the difference shrank during the 90-day study, it remained statistically significant. When the physical form of the presentation of the diet was changed, the differences were abolished.
- Chronic studies:
  - A 2-year feeding study was conducted at 21 and 41% of the diet and a casein control was employed. Growth and reproduction by the F_0 generation was reported to be unaffected. Aside from minor variations, no adverse effects were seen in growth, survival, incidence or onset of tumors, hematology, urinalyses, or histopathology. Pigmentation of organs, seen in the mycoprotein groups, was attributed to the high amount of unsaturated lipid in the mycoprotein.
  - A 1-year study in dogs was performed at doses of 20 and 40% mycoprotein in the diet. No adverse clinical signs or laboratory values were reported, and the dogs showed lower plasma cholesterol and triglycerides. Females had greater thyroid weights.
  - A multigenerational diet was conducted in rats with diets containing mycoprotein at levels of 12.5, 25, and 50%. A 50% casein control was included, in addition to a laboratory diet control. No effects on fertility, reproductive function, or gross or microscopic pathology were reported.
- Teratology and embryotoxicity tests were performed in rats and rabbits, and no effects were noted.

- Humans were fed mycoprotein in amounts of 10 to 40 g/day. Subjects reporting adverse reactions were retested, whereupon either no effect was observed or, in some cases, allergies to other fungi were noted. The authors concluded that mycoprotein was well tolerated with little potential for allergic reaction.

Technical Effect

The authors claim this mycoprotein to be a good source of protein and fiber, and the fat content is typically 2 to 3.5%, with a composition more like vegetable than animal fat. A detailed composition is provided in the review (Rodger, 2001). Thus, the mycoprotein is suggested for use in foods as a replacement for muscle fiber, fats, and cereals.

Exposure

The estimated daily intakes of mycoprotein for the general population is 0.01 to 0.18 mg/kg/day and for vegetarians, 0.24–0.46 mg/kg/day.

Illustration of the Concept of Substantial Equivalence for Mycoprotein (*Fusarium graminearum*) (Jonas, 1993)

This is an example of the decision process involving an organism with no traditional comparator in nature.

Traditional versus Novel Products: Two Classes

Organisms are considered to be "traditional" (*i.e.*, not novel) if the levels of expression are comparable to those of traditional organisms. Included in this definition are lactic acid bacteria with cloned homologous genes or cloned genes derived from other lactic acid bacteria, such as those encoding proteolytic enzymes and enzymes involved in sugar metabolism, nisin production, or resistance to bacteriocins and bacteriophages.

Novel organisms are those whose gene products have not been actively synthesized in dairy products, although they may have been added (such as egg white lysozyme or chymosin). Included here are lactic acid bacteria carrying heterologous genes: for example, genes encoding fimbriae from other prokaryotic sources (bacteria), genes encoding egg white lysozyme or prochymosin, and genes encoding proteolytic enzymes from eukaryotic plant sources.

Manufacturing Information

- The process involves use of a carbohydrate-based culture medium plus micronutrients, all of which must meet specific criteria. The inoculum is maintained under aseptic conditions, and tests for contamination and strain stability are conducted throughout the process. Fermentation is carried out under aseptic conditions, and reaction contents and conditions are continuously monitored.
- RNA content is reduced with the use of thermal shock.
- The mycoprotein is obtained by filtration and consists of about 30% solids.

Evaluation Procedures

- Information on the organism's taxonomy and potential for mycotoxin was supplied by the petitioner. No mycotoxin was detected under conditions used for production, nor under test conditions where other strains of *F. graminearum* were known to produce mycotoxin.
- Analyses included nitrogenous material, amino acids, carbohydrates, fiber, lipids, minerals, and vitamins. No usual amino acids (*e.g.*, D-amino acids) were found, and chitin was the primary carbohydrate.
- Nutritional tests were conducted, and the mycoprotein was judged to be a good source of protein. No antinutritional factors were found.
- A battery of animal tests was carried out, and the only findings were those associated with a high protein diet. Human tests were uneventful, and there were no indications of allergic reactions. The mycoprotein was test-marketed to approximately 4000 people, and no adverse effects were noted.

Rejection of a Petition for Single-Cell Protein by the British Advisory Committee on Novel Foods and Processes

This case study is an example of a substance that was rejected by a national regulatory body, the Advisory Committee on Novel Foods and Processes (ACNFP), ruling through its Committee on Toxicity of Chemicals in Food, Consumer Products, and the Environment (COT, 1996). The subject matter was a petition to this agency for approval of a single-cell protein produced by bacterial fermentation to be used for the production of protein hydrolysates and autolysates.

The COT determined that the test animals may have been overloaded with protein, obscuring possible effects, and there was no demonstration of lack of an effect on reproduction. The COT recommended the following:

- A 90-day study in which the novel protein is added to the diet, with the highest dose constituting the sole source of protein. Further, particular attention should be paid to mineral balance.
- A teratology study and a single-generation reproduction study.
- Additional *in vitro* mutagenicity tests using both polar and nonpolar extracts of the mycoprotein to determine the effects of the nonprotein components.
- Additional allergy testing.

The COT requested that a test battery be conducted on both nucleic acid reduced forms and non–nucleic acid forms, since both forms were requested for clearance.

CONCLUSION

Biotechnologically produced foods and food ingredients hold the promise of alleviating hunger and enhancing nutrition in developing nations. In the western nations, biotechnology holds the promise of reduction in the cost of food as a percentage of household expenses as the result of decreased cost of production, and foods will be safer[22] than conventional foods. Despite these advantages, the "learning curve" for science has gotten significantly ahead of regulations and public confidence. While the American public still has a high degree of confidence in the FDA, European fears stemming from misinformation about bovine spongiform encephalopathy and blood products tainted with HIV have eroded the confidence of the European public in their institutions. Because erosion of confidence acts at a higher energy level than historical credibility, science and regulation must move rapidly to deliver workable solutions and a transparent approval process acceptable to the public.

Guidelines have been developed to address the key issues: characterization of the organism, pathogenicity, toxigenicity, and antinutritive effects, substantial equivalence, and gene transfer, marker genes, and antibiotic resistance. Issues left unresolved include the extent and type of testing required, basis for selection of comparator, and the relative importance of antibiotic resistance versus overuse of antibiotics in the treatment of humans and livestock for infections and the use of antibiotics in feed.

The most valuable asset in determining the reasonable certainty of no harm rests in the judgment of experienced scientists, whether involved in a GRAS process or as regulators assigned to assess a petition. Reasonable, fact-based judgments, in a transparent process, will maintain public confidence.

[22]Associated Press (2001) EU Says Biotech Foods May be Safer. October 9, 2001.

REFERENCES

Bergmans, H. and Knudsen, I. (1993). Lactic acid bacteria, in, *Safety Evaluation of Foods Derived by Modern Biotechnology: Concepts and Principles*, pp. 31–35. Organisation for Economic Co-operation and Development, Paris.

Borzelleca, J. F. (1992). Macronutrient substitutes: Safety evaluation. *Regul. Toxicol. Pharmacol.* **16**, 253–264.

Burdock, G. A. (2000). Dietary supplements and lessons to be learned from GRAS. Regul. Toxicol. Pharmacol., **31** 68–76.

Burdock, G. A. (2001). Flavor regulation, in *Nutritional Toxicology, Target Organ Toxicology Series*, F. Kotsonis and M. Mackey, eds., pp. 316–339. Taylor and Francis, New York.

Carabin, I. G., and Flamm, W. G. (1999). Evaluation of safety of inulin and oligofructose as dietary fiber. *Regul. Toxicol Pharmacol.* **30**, 268–282.

Codex Alimentarius. (2001). http://www.codexalimentarius.net/biotech/en/UnexEf.htm (Site visited August 9, 2001.)

COT. (1996). Committee on Toxicity, Scientific Committee for Food Guidelines on the Assessment of Novel Foods. *Toxicity, Mutagenicity and Carcinogenicity Report, 1996.* http://www.official-documents.co.uk/document/doh/toxicity/chap-1b.htm Site visited September 16, 2001.

Dröge, M., Pühler, A., and Selbitschka, W. (1998). Horizontal gene transfer as a biosafety issue: A natural phenomenon of public concern. *J. Biotechnol.* **64**, 75–90.

EC. (1997a). *Regulation (EC) No 258/97 of the European Parliament and the Council of 27 January 1997 Concerning Novel Foods and Novel Food Ingredients.* Official Journal L 043, 14/02/1997, pp. 001–007. http://europa.eu.int/eurlex/en/fif/dat/1997/en_397R0258.html Site visited October 25, 2001.

EC. (1997b). *97/618/EC Commission Recommendation of 29 July 1997 Concerning the Scientific Aspects and the Presentation of Information Necessary to Support Applications for the Placing on the Market of Novel Foods and Novel Food Ingredients and the Preparation of Initial Assessment Reports Under Regulation (EC) No. 258/97 of the European Parliament and the Council.* Official Journal L253, 16/09/1997, pp. 1–36.

EC. (2001). *Guidance on Submissions for Food Additive Evaluations by the Scientific Committee on Food* (opinion expressed on July 11, 2001). Scientific Committee on Food. Health and Consumer Protection Directorate-General. SCF/CS/ADD/GEN/26 Final, July 12, 2001. http://www.europa.eu.int/commdg24/health/sc/scf/index_en.html Site last visited October 24, 2001.

U.S. FDA. (1996). Safety assurance of foods derived by modern biotechnology in the United States. Presentation at the BioJapan '96 Symposium. Food and Drug Administration, Center for Food Safety and Applied Nutrition, Washington, DC.

U.S. FDA. (1998). [Draft] *Guidance for Industry: Use of Antibiotic Resistance Marker Genes in Transgenic Plants.* Food and Drug Administration, Center for Food Safety and Applied Nutrition, Office of Premarket Approval Washington, DC. http://vm.cfsan.fda.gov/~dms/opa-armg.html Site visited April 2, 2000.

U.S. FDA (2000a). *Toxicological Principles for Evaluating the Safety of Food Ingredients.* Office of Premarket Approval, Center for Food and Applied Nutrition, Food and Drug Administration, Washington, DC.

U.S. FDA (2000b). *Agency Response Letter GRAS Notice No. GRN 000043.* Food and Drug Administration, Washington, DC. http://www.cfsan.fda.gov/~rdb/opa-g043.html Site visited August 9, 2001.

U.S. FDA (2001a). Premarket notice concerning bioengineered foods. *Fed. Regist.*, **66**(12), 4706–4738.

U.S. FDA (2001b). *Summary of all GRAS Notices.* Food and Drug Administration, Washington, DC. http://cfsan.fda.gov/~rdb/opa-gras.html Site visited September 30, 2001.

U.S. FDA (2001c). *Agency Response Letter GRAS Notice No. GRN 000072.* Food and Drug Administration, Washington, DC. http://www.cfsan.fda.gov/~rdb/opa-g072.html Site visited August 10, 2001.

FCC (1996). *Food Chemicals Codex.* 4th edn. National Academy Press, Washington, DC.

Gasson, M. J. (2000). Gene transfer from genetically modified food. *Cur. Opin. Biotechno.* **11**, 505–508.

Greenough, R. J., Perry, C. J., and Stavnsbjerg, M. (1996). Safety evaluation of a lipase expressed in *Aspergillus oryzae. Food Chem. Toxicol.* **34**(2), 161–166.

Hammond, B., Rogers, S. G., and Fuchs, R. L. (1996). Limitations of whole food feeding studies in food safety assessment, in *Food Safety Evaluation*, pp. 85–97. Organisation for Economic Co-operation and Development, Paris.

HPB. (1994). *Guidelines for the Safety Assessment of Novel Foods*, Vol I and II. Food Directorate, Health Protection Branch, Health Canada. 32 pp. JECFA. (1990). Evaluation of Certain Food Additives and Contaminants. WHO Technical Report Series, No. 789. Joint (WHO/FAO) Expert Committee on Food Additives, Geneva.

Jonas, D. A. (1993). Myco-protein, in *Safety Evaluation of Foods Derived by Modern Biotechnology: Concepts and Principles*, pp. 41–44. Organisation for Economic Co-operation and Development, Paris.

Kotsonis, F., Burdock, G. A., and Flamm, W. G. (2001). Food toxicology, in *Toxicology: The Basic Science of Poisons* C. D. Klaassen, 6thed. (p.1050), Pergamon Press, New York.

Lorenz, M. G., and Wackernagel, W. (1994). Bacterial transfer by natural genetic transformation in the environment. *Microbiol. Rev.* **58**, 563–602.

Love, S. L. (2000). When does similar mean the same: A case for relaxing standards of substantial equivalence in genetically modified food crops. *HortScience*, **35**(5) 803–806.

Mackey, M., and Santerre, C. (2000). Biotechnology and our food supply. *Nutr. Today*, **35**(4), 120–128.

Madden, D. (1995). *Food Biotechnology, An Introduction.* International Life Sciences Institute. ILSI Press. Washington, DC.

MAFF. (1999). Toxicological assessment of novel (including GM) foods. Advisory Committee on Novel Foods and Processes. Ministry of Agriculture, Fisheries, and Foods (U.K.), Department of Health and the Scottish Executive (before April 1, 2000, when the Food Standards Agency was established). http://www.foodstandards.gov.uk/maff/archive/food/novel/toxrey. htm Site visited September 16, 2001.

Miller, S. A., and Dwyer, J. T. (2001). Evaluating the safety and nutritional value of mycoprotein. *Food Technol.*, **55**(7), 42–50.

Millstone, E., Brunner, E., and Mayer, S. (1999). Beyond 'substantial equivalence.' *Nature*, **401**, 525–526.

Morrison, M. (1996). Do ruminal bacteria exchange genetic material? *J. Dairy Sci.* **79**, 1476–1486.

Nielsen, K. M., Bones, A. M., Smalla, K., and van Elsas, J. D. (1998). Horizontal gene transfer from transgenic plants to terrestrial bacteria—A rare event? *FEMS Microbiol. Rev.* **22**(2), 79–103.

OECD. (1993). *Safety Evaluation of Foods Derived by Modern Biotechnology: Concepts and Principles.* Organisation for Economic Co-operation and Development, Paris. 79 pp.

OECD. (2000). Report of the Task Force for the Safety of Novel Foods and Feeds, C(2000)86/ADI. Organisation for Economic Co-operation and Development, Paris.

Ow, D. (2000). *Marker Genes*. Biotech 00/14, Joint FAO/WHO Expert Consultation on Foods Derived from Biotechnology, Geneva. 6 pp.

Pariza, M. W., and Foster, E. M. (1983) Determining the safety of enzymes used in food processing. *J. Food Prot.* **46**(5), 453–468.

Pedersen, J. (2000). *Application of Substantial Equivalence Data Collection and Analysis*. Biotech 00/04, Joint FAO/WHO Expert Consultation on Foods Derived from Biotechnology, Geneva. 5 pp.

Rodger, G. (2001). Production and properties of mycoprotein as a meat alternative. *Food Technol.*, **55**(7), 36–41.

Sayles, A. (2001). Genetically engineered foods: Safety issues associated with antibiotic resistance genes. Reservoirs of Antibiotic Resistance Network. http://www.heltsci.tufts.edu/apua/ROAR/salyersreport.htm (Site visited August 11, 2001).

Scarbrough, F. E. (1999). Letter from U.S. Codex Manager regarding elaboration of standards, guidelines or other principles for foods derived from biotechnology. http://www.cefsan.fda.-gov/~dmsbioresp.html (Site visited October 3, 2001).

Smith, J. (1994). New opportunities in food biotechnology. *Food Aust.* **46**(6), 262–265.

Thompson, J. (2000). *Gene Transfer: Mechanisms and Food Safety Risk*. Biotech 00/13, Joint FAO/WHO Expert Consultation on Foods Derived from Biotechnology, Geneva. 12 pp.

Tomlinson, N. (2000). *The Concept of Substantial Equivalence, Its Historical Development and Current Use*. Biotech 00/08, Joint FAO/WHO Expert Consultation on Foods Derived from Biotechnology, Geneva. 6 pp.

U.S. FDA. (1992). Statement of [Food and Drug Administration] Policy: Foods derived from new plant varieties. *Fed. Regist.*, **57**(104): 22984–23005.

Walker, R. (2000). *Safety Testing of Food Additives and Contaminants and the Long-Term Evaluation of Foods Produced by Biotechnology*. Biotech 00/08, Joint FAO/WHO Expert Consultation on Foods Derived from Biotechnology, Geneva. 8 pp.

WHO (1991). *Strategies for Assessing the Safety of Foods Produced by Biotechnology. Report of a Joint FAO/WHO Consultation*. World Health Organisation, Geneva. 59 pp.

WHO. (1996). *Report of a Joint FAO/WHO Consultation, Rome: September* 30-October 4, *1996*. FAO Food and Nutrition Paper 61. World Health Organisation. Rome. 26+ p.

Chapter 4

Food Safety Assessment of Current and Future Plant Biotechnology Products

Bruce M. Chassy
College of Agricultural, Consumer, and Environmental Sciences
University of Illinois
Urbana-Champaign, Illinois

Agricultural biotechnologists have used the recently developed techniques of molecular biology to breed varieties of plants into which have been inserted specific genes that confer desired traits such as tolerance to herbicides and protection from insects. Since the products of this new technology were to be consumed as food and feed, it seemed reasonable to ask whether the products present any new or different food or environmental safety risks. Numerous international scientific organizations have concluded and scientific assessments have confirmed that the risks were no different from those associated with new varieties of plants produced through conventional plant breeding. A regulatory review paradigm evolved that focuses on an evaluation of the safety of the product on a case-by-case basis, rather than on the safety of the new process used to create the product. The U.S. FDA has primary responsibility for premarket food safety assessment through a voluntary premarket notification that will soon become mandatory. The substantial equivalence concept provides a useful framework for safety assessment. In this analysis each product is compared with its conventional counterpart, and similarities and differences are identified. Potential safety implications associated with the differences are then made the focus of the evaluation. Each new product is subjected to detailed molecular characterization to analyze the inserted DNA, composition analysis, toxicological testing, evaluation of allergenicity, and studies of structure and function of the introduced protein(s). It is concluded that this science-based risk analysis paradigm has worked effectively and has the flexibility to be used for the evaluation of novel products in the future.

INTRODUCTION

The Development of Agricultural Biotechnology

Society and agriculture have coevolved over the last 10–15 millennia. The emergence of food and agricultural biotechnology was an integral part of the development of modern agriculture. Plant breeding led to the successful selection of wheat, barley, and millet from large seeded grasses in Eurasia and maize from teosinte, its grasslike ancestor, in meso-America (Diamond, 1997). These domesticated crops also represented the first examples of genetic modification applied to agricultural biotechnology. The use of microbes in cheese making, in the production of fermented beverages, in pickling, and in bread making were also among the earliest applications of biotechnology to foods.

During the last century, food and agricultural biotechnology were the focus of intensive scientific research. Particular attention was devoted to the development of more powerful, more rapid, and more specific methods for genetic modification to assist in plant breeding. Research into the genetics of inherited traits led to the unraveling of DNA structure, and molecular understanding of the functioning of genes that became the foundation of molecular biology. Techniques for the isolation of genes, for the production of recombinant DNA molecules, and for the transfer of DNA into plants developed in the latter decades of the twentieth century made it possible for the first time for plant breeders to preselect a gene from virtually any organism and relatively quickly introduce it into a target crop plant.

The Development of a Regulatory Paradigm for Biotechnology-Derived Foods

There exists no written record, but it is reasonable to assume that, historically, the safety of new products of food and agricultural biotechnology was established by trial and error. The foods consumed today are generally viewed as safe, based on their long history of safe use. In the United States, whole foods that are new to the marketplace are not usually reviewed for safety. Often they have a prior history of safe use somewhere in the world, or they appear to be comparable to other foods. Wholly novel foods or foods that are new to the human diet can be required to undergo premarket review. For example, the proposed marketing in the United States of Quorn, a mycoprotein food product, required a petition to the Food and Drug Administration (FDA) (Miller and Dwyer, 2001). Crop plants that are genetically modified by conventional or traditional breeding techniques are not required to be subjected to a premarket safety review. The developers of a new variety will, however, often voluntarily perform limited compositional and/or nutritional analysis. If a plant contains compounds that are known to be toxic or have non-nutritive health benefits, additional analysis may be performed. Toxic glycoalkaloids such as solanine, found in potatoes and tomatoes, are good examples of such potential toxins (Kuiper *et al.*, 2001).

The Federal Food, Drug, and Cosmetic Act (FFDCA) charged the FDA with primary responsibility for overseeing the safety of the U.S. food supply. The broad charge given to FDA was to ensure that foods would not be *ordinarily* injurious to health. A major focus of FDA regulation under the FFDCA has been premarket review of new food additives and food ingredients through a petition process. Food additives must meet a more stringent safety standard than whole foods. Food additives must be assessed against the possibility that they "may" cause effects that are "injurious to health." FDA has concluded that since products designed for human consumption that are derived from modern biotechnology are foods, they should be assessed against the standard applied to other foods. The developer that proposes a new ingredient or additive must satisfy the FDA that its use conforms to the FDA interpretation of the standard for food safety set by the FFDCA of a reasonable certainty of no harm. Neither the legislation itself nor the legislative history of the FFDCA defines either "reasonable certainty" or "no harm." It is clear, however, that Congress recognized that "zero risk" and "absolute safety" are impossible to achieve in the food system (CAST, 2001). Many foods can easily be rendered unsafe by abuse in storage, preparation, or consumption; and food allergies, food intolerances, food sensitivities, and idiosyncratic responses to food are both common and well documented.

The ability to introduce new traits encoded by genes isolated from an unrelated organism provided plant breeders with a powerful tool with which to produce genetically modified crop plants. Prior to commercialization of crops developed by using this technology, scientists, regulators, and policymakers realized that there were two significant differences between this new form of genetic modification and the conventional approach to plant breeding: the organism from which the genetic material could be selected was virtually unrestricted, and the transformation process did not allow control of the locus at which the incoming genetic trait would be integrated into the chromosome. It seemed logical to ask whether food and environmental safety could be adversely affected by these two characteristics of the new technology. In the United States, the White House Office of Science and Technology Policy chaired a series of meetings that led to the publication, on June 26, 1986, of the *Coordinated Framework for Regulation of Biotechnology* This document assigns responsibility for safety to three lead agencies: the U.S. Department of Agriculture (USDA), the Food and Drug Administration (FDA), and the Environmental Protection Agency (EPA). It also continued the role of the National Institutes of Health (NIH) in providing guidance on laboratory and greenhouse research using recombinant DNA (Coordinated Framework, 1986; CAST 2001).

In a parallel process, the U.S. National of Academy of Sciences (NAS) focused attention on the scientific issues raised by the new technology. An NAS report concluded in 1987 that there were no unique hazards associated with the movement of genes between unrelated organisms or in the use of recombinant DNA techniques (NAS, 1987). The NAS study further concluded the risks associated with the introduction of organisms derived from recombinant DNA biotechnology are no different from those associated with the introduction of unmodified organisms and organisms modified by other methods. A follow-up white paper by the National Research Council concluded that "*no conceptual distinction exists between genetic modification of plants and micro-organisms by classical methods or by molecular techniques that modify DNA and transfer genes*" (NRC, 1989). The conclusions outlined in these reports were reaffirmed by the NAS (NAS, 2000) and by two international bodies (FAO/WHO, 2000), (OECD, 2000).

In 1992 FDA provided a general outline in the form of decision trees for the safety assessment of biotechnology-derived food products based on risk analysis related to the characteristics of the products (FDA, 1992). FDA concluded that the source or focus of a food safety evaluation should not be the method of preparation of a food or ingredient, but the traits and properties of the food or ingredient that affected safety. FDA directed that a science-based risk assessment of the properties and safety of each new product be conducted by developers. *It is emphasized that since it is the safety of the product rather than the process used to produce it that is being assessed, each new product must be evaluated on a case-by-case basis.* FDA established guidelines for a voluntary premarket consultation and reserved the authority to require a Food Addition Petition if it were judged that a biotechnology-derived food product could be considered to contain a wholly new food ingredient or additive. All 49 of the plant products that have been approved for commercialization by the USDA have undergone a premarket consultation with the FDA. Nonetheless, in response to concerns raised about the voluntary nature of this process, FDA has proposed that premarket consultation be mandatory in the future.

The EPA assumes responsibility granted it under the Federal Insecticide, Fungicide, and Rodenticide Act (FIFRA) to regulate the food safety of biotechnology-derived plants that confer pesticidal properties (CAST, 2001). Plants such as *Bt* corn contain pesticidal substances, or plant pesticides, that are regulated by the EPA. In addition, EPA has adopted the term "plant-incorporated protectants" to more accurately describe plant protection system properties that bear no resemblance to chemical pesticides. The EPA applies to plant-incorporated protectants the same safety standard used for pesticide residues in food, regardless of their source. EPA also uses a science-based risk assessment procedure that evaluates the safety of the product, rather than the process used to produce the product. EPA is to evaluate and subject to premarket consultation with FDA the food safety of the "pesticidal" component introduced into a bioengineered plant that has a plant-incorporated protectant as well any marker, such as an antibiotic resistance marker gene, that was introduced to facilitate the introduction of the pesticidal component (CAST, 2001).

Scope of This Chapter

Since biotechnology-derived plants were first approved for commercialization in the United States in 1994, herbicide-tolerant soybeans (> 60% of plantings) and *Bt* corn (> 20% of plantings) have rapidly gained acceptance by farmers (James, 2000). The harvests from these bioengineered seeds have entered the commodity stream and have moved into the food and feed supply. This chapter describes the food safety evaluation that has been applied to these crops.

The underlying approach of the food safety assessment paradigm applied thus far is based on the concept of substantial equivalence, as described in the following section. With but two exceptions, all the crops introduced to date are virtually identical in composition to their conventional counterparts and are thus well suited to an assessment guided by the substantial equivalence concept. Since they are indistinguishable from their conventional counterparts, these crops are intended to enter their respective commodity streams. Each of the scientific questions asked in the safety assessment will be described and representative data will be shown.

It is known that many new biotechnology-derived crops are in various stages of development (see Mackey & Fuchs, Chapter 5). Some of these new crops will lend themselves to the same safety assessment paradigm that has been applied to date. Many, however, will be specifically designed to contain newly introduced compounds for which health benefits are claimed, elevated levels of specific nutrients, or altered nutritional composition. Specific components such as food allergens or natural toxicants (e.g., cyanide in cassava) might also be eliminated. It is necessary to ask whether the food safety paradigm that has been used thus far can be extended to the evaluation of bioengineered plants with significantly altered compositions. One can also ask whether any new strategies and/or new technologies can or should be used to enhance the safety evaluation.

It is fair to say that while there is not a single documented case of harm to humans or animals arising from the consumption of biotechnolgy-derived crops, there remain doubters and critics who question their safety. Some, for example, assert that it is fundamentally risky to introduce a gene from an organism with which the plant would not normally exchange genes and that the insertion of DNA into the chromosome could produce unintended negative effects. They define the newly transformed plants as "wholly new lifeforms" that by their very nature cannot be called "substantially equivalent." Critics also often argue that "no research" has been performed on these new crops. They often point specifically to the lack of a requirement for the long-term human feeding studies they claim are necessary to demonstrate food safety. And, they submit that the precautionary principle should be applied, which translates into a demand for proof that absolutely no harm will result from the consumption of foods and food ingredients derived from them. This chapter provides a reference against which these claims can be judged.

SAFETY EVALUATION AND THE SUBSTANTIAL EQUIVALENCE PARADIGM

Rationale for Risk Assessment–Based Food Safety Evaluation

The strategy of the risk assessment process that is applied to biotechnology-derived crops is based directly on the changes that have been made in the plant. The hazard identification process reveals that there are three principal issues that merit further risk assessment:

- The safety of the inserted DNA
- The safety of the newly introduced component(s)
- The safety of the balance of the whole food

Whole foods do not lend themselves to risk analysis by means of the toxicological techniques that are applied to single chemicals, nutrients, or additives (Table 1). For that reason, a comparative approach is used in which the plant, or a food derived from it, is compared with its conventional counterpart. In each case, the conventional counterpart is an existing food that has a history of safe use. This comparative approach is the essence of the principle of substantial equivalence assessment proposed by OECD (OECD, 1993). The substantial equivalence concept also embodies the concept that a food derived from a genetically modified plant or microorganism should be *as safe as* its traditional counterpart (FAO/WHO, 2000). In practice, similarities and differences between a food derived from a genetically modified plant and its traditional counterpart are identified so that they can be subjected to safety

Table 1

Applicability of Toxicological Testing to Whole Foods In Animals

Chemical toxicology	Testing whole foods
Single chemical	Complex mixture
Highest dose level should produce an adverse effect	Highest dose that does not cause rejection or nutritional imbalance
Low doses, usually > 1% of the diet	High doses, usually > 10% of the diet
Easy to achieve a dose high enough to assure an adequate safety factor (> 100 normal human intake)	Difficult or impossible to achieve doses more than a few multiples of human intake; therefore, conventional toxicological safety factor cannot be assigned
Acute effects obvious	Acute effects, other than those caused by nutritional imbalance, nearly always absent
Possible to study specific routes of metabolism and excretion of toxic compound(s)	Complex metabolism of many ingredients, some unidentified
Cause/effect relatively clear	Effects usually absent or, if observed, confounded by multiple possible causes

Source: Adapted from IFT (2000, Table 1, p. 20).

evaluation. The concept of substantial equivalence has come under criticism by Millstone *et al.* (1999), who claim that regulators have defined as "sub stantially equivalent" crops that are nonequivalent by their very nature to justify a determination that they are safe. Critics argue that unintended and undetected changes may have occurred that justify long-term human feeding trials to demonstrate safety. In fact, "substantial equivalence" is not the *conclusion* of a safety evaluation. It is the *starting point* for a variety of safety assessments that are described in the following sections. A recent expert review concluded that "substantial equivalence" as applies in safety evaluations is a useful, robust, and flexible paradigm that has been improved and refined over the years (FAO/WHO, 2000). In fact, the substantial equivalence concept can be applied to the safety assessment of any new food or food ingredient that has a conventional counterpart.

Bioengineered plants, and the food products derived from them, can be divided into three general categories: essentially identical in composition to the conventional counterpart, identical in composition with the exception of an added new component, and having significant changes in composition and/or content. This can best be illustrated by considering three bioengineered varieties of the same commodity. Soybean oil derived from herbicide-tolerant soybeans is indistinguishable from the oil derived from conventional soybeans. On the other hand, soybean meal derived from herbicide-tolerant soybeans is identical except for the addition of minor quantities of one new protein to the soybean. Soybean oil engineered to have the same composition as olive oil would not be identical to conventional soybean oil and could not be called its equivalent. An oil such as olive oil that is high in oleic acid may be a more suitable comparator.

In practice, the specific questions that must be answered in the safety assessment are as follows:

- Is the transferred DNA safe to consume?
- If an antibiotic resistance marker is used, is it safe?
- Are the newly produced proteins safe to consume?
- Have potential allergens been introduced into the food?
- Are the composition and nutritional value changed?
- Are there changes in the content of important substances (e.g., toxicants, antioxidants, phytochemicals)?
- In what forms will the food or food products isolated from it be consumed?
- Do the newly introduced substances survive processing, shipment, storage, and preparation?
- What is the expected human dietary exposure?

The following sections explore all these issues except for evaluation of potential food allergenicity, which is discussed in Chapter 11.

HOW ARE GENETICALLY MODIFIED FOODS EVALUATED FOR FOOD SAFETY?

SAFETY OF THE NEWLY INSERTED DNA

The first step in evaluating genetically modified plants is to assess the safety of the genetic material introduced into a plant. All characteristics of the genetic insert must be known, including the identity of the source of the genetic material, the nucleotide sequence of the DNA construct being inserted, the number of insertion sites, and the stability of the insertion in the plant genome. The DNA sequence of the inserted gene and flanking regions found in the new variety may be determined to confirm that the insertion occurred as planned. (Typically, however, this is neither done nor considered necessary.) If the newly introduced genetic material is derived from a pathogenic source, or a known allergenic or toxin-producing source, special care must be taken to assure that the injurious traits have not been transferred.

Risks associated with DNA consumption per se are the same for DNA derived from both conventional and biotechnology-produced plants (Beever and Kemp, 2000). The FDA has concluded that DNA is *generally regarded as safe* (GRAS), independent of its source. DNA is always composed of the same four components, which are normal constituents in raw or whole foods (U.S. National Biotechnology Policy Board, 1992; FDA, 2001). Although we consume large quantities of DNA in our diets, there is no evidence that dietary DNA has ever produced any adverse effects. The great majority of ingested DNA is rapidly digested to its constituent nucleotides.

Concerns have been raised about the safety of promoter sequences such as the cauliflower mosaic virus S35 promoter that are inserted to control the expression of inserted genes. It has been suggested that these promoters might become incorporated into human or animal cells and then function as "mutational hot spots." It is important to recognize that large quantities of plant viruses that include S35–like promoters are commonly consumed in the human diet. There is no evidence that such DNA has ever become functionally incorporated in the human genome upon consumption. In fact, despite years of research, there has emerged no solid scientific evidence to date that demonstrates the incorporation of food-derived DNA into mammalian cells, gastrointestinal bacteria, or soil bacteria. There has been a report by Schubert and coworkers (Schubert *et al.*, 1998) that orally administered bacterial DNA can be absorbed and subsequently found incorporated into mouse cells. There are, however, difficulties with the interpretation of this experiment (Beever and Kemp, 2000). It is also not clearly established that infrequent horizontal transfer into individual somatic cells would be deleterious to the host if it were to occur. It is important to note that the newly introduced DNA, or transgene DNA, represents approximately 0.0025–0.005% of the DNA in the plant, and a specific commodity produced from a bioengineered plant may constitute only a tiny fraction of the total dietary DNA. A highly unlikely series of events would be required for this small fraction of plant DNA to transfer genes horizontally into mammalian or bacterial genomes.

To be effective, transgene DNA must do the following:

- Survive harvest, drying, storage, and milling
- Survive food processing
- Be present in the fraction of the plant that is consumed
- Survive acid pH and nucleases in the mammalian gastrointestinal tract
- Compete for uptake with a large excess of dietary DNA
- Survive host nucleases
- Stably integrate into host chromosome
- Express in new host

Safety of the Antibiotic Resistance Marker

The safety evaluation of the newly inserted DNA primarily focuses on potential risks associated with acquisition and expression of the specific genetic information inserted into the plant if the DNA were to transfer to human, animal, or bacterial cells. The safety of genes that encode antibiotic resistance and their potential for transfer often is a major consideration. To date, there has been no observation in nature showing evidence for the transfer of antibiotic resistance marker genes from plants to other organisms (FAO/WHO, 2000). In addition, approval of the use of antibiotic resistance genes has been restricted to those for which resistance to the antibiotic is already widespread in nature. It is also recommended that the gene selected not provide resistance to an antibiotic used to a significant extent in human or veterinary medicine.

The Flavr Savr tomato was the first biotechnology-derived food to be approved by the FDA. Interestingly, the 1994 approval was based on review of a food additive petition for the kanamycin resistance gene product, the antibiotic inactivating enzyme neomycin phosphotransferase, that had been introduced into the tomato (FDA, 1994). Subsequent approvals have not elected to use the Food Additive Petition process, nor have they been required to do so by the FDA. It should be noted that some biotechnology-derived crops do not contain antibiotic resistance marker genes and that new marker genes and selection strategies that reduce the need for antibiotic resistance markers are being developed (FAO/WHO, 2000; Kuiper *et al.*, 2001).

THE SAFETY OF NEWLY INTRODUCED COMPONENT(S)

The second element of the evaluation is the safety assessment of the newly introduced trait(s) or expressed product(s). The newly introduced product is typically a single new protein that is encoded by the inserted DNA, or two new proteins if a marker gene has been used. The association of the protein with a desired property or trait such as herbicide tolerance or insect protection is usually well documented in the literature. The safety assessment begins with a thorough understanding of the history of safe consumption of the introduced new protein—if any exists. This would include detailed information regarding the physiological and biochemical function of the protein, and the relatedness of the protein to other proteins that have a history of safe consumption. The protein is also evaluated at the level of sequence for similarity to toxicants or allergens whose sequence is known. Often these factors are evaluated before product development begins, or at least very early in the development cycle.

The level of expression of the introduced protein(s) in the edible portion of the new plant variety must be estimated. It is also necessary to evaluate whether the protein synthesized in the bioengineered plant is equivalent to that found in the parent organism. Biochemical function or physiological activity is usually demonstrated as well. If there are differences in structure or function of the protein(s) in the new host, the basis of the differences are normally determined. Large quantities of the protein must then be isolated and purified for analytical and animal testing.

The potential toxicity of an inserted protein is usually easily assessed, since proteins that have been shown to be toxic are acutely toxic. Acutely toxic proteins elicit their toxic effects almost immediately upon consumption. Guidelines for conducting such tests are available from numerous sources (FAO, 1995; FAO/WHO, 2000). In a review on food safety evaluation, Kuiper *et al.* (2001) reported 25 references describing toxicity studies on 21 distinct proteins expressed in commercial plant varieties. Several examples of toxicological studies on specific plant recombinant proteins are given in Table 2. Note that none of the proteins produced an adverse effect at the highest concentration tested.

Human consumption of newly introduced DNA and proteins found in biotechnology-derived crops is expected to be quite low. Well over half the field corn and soybeans grown in the United States is consumed as animal feed. A large portion of the remaining soybean and corn is consumed by humans in the form of products containing soybean oil, starch, and corn oil that are virtually devoid of DNA and protein. A small quantity of soybean finds its way into alkali-denatured protein fractions such as soy isolate, and specialty products such as tofu, soy milk, and soy

Table 2

Acute Toxicity Evaluation of Proteins Introduced into Commercial Crops

Protein studied	No observable adverse effect level (mg/kg/day)	Stable to digestion?	Stable to processing?
Cry1Ab	>4000	No (30 s)	No
Cry1Ac	>5000	No (30 s)	No
Cry2Aa	>4011	No (30 s)	No
Cry2Ab	>1450	No (30 s)	No
Cry3A	>5220	No (30 s)	No
Cry3Bb	>3780	No (30 s)	No
Cry9C	>3760	± (30 min)	Partial?
NPTII	>5000	No	No
CP4 EPSPS	>572	No	N.A.
GUS	>100	No	N.A.

Source: Betz *et al.* (2000), Astwood *et al.* (2001), EPA Biopesticide Fact Sheets
http://www.epa.gov/pesticides/biopesticides/factsheets/fs006466t.htm

yogurt. Approximately 1% of the corn crop is used as dry-milled whole corn in products such as tortillas and corn chips. The estimated dietary intake is further reduced by milling, drying, and processing operations (see later).

Toxicological risk is proportional to dose × exposure. Therefore, after expression levels of the gene product have been evaluated, the estimated dietary intake is determined. In the United States, estimated intakes can be calculated from human dietary intake data that can be found in the National Health and Nutritional Examination Survey (NCHS, 2001) and Common Food and Shelter databases (FSRG, 2001). Dietary intakes should be estimated for various criteria and demographic groups, including age, gender, socioeconomic status, geography, and ethnicity, to reveal group-related variations in estimated daily intakes. In commodities that are comparable to their conventional counterparts such as *Bt* corn, newly introduced proteins are often expressed in the range of 0.01–0.1% of the total protein content of a plant (Betz, 2000). Average, estimated daily dietary intakes of a newly inserted protein are often 1–10 μg/day and would be unlikely to exceed the range of 10–100 μg/day. In the case of biotechnology-derived foods and food ingredients, it is therefore often possible to conduct acute toxicity studies in rodents using doses that are thousands to a million times higher than what a human would normally be expected to consume.

Toxicologists define an acceptable daily intake (ADI) as a dietary level of 0.01 of the highest concentration at which no adverse effects are observed (NOAEL). The lack of effects observed in representative toxicity studies performed at high dietary concentrations of several purified proteins (Table 2) that have been inserted into commercial plant varieties translates to safety factors many times in excess of calculated ADI values. The observed safety factors range from a thousand fold to more than a million fold above the anticipated dietary intake of these proteins. It should also be noted that the maximum level tested in these studies was often artificially low owing to the scarcity of purified recombinant proteins and the relatively large quantities that are required for *in vivo* toxicity studies. The results of the digestion and processing studies described in the following paragraph are consistent with the conclusion that feeding higher levels of protein would not have evoked an adverse effect.

Additional insight into the potential biological activity (i.e., toxicity or allergenicity) of a protein can be gained through an analysis of its digestibility and stability to processing. Digestibility is determined by incubation of the protein in simulated gastric fluid (SGF). Almost all the proteins that have been inserted to date in biotechnology-derived crop plants are rapidly digested and are therefore highly unlikely to retain any residual biological activity (Table 2). Cry9C, the *Bt* protein found in the controversial *Bt* corn variety Starlink, is partially stable to SGF. Unlike the other proteins shown in Table 2, Cry9C is also partially stable to some food processing operations. Although Cry9C is not toxic to laboratory animals, the properties of partial digestibility and partial processing stability make it difficult to absolutely preclude the possibility that it could act as a food allergen (see Taylor & Hefle, Chapter 11). More complete information regarding the properties of Cry9C can be found at EPA websites:
http://www.epa.gov/scipoly/sap/
http://www.epa.gov/pesticides/biopesticides/factsheets/fs006466t.htm.
Inserted proteins, like most dietary proteins, are usually degraded and/or denatured by thermal processing operations (baking, extrusion, frying, microwave, etc.) and treatments with strong acid or alkali such as corn wet milling (Table 2).

Safety of the Balance of the Whole Food

The substantial equivalence concept is used to focus investigation on any differences that might exist between the new variety being tested and its conventional counterpart, provided a conventional counterpart can be identified. Changes in composition that may have occurred in the remaining edible portions of a genetically modified plant must, therefore, be assessed. These analyses are performed with foods derived from both biotechnology-derived and conventional crops that have been grown side by side under a diversity of environmental conditions that are representative of the conditions under which these crops will be grown commercially. Comparisons are made first between the levels of the selected components in the biotechnology-derived and conventional crops. If differences are observed, the differences are assessed in relation to the range of that specific component normally found in that crop to determine whether that change is biologically significant.

The next step is a comparative evaluation of the concentrations of macro- and microconstituents of the edible portion of the biotechnology-derived food, including profiles of major constituents such as fats, proteins, and carbohydrates, and minor constituents such as vitamins and minerals. For example, it has been reported that a variety of *Bt* corn has a macronutrient composition that lies within the range of literature values reported for various varieties of conventional corn (Table 3). Similar comparisons have made for herbicide-tolerant soybeans (Table 4). Two varieties of herbicide-tolerant soybeans have compositions that closely resemble that reported for the conventional soybean control.

Amino acid analysis of the same variety of *Bt* corn just discussed shows that most amino acids are present within the range of values reported in the literature with only minor differences from values reported for conventional corn (Table 5). A comparison of the amino acid content of two varieties of herbicide-resistant soybeans with their conventional counterpart showed striking similarities in the content of the 18 amino acids reported (Table 6). The values also fell into the range of amino acid content reported in the literature for several soybean varieties.

Similarly, lipid analysis shows that the fatty acid composition of *Bt* corn lies within the range of values reported in the literature for conventional

Table 3

Composition (% dry wt)

Nutrient	*Bt*-Corn	Corn (Literature)
Protein	13.1	6.0–12.0
Fat	3.0	3.1–5.7
Fiber	2.6	2.0–5.5
Ash	1.6	1.1–3.9
Carbohydrate	82.4	N.A.

Source: Astwood *et al.* (2001).

Table 4

Glyfosate-Tolerant Soybean Composition Measured at Nine Sites in 1992

	Composition (% dry wt)			
		Glyphosate-tolerant soybean mean		
Component	Control soybean mean	Variety 1	Variety 2	Literature Range
Protein	41.6	41.4	41.3	36.9–46.4
Ash	5.04	5.24	5.17	4.61–5.37
Moisture, g/100 g fresh wt	8.12	8.12	8.20	7–11
Fat	15.52	16.28	16.09	13.2–22.5
Fiber	7.13	6.87	7.08	4.7–6.48
Carbohydrates	38.1	37.1	37.5	30.9–34.0

Source: Data from Padgette *et al.* (1996).

Table 5

Amino Acid Composition of *Bt* Corn and Conventional Corn (values from the literature)

Amino acid	Literature low	*Bt* Corn	Literature high
Alanine	6.4	8.2	9.9
Arginine	2.9	4.5	5.9
Aspartic acid	5.8	7.1	7.2
Cysteine	1.2	2.0	1.6
Glutamic acid	12.4	21.9	19.6
Glycine	2.6	3.7	4.7
Histidine	2.0	3.1	2.8
Isoleucine	2.6	3.7	4.0
Leucine	7.8	15.0	15.2
Lysine	2.0	2.8	3.8
Methionine	1.0	1.7	2.1
Phenylalanine	2.9	5.6	5.7
Proline	6.6	9.9	10.3
Serine	4.2	5.5	5.5
Threonine	2.9	3.9	3.9
Tryptophan	0.5	0.6	1.2
Tyrosine	2.9	4.4	4.7
Valine	2.1	4.5	5.2

Source: Astwood *et al.* (2001).

Table 6
Amino Acid Composition of Glyphosate-Tolerant Soybeans

	Composition (% dry wt)			
	Glyphosate-tolerant soybean mean			
Amino acid	Control soybean mean	Variety 1	Variety 2	Literature Range
Aspartic acid	4.53	4.42	4.48	3.87–4.98
Threonine	1.60	1.56	1.58	1.33–1.79
Serine	2.10	2.04	2.07	1.81–2.32
Glutamic acid	7.34	7.10	7.26	6.10–8.72
Proline	2.03	1.98	2.02	1.88–2.61
Glycine	1.72	1.67	1.69	1.88–2.02
Alanine	1.71	1.67	1.69	1.49–1.87
Valine	1.85	1.80	1.83	1.52–2.24
Isoleucine	1.78	1.73	1.76	1.46–2.12
Leucine	3.05	2.97	3.03	2.71–3.20
Tyrosine	1.45	1.40	1.43	1.12–1.62
Phenylalanine	1.97	1.90	1.95	1.70–2.08
Histidine	1.06	1.03	1.04	0.89–1.08
Lysine	2.61	2.56	2.58	2.35–2.86
Arginine	2.94	2.85	2.90	2.45–3.49
Cysteine	0.60	0.62	0.60	0.56–0.66
Methionine	0.55	0.55	0.54	0.49–0.66
Tryptophan	0.59	0.59	0.58	0.53–0.54

Source: Data from Padgette *et al.* (1996).

varieties of corn (Table 7). A more striking comparison was observed upon comparison of two varieties of herbicide-resistant soybeans with their conventional counterpart. None of the 12 fatty acids measured differed by a statistically significant value in the content measured for the three varieties (Table 8).

The content of known endogenous toxins or antinutritional factors characteristic of the food crop in question is also evaluated. Typically, at least 50, and sometimes hundreds, of key nutrients, phytochemicals, and antinutrients are assessed either in raw materials (e.g., grains) or, where appropriate, in food products that are derived. For example, a health claim has been approved in the United States for soybean isoflavones that have been shown to have effects beneficial to health. Isoflavone content is therefore routinely evaluated in new biotechnology–derived soybean varieties (Padgette *et al.*, 1996; Taylor *et al.*, 1999). Pagette *et al.* also report values for quality

Table 7

Fatty Acid Composition of *Bt* Corn

	Composition (% of total)		
Literature	Literature low	*Bt* con	Literatures high
Palmitic acid	7	10.5	19
Stearic acid	1	1.9	3
Oleic acid	20	23.2	46
Linoleic acid	35	62.6	70
Linolenic Acid	0.8	0.8	2

Source: (Astwood, 2001).

Table 8

Fatty Acid Composition of Glyphosate-Tolerant Soybeans

	Composition (% dry wt)			
		Glyphosate-tolerant soybean mean		
Fatty acid	Control soybean mean	Variety 1	Variety 2	Literature range
6:0	0.11	0.11	0.11	
16:0	11.19	11.21	11.14	7–12
17:0	0.13	0.13	0.13	
18:0	4.09	4.14	4.05	2–5.5
18:1 *cis*	19.72	19.74	19.81	20–50
18:2	52.52	52.31	52.48	35–60
18:3	8.02	8.23	8.12	2–13
20:0	0.36	0.37	0.35	
20:1	0.17	0.17	0.17	
22:0	0.50	0.53	0.49	
24:0	0.18	0.19	0.18	
Unknown	2.63	2.48	2.59	

Source: Data from Padgette *et al.* (1996).

determinants such as phytate, stacchyose, raffinose, urease, trypsin inhibitor, and lectin. One research report erroneously claimed that isoflavone content was lower in herbicide-tolerant soybeans than in conventional soybeans (Lappé *et al.*, 1999). The investigators failed to compare isogenic strains and the reported isoflavone content that was within the three-to fivefold range in concentrations that exists between conventional soybean varieties (Wang and Murphy, 1994). Kuiper *et al.* (2001) have tabulated references for the compositional analysis of 21 genetically enhanced plant varieties.

All varieties of plants – including biotechnology-derived varieties – that are intended for commercialization must meet rigorous agronomic specifications (Astwood *et al.*, 2001). Commercial varieties are also often the progeny of numerous backcrosses with elite varieties that have the desired agronomic traits. These two processes would be expected to strongly select against changes in composition, forcing the selection of only plants with great conformity. In the selection process unintended effects should be largely eliminated. It should not be surprising to find, therefore, that most of biotechnology-derived varieties is use today are virtually identical in composition to their conventional counterparts.

The finding that two plants have the same composition has profound biochemical and nutritional implications that are often overlooked or at least underestimated. Equivalent composition means that the plant will probably contain the same genes coding for transport, metabolic, and biosynthetic pathways, and that the genes are probably being expressed in the same temporal sequence to an equivalent level of transcription. It further suggests that the concentrations of the key metabolites in the cell are equivalent, and that the kinetic flux through pathways is virtually identical. It cannot be stressed strongly enough that the thousands of biochemical reactions that take place in a cell are interconnected in an interactive network that is mediated by small-molecule metabolites that are shared in common between two or more pathways. As a result, the observation that 50 or 100 diverse metabolites have the same concentration in two samples from closely related varieties strongly supports the conclusion that virtually all cellular metabolites will be present at comparable concentrations. Thus, there would be no material difference between the two plants when they were used as a food or feed.

The conclusion regarding general comparability stated in the preceding paragraph should not be taken to mean, however, that unintended and undesired changes have not occurred in the development of new varieties (Kuiper *et al.*, 2001). For example, the expression of yeast invertase in potatoes resulted in a reduction in glycoalkaloid content (Engel *et al.*, 1998), while expression of soybean glycinin in potatoes had the unexpected reverse effect of raising the glycoalkaloid content (Hashimoto *et al.*, 1999a,b).

These reports also demonstrate that unintended changes can be detected early in the development cycle.

Animal Testing of Whole Foods

It was stated earlier without explanation that whole foods do not lend themselves to the toxicological safety evaluation process that is applied to food additives and other single or defined chemicals found in food. Tests of whole foods are very difficult to conduct in animals because feeding diets that contain large amounts of single whole foods has a high potential to induce nutrient imbalances and possible secondary adverse effects that may be interpreted as toxicity (LSRO, 1998). Furthermore, the safety factors used for single compounds cannot be achieved with whole foods. The differences between toxicological testing of chemicals or food additives and whole foods were summarized in Table 1.

Kuiper *et al.* (2001) identified 12 published studies in which whole-food toxicological studies attempted to overcome the challenges noted in Table 1. The problem with conducting such studies is exemplified by the experience reported by MacKenzie (1999). When rats were fed freeze-dried tomato extract equivalent to 13 tomatoes a day to compare the toxicity of biotechnology-derived and conventional tomatoes, both groups developed electrolyte imbalances. Nonetheless, toxicologists concluded that insufficient levels of tomato had been fed to make a meaningful conclusion about toxicity. FAO/WHO (2000) has recognized that the practical difficulties associated with whole-foods testing preclude its use as a routine testing technique.

Poorly performed animal studies can lead to erroneous conclusions that are reported in the media, frightening the general public and consequently raising major challenges to regulators and policymakers. An unfortunate example of just such an occurrence was a claim widely announced in the media by two researchers who studied potatoes into which had been inserted genes encoding a lectin. The study was later published in *Lancet* in spite of reviewers' rejections with the justification that because of the publicity associated with the research, the public had a right to see the study (Ewen and Pusztai, 1999).

What did the study evaluate and report? Since it was known that certain lectins have antinutritional properties and can be toxic, the researchers fed three groups of rats a diet that contained, respectively, raw potatoes, raw potatoes plus added lectin, or potatoes that contained the inserted lectin gene. They reported that they saw proliferative and antiproliferative changes in gastrointestinal epithelial cells in rats fed the bioengineered potatoes containing the lectin gene. They concluded that the observed changes were caused by the genetic modification per se, not the toxic lectin, since controls containing added lectin displayed no such effect. Their conclusions should be viewed in light of the facts that raw potatoes are toxic, that the diets caused protein deficiency that prompted early termination of the protocol, that the diets used were unmatched, and that the GM and control potatoes were quite different in composition because isogenic comparator potatoes were not used; moreover, the number of animals used in each group was too small to allow statistically sound conclusions to be drawn. It was concluded after review of evidence and testimony before a royal commission that a meaningful scientific conclusion cannot be drawn from the study (Royal Society, 1999). This study illustrates the difficulties inherent in whole-food testing in animals. It has been suggested that whole-food animal toxicity studies that are performed should be conducted for a minimum of 90 days (FAO/WHO, 2000).

Animal Performance Trials

Since many of the genetically modified crops presently on the market (e.g. corn and soybean) are used as animal feeds, animal feeding studies with several species that demonstrate their equivalent performance have been performed. These studies utilize nutritional performance end points rather than toxicological end points. Time-to-market weight, rate of weight gain, general health, and feed consumption are commonly evaluated. A review of more than 25 published studies led to the conclusion that no changes in animal performance are associated with high-level intake of biotechnology-derived feed crops by domestic animals (Clark and Ipharragnene, 2001). Nineteen animal performance studies have been identified and reviewed by Kuiper *et al.* (2001). It has also been demonstrated that DNA and proteins from the genetically modified crops are not present in products such as meat, eggs, and milk derived from animals that have been fed grain from genetically modified crops (Faust, 2000).

Safety Evaluation of Products with Altered Composition

There are two known examples of products from crops developed using biotechnology that have significant compositional differences from their traditional counterparts. A canola oil high in lauric acid (C12:0), a fatty acid not normally found in canola oil, has been developed to serve as a substitute for tropical oils in certain food applications (FAO/WHO, 1996), and a soybean oil has been developed to have high levels of oleic acid (C18:1) at the expense of linoleic acid (C18:2 *n*–6) (OECD, 1998). The difference in the composition of these products from ordinary canola or soybean oil does not cause a problem for the substantial equivalence concept. The concept can be applied in assessing the safety of these components in the processed food (i.e., oil) by comparing the composition of the new foods with those of similar processed products for which they substitute in the diet and by comparing their anticipated intake against the intake of the component from other foods in the diet. For example, in the case of high laurate canola oil for which the comparator is tropical oils, it was demonstrated that the total intake of lauric acid in the diet would not be changed significantly by the substitution of this product for tropical oils in food applications. Additionally, it is well recognized that tropical oils have a long history of safe use. Similarly, soybean oil high in oleic acid most closely resembles olive oil, so olive oil can serve as a comparator. Once again, the impact of substitution of a new oil for an existing dietary oil on human dietary intake is the important consideration. This example illustrates how the concept of substantial equivalence can be applied to the evaluation of the safety of whole foods or food components that have been modified in composition. The traditional counterpart of a food from a biotechnology-derived crop need not be the chosen comparator. The product for which it substitutes in the diet is often a more relevant comparator.

FUTURE TRENDS IN FOOD SAFETY ASSESSMENTS

Evaluating New Products of Biotechnology

Many future products developed through biotechnology will be intentionally designed not to be equivalent in composition or nutritional content with their conventional counterparts. These will often be products that are intended to directly provide consumer benefits through enhanced nutrition and health. Others of these products are designed to address food security and/or nutritional adequacy in developing countries. Still other products are being developed to improve animal feed. The differences between these new products and their conventional counterparts could be as simple as a change in one component of a food's composition (such as a higher level of one amino acid) or as complex as the introduction of large quantities of one or more new proteins, accompanied with a concomitant decrease in other components. Can the paradigm used thus far for food safety evaluations be applied to products that not only are significantly different in composition from their parent varieties but have no conventional comparator?

The answer is that the fundamental issue comprises the dietary and health implications and impact of the changes. The finding that changes have taken place does not per se constitute a basis for concern from a scientific or regulatory perspective. It must be remembered that human dietary intake of any nutrient will vary greatly among individuals. There are many patterns of human dietary consumption, and these patterns can change frequently throughout the life cycle. A few hypothetical examples illustrate some possible approaches to the safety evaluation of new nonconventional foods.

Case 1. Golden rice is a rice variety into which have been inserted genes that encode for the synthesis of β-carotene, a vitamin A precursor (Ye *et al.*, 2000). Genes have also been added that produce small amounts of proteins that increase dietary iron bioavailability and uptake (Goto *et al.*, 1999; Potrykus, 2001). The genes and gene products that were used in the constructions have closely related analogues that are commonly found in the human diet, which implies a prior history of safe use. Golden rice is not nutritionally equivalent to conventional rice, but if conventional rice is used as a comparator, it is likely that the composition and nutritional value will in all other ways be essentially the same. ■

Perhaps the more relevant nutritional questions are whether the iron and vitamin A are bioavailable and whether these compounds overcome deficiencies in the target rice-eating population, which is deficient in vitamin A and iron. The set of evaluations described under "Safety Evaluation and the Substantial Equivalence Paradigm" can be applied to golden rice, since a clear comparator, conventional rice, is available.

Case 2. A research group developed a soybean variety that contained a 2S albumin gene isolated from Brazil nut. The gene insertion produced a significant amount of a sulfur-rich albumin that improved the total essential amino acid content of soy protein. However, the safety assessment revealed that the 2S albumin is one of the food allergens responsible for Brazil nut allergies, and the research was halted (Nordlee *et al.*, 1996). ■

The question remains, How would the food safety of this new variety of soybean have been established? The example could in fact be generalized to any conventional food such as corn, wheat, or rice that contains no new components but has had its composition significantly altered with respect to major macronutrients—starch, protein, and oil. These new products would need to be assessed in terms of impact on total human dietary intake of macronutrients, but there is no apparent reason that the assessments described under "Safety Evaluation and the Substantial Equivalence Paradigm" cannot be applied successfully.

Case 3. A coffee variety is developed through biotechnology that contains a gene whose product is an almost noncaloric protein sweetener with no history of safe use in foods. A gene encoding an enzyme that quantitatively and completely converts caffeine to a phytochemical antioxidant is also added. The product is intended to offer consumers a health-beneficial antioxidant in a caffeine-free coffee that contains a non caloric sweetener. There are no additional compositional changes in the coffee. In the United States, the FDA would require that a Food Additive Petition be filed for the sweetener, since it is a new food additive. The safety assessment would require purification of the protein sweetener additive from the bioengineered coffee plant, followed by testing in accordance with the regulatory requirements applied to new food additives. ■

What is less than clear is how the phytochemical antioxidant portion would be evaluated. Since it is thought to affect human health, it could be subjected to rigorous toxicological testing under the Food Additive Petition process. The same antioxidant is, however, sold over the counter as a dietary supplement in the United States. Under the Dietary Supplement and Education Act of 1996, no premarket safety review is required for dietary supplements. Could the developer offer this product as a dietary supplement? It is likely that the safety review process applied to plants that are derived through biotechnology would take precedence and that a Food Additive Petition for the antioxidant would be required even though no premarket safety review was conducted for its over-the-counter analogue. There seems, however, to be no compelling reason not to apply the safety assessment paradigm described under "Safety Evaluation and the Substantial Equivalence Paradigm" to the balance of the whole food—in this case, a coffee bean. It is, however, probably inappropriate to attempt whole-animal feeding studies with the raw coffee beans, since test animals do not typically consume coffee beans nor tolerate them well.

A few conclusions emerge from these simple cases:

1. The general issues that need to be addressed in a food safety assessment are much the same for all crops.
2. New crop foods or food ingredients should be assessed on a case-by-case basis so that the most appropriate approach is used to ensure that each final food product is safe for consumption.
3. The paradigm used for assessing food safety today can be directly extended to include foods that are by design not compositionally equivalent to their conventional counterparts.
4. The estimated dietary intake and effect on total nutrient intake are important additions to the safety assessment paradigm for nonequivalent foods.
5. Better data on dietary intake, nutrigenomic differences between individuals, and individual differences in response to nutrients need to be gathered.

6. A thorough understanding of the mode of action and potential health effects of one or more phytochemicals needs to be available if a safety assessment is to be applied to a plant in which the phytochemical content has been altered.

The Role of New Technology

Is there new technology under development that will facilitate the safety assessment of new varieties? There is research on the development of new methods that allow for the simultaneous screening of differences between the modified organism and its conventional counterpart. Methods are being developed that would allow comparison of the genome (DNA), gene expression patterns (mRNA), protein synthesis (translation), and small-molecule metabolites (Kuiper *et al.*, 2001). These are generally very rapid and sensitive methods that allow large sample throughput at relatively low cost. The major limitation beyond methodological development is that databases have not yet been developed that will allow an assessment of the significance of detected differences. It is, therefore, at present difficult to assess whether an observed difference between two varieties reflects biologically meaningful differences. Current research is directed at refining the methods, collecting baseline data on plants, and developing the information handling systems that will be necessary (Kuiper *et al.*, 2001). The methods have already proven to be valuable research tools, but it is not yet clear whether they will be generally applicable to safety assessments or will remain reserved for more complex or difficult cases. Perhaps the development of accurate rapid methods for parallel determination of large numbers of cellular analytes will prove to be the most useful of these technologies. There is as yet no compelling reason to believe that detailed knowledge of the comparative concentration of cellular metabolites is not a sufficiently powerful measure of equivalence and consequently of safety.

CONCLUSIONS AND FUTURE PROSPECTS

The concept of substantial equivalence has served as an effective, science-based framework for identifying the similarities and differences between a bioengineered crop and its conventional comparator. Biotechnology-derived crops are evaluated for potential toxicity, food allergy potential, composition, and nutritional value. Whole-food animal studies and animal feed performance studies may supplement the analysis. Human feeding trials from which sound, hypothesis-driven conclusions could be reached would be exceedingly difficult, if not impossible, to design and conduct, as well as extremely expensive. This is largely because it is difficult to propose a plausible biochemical or physiological mechanism by which harmful effects might occur that has not been investigated and excluded. As it happens, human feeding trials are not necessary to provide consumers a *reasonable certainty of no harm*. No amount of experimentation can provide an absolute certainty of no harm, so a precautionary approach that demands proof of no harm cannot be satisfied. Even in the United States, where the product not the process is evaluated for safety, the products of biotechnology are evaluated much more stringently than conventional products, which often receive no premarket review. Why is this so?

Science-based risk–cost–benefit analysis is the appropriate standard for the safety analysis of foods, since foods are generally safe for human consumption. It is also important that the stringency, rigor, and cost of the safety evaluation system be commensurate with the associated risks. No data have been advanced that demonstrate an inherent danger in the use of biotechnology, and in years of experience with production and consumption of these crops, no harm has been documented. Scientific analysis has consistently concluded that these new foods are as safe as, or safer than, foods produced from conventional crops. Risks must be viewed from a proper frame of reference for comparison. What is important is not that these new crops be demonstrated to be risk free, but that the public have assurance that the risks associated with them are no different from are those presented by conventional crops. The creation of regulatory systems that demand extensive and arduous assessment followed by a lengthy decision-making process do more to serve as barriers to trade and development than they do to ensure safety for consumers.

New bioengineered crop varieties that offer clear consumer benefits and choices are being developed. Many of these crops will by their very design not be comparable in composition to their conventional counterparts. As we have seen in the three case studies, however, the principles that are used today for the food safety assessment of bioengineered plants can in general be applied to extensively modified future crops by invoking the substantial equivalence concept. Improved analytical methods such as profiling techniques may in the future facilitate the analysis of crops that have significantly altered content, but these techniques will not change the fundamental underlying questions that must be answered in any food safety assessment. The robustness and flexibility of the science-based risk assessment of product rather than process will prove useful in the evaluation of the newly developed crops in the future.

If products with strikingly altered composition or elevated levels of putative health-beneficial phytochemicals are developed, the analysis of their safety will depend more on enhanced understanding of the role of specific nutrients and phytochemicals in the diet than on the safety per se of the new variety. Knowledge of diet and health relationships and of the potential health-protective or health-beneficial roles of nonnutritive components is incomplete at best. This, coupled with the fact that humans have distinct genetic makeups that will respond uniquely to specific dietary components, may move the focus of analysis of new plants into the arena in which all foods should be judged: How do they impact composite dietary intake, and how does the intake pattern in turn affect health? Perhaps the greatest challenge for bioengineered crops that are intended to provide direct consumer benefits will be to demonstrate direct human health benefits, to permit verification of appropriate and important health claims.

It is predicted that sound scientific methods for food safety assessment will continue to ensure that biotechnology-derived foods are acceptably safe. These methods will doubtless be improved over time and will provide an even greater confidence in safety. Expanded research on analysis, profiling and screening methods, and food allergy could contribute to a fuller understanding and enhanced management of risk. As noted earlier, risk should be placed in the proper perspective. Attention and resources should be directed at major food safety threats such as those posed by microbial pathogens, bovine sponginform encephalopathy, and now—sadly—bioterrorism. Food-borne illness is a proven killer that poses the greatest food safety risks. Diet is the second greatest health threat associated with food. Food sufficiency, nutritional adequacy, and overnutrition are different facets of diet that affect the health of billions of people worldwide, often causing death or illness. It is tragic that for many of the world's children the greatest food risk is that there is no food. The benefits of scientific attention and allocation of resources to these topics could have an impact on diet and health that far exceeds safety issues that are today associated with biotechnology. It may even come to pass that agricultural biotechnology will be a key in reducing and managing hunger, food security, and food safety risks.

REFERENCES

Astwood J., *et al.* (2001). Status and safety of biotech crops, in *Agrochemical Discovery, Insect, Weed and Fungal Control*, ACS Symposium Series 774, American Chemical Society, Washington, DC, pp. 152–164.

Beever, D. E., and Kemp, C. F. (2000). Safety issues associated with the DNA in animal feed derived from genetically modified crops. A review of scientific and regulatory procedures. *Nutri. Abstr. Rev.*, **70**(3), 175–182.

Betz, F. S., Hammond, B. G., and Fuchs, R. L. (2000.) Safety and advantages of *Bacillus thuringiensis*-protected plants to control insect pests. *Regul. Toxicol. Pharmacol.*, **32**, 156–173.

CAST (2001). *Evaluation of the U.S. Regulatory Process for Crops Developed through Biotechnology*, Council for Agricultural Science and Technology, Washington, DC

Clark, J. H., and Ipharraguerre, I. R. (2001). Livestock performance: Feeding biotech crops. *J. Dairy Sci.* **84** (suppl.), E9–E18.

Coordinated Framework for Regulation of Biotechnology. (1986). *Fed. Regist.*, **51**, 23302–23347, June 26.

Diamond, J. (1997). *Guns, Germs, and Steel: The Fate of Human Societies*, Norton, New York

Engel, K. H., Gerstner, G., and Ross, A. (1998). Investigation of glycoalkaloids in potatoes as example for the principle of substantial equivalence, in *Novel Food Regulation in the EU – Integrity of the Process of Safety Evaluation*, pp. 197–209. Federal Institute of Consumer Health Protection and Veterinary Medicine.

Ewen, S. W. B., and Pusztai, A. (1999). Effect of diet containing genetically modified potatoes expressing *Galanthus nivalis* lectin on rat small intestine. *Lancet*, **354**, 1353–1354.

FAO. (1995). *Report of the FAO Technical Consultation on Food Allergies*, Food and Agriculture Organisation of the United Nations, Rome, November 13–14, 1995.

FAO/WHO. (1996). *Biotechnology and Food Safety*, report of a joint FAO/WHO consultation, Rome, September 30–October 4, 1996. FAO Food and Nutrition Paper 61. Food and Agriculture Organisation of the United Nations, Rome. http://www.fao.org/es/esn/gm/biotec-e.htm

FAO/WHO (2000). *Safety Aspects of Genetically Modified Foods of Plant Origin*, report of a joint FAO/WHO expert consultation on foods derived from biotechnology, May 29–June 2, 2000. World Health Organization (WHO), in collaboration with the FAO (Food and Agriculture Organisation of the United Nations). WHO Headquarters, Geneva, Switzerland.

Faust, M. A. (2000). Livestock products: Composition and detection of transgenic DNA/proteins, in *Selected Proceedings from the Agricultural Biotechnology in the Global Marketplace Symposium*, Baltimore, July 24, 2000&I.; American Society of Animal Science, Savoy, IL.

FDA (1992). Statement of policy: Foods derived from new plant varieties. *Fed. Regist.* **57**, 22984.

FDA (1994). Secondary direct food additives permitted in food for human consumption; food additives permitted in feed and drinking water of animals; amino glycoside 3′-phosphotransferase II. *Fed. Regis.* **59**, 26700–26711.

FDA (2001). Premarket notification concerning bioengineered foods. *Fed. Regist*, **66**, 12,4706–12,4738, January 18.

FSRG (2001). Food Surveys Research Group), U.S. Department of Agriculture, Beltsville, MD. http://www.barc.usda.gov/bhnrc/foodsurvey/home.htm

Goto, F., Yoshihara, T., Shigemoto, N., Toki, S., and Takaiwa, F. (1999). Iron fortification of rice seed by the soybean ferritin gene. *Nat. Biotechnol.*, **17**, 282–286.

Hashimoto, W., Momma, K., Katsube, T., Ohkawa, Y., Ishige, T., Kito, M., Utsumi, S., and Murata, K. (1999a). Safety assessment of genetically engineered potatoes with designed soybean glycinin: Compositional analyses of the potato tubers and digestibility of the newly expressed protein in transgenic potatoes. *J. Sci. Food Agric.*, **79**, 1607–1612.

Hashimoto, W., Momma, K., Yoon, H.-J., Ozawa, S., Ohkawa, Y., Ishige, T., Kito, M., Utsumi, S., and Murata, K. (1999b). Safety assessment of transgenic potatoes with soybean glycinin by feeding studies in rats. *Biosci. Biotechnol. Biochem.*, **63**, 1942–1946.

James, C. (2000). Global status of commercialized transgenic crops: 2000. ISAAA Briefs no. 21: Preview. International Service for the Acquisition of Agri-biotech Applications, Ithaca, NY. http://www.isaaa.org/publications/briefs/Brief_21.htm (site visited September 12, 2001.)

Kuiper, H. A., Kleter, G. A., Noteborn, H. P. J. M., and Kok, E. J. (2001). Assessment of the food safety issues related to genetically modified foods. *Plant J.* **27**(6), 503–528.

Lappé, M. A., Bailey, E. B., Childress, C., and Setchell, K. D. R. (1999). Alterations in clinically important phytoestrogens in genetically modified, herbicide–tolerant soybeans. *J. of Med. Food*, **1**(4), 241–245.

LSRO (1998). *Alternative and Traditional Models for Safety Evaluation of Food Ingredients*, Life Sciences Research Office., American Society for Nutritional Sciences, Bethesda, MD. FDA contract 223-92-2185.

MacKenzie, D. (1999). Unpalatable truths. *New Scie.*, April 17, pp. 18–19.

Miller, S. A., and Dwyer, J. T. (2001). Evaluating the safety and nutritional value of mycoprotein. *Food Technol.*, **55**, 42–46.

Millstone, E., Brunner, E., and Mayer, S. (1999). Beyond substantial equivalence. *Nature*, **4015** 525–526.

National Center for Health Statistics. Centers for Disease Control and Prevention, Hyattsville, MD.
http://www.cdc.gov/nchs/nhanes.htm

NAS (1987). *Introduction of Recombinant DNA–Engineered Organisms into the Environment: Key Issues*, National Academy of Sciences, National Academy Press, Washington, DC.

NAS (2000). *Genetically Modified Pest-Protected Plants: Science and Regulation*, National Academy Press, Washington, D.C.

Nordlee, J. A., Taylor, S. L., Townsend, J. A., Thomas, L. A., and Bush, R. K. (1996). Identification of a Brazil-nut allergen in transgenic soybeans. *New Engl. J. Med.*, **334**, 688–692.

NRC (1989). *Field Testing Genetically Modified Organisms: Framework for Decisions*, National Research Council, National Academy Press, Washington, DC.

OECD (1993). *Safety Evaluation of Foods Derived by Modem Biotechnology: Concepts and Principles*, Organisation for Economic Co-operation and Development, Paris.

OECD (1998). *Report of the OECD Workshop on the Toxicological and Nutritional Testing of Novel Foods*, Aussois, France, March 5–8, 1997. Organisation for Economic Co-operation and Development, Paris.
http://www.oecd.org/ehs/ehsmono/aussoidrEN.pdf

OECD (2000). *Report of the Task Force for the Safety of Novel Foods and Feeds*, Organisation for Economic Co-operation and Devel-opment, Paris.

Padgette, S. R., *et al.* (1996). The composition of glyphosate-tolerant soybean seeds is equivalent to that of conventional soybeans. *J. Nutr* **126**, 702–716.

Potrykus, I. (2001). Golden rice and beyond. *Plant Physiol.*, **125**(3), 1157–1161.

Royal Society. (1999). Review of Data on Possible Toxicity of GM Potatoes (Ref. 11/99), The Royal Society, London.
http://www.royalsoc.ac.uk/templates/statements/statementDetails.cfm?StatementID=29

Schubert, R., Hohlweg, U., Renz, D., and Doerfler, W. (1998). On the fate of orally ingested DNA in mice: Chromosomal association and placental transmission to the fetus. Mol. Gen. Genet., **259**, 569–576.

Taylor; N. B., Fuchs; R. L., MacDonald; J., Shariff, A. R., and Padgette, S. R. (1999). Compositional analysis of glyphosate-tolerant soybeans treated with glyphosate. *J. Agric. Food Chem.*, **47**(10), 4469–4473.

U.S. National Biotechnology Policy Board. (1992). *Report.* National Institutes of Health, Office of the Director, Bethesda MD.

Wang, H., and Murphy, P. A. (1994). Isoflavone (phytoestrogen) composition of American and Japanese soybeans in Iowa: Effects of variety, crop year, and location. *J. Agric. Food Chem.*, **42**, 1674–1677.

Ye, X., Al Babili, S., Kloeti, A., Zhang, J., Lucca, P., Beyer, P., and Potrykus, I. (2000). Engineering the provitamin A (beta-carotene) biosynthetic pathway into (carotenoid-free) rice endosperm. *Science*, **287**, 303–305.

Chapter 5

Plant Biotechnology Products with Direct Consumer Benefits

Maureen A. Mackey
Monsanto Company
St Louis, Missouri

Roy L. Fuchs
Monsanto Company
St Louis, Missouri

Foods from crops enhanced via biotechnology first appeared on the market in the United States in 1994. With few exceptions the crops introduced to date have been enhanced to have traits that primarily benefit farmers. Over the next several years, it is anticipated that crops will be enhanced to have nutritional and food quality traits that will directly appeal to consumers. Oilseed crops can be modified to have more healthful fatty acid compositions, and the protein quality of grains may be enhanced by the insertion of genes for proteins that provide increased levels of essential amino acids. The micronutrient composition of foods also can be enhanced; rice has already been modified to provide β-carotene, and the iron content and bioavailability of grains also can be increased. Efforts have been initiated to express human milk proteins in plants so that plant-based infant formulas can contain these important proteins, and in the future, vaccines may be expressed in foods to facilitate administration. Crop biotechnology also may lengthen the shelf life of fruits and vegetables, prevent browning of apples and potatoes, and reduce or even eliminate the allergens in plant foods. Research efforts span areas from basic studies of plant physiology and metabolism to safety assessments for product development and regulatory acceptance. Over the next few years, consumers can expect to see several new products resulting from these extensive research programs.

INTRODUCTION

Foods from crops enhanced via biotechnology have been on the market in the United States since 1994. The majority of these enhancements have provided agronomic traits, such as resistance to a variety of insect or viral pests and tolerance to herbicides. While the benefits of these crops to farmers are evident in that they chose to increase plantings every year over the past six years (James, 2000), the benefits to consumers have not been as obvious. Significant reductions in use of chemical insecticides and herbicides (Heimlich *et al*, 2000; Carpenter and Gianessi, 2001) have been realized as a result of adoption of these genetically enhanced crops, which should be welcome news to consumers.

Researchers are exploring the opportunities to genetically enhance crops with attributes such as improved nutritional value and food quality, which more directly benefit consumers. A few products have already appeared on the market, and more are expected over the next several years. This chapter summarizes the research reported in the scientific literature on potential new crop products with enhanced nutritional value or quality developed via biotechnology. A wide array of possibilities has been reported, demonstrating the potential and adaptability of the technology to produce successful products. Many of the studies reviewed here are in the discovery phase, and several years of product development and regulatory review will be necessary before a product can be commercialized. Other studies report characterization of a plant's physiology and metabolism as a first step so that multitrait modifications to metabolic pathways can be attempted. As with all types of new product development, many projects successful at the research stage will not be successful in the development stage. Thus, it is realistic to expect that only some of the projects reported here will result in commercial products.

Much research is ongoing around the world to improve the yields of staple crops such as rice, wheat, and corn, upon which most of the population relies for energy. To the extent that crop biotechnology can improve crop yields, it will improve global nutrition by assuring food security for the world's growing population. Reviews of these efforts are not covered here; they can be found in Dunwell (2000) and Khush (2001), and at the websites of the Consultative Group on International Agricultural Research (http://www.cgiar.org) and the International Service for the Acquisition of Agri-biotech Applications (http://www.isaaa.org).

IMPROVED NUTRITIONAL QUALITIES

MACRONUTRIENTS

Fats

It has long been recognized that the fatty acid composition of dietary fats impacts human health. Saturated fatty acids and trans fatty acids can elevate blood cholesterol, increasing risk for cardiovascular disease, while mono- and polyunsaturated fatty acids have the opposite effect (AHA, 2001). Thus, dietary guidelines advise a limit on intake of foods high in saturated and trans fatty acids and a preference for foods containing mono- and polyunsaturated fatty acids (USDA, 2000). These recommendations have spurred research and development of vegetable oils genetically enhanced to have more health-promoting fatty acid profiles. Another motivation has been the development of oils with improved processing qualities, which are discussed later in this chapter.

Several examples of fatty acid modifications have been reported in the literature: for example, soybean, canola, and sunflower oils with high levels of oleic acid (Kinney, 1996; Voelker, 1997) and canola and soybean oils that are low or even free of saturated fatty acids. More recently, researchers at the Commonwealth Scientific and Industrial Research Organization in Australia reported that cottonseed oil has been modified to contain increased levels of oleate and decreased levels of palmitate, thereby creating an oil with enhanced heart health characteristics (Liu, *et al.* 2000). Another product with unique health benefits is canola oil that contains high levels of an ω-3 fatty acid, stearidonic acid (SDA, 18:4 *n*–3) (Knutzon, 1999) that is the precursor of long-chain ω-3 fatty acids, such as docosahexaenoic acid (DHA, 22:6 *n*–3). Since SDA contains only four double bonds and is the precursor of long-chain ω-3 fatty acids with five or six double bonds, an oil with elevated levels of SDA not only could help meet the dietary needs for long-chain ω-3 fatty acids but also would be less susceptible to oxidative rancidity.

Palm oil is widely used in tropical regions, such as Malaysia, West Africa, and Central and South America. This oil, however, is rich in palmitic acid, a pro-atherogenic saturated fatty acid, and thus is not attractive to populations of the developed world. Efforts are under way in Malaysia to increase the conversion of palmitate to oleate in palm by suppressing the expression of the gene for palmitoyl ACP thioesterase and promoting the expression of the gene for β-keto acyl ACP synthase II (Jalani *et al.*, 1997). Because palm oil is rich in carotenoids, tocopherols, tocotrienols, and sterols, an enrichment in its oleate content would further enhance its nutritional value and improve acceptance by western populations.

Protein

Although the diets of people in the developed world have plenty of high-quality animal protein, people in the developing world still largely rely on plants for this major nutrient. Plant proteins can be deficient in one or more essential amino acids, which means that for adequate nourishment, the diet must contain a mixture of plant proteins whose levels of specific amino acids complement each other, or there must be some animal protein to balance the dietary protein quality. Economically depressed populations may have little access to animal proteins and may have diets consisting of primarily one food, such as rice, corn, cassava, or potatoes. Thus, improving the content of essential amino acids in these subsistence foods would help improve the nutritional status of these populations.

Zheng and coworkers (1995) reported success in inserting the gene for the lysine-rich protein, β-phaseolin, from beans into rice. Like other cereal grains, rice is deficient in the essential amino acid lysine. The stable insertion of the β-phaseolin gene resulted in the expression of the protein at as high as 4% of total endosperm protein, which was estimated to increase lysine content significantly. Since rice forms the basis of the diet for about one-third of the world's population, and many in this group have marginal diets, an increase in protein quality of this grain makes good nutritional sense.

Soybeans are deficient in the essential amino acid methionine. One of the major storage proteins in soybeans, glycinin (11S, globulin), contains more methionine than the other major storage protein, β-conglycinin (7*S* globulin). To improve the protein quality of soybeans via biotechnological techniques, workers at the Research Institute for Food Science of Kyoto University (Kim *et al.*, 1990) first deleted certain variable nucleotide sequences in the CDNA coding for proglycinin and inserted synthetic DNA encoding four methionines into these regions of the cDNA. Plasmids containing the cDNA were expressed in *E. coli*, and the modified proteins accumulated in the cells. The resulting methionine-enriched glycinin exhibited improved emulsification and gelation properties in comparison to native glycinin.

Having succeeded in enriching the methionine content of glycinin, these workers next demonstrated that the modified glycinin was appropriately expressed and processed in tobacco (a model plant) (Utsumi *et al.*, 1997). The next step was to introduce the methionine-rich glycinin gene into an important food crop. Rice was chosen because its protein complements that of soy. These workers found that not only was the glycinin expressed in rice, but also that the rice protein, glutelin, assembled with glycinin to form higher order protein structures (Katsube *et al*, 1999). Further evaluation of the rice's composition and digestibility demonstrated that the rice containing soybean glycinin was compositionally similar to control rice except for the protein and amino acid levels, which were 20% higher in the enhanced rice (Momma *et al.*, 1999). Digestibility of glycinin as determined by simulated gastric and intestinal fluids showed that glycinin was completely digested within 10 minutes in gastric fluid and within 30 minutes in intestinal fluid. Time points between zero and 10 or 30 minutes, respectively, were not reported. Rat feeding studies demonstrated that the enhanced rice was as safe as nontransformed rice (Momma *et al.*, 2000). These studies show that it is possible to enhance the protein quality and nutritional value of crops via biotechnology.

Work also is being done to improve the protein content of sweet potatoes, a staple crop in Africa. Prakash and coworkers (1997) have inserted a gene that codes for a storage protein rich in essential amino acids into sweet potatoes. These potatoes had a fivefold increase in total protein vs controls, and levels of the essential amino acids methionine, threonine, tryptophan, isoleucine, and lysine were significantly increased. While there were no adverse phenotypic effects, the transformed tubers appeared to produce storage roots more slowly than untransformed controls. In further studies it was shown that the protein efficiency ration (PER) of the transformed potatoes was 3.71, essentially the same as soy protein (PER = 3.72) vs 2.57 for control sweet potato (Egnin *et al.*, 1999).

Carbohydrates

Oligosaccharides such as fructans have been recognized as having health benefits via maintenance of healthy microbial flora in the digestive tract (Roberfroid and Delzenne, 1998). These carbohydrates are present in low concentrations in a few foods, such as onions, Jerusalem artichokes, and chicory, and they can be manufactured from sucrose via enzymatic hydrolysis for addition to other foods. Since this hydrolysis process is expensive, others have sought genetic modification of plants to increase the levels of oligosaccharides. For example, Sevenier and coworkers (1998) have isolated the gene for 1-sucrose:sucrose fructosyl transferase, the enzyme that converts sucrose into fructans, from Jerusalem artichoke and inserted it into sugar beets. The resulting fructan beets had no visible or measurable effects on plant phenotype or on rate of dry weight accumulation, but they converted almost all their sucrose into fructans, yielding 40% of their dry weight as fructans. Such genetically enhanced beets could provide a low-cost alternative to fructans produced via enzymatic hydrolysis. Vijn and coworkers (1997) reported success in introducing into chicory a gene from onions that codes for production of fructan:-fructan 6G-fructosyltransferase. This enzyme catalyzed the formation of fructans of the inulin neoseries while not affecting the production of linear inulin. These new fructans are more polymerized than inulin and thus have distinct functional properties that could be useful to food manufacturers.

MICRONUTRIENTS: VITAMINS AND MINERALS

Nutrient deficiencies still persist in the human diet, particularly in the least developed countries. The most serious and widespread deficiencies are in Vitamin A, iron, and iodine, but several other nutrients are present in inadequate amounts as well. The traditional means of providing these nutrients, fortification of basic foodstuffs and supplementation, have resulted in important improvements but have not solved the problem completely. Genetic modification of crops to enhance levels of micronutrients provides another option to meet the nutritional needs of the world's population.

One of the most serious and prevalent deficiencies is in vitamin A. This vitamin is provided in animal products, and its precursors, certain members of the carotenoid family, are provided in fruits and vegetables. Staple foods, such as wheat and rice, generally are low in carotenoids. Populations whose diets consist primarily of grains with few fruits and vegetables and little animal-derived food have high rates of vitamin A deficiency. These populations include many groups in India, Southeast Asia, and Africa. Vitamin A plays a critical role in maintaining the integrity of the eye and of vision; it also is important in the immune response and in the maintenance of the epithelium. A prolonged and significant deficiency of vitamin A results in poor night vision and eventually xerophthalmia and blindness (Food and Nutrition Board, 2001).

In the 1990s, scientists in Switzerland and Germany recognized the need for a widely cultivated and consumed food that could be developed to provide vitamin A for deficient populations. Using primarily funds from the Rockefeller Foundation, Drs. Ingo Potrykus and Peter Beyer inserted into rice genes from the daffodil flower and from a bacterium that provide the necessary enzymes to enable the rice to synthesize β-carotene, the precursor of vitamin A. The initial genetic transformation resulted in "golden rice" that provides 1.6 μg of β-carotene per gram, and concentrations severalfold higher than this are expected (Burkhardt *et al.*, 1997; Ye *et al.*, 2000). The goal is to develop rice that provides about one-fourth to one-half of the daily requirement of vitamin A in a typical daily intake (i.e., a few hundred grams of rice per day). Research is ongoing at the International Rice Research Institute in the Philippines and at other national agricultural research centers to transfer these traits to local varieties of rice in India, Southeast Asia, China, Africa, and Latin America. Several organizations and biotechnology companies have provided Potrykus and coworkers free access to their intellectual property and patented technologies to enable golden rice to be developed and provided to poor farmers free of charge (Potrykus, 2001).

While some have recognized golden rice as an important component of the solution to vitamin A deficiency (Chassy, 2001), others have voiced skepticism that this rice will make any difference (Nestle, 2001). Concerns have been raised that golden rice cannot solve vitamin A deficiency because there will not be enough β-carotene in the product to provide the recommended daily allowance (RDA) of the vitamin in a reasonable daily intake. Questions also have been raised about the bioavailability of the β-carotene in the rice and whether it can be absorbed when the diet is deficient in other nutrients. In an analysis made available on the Internet Robertson and coworkers (2001) estimated the impact of the substitution of golden rice for regular rice on vitamin A intake of children in a population in the Philippines. Assuming that rice was the major staple of the diet, that golden rice contained 2 μg of β-carotene per gram, and that all rice was replaced by golden rice, Robertson concluded that the vitamin A intake of children could increase by 76 (rational equivalents) RE (12.7% of the RDA.) When put in the context of the current diet, which may provide only 100–150 RE/day (about one-third the RDA for children), the substitution of golden rice for regular rice could add significantly to vitamin A intake. In populations with low intakes of vitamin A, even an additional fraction of the RDA is useful for preventing xerophthalmia and blindness.

Vitamin A is a fat-soluble vitamin, and some fat is needed in the diet to promote absorption. Thus, providing vitamin A in an oily vehicle makes good nutritional sense. Researchers at Calgene have developed a canola oil with high levels of carotenoids. Via biotechnology, canola plants (*Brassica napus*) were equipped with a bacterial gene for the enzyme phytoene synthase, which enables the plant to synthesize as much as a 50-fold increase in carotenoids, primarily α- and β-carotene. The resulting oil contained about 2000 μg of carotenoids per gram, whereas the next richest source of carotenoids, red palm oil, contains about 600 μg/g. A teaspoon of the canola oil would provide about 4000 μg of β-carotene, or [assuming an RE factor of 2:1 for β-carotene in oil (Food and Nutrition Board, 2001)], 66% of the RDA (Dhawan, 2001).

Oil from the mustard plant, a relative of canola, is used in cooking in India and could be a useful vehicle to provide additional vitamin A to the diet. A joint research project has been established among the Tata Energy Research Institute, a not-for-profit Indian research institute in Delhi, Michigan State University's Agricultural Biotechnology Support Project, and the Monsanto Company, with support from the U.S. Agency for International Development, to use the technology developed for canola in producing mustard oil that is high in β-carotene (Dhawan, 2001). Like all products developed via biotechnology, "golden" mustard oil will be evaluated for safety and efficacy and must obtain regulatory clearances prior to introduction on the market. The National Institute of Nutrition in India will participate in these evaluations and further will develop strategies to introduce the product to farmers and ultimately the population.

Other strategies to alter the carotenoid levels in foods have focused on fruits and vegetables, which may already be good sources of carotenoids. For example, Hauptmann and coworkers (1997) claimed to have produced a two- to five fold increase in total carotenoid content by inserting into carrots a bacterial (*Erwinia herbicola*) gene for phytoene synthase, the enzyme that catalyzes the conversion of geranylgeranyl pyrophosphate to phytoene. Increases reported in α- and β-carotene, and in zeaxanthin or lutein suggested that enhanced synthesis of one carotenoid was not done at the expense of another. Hauptmann proposed that this transformation could be accomplished in a variety of fruits and vegetables, including potatoes, melon, squash, corn, and tangerines, to name a few. When Romer and coworkers (2000) inserted the bacterial (*Erwinia uredovora*) gene for the enzyme phytoene desaturase into tomatoes, there was a doubling in concentration of β-carotene such that one fruit provided 42% of the RDA for vitamin A while the control fruit provided 23%. However, the lycopene content of the modified tomato decreased by two-thirds and total carotenoid content decreased by half. Since tomatoes are one of the best sources of lycopene in the diet, further work is needed to assure that efforts to enhance particular carotenoids do not compromise the content of others.

Iron is another essential nutrient that is widely deficient in the diets of people of both the developed and developing world. Iron deficiency is severe enough in the developing world to result in widespread anemia. Again, the problem arises from the lack of adequate animal protein and reliance on plant foods that are low in bioavailable iron and may have phytate, which interferes with iron absorption. Fortification of foods with iron salts is technically difficult, and providing supplements is expensive and inefficient. Thus, enhancing the content and bioavailability of iron via biotechnology in foods consumed by poor populations has been pursued.

In 1999 Goto and coworkers reported that they had succeeded in inserting the gene for ferritin, the iron storage protein, from soybeans into rice. Ferritin stably accumulated in the endosperm and resulted in a tripling in iron content of rice. Thus, a 150-g serving of rice would provide 5–6 mg of iron, or about one-third the daily requirement.

The same workers who developed golden rice also have enhanced the iron content of rice by a multipronged strategy. Not only did they insert into rice the ferritin gene from the bean *Phaseolus vulgaris*, they also overexpressed the gene that codes for a metallothionein protein high in cysteine because cysteine-rich proteins enhance iron absorption. In addition, they inserted the gene for phytase so that the rice's content of the iron absorption inhibitor phytic acid would be reduced (Lucca *et al.*, 2001). The three transgenic rice lines were then crossed to combine the three qualities into one rice line, and also were crossed with the golden rice line. According to the Rockefeller Foundation

(1999), which funded work on golden rice, this rice will be distributed to local rice breeders free of charge by the International Rice Research Institute and other national agricultural research centers in developing countries. The breeders can then cross the enhanced rice with local varieties and make them available to farmers in these regions.

As has been demonstrated with carotenoids, inserting a gene coding for a key enzyme in a synthetic pathway enables the alteration of metabolic pathways involved in synthesis of nutrients in plants. Similar work is being pursued to enhance the content of the α form of vitamin E, the most biologically active form for human nutrition, in oilseeds. Oilseeds are rich in total tocopherols, but predominantly in the γ form, which is only one-tenth as active as the α form in humans. Shintani and Della Penna (1998) have demonstrated that the α-tocopherol content of *Arabidopsis* can be increased at the expense of γ-tocopherol by inserted and overexpressing *Arabidopsis* cDNA for γ-tocopherol methyltransferase, the enzyme that catalyzes the conversion of γ- to α-tocopherol. The content of α-tocopherol in the transformed *Arabidopsis* oil was ninefold higher than that in the wild *Arabidopsis* oil, while the total tocopherol content was unchanged. If this transformation could be accomplished in commercially important oilseeds, such as corn, canola, cotton, or soy, the nutritional value of these oils would be greatly improved (Grusak and Della Penna, 1999).

In addition to the work described for enhancement of iron content and bioavailability in plants, efforts are being undertaken to accomplish the same for zinc and calcium. Increasing the ability of plants to take up and store these minerals may require the coordinated expression of cellular mineral transporters and storage components (e.g., pectin). Other strategies include raising the levels of absorption-enhancing components, such as ascorbic acid, and decreasing the level of phytic acid, which interferes with mineral absorption (Frossard *et al.*, 2000)

Other Functional Components

In addition to the classical vitamins and minerals, nutritionists now recognize a number of components in plants that may have health-enhancing properties. Collectively called phytochemicals or phytonutrients, these components include isoflavones, sterols, phytoestrogens, and anthocyanins. While much remains to be learned about the effects of these components on health and mitigation of disease, efforts are now under way to understand better how plants synthesize these components so that their levels can be manipulated. The idea that multienzyme plant pathways yielding nutritionally important components can be elucidated and controlled or reproduced in other plants has been called "nutritional genomics" or "metabolic engineering" (Della Penna, 1999, 2001). For example, it has been hypothesized that glucoraphanin, a glucosinolate found in *Brassica* vegetables, may have cancer-preventing effects in humans via its induction of liver detoxification enzymes. Thus, enhancing the expression of glucoraphanin in other, more acceptable foods may be an appealing target for biotechnology. However, even if the hypothesis that glucoraphanin is sufficient for exerting the anticancer effect proves true, the task of selectively over expressing one glucosinolate in a complex pathway likely will present significant challenges.

Flavonoids are a class of compounds that include isoflavones, anthocyanins, rutin, quercitin, and kaempferol. Synthesized from phenylalanine via chalcone synthase and chalcone isomerase, these compounds are present in fruits, vegetables, nuts, and seeds and have antioxidant activity. Although definitive evidence is still needed, it is hypothesized that the health benefits associated with diets rich in fruits and vegetables derive from intake of components such as flavonoids. Since many people, even in developed countries, do not eat enough fruits and vegetables, enhancing the levels of components such as flavonoids in fruits and vegetables via biotechnology may help increase their intake. Jung and coworkers (2000) expressed the gene for soybean isoflavone synthase in the laboratory model plant *Arabidopsis thaliana* in such a way that this nonleguminous plant produced genestein. Taking the process the next step, Muir and coworkers (2001) reported that they genetically enhanced tomatoes to overexpress the enzyme chalcone isomerase, obtained from petunia. Paste processed from the transformed tomatoes contained up to 1.9 mg of flavonols per gram dry weight, a 21-fold increase over wild controls. Further, there were no negative effects on the plants, and no undesirable flavors developed in the paste product.

Plant phytosterols such as β-sitosterol were once used as cholesterol-lowering drugs. Now functional food products, such as margarines containing phytosterols or phytostanols, are available to consumers in Europe and the United States. The sources of these compounds include pine tree resin and soybeans. Venkatramesh and coworkers (2000) claimed to have increased the content of phytosterols in oilseeds via biotechnology. Such an improvement could increase the availability of this new functional component so it could be included in more foods.

Special Products

Vaccines are one of the most cost-effective health care measures, but with few exceptions they require administration via injection. This means that delivery of vaccines is complicated by the need for refrigeration, sterile equipment, and trained personnel to administer the injections, requirements that add prohibitive costs for public health agencies in the developing world. Over the past several years, the possibility that edible vaccines might be expressed in common foods has come nearer to reality (Langridge, 2000). Dr. Charles Arntzen and coworkers have engineered vaccines against *E. coli* enterotoxin B and the Norwalk virus into potatoes and have elicited immune responses in mice and humans (Mason *et al.*, 1998; Tacket *et al.*, 2000). Arawaka and coworkers (1998) also transformed potatoes that induced immunity against cholera toxin in mice. These viruses are responsible for many cases of food-borne illness throughout the world. Arntzen's team also developed and tested an oral immunization against hepatitis B in mice (Richter *et al.*, 2000) and is now developing this vaccine in bananas and tomatoes.

Other workers (Sandhu *et al.*, 2000) have developed transgenic tomatoes bearing a vaccine against respiratory syncytial virus, a common pathogen that causes serious lower respiratory tract disease in infants and children. When fed to mice, it elicited serum IgG and IgA responses and antibodies against the virus. Future studies are needed to evaluate the ability of this food-borne vaccine to protect against expression of the disease. Yu and Langridge (2001) have developed transgenic potatoes with multicomponent vaccines against cholera, rotavirus enterotoxin, and enterotoxigenic *E coli* and shown them to be effective in mice. For this technology to have practical application, it must be found possible for the vaccine(s) to be developed and standardized in foods in a consistent manner, and policies and procedures must be developed to assure that the vaccine-bearing foods are segregated, distributed, and handled appropriately.

The development of infant formulas is guided by the composition of human milk. Currently, the proteins in infant formulas are derived from cow's milk or soy and thus do not provide the infant with benefits ascribed to the proteins in human milk. Biotechnology may change this in the future. Work is being done to express in plants human milk proteins that can be recovered for inclusion in infant formulas. For example, Chong and coworkers (1997) reported that they expressed the human milk protein β-casein in transgenic potato plants. While these initial efforts produced only about 0.01% of the total soluble protein as β-casein, the researchers expect to be able to increase expression of β-casein to as high as 2% via the use of promoter genes. At these levels, an acre of potatoes would yield enough β-casein to supply over 7000 liters of formula. Takase and Hagiwara (1998) expressed human α-lactalbumin in tobacco, but again the concentrations achieved (5 μg/g fresh leaves) were very low, even with the use of a promoter gene. More recently, Rodriguez and coworkers (2000) reported that they had expressed several human milk proteins, lactoferrin, lysozyme and α_1-

antitrypsin in rice. The presence of these disease-fighting proteins in rice could enhance the nutritional value of infant formulas and foods.

The expression of human proteins in plants for nutritional purposes raises several questions. From a technical perspective, can enough protein be expressed and recovered in the plant to be significant and cost-effective? Will the high expression of human proteins in plants negatively affect their growth and reproduction? From the public policy perspective, will consumers accept the insertion of human genes and the expression of human proteins in plants? If all these questions are answered in the affirmative, what control processes need to be in place to manage the cultivation of such plants so that they are effectively segregated from plants intended for the general food supply? Although the notion of expressing animal or human genes in plants may be exciting from a scientific point of view, consumers will need to understand and value the benefits derived from such modifications before such products are pursued.

PRODUCTS WITH ENHANCED QUALITY TRAITS

Biotechnology is being used to improve the quality of fruits, vegetables, and cereal grains. Freshness or shelf life can be extended, flavor and sweetness can be enhanced, and other functional traits can be modified to suit the needs of food processors and consumers.

Enhancing Freshness

Prolonging the time during which fruits and vegetables stay fresh after harvesting can have important consumer benefits. First, fruits and vegetables could be shipped longer distances so consumers could enjoy a greater variety of fresh produce all year long. Second, consumers could reduce spoilage and waste at home, since even after purchase, these products stay fresher longer. Crop biotechnology has already been used to delay the ripening of tomatoes, and other strategies are being explored to prevent other aspects of senescence.

The process of softening in tomatoes is controlled by the enzyme polygalacturonase, which breaks down pectin. Through genetic modification it is possible to "switch off" this enzyme. The first commercialized product of crop biotechnology, Calgene's Flavr Savr™ tomato, was developed by excising and reversing the orientation of the gene coding for this enzyme. Tomatoes with this "antisense" gene can ripen on the vine to develop flavor and color before the softening process starts. Even after picking, the tomatoes have extended shelf lives. Calgene marketed fresh Flavr Savr™ tomatoes to supermarkets; a similar tomato was developed by Zeneca for use in processed tomato products in Europe (Roller and Harlander, 1998). Owing to supply and quality issues, however, neither product was a commercial success.

Another approach to prolong shelf life of fruits and vegetables is to inhibit the expression of ethylene. Ethylene is synthesized from *s*-adenosyl-L-methionine via two key enzymes, aminocyclopropane carboxylate (ACC) synthase and ACC oxidase. Ethylene regulates the expression of many ripening-related genes as well as those responsible for senescence (King and O'Donoghue, 1995). Workers in France (Ayub *et al* 1996; Guis *et al.*, 1997) have developed cantaloupe melons that express an antisense gene to ACC oxidase. The production of ethylene and the ripening process were blocked; however, ripening could be induced by the application of exogenous ethylene. These melons were able to remain on the vine longer to accumulate higher amounts of soluble sugars and improve in taste without the other aspects of ripening, such as softening, that shorten the product's shelf life. Similar work has been reported by Henzi and coworkers (1998), who suppressed ACC oxidase expression in broccoli. Alternatively, Theologis and Sato (2000) claimed to have developed a method to inhibit the expression of the ACC synthase gene that could be used to prolong freshness and flavor in a variety of fruits and vegetables. A number of biotechnology companies have attempted to delay the ripening of tomatoes by controlling ethylene, but additional work will be needed to produce products that meet or exceed consumers' and processors' expectations.

Cytokinin is a plant hormone associated with plant senescence; when levels of this hormone decline, senescence is signaled in the plant. Thus, strategies to maintain cytokinin levels should delay senescence. Gan and Amasino have shown that it is possible to maintain cytokinin levels in plants by controlling the expression of the gene coding for isopentenyl transferase, the enzyme that catalyzes the rate-limiting step in cytokinin biosynthesis, with a senescence-specific promoter. If cytokinin levels in plants can be maintained, the plant will grow longer and produce more seed or fruit than is possible for a nontransgenic plant (Gan and Amasino, 1995; Amasino and Gan, 1996).

Still another way to delay senescence is via inhibition of farnesyl transferase activity. McCourt and coworkers (1999) claimed to have constructed and inserted into *Arabidopsis* an antisense gene to the gene encoding farnesyl transferase. The transformed plants remained green and viable long after wild control plants had died. Producing such a modification in crops could improve harvest quality, keep produce immature longer, permit shipping without wilting, and require less misting or waxing by the grocer to maintain a fresh appearance.

Improving Flavor and Sweetness

On average, consumption of fruits and vegetables by Americans is well below the recommended five servings per day. This is particularly true for children, who generally dislike the taste of many fruits and vegetables. Making these products sweeter might increase their acceptance by consumers. The delayed ripening strategies just described allow fruits and vegetables to increase sugar and flavor levels while other aspects of ripening, such as softening, are delayed. Another approach to enhance sweetness is to enable plants to synthesize naturally sweet proteins. For example, the tropical plants *Dioscoreophyllum cumminsii* and *Thaumatococcus danielli* produce intensely sweet proteins, monellin and thaumatin, respectively, in their fruit. Thaumatin has been extracted from the fruit and developed as a food additive, but it is prohibitively expensive and not widely used. Researchers have inserted the gene encoding monellin into tomatoes and lettuce (Penarrubia *et al.*, 1992), and the gene encoding the precursor protein for thaumatin into potatoes (Zemanek and Wasserman, 1995) and cucumbers (Szwacka *et al.*, 1999) and observed enhanced sweetness. In describing the insertion of either monellin or thaumatin into fruits and vegetables, Fischer *et al.* (1998) claim that it may be possible to reduce the amount of added sugars in the recipes of certain foods, such as pumpkin pie filling and applesauce, because the pumpkin and apples are already increased in sweetness. Whether this approach will be used to enhance sweetness of fruits and vegetables remains to be seen. Of course, these proteins and the derived products would have to be determined to be safe for consumption.

Another protein-based sweetener, brazzein, is being developed by a joint venture between ProdiGene and NeKtar Worldwide (Anon., 2000). The approach is to express the protein in corn and recover it during conventional milling for use as a high-intensity sweetening ingredient. Because the protein is 500–2000 times sweeter than sucrose, 0.1 ton of brazzein corn is estimated to yield the sweetening equivalent of 0.5 metric ton of sugar, while preserving the value of other corn-derived ingredients. However, while all these taste-modifying proteins are used in West Africa to suppress bitterness and improve flavor of foods, the acceptance of their sensory characteristics by westernized populations will need to be determined.

Preventing Enzymatic Browning

Enzymatic browning is a familiar reaction observed when produce (e.g., apples, potatoes) is sliced and exposed to the air. Mechanical harvesting, transport, and processing of these products also can cause bruising and the development of black spots, which decrease consumer appeal. The reaction is catalyzed by the action of polyphenol oxidase (PPO) on phenolic compounds, which polymerize to form dark pigments. Research has shown that the activity of PPO can be suppressed in both apples (Murata *et al.*, 2000) and potatoes (Bachem *et al.*, 1994; Coetzer *et al.*, 2001) by insertion of the antisense gene for PPO. The development of nonbrowning apples and potatoes would be particularly useful to the processing industry, where browning of juices and French fries results in processor waste. Similarly, if white grapes used for juices and wine could be prevented from undergoing this browning reaction, the need to use antioxidant sulfites could be reduced or eliminated.

Improving Functional Characteristics

Wheat, one of the major staple crops for the world's population, is used to make a variety of food products, including bread, pasta, cakes, and cereals. The gluten proteins in wheat impart elasticity and extensibility, which are key to the successful use of this grain in food making. The family of wheat glutens can be classified as gliadins and glutenins and also by their molecular size. High molecular weight (HMW) glutenins are particularly important in determining elasticity, an important characteristic in bread dough development. Thus, efforts are being directed toward the genetic enhancement of HMW wheat glutenins to improve bread-making qualities (Shewry *et al.*, 1995; Shewry and Tatham, 1997; Vasil and Anderson, 1997). One of the ways this can be accomplished is by increasing the number of expressed HMW glutenin genes. Blechl and Anderson (1996) constructed a hybrid HMW glutenin gene from two wheat genes coding for other glutenins. The resulting hybrid HMW glutenin accumulated to levels comparable to those of other glutenins. Altpeter and coworkers (1996) achieved a similar outcome—increased level of HMW glutenin—by inserting into a variety of wheat a particular HMW glutenin gene that it lacked. The next step was accomplished by Barro *et al.*, (1997), who not only transformed wheat with genes for HMW glutenin, but also tested the product for functional properties. Dough from wheat in which one or two additional HMW glutenins was expressed also showed improved elasticity and strength. These achievements demonstrate how it is possible to improve wheat varieties via biotechnology. In the future, wheat varieties that grow well in certain geographies but have suboptimal food-making qualities could be improved to meet local consumer needs.

Starch is another target of genetic enhancement of food crops. Cornstarch is one of the primary ingredients used in the United States to thicken or gel foods, but potato starch is widely used in Europe. The ratio of amylose to amylopectin in starch greatly influences its functional and sensory properties, and attempts have been made to alter the ratio of these two fractions via traditional breeding and identification of mutants. In addition, starch is chemically or physically treated to develop various functionalities suitable for food applications. Biotechnological alterations of starch that would reduce the need for chemical or physical treatments would be advantageous to food manufacturers.

Because potatoes are relatively easy to genetically modify, attempts to improve their starch characteristics have already been made (Heyer *et al.*, 1999). For example, antisense technology was used in the development of an amylose-free potato: the activity of granule-bound starch synthase was inhibited (Visser *et al.*, 1991), and the resulting starch produced a strong gel that remained clear under cold temperature conditions, characteristics ideal for certain food applications (Visser *et al.*, 1997). Schwall and coworkers (2000) reported success in genetically modifying potatoes to contain high levels of amylose with insignificant levels of amylopectin. This was accomplished by inserting antisense genes into the two isoforms of starch branching enzymes. The resulting starch may offer novel properties to food product developers. Attempts also have been made to modify the starch characteristics of wheat to produce starches both high and low in amylose, suited for production of certain foods (Baga *et al.*, 1999).

Biotechnological approaches also have been used to improve the quality or amount of starch in whole foods. For example, Shimada and coworkers (1993) reported that they reduced the amylose content of rice from 19% to 6% by inserting the antisense gene for granule-bound starch synthase. The reduced amylose content produces the opaque appearance and sticky consistency preferred in Asian cuisine.

The solids content of potatoes is an important factor in finished product quality. A higher solids content is desirable to the potato processing industry because it reduces cooking time and costs and increases product recovery. In addition, less oil is absorbed by high-solids potatoes, resulting in a product that is lower both in fat and in calories. Stark and coworkers (1992) developed potatoes with 24% more dry matter by inserting a bacterial gene for the enzyme ADP glucose pyrophosphorylase (ADPGPP), that was not sensitive to feedback inhibition by cellular starch accumulation. These potatoes also better survived long-term storage temperatures and were less likely to sprout. However, they were more susceptible to blackspot bruising, and steps will need to be taken to prevent this oxidative browning before these potatoes can become commercially viable (Stark, 1998).

Taking a different approach to the regulation of ADPGPP, Leaver (1998) claimed that the starch content of potatoes could be increased by using antisense technology to reduce the activity of NAD malic dehydrogenase (NAD-ME), the enzyme that catalyzes the oxidative decarboxylation of malate to pyruvate. Inhibition of NAD-ME resulted in the accumulation of glycolytic intermediates (e.g., 3-phosphoglycerate and phosphoenolpyruvate), that stimulate ADPGPP (Jenner *et al.*, 2001). Leaver noted that because of the similarity in enzymes associated with carbohydrate metabolism across a variety of plants, the same transformation is feasible in corn, rice, wheat, peas, soybeans, cassava, other root vegetables, and tomatoes, to name a few.

In the processed potato industry, it is important to limit the postharvest degradation of starch in the tuber to glucose and fructose, because during frying these reducing sugars will undergo a Maillard reaction, resulting in an unacceptable darkening of chips and fries. Secor and coworkers (1997) claim to have genetically modified potatoes to inhibit accumulation of reducing sugars. In this work the expression of the endogenous UDP-glucose pyrophosphorylase (UDPGase) gene was inhibited by the introduction of a polynucleotide from the same potato gene for UDPGase in an antisense orientation. When UDPGase expression is suppressed, it is expected that less sucrose, and thus less glucose and fructose, will be formed from starch.

As described in detail earlier, there has been significant investment in efforts to improve potato quality traits (inhibition of browning and postharvest degradation, improved starch qualities, higher solids) that would appeal to the processing industry. Potatoes genetically enhanced to resist infestation by Colorado potato beetles as well as potato leafroll virus and potato virus Y, all major potato crop pests, were developed and marketed to farmers in the late 1990s. Many potato farmers embraced these pest-resistant varieties because they allowed them to use less pesticide. However, concern about consumer acceptance of foods developed via biotechnology made some major restaurant customers of processed potatoes reluctant to accept the pest-resistant potatoes. They demanded traditional potatoes from farmers, and after the 2001 season, sale of the genetically enhanced potato seeds stopped. Until major customers believe their consumers will accept genetically enhanced potatoes, it is unlikely the potato quality improvements just described will advance to the commercialization stage.

Food manufacturers require a variety of vegetable oils with different functional properties, such as plasticity, melting point, and oxidative stability. Soybean, cottonseed, canola, and corn oils predominantly have unsaturated fatty acids, while tropical oils, such as coconut, palm, and palm kernel, contain high levels of saturated fatty acids. Unsaturated vegetable oils are chemically hydrogenated to produce certain functional characteristics, but at the same time, trans fatty acids are formed. These trans fatty acids have been related to increased risk for coronary heart disease, and recommendations to reduce their intake have been made (USDA, 2000).

The modification of fatty acid composition in vegetable oils can be achieved by modifying the enzymes associated with chain elongation and desaturation. Thus, for example, it is possible to generate oils high in stearate and oleate and low in linoleate and linolenate by suppressing the enzymes associated with desaturation. Oils with shorter chain fatty acids, such as laurate, can be developed by inserting genes for thioesterases, which compete with enzymes involved in chain elongation.

Biotechnology is being used to develop oilseeds with favorable compositional and functional characteristics (Liu and Brown, 1996; Kinney, 1997; Mazur *et al.*, 1999; Riley and Hoffman, 1999). For example, soybean and canola oils have been modified to contain increased levels of oleic acid at the expense of linoleic and linolenic acids. Since oleic acid has only one double bond while linoleic has two and linolenic has three, oils high in oleic acid are more stable to oxidation. Soybean and canola also are being modified to have a higher content of stearic acid, a saturated fatty acid that provides the desired functional properties for shortenings and margarines without the need for chemical hydrogenation and the production of trans fatty acids. Facciotti and coworkers (1999) reported that they were able to increase stearate accumulation in canola from about 2% of fatty acids in nontransgenic controls to 20% in genetically enhanced varieties. And workers in Australia reported having genetically enhanced cottonseed oil to have stearate levels as high as 38%, with this increase coming at the expense of palmitate, oleate, and linoleate (Liu *et al.*, 2000). Such oils could provide options to manufacturers of shortenings and margarines and to manufacturers who use such ingredients in baked goods (List *et al.*, 1996). In addition, a canola high in lauric acid has been developed to serve as a substitute for tropical oils in certain food applications, such as confectionery coatings and coffee whiteners (Del Vecchio, 1996).

Reducing or Eliminating Allergens and Other Undesirable Components

Since allergens in foods are almost always proteins, it is reasonable to expect that genetic enhancement might serve to alter or even eliminate food allergens. The most common plant food allergens exist in nuts, soy, peanuts, and wheat. Less common food allergens can exist in just about any other plant, such as rice, potatoes, and kiwis. While work in this area is at a very early stage, one group of researchers has shown that an allergenic protein in rice can be reduced via genetic modification (Tada *et al.*, 1996). Other workers are using the knowledge of the structure of peanut allergen (Bannon *et al.*, 1999) and potato allergen (Alibhai *et al.*, 2000; Astwood *et al.*, 2000) to modify their allergenic epitopes and hopefully reduce their allergenic potential. Still another approach in being pursued by Buchanan and coworkers (1997,

1999), who have found that thioredoxin, a ubiquitous regulatory disulfide protein, reduces the disulfide bonds in allergens, thereby rendering them less allergenic. By inserting the gene for thioredoxin into allergenic plant foods, Buchanan expects to make them less allergenic. The team's initial focus is on hypoallergenic wheat.

Glucosinolates are a class of compound found naturally in vegetables of the *Brassica* family, such as cabbage, brussels sprouts and broccoli, as well as in the oil-yielding plants rapeseed and mustard. Glucosinolates are hydrolyzed to pungent-tasting isothiocyanates, which affect the acceptability of the food. In addition, some isothiocyanates have toxic effects. Plant-breeding efforts have succeeded in producing varieties of rapeseed and mustard with greatly reduced levels of glucosinolates such that the meal remaining after oil extraction can be used safely in animal feed. However, glucosinolates enable plants to protect themselves against insect pests and pathogens and thus perform vital functions. Efforts are being made to use antisense technology to maintain glucosinolate levels in plant leaves and seed pods but block their accumulation in oilseeds (Vageeshbabu and Chopra, 1997).

Steroidal glycoalkaloids are naturally occurring food toxicants. Among them are α-solanine and α-chaconine in potatoes. Workers in Germany have discovered that by inserting an invertase gene from yeast into potato, it is possible to reduce the content of these toxicants (Engel *et al.*, 1996). Further work is being pursued to determine whether this effect is caused by metabolic interference with the sugar moieties needed for biosynthesis of glycoalkaloids. Another approach is being taken by researchers at the U.S. Department of Agriculture Agricultural Research Service in California: by inserting in antisense orientation a gene that codes for a key enzyme in glycoalkaloid synthesis, potatoes with much lower content of this toxin have been produced (http://nps.ars.usda.gov/menu.htm?newsid=1162). The technology also is being shared with scientists in the Andean region of Peru and Ecuador, the ancestral home of potatoes. There, potatoes must be processed to remove the glycoalkaloids, but protein and vitamins are lost as well. Eliminating the need for this processing step would help preserve the nutritional value of Andean potatoes (http://nps.usda.gov/menu.htm?newsid =1434).

Corn plants can become infected with fungi, such as *Fusarium*, that are carried by insect pests and gain entry into the plant via wounds made by pests. These fungi produce fumonisins, present at low levels in most field-grown corn. Fumonisin has been shown to be acutely toxic to certain livestock and to cause liver cancer in laboratory rats (Wentzel *et al.*, 2001a) and liver and kidney damage in nonhuman primates (Wentzel *et al.*, 2001b.) Epidemiological evidence also relates the consumption of corn with high concentrations of fumonisin to esophageal cancer in humans, and the International Agency for Research on Cancer has categorized fumonisin as a probable human carcinogen. Because fumonisin can spike to high levels in corn, depending on environmental conditions and the genetic susceptibility of the host, control of fumonisin is an important public health goal.

In the late 1990s as farmers in the United States adopted corn genetically enhanced to contain the gene for the *Bacillus thuringiensis* protein that protects corn against certain insect pests, it was discovered that infection by *Fusarium* was greatly reduced. By protecting the plant against these pests and the damage they cause, *Bt* also indirectly protects the corn plant against fungal infections and the formation of fumonisins (Munkvold *et al.*, 1999; Cahagnier and Melcion, 2000; Dowd, 2000). In a 3-year study conducted in northern Italy, where fumonisin contamination of corn is a major problem, fumonisin levels in *Bt* corn were greatly reduced in comparison to traditional hybrids (Pietri and Piva, 2000). Thus, animal feed and foods based on *Bt* corn can be expected to be safer and more wholesome than those containing non-*Bt* corn that has been exposed to these fungi.

The caffeine content of tea and coffee can pose health problems for some consumers, and thus decaffeinated versions of these beverages are produced. However, the process used, supercritical fluid extraction, is expensive and can detract from the flavor of the product. Now, the gene coding for caffeine synthase has been sequenced, opening the door to the possibility of silencing the expression of the enzyme and producing caffeine-free coffee and tea plants (Kato *et al.*, 2000).

CONCLUSIONS

The next decade is expected to bring biotech crops with enhanced nutritional or food quality properties that will benefit both the developed and developing worlds. Some of these products will help correct some of the problems of the food supply, whether they be inadequacies in nutrients or excesses in undesirable components. Others will have prolonged shelf lives and/or improved flavor that will make nutritious foods more available and appealing. While crop biotechnology holds enormous promise, it should be recognized that, like any other new technology, some projects will produce commercial successes, but others will be abandoned along the product development path.

Like all crop products developed via biotechnology, products with enhanced nutritional value or quality will undergo safety assessments to obtain regulatory clearances. Additional components of these assessments may include evaluations of the nutritional impact of the product. For example, it may be appropriate to determine, via modeling of food consumption data, the likely effect of the addition of a product with enhanced vitamin content on the overall vitamin intake of the population as well as of sensitive subpopulations. Additionally, it may be appropriate to conduct human studies to assess the effect of increased (or decreased) intake of a food component on clinical end points, such as blood lipids, particularly if a health claim is to be made about the product. These studies are needed not because the product was developed via biotechnology, but because the product was developed to have enhanced nutritional qualities, which should be assessed empirically. Finally, the development of these products should occur with consumer needs in mind, to assure that such new products will be understood and accepted in the market place.

ACKNOWLEDGMENTS

We wish to thank Bradley Krohn, Elizabeth Owens, and David Stark for their thoughtful review and input to this chapter.

REFERENCES

AHA (2001). American Heart Association scientific statement: Summary of the Scientific Conference of Dietary Fatty Acids and Cardiovascular Health. *J. Nutr.* **131**, 1322–1326.

Alibhai, M., *et al.* (2000). Re-engineering patatin (*Sol t* 1) protein to eliminate IgE binding. *J. Allergy Clin. Immunol.* **105**, S79.

Altpeter, F., *et al.* (1996). Integration and expression of the high-molecular-weight glutenin subunit *1Ax1* gene into wheat. *Nat. Biotechnol.*, **14**, 1155–1159.

Amasino, R. M., and Gan, S. (1996). Transgenic plants with altered senescence characteristics. International Patent Publication WO 96/29858.

Anon. (2000). Corn extract 500X sweeter than sugar. *Food Ingredients News*, **8**.

Arawaka, T. *et al.* (1998). Efficacy of a food plant–based oral cholera toxin B subunit vaccine. *Nat. Biotechnol.*, **16**, 292–297.

Astwood, J., *et al.* (2000). Identification and characterization of IgE binding epitopes of patatin, a major food allergen of potato. *J. Allergy Clin. Immunol.* **105**, S184.

Ayub, R., *et al.* (1996). Expression of ACC oxidase antisense gene inhibits ripening of cantaloupe melon fruits. *Nat. Biotechnol.*, **14**, 862–866.

Bachem, C. W. B., *et al.* (1994). Antisense expression of polyphenol oxidase genes inhibits enzymatic browning in potato tubers. *Bio/Technology*, **12**, 1101–1105.

Baga, M., *et al.* (1999). Wheat starch modification through biotechnology. *Starch*, **51**, 111–116.

Bannon, G. A., *et al.* (1999). Tertiary structure and biophysical properties of a major peanut allergen, implications for the production of a hypoallergenic protein. *Int. Arch. Allergy Immunol.* **118**, 315–316.

Barro F., *et al.* (1997). Transformation of wheat with high molecular weight subunit genes results in improved functional properties. *Nat. Biotechnol.*, **15**, 1295–1299.

Blechl, A. E., and Anderson, O. D. (1996). Expression of a novel high-molecular-weight glutenin subunit gene in transgenic wheat. *Nat. Biotechnol.*, **14**, 875–879.

Buchanan, B. B. (1999). Dr. Bob Buchanan explains how his research using plant biotechnology is removing allergens from existing foods. Statement to the Senate Committee on Agriculture, Nutrition and Forestry, October 6, 1999.

http://aspp.org/pubaff/tesbuch.htm

Buchanan, B. B., *et al.* (1997). Thioredoxin-linked mitigation of allergic responses to wheat. *Proc. Natl. Acad. Sci., USA*, **94**, 5372–5377.

Burkhardt, P. K., *et al.* (1997). Transgenic rice (*Oryza sativa*) endosperm expressing daffodil (*Narcissus pseudonarcissus*) phytoene synthase accumulates phytoene, a key intermediate of provitamin A biosynthesis. *Plant J.*, **11**, 1071–78.

Cahagnier, B., and Melcion, D. (2000). Mycotoxines de *Fusarium* dans les mais-grains à la recolte: Relation entre la présence d'insectes (pyrale, sésame) et la teneur en mycotoxines, in *Proceedings of the International Feed Production Conference*, Piacenza, November 27–28 2000.

Carpenter, J. E., and Gianessi, L. P. (2001). *Agricultural Biotechnology: Updated Benefit Estimates*. National Center for Food and Agricultural Policy, Washington DC. http:www.ncfap.org

Chassy, B. M. (2001). Imagine a healthier world. *Chicago Tribune*, March 11.

Chong, D. K. X., *et al.* (1997). Expression of the human milk protein beta-casein in transgenic potato plants. *Transgenic Res.*, **6**, 289–296.

Coetzer, C., *et al.* (2001). Control of enzymatic browning in potato (*Solanum tuberosum* L.) by sense and antisense RNA from tomato polyphenol oxidase. *J. Agric. Food Chem.* **49**, 652–657.

Del Vecchio, A. J. (1996). High laurate canola. *Int. News Fats Oils Relat. Mater.* **7**, 230–243.

Della Penna, D. (1999). Nutritional genomics: Manipulating plant micronutrients to improve human health. *Science*, **285**, 375–379.

Della Penna, D. (2001). Plant metabolic engineering. *Plant Physiol.*, **125**, 160–163.

Dhawan, V. (2001). Biotechnology and the promise for control of vitamin A deficiency. Presentation at the Twentieth Meeting of the International Vitamin A Consultative Group, February 15, 2001, Vietnam.

Dowd, P. (2000). Indirect reduction of ear molds and associated mycotoxins in *Bacillus thuringiensis* corn under controlled and open field conditions: Utility and limitations. *J. Econ. Entomol.*, **93**, 1669–1679.

Dunwell, J. M. (2000). Transgenic approaches to crop inprovement. *J. Exp. Bot.*, **51**, 487–496.

Egnin, M., *et al.* (1999). Enhanced protein content and quality of sweetpotato engineered with a synthetic storage protein gene. Abstract presented at the Annual Meeting of the American Society of Plant Physiology, Baltimore, July 20–24.

Engel, K. H., *et al.* (1996). Modern biotechnology in plant breeding: Analysis of glycoalkaloids in transgenic potatoes, in *Biotechnology for Improved Foods and Flavors*, G. R. Takeoka *et al.*, eds., pp. 249–260. American Chemical Society, Washington DC.

Facciotti, M. T., *et al.* (1999). Improved stearate phenotype in transgenic canola expressing a modified acyl–acyl carrier protein thioesterase. *Nat. Biotechnol.*, **17**, 593–597.

Fischer, R., *et al.* (1998). Endogenously sweetened transgenic plant products. U.S. Patent 5,739,409.

Food and Nutrition Board. (2001). *Dietary Reference Intakes for Vitamin A, Vitamin K, Boron, Chromium, Copper, Iodine, Iron, Manganese, Molybdenum, Nickel, Silicon, Vanadium, and Zinc*. A report of the panel of micronutrients, subcommittee on upper reference levels of nutrients and of interpretation and use of dietary reference intakes, and the standing committee on the scientific evaluation of dietary reference intakes. National Academy Press, Washington DC.

Frossard, E., *et al.* (2000). Potential for increasing the content and bioavailability of Fe, Zn and Ca in plants for human nutrition. *J. Sci. Food Agric.* **80**, 861–879.

Gan, S., and Amasino, R. M. (1995). Inhibition of leaf senescence by autoregulated production of cytokinin. *Science*, **270**, 1986–1988.

Goto, F. *et al.* (1999). Iron fortification of rice seed by the soybean ferritin gene. *Nat. Biotechnol.*, **17**, 282–286.

Grusak, M. A., and Della Penna, D. (1999). Improving the nutrient composition of plants to enhance human nutrition and health. *Annu. Rev. Plant Physiol. Plant Mol. Biol.*, **50**, 133–161.
Guis, M., *et al.* (1997). Ethylene and biotechnology of fruit ripening: Pre- and postharvest behaviour of transgenic melons with inhibited ethylene production. *Acta Hortic.* **463**, 31–37.
Hauptmann, R., *et al.* (1997). Enhanced carotenoid accumulation in storage organs of genetically engineered plants. U.S. Patent 5,618,988.
Heimlich, R. E., *et al.* (2000). Genetically engineered crops: has adoption reduced pesticide use? *Agric. Outlook* (a publication of the U.S. Department of Agriculture, Economic Research Service), August, pp. 13–17.
Henzi, M. X., *et al.* (1998). Transgenic broccoli (*Brassica oleracea* L. cv. "*Italica*") plants containing an antisense ACC oxidase gene. *Acta Hortic.*, **464**, 147–151.
Heyer, A. G. *et al.* (1999). Production of modified polymeric carbohydrates. *Curr. Opin. Biotechnol.*, **10**, 166–174.
Jalani, B. S., *et al.* (1997). Improvement of palm oil through breeding and biotechnology. *J. Am. Oil Chem. Soc.*, **74**, 1451–1455.
James, C. (2000). Global review of commercialized transgenic crops: 2000 Presentation at the World Food Prize International Symposium Des Moines, IA., October 12, 2000.
Jenner, H. L., *et al.* (2001). NAD malic enzyme and the control of carbohydrate metabolism in potato tubers. *Plant Physiol.*, **126**, 1139–1149.
Jung, W., *et al.* (2000). Identification and expression of isoflavone synthase, the key enzyme for biosynthesis of isoflavones in legumes. *Nat. Biotechnol.* **18**, 208–212.
Kato, M., *et al.* (2000). Caffeine synthase gene from tea leaves. *Nature*, **406**, 956–957.
Katsube, T., *et al.* (1999). Accumulation of soybean glycinin and its assembly with glutelins in rice. *Plant Physiol.* **120**, 1063–1073.
Khush, G. S. (2001). Challenges for meeting the global food and nutrient needs in the new millennium. *Proc. Nutr. Soc.*, **60**, 15–26.
Kim, C. S., *et al.* (1990). Improvement of nutritional value and functional properties of soybean glycinin by protein engineering. *Protein Eng.* no **3**, 725–731.
King, G. A,. and O'Donoghue, E. M. (1995). Unravelling senescence: New opportunities for delaying the inevitable in harvested fruit and vegetables. *Trends Food Sci. Technol.*, **6**, 385–389.
Kinney, A. J. (1996). Designer oils for better nutrition. *Nat. Biotechnol.*, **14**, 946.
Kinney, A. J. (1997). Genetic engineering of oilseeds for desired traits, in *Genetic Engineering*, J. K. Setlow, ed., pp. 149–166. Plenum Press, New York.
Knutzon, D. (1999). Polyunsaturated fatty acids in plants. International Patent Publication WO9964614. Filed June 10.
Langridge, W. H. R. (2000). Edible vaccines. *Sci. Am.*, **283**, 66–77.
Leaver, C. J., *et al.* (1998). Transgenic plants having increased starch content. World Intellectual Property Organization. International Patent Publication WO 98/23757.
List, G. R., *et al.* (1996). Potential margarine oils from genetically modified soybeans. *J. Am. Oil Chem. Soc.*, **73**, 729–732.
Liu, K. and Brown, E. A. (1996). Enhancing vegetable oil quality through plant breeding and genetic engineering. *Food Technol.*, **50**, 67–71.
Liu, Q., *et al.* (2000). Genetic modification of cotton seed oil using inverted-repeat gene-silencing techniques, in *Recent Advances in the Biochemistry of Plant Lipids*, J. L. Harwood and P. J. Quinn, eds., pp. 927–929. Portland Press, London.
Lucca, P., *et al.* (2001). Genetic engineering approaches to improve the bioavailability and the level of iron in rice grains. *Theor. Appl. Genet.*, **102**, 392–397.
Mason, H. S., *et al.* (1998). Edible vaccine protects mice against *Escherichia coli* heat-labile enterotoxin (LT): Potatoes expressing a synthetic LT-B gene. *Vaccine*, **16**, 1336–1343.

Mazur, B., *et al.* (1999). Gene discovery and product development for grain quality traits. *Science*, **285**, 372–375.
McCourt, P., *et al.* (1999). Stress tolerance and delayed senescence in plants. International Patent Publication WO 99/06580.
Momma, K., *et al.* (1999). Quality and safety evaluation of genetically engineered rice with soybean glycinin: Analyses of the grain composition and digestibility of glycinin in transgenic rice. *Biosci. Biotechnol. Biochem.* **63**, 314–318.
Momma, K., *et al.* (2000). Safety assessment of rice genetically modified with soybean glycinin by feeding studies on rats. *Biosci. Biotechnol. Biochem.*, **64**, 1881–1886.
Muir, S. R., *et al.* (2001). Overexpression of petunia chalcone isomerase in tomato results in fruit containing increased levels of flavonols. *Nat. Biotechnol.*, **19**, 470–474.
Munkvold, G. P., *et al.* (1999). Comparison of fumonisin concentrations in kernels of transgenic *Bt* maize hybrids and nontransgenic hybrids. *Plant Dis.*, **83**, 130–138.
Murata, M., *et al.* (2000). Transgenic apple (Malus x domestica) shoot showing low browning potential. *J. Agric. Food Chem.*, **48**, 5243–5248.
Nestle, M. (2001). Genetically engineered "golden" rice unlikely to overcome vitamin A deficiency. *J. Am. Diet. Assoc.*, **101**, 289–290.
Penarrubia, L., *et al.* (1992). Production of the sweet protein monellin in transgenic plants. *Bio/Technology*, **10**, 561–564.
Pietri, A., and Piva, G. (2000). Occurrence and control of mycotoxins in maize grown in Italy. *Proceedings of the Sixth International Feed Production Conference*, Piacenza, November 27–28, 2000.
Potrykus, I. (2001). Golden rice and beyond. *Plant Physiol.*, **125**, 1157–161.
Prakash, C. S., *et al.* (1997). Molecular insights into the biology of sweetpotato (*Ipomoea batatas*), in *Radical Biology: Advances and Perspectives on the Function of Plant Roots*, H. E. Flores *et al.*, eds., pp. 207–219. Vol. 18 in *Current Topics in Plant Physiology*. American Society of Plant Physiologists, Rockville, MD.
Richter, L. J., *et al.* (2000). Production of hepatitis B surface antigen in transgenic plants for oral immunization. *Nat. Biotechnol.*, **18**, 1167–1171.
Riley, P. A., and Hoffman, L. (1999). Value-enhanced crops: Biotechnology's next stage. *Agric. Outlook* (a publication of the U.S. Department of Agriculture, Economic Research Service), March, pp. 18–23.
Roberfroid, M. B., and Delzenne, N. M. (1998). Dietary fructans. *Annu. Rev. Nutr.*, **18**, 117–143.
Robertson, R., *et al.* (2001). Golden rice: What role could it play in alleviation of vitamin A deficiency? http://www.agbioworld.org/articles/goldenrice/html.
Rockefeller Foundation. (1999). New rice may help address vitamin A and iron deficiency, major causes of death in the developing world. Press release, St. Louis, MO, August 3.
Rodriguez, R. L., *et al.* (2000). Functional expression of recombinant human milk proteins in rice. Abstracts of Papers American Chemical Society 219:pBIOT 16.
Roller, S., and Harlander, S. (1998). Modern food biotechnology: Overview of key issues, in *Genetic Modification in the Food Industry. A Strategy for Food Quality Improvement*, S. Roller and S Harlander, eds., pp. 3–26. Blackie Academic & Professional, London.
Romer, S., *et al.* (2000). Elevation of the provitamin A content of transgenic tomato plants. *Nat. Biotechnol.*, **18**, 666–669.
Sandhu, J. S., *et al.* (2000). Oral immunization of mice with transgenic tomato fruit expressing respiratory synctial virus–F protein induces a systemic immune response. *Transgenic Res.*, **9**, 127–135.
Schwall, G. P., *et al.* (2000). Production of very-high-amylose potato starch by inhibition of SBE A and B. *Nat. Biotechnol.*, **18**, 551–554.
Secor, G. A., *et al.* (1997). Modulation of sugar content in plants. U.S. Patent 5,646,023.

Sevenier, R., *et al.* (1998). High level fructan accumulation in a transgenic sugar beet. *Nat. Biotechnol.*, **16**, 843–846.
Shewmaker, C. K., *et al.* (1999). Seed-specific overexpression of phytoene synthase: Increase in carotenoids and other metabolic effects. *Plant J.*, **20**, 401–412.
Shewry, P. R., and Tatham, A. S. (1997). Biotechnology of wheat quality. *J. Sci. Food Agric.*, **73**, 397–406.
Shewry, P. R., *et al.* (1995). Biotechnology of breadmaking: Unraveling and manipulating the multiprotein gluten complex. *Bio/Technology*, **13**, 1185–1190.
Shimada, H., *et al.* (1993). Antisense regulation of the rice waxy gene expression using a PCR-amplified fragment of the rice genome reduces the amylose content in grain starch. *Theor. Appl. Genet.* **86**, 665–672.
Shintani, D. and Della Penna, D. (1998). Elevating the vitamin E content of plants through metabolic engineering. *Science*, **282**, 2098–2100.
Szwacka, M., *et al.* (1999). Transgenic cucumber plants expressing the thaumatin gene, in *Food Biotechnology: Proceedings of an International Symposium, Zakopane*, S. Bielecki *et al.*, eds., pp.43–38. Elsevier, Amsterdam.
Stark, D. M. (1998). Potatoes, in *Genetic Modification in the Food Industry. A Strategy for Food Quality Improvement*, S. Roller and S. Harlander, eds., pp. 214–227. Blackie Academic & Professional, London.
Stark, D. M., *et al.* (1992). Regulation of the amount of starch in plant tissues by ADP glucose pyrophosphorylase. *Science*, **258**, 287–292.
Tacket, C. O., *et al.* (2000). Human immune responses to a novel Norwalk virus vaccine delivered in transgenic potatoes. *J. Infect. Dis.*, **182**, 302–305.
Tada, Y., *et al.* (1996). Reduction of 14–15 kDa allergenic proteins in transgenic rice plants by antisense gene. *FEBS Lett.*, **391**, 341–345.
Takase, K., and Hagiwara, K. (1998). Expression of human alpha-lactalbumin in transgenic tobacco. *J. Biochem.*, **123**, 440–444.
Theologis, A., and Sato, T. (2000). Control of fruit ripening through genetic control of ACC synthase synthesis. US Patent 6,156,956.
USDA (2000). *Dietary Guidelines for Americans*, 5th ed. U.S. Department of Agriculture, Washington, DC.
Utsumi, S., *et al.* (1997). Molecular design of soybean glycinins with enhanced food qualities and development of crops producing such glycinins, in *Food Protein and Lipids*, S. Damodaran, ed., pp. 1–15. Plenum Press, New York.
Vageeshbabu, H. S., and Chopra, V. L. (1997). Genetic and biotechnological approaches for reducing glucosinolates from rapeseed–mustard meal. *J. Plant Biochem. Biotechnol.*, **6**, 53–62.
Vasil, I. K., and Anderson, O. D. (1997). Genetic engineering of wheat gluten. *Trends Plant Sci.*, **2**, 292–297.
Venkatramesh, M., *et al.* (2000). Transgenic plants containing altered levels of sterol compounds and tocopherols. International Patent Publication WO 00/61771.
Vijn, I., *et al.* (1997). Fructan of the inulin neoseries is synthesized in transgenic chicory plants (*Chicorium intybus* L.) harbouring onion (*Allium cepa* L.) fructan:fructan 6G-fructosyltransferase. *Plant J.*, **11**, 387–398.
Visser, R. G. F., *et al.* (1991). Inhibition of the expression of the gene for granule-bound starch synthase in potato by antisense constructs. *Mol. Gen. Genet.*, **225**, 289–296.
Visser, R. G. F., *et al.* (1997). Some physicochemical properties of amylose-free potato starch. *Starch*, **49**, 443–448.
Voelker, T. (1997). Transgenic manipulation of edible oilseeds, in *Functionality of Food Phytochemicals*, T. Johns and J. Romeo, eds., pp. 223–236. Plenum Press, New York.

Wentzel, C. A., *et al.* (2001a). Fumonisin-induced hepatocarcinogenesis: Mechanisms related to cancer initiation and promotion. *Environ. Health Perspect.*, **109** (suppl. 2), 291–300.

Wentzel, C. A., *et al.* (2001b). Toxicity of culture material of *Fusarium verticillioides* strain MRC 826 to nonhuman primates. *Environ. Health Perspect.*, **109**(suppl. 2), 267–276.

Ye, X., *et al.* (2000). Engineering the provitamin A (beta-carotene) biosynthetic pathway into (carotenoid-free) rice endosperm. *Science*, **287**, 303–305.

Yu, J., and Langridge, W. H. R. (2001). A plant-based multicomponent vaccine protects mice from enteric diseases. *Nat. Biotechnol.* **19**, 548–552.

Zemanek, E. C., and Wasserman, B. P. (1995). Issues and advances in the use of transgenic organisms for the production of thaumatin, the intensely sweet protein from *Thaumatococcus danielli. Crit. Rev. Food Sci. Nutr.*, **35**, 455–466.

Zheng, A., *et al.* (1995.) The bean seed storage protein -phaseolin is synthesized, processed, and accumulated in the vacuolar type-II protein bodies of transgenic rice endosperm. *Plant Physiol.*, **109**, 777–786.

Chapter 6

Animal Feeds from Crops Derived through Biotechnology: Farm Animal Performance and Safety

Marjorie A. Faust
Iowa State University
Ames, Iowa

Barbara P. Glenn
Federation of Animal Science Societies
Bethesda, Maryland

Farm animals consume a large proportion of the U.S. grain, forage, and crop by-products. It has been postulated that consumption of these products from biotechnology-derived crops might have an impact on the safety of meat, milk, and eggs for human consumption or on the performance of animals fed these products. Proteins and nucleotides (transgenes) found in biotech crops consumed by farm animals generally have a very high digestibility, with only a limited proportion of undigested residual nitrogenous compounds excreted in the feces. Because of the high digestibility

Biotechnology and Safety Assessment, 3rd edition

and breakdown to amino acids, ammonia, and carbon skeletons, tracking absorbed amino acids from digested dietary protein into meat, milk, or eggs is difficult, especially when these nutrients undergo transamination, urea production, or protein synthesis. To date, commercially available conventional and biotechnology-derived crops, including pest-protected, viral-resistant, herbicide-tolerant, and modified nutritive value varieties of corn, soybeans, canola, beets, flax, cotton, and potatoes, have been compositionally comparable except with respect to their introduced traits. Safety for farm animals has been confirmed by regulatory reviews, and results from a large number of research studies are that performance, health, and nutrient utilization by farm animals has not differed when animals are fed conventional and biotechnology-derived crops and/or their coproducts. Further, for the biotechnology-derived Bt *crops evaluated to date as feeds for farm animals, no biologically relevant differences in composition of animal products including meat and milk have been reported. No intact or immunologically reactive fragments of the transgenic plant proteins or fragments of transgenic plant source DNA have been detected in samples of milk, meat, eggs, lymphocytes, blood, organ tissue, duodenal fluid, and excrement from animals fed* Bt *crops. No biologically relevant differences in composition of animal products including meat and milk have been reported when farm animals are fed commercially available herbicide-tolerant varieties and their conventional genetic counterparts. Further, no transgenic plant proteins or fragments of transgenic plant source DNA have been detected in samples of milk, meat, eggs, skin, duodenum, leukocytes, lymphocytes, blood, organ tissue, duodenal fluid, or excrement when farm animals have been fed commercially available herbicide-tolerant crops. Products produced by farm animals fed biotechnology-derived crops are as wholesome, safe, and nutritious as similar products produced by animals fed conventional crops.*

INTRODUCTION

Animal agriculture is an integral part of food-producing systems, with foods of animal origin representing about one-sixth of human food energy and one-third of human food protein on a global basis (CAST, 1999). Per-capita consumption of meat, milk, and eggs is much higher in developed countries, but the current rapid increase in consumption in many developing countries is expected to continue. Total meat consumption in developing countries is expected to more than double by the year 2020, while projections for developed countries predict an increase in consumption at or below the rate of population growth. Because most of the world's population is in developing countries that are experiencing the most rapid growth rates, global demand for meat is projected to increase by more than 60% of current consumption by 2020.

By the year 2001, biotechnology-derived crops represented a significant portion of feed crops fed to farm animals in the United States and a growing percentage worldwide. The majority of biotechnology-derived crops now consumed by farm animals are pest-protected and herbicide-tolerant varieties of corn, soybeans, canola, and cotton. Future biotechnology-derived crops are expected to have direct benefits for animal agriculture through improved environmental stewardship, production of animal products with enhanced nutritive value and safety for consumers, and improved animal health and efficiency.

Evaluations of safety and nutritive value for biotechnology-derived crops consider whether the new crop is "as safe and nutritious as" its conventional genetic counterpart. In the United States, this evaluation is completed as part of the regulatory process for biotechnology-derived crops by assessing safety for inserted genes and recipient plants, potential for allergenicity, integrity for genes, functionality for novel proteins, agronomic characteristics and environmental safety for novel crops, and composition for key nutrients and toxicants for crops and important products. Similar processes for assessing safety are used by regulatory agencies globally. Through these regulatory evaluations, the currently available biotechnology-derived crops have been found to be as safe and nutritious as conventional non-biotechnology-derived counterparts. In addition, numerous independent studies have corroborated the safety and wholesomeness of biotechnology-derived crops as feeds for farm animals.

To further understand the digestive fate of biotechnology-derived crops for farm animals and the potential impacts for characterizing animal products as a result of feeding these crops, several studies have been conducted to determine whether the novel genes and proteins in feeds containing biotechnology-derived crops can be detected in products and tissues from farm animals. No proteins or DNA introduced from biotechnology have been detected to date in tissues or fluids from farm animals fed biotechnology-derived crops or their by-product feeds.

Objectives for this discussion are to provide an overview of crops commonly used as feeds for livestock with emphasis on biotechnology-derived crops, to introduce background discussion of digestion in farm animals as the underlying basis of the regulatory process used to evaluate the safety and nutritive value of biotechnology-derived crops as feeds for farm animals, to review studies that have used biotechnology-derived crops as feeds for farm animals, to examine studies designed to detect biotechnology-derived proteins and DNA in animals, and to introduce concepts of future biotechnology-derived crops with specific benefits for animal agriculture.

CROPS FED TO FARM ANIMALS IN THE UNITED STATES

Worldwide, diets for farm animals consist almost exclusively of plant foods, including human-edible and human-inedible products. Diets of ruminants, such as beef cattle, dairy cattle, sheep, and goats, comprise as much as 70% of materials that are inedible for humans such as forages (hay and silage) and crop residues. Swine and poultry diets typically contain 50 to 70% of human-edible feedstuffs, and worldwide, farm animals consume one-third of the global cereal grain supply. Additionally, farm animals are able to use food and fiber by-products such as soybean meal, canola meal, cottonseed hulls, and corn distillers' dried grains. In the United States, farm animals consume a large proportion of the grain, forage, and by-product crops produced, and biotechnology-derived crops now constitute a significant proportion of many animal feed crops.

The U.S. production for crops used commonly as feedstuffs for farm animals is in Table 1. In the United States, corn (maize), soybeans, and their associated by-product feeds are important feedstuffs for farm animals. In fact, approximately 80% of the U.S. yield of soybeans is consumed by animals, and during 2000 and 2001, 54 and 68%, respectively, of this crop (USDA, 2001; Table 2) was derived from biotech seed. The vast majority of biotechnology-derived crops now fed to farm animals include pest-protected and herbicide-tolerant varieties of corn, soybeans, cotton, and canola. Future biotechnology-derived crops are expected to provide unique benefits to animal agriculture. Thus, it is important to consider the safety of biotechnology-derived crops when fed to farm animals for the production of meat, milk, and eggs.

CHARACTERISTICS OF INTAKE AND DIGESTION BY FARM ANIMALS

SUMMARY

Evaluating the safety of biotechnology-derived crops as feedstuffs for farm animals requires an understanding of the digestive process for various farm animal species, changes made to plants through the biotechnology process, and the ultimate outcome of these plant changes during digestion. To date, biotechnology-derived crops include unique traits conferred to plants as a result of the insertion of novel DNA that codes for unique proteins. For the vast majority of commercially available biotechnology-derived crops, the unique proteins confer directly the desired traits to the plants (i.e., pest protection and herbicide tolerance). Consequently, the ensuing discussion about the digestive process in farm animals focuses on the fate of ingested proteins and DNA (nucleotides). Overall, the proteins and nucleotides (transgenes) found in biotech crops are highly digestible, with only a limited proportion of undigested residual nitrogenous compounds excreted in the feces.

Table 1

Production of Feed Crops in the United States for 2000

Crop	Harvested (ha)	Yield (metric tons/ha)	Production (metric tons)
Barley	2,104,790	3.29	6,920,690
Corn for grain[a]	29,433,910	8.6	253,207,960
Corn for silage	2,374,720	37.64	89,392,170
Hay			
Alfalfa	9,339,030	7.80	72,899,570
Others	14,883,280	4.38	65,168,520
Oats	940,500	2.30	2,165,560
Proso millet	149,740	1.11	166,010
Rice	1,229,850	7.04	8,657,810
Rye	122,220	1.79	218,930
Sorghum for grain[a]	3,125,420	3.82	11,940,330
Sorghum for silage	107,240	24.22	2,597,270
Wheat, all	21,459,900	2.82	60,512,120
Oilseeds			
Canola	610,680	1.50	914,870
Cottonseed			5,838,280
Flaxseed	209,220	1.30	272,550
Peanuts	540,670	2.74	1,481,210
Rapeseed	1,580	1.65	2,610
Safflower	79,720	1.61	128,160
Soybeans for beans	29,428,250	2.56	75,377,930
Sunflower	1,063,930	1.53	1,625,830
Cotton			
Upland	5,214,030	0.70	3,657,590
Amer-Pima	68,390	1.24	84,720
Sugar beets	556,170	52.91	29,425,440
Dry peas, beans, lentils	810,790		1,529,550
Hops	14,620	2.10	30,650
Potatoes	546,980	42.80	23,409,130

[a]Area planted for all purposes.
Source: USDA (2001).

Table 2

Percentages of Acreage for Major U.S. Feed Crops from Plantings of Biotechnology-Derived Seed[a]

	Planting year	
	2000	2001
Corn for grain	25	26
Soybeans for beans	54	68
Upland cotton	61	69
Potatoes	2 – 3[2]	Data not available

[a]Carpenter and Gianessi (2001).
Source: USDA (2001), (except as noted).

Intake Effects

Intake of feed and the digestibility of nutrients in the feed are the drivers of animal growth, milk production, egg production, reproduction, and maintenance (Maynard and Loosli, 1979). Maintenance is the level of energy required to maintain cellular and tissue level functions. Nutrients consumed in excess of those required for maintenance are available to animals for supporting a fetus during gestation and for the production of meat, milk, eggs, and fiber. For example, dairy cows that consume twice maintenance as feed dry matter produce considerably less milk than comparable cows consuming four times maintenance as feed dry matter. However, because feeding for maximum production typically is not economically beneficial, the general goal is to achieve economically optimum feed intake and production.

Characteristics of feedstuffs can influence feed intake by farm animals, and these include palatability or acceptability and especially for ruminants, the content of different fiber fractions. The presence of some mycotoxins can reduce intake of feed by farm animals, and to some degree this effect is mediated by reduced palatability of the feed. For high-producing ruminants, such as dairy cows, feed intake often is limited by gut fill, and contents of individual fiber components in roughage feeds are important indices of their impact for intake. Consequently, feeding studies for farm animals typically evaluate performance and feed intake.

Anatomy of Digestive Tracts for Farm Animals

The digestive tracts of farm animals differ (Table 3); consequently modes of digestion differ. Digestive tracts for primary fiber digesters (cattle, sheep) are

Table 3
Comparative Measurements of the Gastrointestinal Tract

	Relative capacity (%)			
Animal	Stomach	Small intestine	Cecum	Colon and rectum
Ox	71	18	3	8
Sheep and goat	67	21	2	10
Horse	9	30	16	45
Pig	29	33	6	32
Dog	63	23	1	13
Cat	69	15	–	16

Source: Adapted from *Dukes' Physiology of Domestic Animals* (1993).

considered to be more complex than tracts for carnivores and omnivores (pigs, poultry). Ruminants such as cattle and sheep have four stomach compartments –the rumen, the reticulum, the omasum, and the abomasum. The rumen is the largest compartment and contains a vast symbiotic microbial population consisting of bacteria and protozoa that enable ruminants to utilize large amounts of fiber. Chickens, turkeys, and ducks have a crop, an organ where food is stored and soaked after intake, a proventriculus, which is similar to a true stomach in that it produces digestive juices, and a gizzard, which aids in grinding feed. The digestive system of pigs and other omnivores is considerably less complex than digestive systems for fiber digesters such as ruminants (Table 3). Interestingly, the digestive system of pigs is quite similar to that of humans.

Digestion Differences

Digestion involves a series of processes in the digestive tract during which feeds are broken down in particle size and finally rendered soluble so that absorption is possible (Maynard and Loosli, 1979). These processes are mechanical and enzymatic; further, microoganisms aid digestion by providing important enzymes not secreted by mammalian tissues.

Monogastric animals such as swine and poultry use enzymatic digestion from the endogenous production of digestive enzymes. For monogastrics, no enzymes are secreted in the alimentary tract to digest cellulose or other higher polysaccharides, thus these farm animals cannot use forages efficiently. In monogastric animals, ingested proteins are acted on by enzymes secreted by the stomach, pancreas, and small intestine. Monogastric animals rely directly on ingested protein for sources of amino acids and peptides.

In contrast to monogastrics, ruminants are able to utilize large quantities of roughage feeds by relying heavily on populations of bacteria and protozoa in the rumen for digestion of plant cellulose. Further, unlike monogastric animals, which rely directly on ingested protein for amino acids, rumen microorganisms utilize ingested nitrogen sources to provide the majority of necessary amino acids used by ruminants. Reliance on microorganisms for primary protein synthesis allows ruminants to utilize nitrogen from diverse sources including dietary protein, ammonia, amides, and even nitrates. Ultimately, proteins synthesized by these microorganisms are digested in the stomach and intestines. Ruminants obtain a smaller percentage of necessary amino acids from proteins that "escape" digestion by microorganisms and are digested directly by the animal in the stomach and intestine. In addition, rumen microorganisms are responsible for synthesis of other essential nutrients such as the B vitamins.

Digestion of Protein

Enzymes responsible for digestion of proteins in ruminant and monogastric animals are listed in Table 4. Pepsin released from stomach mucosa splits

Table 4
Enzymatic Digestion of Protein

Substrate	Enzyme	Origin	End products
Protein	Pepsin	Gastric mucosa	Proteoses Peptones Polypeptides
⇓			
Products of gastric digestion	Trypsin Chymotrypsin Carboxypeptidase	Pancreas	Peptides Amino acids Nucleoproteins
⇓			
Products of pancreatic digestion	Dipeptidase Nucleotidase Nucleotides	Small intestine	Amino acids Nucleosidase Nucleosides Purines Phosphoric acid

Source: Adapted from *Dukes' Physiology of Domestic Animals* (1993); Maynard and Loosli (1979).

proteins into polypeptides, proteoses, and peptones (Table 4, *Dukes' Physiology of Domestic Animals*, 1993). These polypeptides are further digested to peptides and amino acids by trypsin, chymotrypsin, and elastase from the pancreas. Also, carboxypeptidase from the pancreas and aminopeptidase and dipeptidase from the small intestine break polypeptides down into amino acids. Nucleoproteins such as nucleic acids are digested to nucleotides, nucleosides, purines, and phosphoric acid by nucleotidase in the small intestine. The primary nitrogenous end products of digestion are amino acids that enter the bloodstream from the intestine. Ammonia and simple peptides are absorbable, also. Under normal conditions, 98% of all proteins ultimately are broken down to amino acids (Guyton, 1976). Similarly, others reported that when ovalbumin was administered orally to humans, 0.007 to 0.008% of the intact ingested protein was detected in circulation (Tsume *et al.*, 1996).

Digestion of DNA and Nucleotides

Deoxyribonucleic acid (DNA) provides the genetic coding for all plants, animals, bacteria, and many viruses (Beever and Kemp, 2000). Typically, in plants and animals each chromosome contains a single, long molecule of DNA. The DNA molecule is double stranded, and each strand comprises smaller units or nucleotides that are arranged in linear sequences. Sequence of the four different nucleotides ultimately determines the instructions conferred by segments of DNA. Corresponding DNA strands are linked by hydrogen bonds between complementary pairs of nucleotides on the two strands—adenylic acid is complementary to thymidylic acid, and guaninylic acid and cytosine form the second complementary pair. Bonds within and between strands cause the DNA molecule to form an antiparallel double-helical structure. Linear groups of 1000 or more nucleotides act together as functional units known as genes. An average plant species contains 20×10^6 to 50×10^6 different genes.

Most foodstuffs contain a complex mixture of proteins, lipids, carbohydrates, minerals, vitamins, and nucleic acids. The relative proportion of individual constituents varies widely; however, the quantity of DNA in most food crops generally is less than 0.02% (dry matter basis) (Beever and Kemp, 2000). Virtually all feedstuffs contain DNA. In addition, farm animals are exposed to other sources of DNA in the gut, including shed epithelial cells, white blood cells, and bacteria and protozoa resident in the gut. Consequently, exogenous DNA is present constantly within the gastrointestinal tract of farm animals and humans.

The transgenic DNA in biotechnology-derived crops consists of the same four nucleotides found in host plant DNA, and thus, is structurally identical to endogenous plant DNA. Further, for the currently available biotechnology-derived crops, transgenic DNA constitutes a small proportion of total plant DNA (< 0.0004% of total plant DNA: Glenn, 2001) and a small proportion of DNA ingested by farm animals. Beever and Kemp (2000) estimated that a 600-kg lactating dairy cow consumes approximately 608,000 mg of DNA daily; when fed a standard diet containing *Bt* corn, the daily intake of transgenic DNA of such an animal would be approximately 0.00024% of the total DNA ingested (~1.5 mg).

Within the digestive tract, DNA is hydrolyzed by high concentrations of DNase I, the endonuclease enzyme, that disrupts the double-stranded DNA. The DNase endonuclease is produced and secreted by the salivary glands, the pancreas, the liver, and the Paneth cells of the small intestine (Beever and Kemp, 2000). In addition, the recently characterized endonuclease called DNase II, found in lysosomes within phagocytes, is involved in the catabolism of DNA. McAllan (1982) estimated that more than 85% of the plant DNA consumed by ruminants is reduced to nucleotides or smaller constituents before entering the duodenum, with most of the larger nucleic acid fragments in small intestinal contents arising from rumen microbes. Further, these workers reported that ingested DNA was degraded within 4 hours to mononucleotides.

Horizontal Gene Transfer

It has been postulated that plant DNA (endogenous or inserted) may be incorporated into the genome of farm animals that consume biotechnology-derived crops or into that of microorganisms. For farm animals, horizontal gene transfer is unlikely because DNA is digested completely in a relatively short time. In fact, Beever and Kemp (2000) reviewed the literature and concluded that no plant gene or plant gene fragment has ever been detected in the genome of animals or humans, and further that there is no evidence that tissues from animals consuming plant material express any plant proteins.

Several of the early insect-resistant corn varieties incorporated an antibiotic resistance gene as a marker for the specific transgene conferring insect resistance, and some have considered whether the feeding of biotechnology-derived plants may lead to the development of new antibiotic resistant microorganisms in the digestive tract of farm animals. This question is extremely important to animal agriculture because the extensive use of antibiotics for growth promotion and therapeutic use in farm animal production has resulted in multiple antibiotic-resistant microbes.

Transmission of antibiotic resistance among bacterial strains is well documented and occurs by transfer of plasmids, small circular extrachromosomal pieces of DNA or, in some bacterial species, by insertion of intact antibiotic-resistant genes from genomic DNA of one bacterium to that of another (Beever and Kemp, 2000). However, one species of bacteria (*Acinetobacter* sp. BD413) has been shown to have incorporated a fragment of plant DNA (deVries and Wackernagel, 1998). It has been shown that plant DNA containing the *npt*II gene, which encodes resistance to neomycin and kanamycin, can at low frequency, rescue *Acinetobacter* sp. that already have an *npt*II gene containing a small deletion in the gene (Gebhard and Smalla, 1998). However these studies did not demonstrate the uptake and function of a complete plant *npt*II gene, suggesting that de novo acquisition of complete genes from plants is extremely unlikely even in the presence of antibiotics providing selection pressure for recombinants. Although Acetinobacter sp. BD413 can be induced by nutrients to acquire competence under soil conditions, a study involving the field release of transgenic sugar beet containing the *npt*II gene failed to demonstrate horizontal transfer of this gene from sugar beet to soil microorganisms (Gebhard and Smalla, 1999).

Results obtained to date indicate that horizontal gene transfer from plants to microorganisms is an extremely rare evolutionary event (Nielsen *et al.*, 1998). Further, the antibiotic resistance genes used initially to identify biotechnology-derived crops are relatively unimportant antimicrobials for farm animal production. Equally important, the use of antibiotic resistance genes as markers in agricultural biotechnology is being discontinued as a result of heightened awareness.

PERFORMANCE, HEALTH, AND NUTRIENT UTILIZATION FOR FARM ANIMALS CONSUMING BIOTECHNOLOGY-DERIVED CROPS

SUMMARY

Commercially available conventional and biotechnology-derived crops, including pest-protected, viral-resistant, herbicide-tolerant, and modified nutritive value varieties of corn, soybeans, canola, beets, flax, cottonseeds, and potatoes, are compositionally comparable except with respect to their introduced trait. Safety for farm animals as confirmed by regulatory reviews has been sufficient, and results from research studies indicate that animal performance, health, and nutrient utilization by farm animals has not differed when animals are fed conventional and biotechnology-derived crops. Further, genetic differences among varieties appear to be considerably more important than the presence of the tested transgenes for influencing animal performance. Under conditions that favor mycotoxin development, lower levels of mycotoxins for pest-protected *Bt* crops may result in improved health and performance for farm animals fed these grains as opposed to those fed conventional non-pest-protected grains with high levels of mycotoxins.

Introduction

Development of new varieties of feed crops has been prolific, and crops with improved agronomic properties, nutritive characteristics, and other properties are now available. These improved crops have been developed by using conventional plant breeding and technological methods, including biotechnology. The scope of the current discussion is limited to biotechnology-derived crops and their realized safety and nutritive characteristics for farm animals, and one important outcome for research in this field is the indirect assessment of unintended changes for the new varieties. It is important to note that unintended changes to crops can result from using conventional methods and biotechnology and that most changes to plants do not affect the safety for these new plants. In rare instances, however, new plants may be less safe than their parent varieties; a U.S. National Academy of Sciences committee cited two crops developed by means of conventional breeding methods for which unintended changes were significant and detrimental (NAS, 2000). Consequently, to ensure the safety of new varieties developed by means of conventional and biotechnology methods, NAS (2000) recommended developing comprehensive databases of endogenous levels of potential nutritional and antinutritional compounds in plants and the plant tissues in which they are present.

Biotechnology-derived crops that are available commercially include those with protection from insects, tolerance to herbicides, sterility/fertility characteristics, modified seed fatty acid profiles, incorporation of the phytase enzyme, and resistance to viruses (AGBIOS Inc., 2001; U.S. FDA, 2001d; see Table 5). Available crops have completed regulatory reviews in many countries aimed at assessing the safety and nutritive value of these crops for farm animals. In the United States, Canada, and elsewhere worldwide, these reviews include evaluations of toxicity, compositional equivalence, bioavailability of important nutrients, and level of intake for novel constituents (U.S. FDA, 1992; MacKenzie, 2000).

Evaluations for these food and feed safety and nutritional factors include compositional comparisons for the resulting crop and commonly produced coproducts, as well as oral toxicity testing, functional assessment, and historical safety of the novel components. In addition, these evaluations consider the relative consumption levels of the new crop and for humans the potential for allergenicity is also considered. Results of acute oral toxicity testing

Table 5

Traits Expressed and Associated Genes That Have Been Incorporated into Crops Used Commonly as Animal Feeds and Have Been Commercialized[a]

Trait	Genetic element(s)	Gene source
Degradation of phytate in animal feed	Phytase	*Aspergillus niger* van Tieghem
Fertility restorer	Barstar	*Bacillus amyloliquefaciens*
Glufosinate herbicide tolerance	Phosphinothricin *N*-acetyltransferase (PAT)	*Streptomyces hygroscopicus* or *S. viridochromogenes*
Glyphosate herbicide tolerance	5-Enolpyruvylshikimate-3-phosphate synthase (EPSPS)	*Agrobacterium tumefaciens* strain CP4 or modified endogenous maize enzyme
	Glyphosate oxidoreductase	*Ochrobactrum anthropi* or *Achromobacter* sp. strain LBAA
Insect resistance	*cry*IAb, *cry*IAc, *cry*IIIA, *cry*IF, *cry*9C	*Bacillus thuringiensis*
Male sterility	Barnase ribonuclease	*Bacillus amyloliquefaciens*,
	DNA adenine methylase (DAM)	*Escherichia coli*
Modified seed fatty acid profile	δ-12 Desaturase	Soybean; coordinate suppression of endogenous gene
	12:0 Acyl carrier protein thioesterase	*Umbellularia californica* (California bay)
Oxynil herbicide tolerance	Nitrilase	*Klebsiella ozaenae* subsp. *ozaenae* or *Klebsiella pneumoniae* subsp. *ozaene*
Sulfonyl urea herbicide tolerance	Variant form of acetolactate synthase	*Arabidopsis thaliana* or *Nicotiana tabacum* (tobacco)
Virus resistance	Coat proteins Helicase/replicase	Potato virus Potato leafroll virus

Sources: AGBIOS Inc. (2001) and U.S. FDA (200d), by permission of AGBIOS Inc.

for a variety of novel proteins expressed by biotechnology-derived plants (Table 6) indicate no detrimental effects for mice that received levels of these proteins that far exceed normal consumption levels for farm animals (AGBIOS Inc., 2001). For example, testing levels in Table 6 for the CP4 5-enolpyruvylshikimate-3-phosphate synthase (EPSPS) and crystalline IIIA

Table 6

Results of Acute Oral Toxicity Studies for Novel Proteins Expressed in Biotechnology-Derived Plants: No Observed Effect Level (NOEL)[a]

Protein[a]	Crop(s)	NOEL (mg/kg body weight)
CryIAb	Corn	4000
CryIAc	Cotton, tomato	4200
CryIIAa	Cotton	3000
CryIIAb	Corn, cotton	3700
CryIIIA	Potato	5200
CP4 EPSPS	Canola, corn, cotton, soybean, sugar beet	572
mzEPSPS	Corn	350
NPTII	Cotton, potato, tomato	5000
GUS	Sugar beet	100
GOX	Canola	100
ACC deaminase	Tomato	602

[a]Cry, crystalline protein typically derived from *Bacillus thuringiensis*; EPSPS, 5-enolpyruvylshikimate-3-phosphate synthase; mzEPSPS, modified glyphosate-tolerant form of maize EPSPS; NPTII, neomycin phosphotransferase II; GUS, β-D-glucuronidase; GOX, glyphosate oxidase; ACC, 1-aminocyclopropane-1-carboxylic acid.
Source: AGBIOS Inc. (2001), by permission.

(CryIIIA) protein exceeded human consumption levels by 1000 and 1,000,000 times, respectively (AGBIOS Inc., 2001). To date, these regulatory evaluations have provided a comprehensive assessment of the safety and nutritional value to farm animals for biotechnology-derived crops.

As part of the process of scientific inquiry and product stewardship, studies have been completed to assess the nutritive value for livestock of biotechnology-derived crops. These studies have been completed for pest-protected crops such as those containing genes from *Bacillus thuringiensis* (*Bt*), herbicide-tolerant crops, and crops with modified compositional characteristics. Clark and Ipharraguerre (2001) provided an early review of studies completed using biotechnology-derived crops as feeds for livestock.

Pest-Protected Plants: *BT*

Pest-protected *Bt* plants developed by means of agricultural biotechnology and used commonly as feedstuffs for farm animals include *Bt* corn, *Bt* soybeans, *Bt* cotton, and *Bt* potatoes. All these biotechnology-derived crops except *Bt* potatoes have been compared with their conventional counterpart varieties as feeds for farm animals in studies reported in the scientific literature. Table 7 provides a summary of these reported studies, which have included broiler chickens, laying hens, beef cattle, dairy cattle, swine, and sheep that were fed *Bt* and conventional corn as grain, silage, whole-plant green chop, and stover; *Bt* soybeans as soybean meal; and *Bt* cottonseeds. The *Bt* varieties evaluated contained the following *cry* genes from *Bacillus thuringiensis*: *cry*IAb, *cry*IAc, *cry*IIAb, and *cry*9C. For some of these studies, several different varieties or sources of commercially available conventional feedstuffs were included as diet treatments to provide an appropriate context in which to evaluate animal performance for the biotechnology-derived varieties and their conventional genetic counterparts.

Performance, health, and nutrient utilization characteristics evaluated collectively in these studies include animal growth, intake of feed, survival and livability, gross efficiency of feed utilization, and digestibilities of energy, nitrogen components, organic matter, nitrogen-free extracts, fat, and fiber components. Conclusions for all these studies indicate no detrimental effects to performance, health, and nutrient utilization for farm animals consuming the tested biotechnology-derived crops (Faust and Miller, 1997; Aulrich *et al.*, 1998; Brake and Vlachos, 1998; Halle *et al.*, 1998; Daenicke *et al.*, 1999; Faust, 1999; Mayer and Rutzmoser, 1999; Faust and Spangler, 2000; Folmer *et al.*, 2002; Hendrix *et al.*, 2000; Mirales *et al.*, 2000; Russell *et al.*, 2000a, b; Weber *et al.*, 2000; Anonymous, 2001; Barrière *et al.*, 2001; Castillo *et al.*, 2001a,b; Gaines *et al.*, 2001a,b; al., 2001; Kerley *et al.*, 2001; Petty *et al.*, 2001a; Piva *et al.*, 2001a, b; Reuter *et al.*, 2001; Russell *et al.*, 2001a,b; Taylor *et al.*, 2001a; Weber and Richert, 2001; see Table 7). In addition, for several studies that also included a variety of sources for conventional varieties, researchers reported no differences in performance for animals fed different conventional sources of feedstuffs (Castillo *et al.*, 2001a,b; Kan *et al.*, 2001; Taylor *et al.*, 2001a). However, others found differences in animal performance for animals fed different conventional sources and no consistent differences for animals fed the biotechnology-derived and control varieties from the same genetic background (Folmer *et al.*, 2002; Weber *et al.*, 2000; Barrière *et al.*, 2001; Gaines *et al.*, 2001a,b; Weber and Richert, 2001). This set of finding implies that genetic differences among varieties are considerably more important for influencing animal performance than is the presence of the tested transgenes.

For three studies using pest-protected *Bt* crops for broiler chickens and swine, scientists reported slight advantages for growth parameters when animals were fed *Bt* instead of *Bt* near-isogenic counterparts (Brake and Vlachos, 1998; Piva *et al.*, 2001a,b; also see Table 7). These scientists also reported that levels for at least one mycotoxin (fumonisin) known to have

Table 7

Research Results for Completed Farm Animal Studies of Insect-Protected Crops Developed with Biotechnology

Crop	Event or variety studied[a]	Animal parameters evaluated	Results	Reference
Broiler chickens: corn grain, soybean meal				
Corn grain	*Bt*176/CryIAb	Performance, carcass characteristics	No differences: live weight, survival, yield of neck, fat pad, legs, thighs, wings, *pectoralis major*, ribs, and back. Slight advantage for *Bt*-fed birds: feed efficiency, yields of breast skin, *pectoralis minor* – likely due to lower mycotoxin levels in *Bt* corn.	Brake and Vlachos (1998).
Corn grain	*Bt*176/CryIAb	Performance, digestibility	No differences: final weight, gain, feed intake, feed efficiency, digestibility of protein.	Halle *et al.* (1998).
Corn grain	*Bt* – unspecified	Performance, digestibility	No differences: weight gain, feed efficiency, digestibility of true metabolizable energy and amino acids.	Mirales *et al.* (2000).
Corn grain	StarLink[b] *Bt* + glufosinate-tolerant/Cry9C + PAT	Detection of transgenic DNA proteins	No abnormalities: no transgenic DNA or proteins detected in samples of blood, liver, and muscle.	Anonymous (2001).
Corn grain	*Bt*176/CryIAb	Detection of transgenic DNA	No transgenic DNA: liver spleen, kidney, leg, and breast muscle.	Einspanier *et al.* (2001).
Corn grain	MON810 *Bt*/CryIAb	Performance, digestibility	Differences: average daily feed intake, gain/feed between corn genetic backgrounds, but not between *Bt* and non-*Bt* hybrids with the same genetic background. No differences: diet digestible energy.	Gaines *et al.* (2001a)

(continues)

Table 7 (*continued*)

Corn grain	MON810 *Bt*/CryIAb	Performance	No differences: daily gain, feed intake, feed efficiency. Slight advantage for *Bt*-fed birds (heavier live weight) attributed to lower fumonisin B_1 levels in *Bt* corn.	Piva *et al.* (2001b)
Corn grain	MON810 *Bt*/CryIAb, MON810 *Bt* + Roundup Ready®[c]/CryIAb + mzEPSPS	Performance, carcass characteristics	No differences: live weight, feed intake, feed efficiency. No biologically relevant differences: carcass yields, composition.	Taylor *et al.* (2001a)
Soybean meal	*Bt* – unspecified	Performance, carcass characteristics	No differences: final weight, feed conversion, carcass weight, carcass yield, composition.	Kan *et al.* (2001).
Laying hens: corn grain				
Corn grain	*Bt*176/CryIAb	Digestibility	No differences: digestible organic matter, crude protein, metabolizable energy.	Aulrich *et al.* (1998).
Corn grain	*Bt*176/CryIAb	Detection of transgenic DNA	No transgenic DNA: leg muscle, liver, spleen, kidney, egg, excrements.	Einspanier *et al.* (2001).
Swine: corn grain				
Corn grain	MON810 *Bt*/CryIAb	Performance, carcass characteristics, detection of transgenic DNA, protein	No differences: daily gain, daily feed intake, feed efficiency. No transgenic or endogenous plant source DNA or protein detected in loin tissue samples. No differences: *Bt* and isoline non-*Bt* corn for hot carcass weight, dressing percentage, carcass lean, loin eye area, color, marbling, firmness. Minor differences for alternative source of conventional corn: for some.	Weber *et al.* (2000); Weber and Richert 2001

(*continues*)

Table 7 (*continues*)

Crop	Event or variety studied[a]	Animal parameters evaluated	Results	Reference
Corn grain	MON810 *Bt*/CryIAb	Performance, digestibility	No differences: diet digestible energy coefficients between *Bt* and non-*Bt* hybrids, but coefficients differed between corn genetic backgrounds.	Gaines *et al.* (2001b)
Corn grain	MON810 *Bt*/CryIAb	Performance	No differences: feed intake, feed efficiency. Slight advantage for *Bt*-fed pigs: heavier live weight and greater daily gain, attributed to lower fumonisin B1 and deoxynivalenol levels in *Bt* corn.	Piva *et al.* (2001a).
Corn grain	*Bt* – unspecified	Performance, digestibility	No differences: daily gain, feed intake, feed efficiency, digestibilities of crude protein, nitrogen-free extracts, metabolizable energy.	Reuter *et al.* (2001).
Dairy cattle: freshly chopped whole plant, corn silage, corn grain, cottonseeds				
Freshly chopped whole plant	*Bt*176/CryIAb, *Bt*11/CryIAb	Detection of transgenic proteins, DNA; performance; milk composition.	No transgenic proteins or DNA detected. No differences: milk yield, composition, somatic cell count, intake.	Faust and Miller (1997); Faust (2000).
Corn silage	Pactol GB *Bt*	Performance, milk composition	No differences: milk and component yields, composition, somatic cell count, intake.	Mayer and Rutzmoser (1999).
Corn silage and grain	*Bt*11/CryIAb	Performance, digestibility, milk composition	No differences: feed intake, milk yield, composition, feed efficiency, ruminal pH, acetate-to-propionate ratio, digestion kinetics	Folmer *et al.* (2002).

(*continues*)

Table 7 *(continued)*

Corn silage	*Bt*176/CryIAb	Performance, milk composition,	No differences: milk yield, body weight gain, content of milk proteins, fatty acids, urea, cheese-making characteristics.	Barrière *et al.* (2001).
Corn grain	*Bt*176/CryIAb	Detection of transgenic DNA	No transgenic DNA: lymphocytes, blood, milk, chyme, excrements.	Einspanier *et al.* (2001).
Corn grain	MON810 *Bt*/CryIAb	Detection of transgenic DNA	No transgenic DNA detected (fragments ≥ 200 base pairs).	Phipps *et al.* (2001).
Cottonseeds	BollGard®[d] *Bt*/CryIAc, BollGardII®[d] *Bt*/ CryIAc + CryIIAb	Performance, milk composition	No differences: milk yield, body condition score, dry matter intake, intake of cottonseeds, and milk composition for fat, protein, lactose, nonfat solids, and urea.	Castillo *et al.* (2001a)
Cottonseeds	BollGard®[d] *Bt* + Roundup Ready®[c]/CryIAc + CP4 EPSPS	Performance, milk composition	No differences: milk yield, body condition score, dry matter intake, intake of cottonseeds, and milk composition for fat, protein, lactose, nonfat solids, and urea.	Castillo *et al.* (2001b)
Beef cattle: corn silage, corn grain, corn stover				
Corn silage	*Bt*176/CryIAb	Performance, carcass characteristics	No differences: gain, feed efficiency, intake of silage and protein, final weight, carcass weight, carcass yield, abdominal fat. Slight advantage for steers fed *Bt* silage: total intake of dry matter daily was lower.	Daenicke *et al.* (1999).
Corn silage	*Bt*11/CryIAb	Performance	No consistent differences in performance due to *Bt* gene. Hybrid genotype may influence performance.	Folmer *et al.* (2002).
Corn silage and grain	MON810 *Bt*/CryIAb	Performance, carcass characteristics	No differences: daily gain, intake, overall feed efficiency, carcass characteristics.	Hendrix *et al.* (2000), Petty *et al.* (2001a).

(continues)

Table 7 (*continues*)

Crop	Event or variety studied[a]	Animal parameters evaluated	Results	Reference
Corn grain	*Bt*176/CryIAb	Detection of transgenic DNA	No transgenic DNA: muscle, blood, liver, spleen	Einspanier *et al.* (2001).
Corn grain	StarLink[b] *Bt* + glufosinate-tolerant/Cry9C + PAT	Performance, carcass characteristics	No differences: daily gain, feed intake, feed efficiency, carcass yield grade, quality grade.	Kerley *et al.* (2001).
Corn stover	*Bt*11/CryIAb	Performance, grazing preference	No differences: daily gain, grazing preferences.	Folmer *et al.* (2002).
Corn stover	MON810 *Bt*/CryIAb	Grazing preference	Grazing patterns varied greatly: 46 and 54% of cows observed grazing *Bt* and control fields, respectively.	Hendrix *et al.* (2000).
Corn stover	*Bt*176/CryIAb, *Bt*11/CryIAb, MON810 *Bt*/CryIAb	Overwintering, grazing selectivity (fistulated steers)	No differences due to corn transgenes: hay required to maintain body condition score, nutritive value of crop residues. No differences (fistulated steers): fiber fractions, digestibility of forage selected.	Russell *et al.* (2000a,b; 2001a,b).
Sheep: corn silage				
Corn silage	*Bt*176/CryIAb, Pactol GB *Bt*	Digestibility	No differences: digestibility of organic matter, protein, fat, fiber, energy	Daenicke *et al.* (1999); Mayer and Rutzmoser, (1999).
Corn silage	*Bt*176/CryIAb	Digestibility	No differences: intake, digestibility of organic matter, crude fiber, neutral detergent fiber, energy, protein between *Bt* and non-*Bt* hybrids, but parameters differed between corn genetic backgrounds.	Barrière *et al.* (2001).

(*continues*)

Table 7 *(continued)*

In vitro composition – *corn silage*				
Corn silage	MON810 *Bt*/CryIAb	*In vitro* composition	No differences: 19 analytes, including measures of nitrogen, minerals, energy, fiber, and digestibility.	Faust (1999); Faust and Spangler (2000).

[a]Cry, crystalline protein typically derived from *Bacillus thuringiensis* (*Bt*); EPSPS, 5–enolpyruvylshikimate-3-phosphate synthase; mzEPSPS, modified glyphosate-tolerant form of maize EPSPS; PAT, phosphinothricin acetyltransferase.

[b]StarLink is a trademark of Aventis CropSciences.

[c]Roundup Ready is a registered trademark of Monsanto Technology LLC.

[d]Bollgard is a registered trademark of Monsanto Technology LLC.

detrimental effects on livestock performance and/or health were lowest for the biotechnology-derived variety. They concluded that performances advantages detected for animals fed *Bt* crops likely were due to this difference for mycotoxins. Others have reported lowest levels of *Fusarium* fungi and the fumonsin mycotoxin that they produce for corn grain from hybrids that express *Bt* proteins in kernels (Munkvold *et al.*, 1997, 1999). Mean total fumonisin concentrations for *Bt* varieties were 78 to 87% lower than levels for conventional genetic counterpart varieties when plants were infested manually with neonatal larvae of European corn borers (Munkvold *et al.*, 1999). For farm animals, the detrimental effects of mycotoxins can be severe and even fatal, ranging from nervous symptom disorders to gangrene in extremities to immune suppression to hepatotoxicity. In addition, animal species differ in their susceptibility to the effects of individual mycotoxins (Merck & Co., Inc., 1998). Thus, it is reasonable to expect improved health and performance for farm animals fed feeds containing lowest levels of mycotoxins for which they are susceptible as indicated for several biotechnology-derived crop studies (Brake and Vlachos, 1998; Piva *et al.*, 2001a,b). Under conditions that favor mycotoxin development, lower levels of mycotoxins for pest-protected *Bt* crops may be an added health and performance benefit when these grains are fed to farm animals.

Pest-Protected Plants: Viral Resistant

It is expected that a wider variety of viral-resistant crops developed by means of biotechnology will become available in the near future. Currently, biotechnology-derived potatoes are the example of viral-resistant crops used most frequently as feeds for farm animals. Potatoes represent an important staple crop worldwide. To date, viral-resistant potatoes have not been evaluated specifically as feeds for farm animals. Viral-resistant crops such as potatoes commonly incorporate genes for viral coat proteins. These same coat proteins are present naturally in virally infected crops, and animals that consume these infected crops are exposed to these gene products. Thus, no specific safety-related issues are expected for farm animals that are fed biotechnology-derived crops that include these viral coat proteins, and these viral-resistant crops are expected to be as safe as their conventional counterparts that have been infected by the target viruses. The National Academy of Sciences committee on genetically modified pest-protected plants has reached similar conclusions of comparable safety for viral coat proteins present in biotechnology-derived and conventional crops (NAS, 2000).

Herbicide-Tolerant Plants

Biotechnology-derived plants that are tolerant of herbicides and are used commonly as feedstuffs for farm animals include glufosinate-tolerant canola, corn, soybeans, and sugar beets; glyphosate-tolerant canola, corn, cotton, soybeans, and sugar beets; canola and cotton that are tolerant to the herbicide Oxynil; and cotton tolerant to the herbicide sulfonyl urea. Table 5 provides additional details about sources for these traits in biotechnology-derived crops (AGBIOS Inc., 2001; U.S. FDA, 2001d). Sources for several of the genes that confer herbicide tolerance are bacteria; however, other varieties were developed from a modified version of a glyphosate tolerance gene that occurs naturally in some varieties of maize (U.S. FDA, 2001d; see Table 5).

Studies to evaluate the wholesomeness of herbicide-tolerant crops as feeds for farm animals have been completed and reported for broiler chickens (glyphosate-tolerant corn and soybean meal), swine (glufosinate-tolerant corn and sugar beets and glyphosate-tolerant corn), lactating dairy cattle (glyphosate-tolerant corn, cottonseeds, and soybeans), beef cattle (glyphosate-tolerant corn), sheep (glyphosate-tolerant sugar beets, fodder beets, and beet pulps), and catfish (glyphosate-tolerant soybean meal). Details of these studies are in Table 8. Several studies listed in Table 8 have used crops that are both herbicide tolerant and pest protected (Anonymous, 2001; Castillo *et al.*, 2001b; Kerley *et.al.*, 2001, Taylor *et al.*, 2001b). Factors for farm animals that were studied included growth rates, intake of feed, survival and livability, gross efficiency of feed utilization, ruminal volatile fatty acids, and digestibilities of energy, nitrogen components, organic matter, nitrogen-free extracts, fat, and fiber components. Reports from one study indicated that at least one performance measure was highest for lactating dairy cattle and catfish fed glyphosate-tolerant soybeans and meal in comparison to animals fed conventional genetic counterpart varieties (Hammond *et al.*, 1996). However, in this same study, the overwhelming majority of parameters were not different for biotechnology-derived and control-fed groups, and in total, evidence suggests that these results did not reflect biologically important differences for the soybeans.

Other researchers noted differences in diet digestible energy coefficients between swine fed corn from different genetic backgrounds, but not between those fed biotechnology-derived and conventional genetic counterpart varieties from the same genetic background (Gaines *et al.*, 2001b). The overall conclusion from all performance, health, and nutrient utilization studies for farm animals is that for the varieties studied, there are no biologically relevant differences between herbicide-tolerant crops and their non-biotechnology-derived genetic counterpart varieties as feeds for farm animals (Hammond *et al.*, 1996; Padgett *et al.*, 1996; Böhme and Aulrich, 1999; Donkin *et al.*, 2000; Sidhu *et al.*, 2000; Anonymous, 2001; Castillo *et al.*, 2001b; Cromwell

Table 8

Research Results for completed Farm Animal Studies of Herbicide-Tolerant Crops developed with Biotechnology

Crop	Event or variety studied[a]	Animal parameters evaluated	Results	Reference
Broiler chickens: corn grain, soybean meal				
Corn grain	GA21 glyphosate-tolerant/mzEPSPS	Performance	No differences: final weight, feed efficiency, fat pad weight.	Sidhu *et al.* (2000).
Corn grain	StarLink[b] *Bt* + glufosinate-tolerant/Cry9C + PAT	Detection of transgenic DNA, proteins	No abnormalities. No transgenic DNA or proteins detected in samples of blood, liver, and muscle	Anonymous (2001).
Corn grain	Roundup Ready®[c] glyphosate-tolerant/mzEPSPS	Performance, digestibility	No differences: daily gain, daily feed intake, feed efficiency, apparent metabolizable digestibility coefficients.	Gaines *et al.* (2001a).
Corn grain	MON810 *Bt* + Roundup Ready®[c]/CryIAb + mzEPSPS, NK603 glyphosate-tolerant/CP4 EPSPS	Performance, carcass yields, composition	No differences: live weight, feed intake, feed efficiency. No biologically relevant differences: carcass yields, composition.	Taylor *et al.* (2001a,b).
Soybean meal	Glyphosate tolerant/CP4 EPSPS	Performance, carcass composition	No differences: daily gain, daily feed intake, feed efficiency, livability, breast and fat pad weights	Hammond *et al.* (1996).
Soybean meal	Roundup Ready®[c] glyphosate-tolerant/CP4 EPSPS	Detection of transgenic DNA	No transgenic DNA detected: samples of meat, skin, duodenum, liver.	Khumnirdpetch *et al.* (2001).

(*continues*)

Table 8 *(continued)*

Chicken (laying hens): soybeans, soybean meal				
Soybeans, meal	Roundup Ready®[c] glyphosate-tolerant/CP4 EPSPS	Detection of transgenic protein	No transgenic protein detected: samples of whole egg, egg white, liver, feces. Transgenic protein detected in samples of raw soybeans, soybean meal, complete diet.	Ash *et al.* (2000).
Swine: corn grain, sugar beets				
Corn grain	Glufosinate-tolerant/PAT	Digestibility	No differences: digestibilities of organic matter, crude protein, nitrogen-free extract.	Böhme and Aulrich (1999).
Soybean meal	Roundup Ready®[c] glyphosate- tolerant/CP4 EPSPS	Performance, carcass characteristics, detection of transgenic protein	No differences: daily gain, daily feed intake, feed efficiency, calculated percentage carcass lean, scanned backfat and longissimus area. No differences: carcass characteristics. No transgenic protein detected in samples of loin tissue.	Cromwell *et al.* (2001).
Corn grain	Roundup Ready®[c] glyphosate-tolerant/ mzEPSPS	Performance, digestibility	Differences: diet digestible energy coefficients between corn genetic backgrounds, but not between biotech and non biotech hybrids with the same genetic background.	Gaines *et al.* (2001b).
Corn grain	GA21 glyphosate-tolerant/ mzEPSPS	Performance, carcass composition	No differences: daily gain, daily feed intake, feed efficiency, chemical composition of muscle.	Stanisiewski *et al.* (2001).
Sugar beets	Glufosinate-tolerant/PAT	Digestibility	No biologically relevant differences: digestibilities of organic matter, crude protein, nitrogen-free extract.	Böhme and Aulrich (1999).

(continues)

Table 8 (*continued*)

Crop	Event or variety studied[a]	Animal parameters evaluated	Results	Reference
Dairy cattle: corn silage, corn grain, soybeans, soybean meal				
Corn silage and grain	Roundup Ready®[c] glyphosate-tolerant	Performance, milk composition	No differences: dry matter intake, milk and component yields, composition, feed efficiency, somatic cell count.	Donkin *et al.* (2000).
Cottonseeds	BollGard®[d] *Bt* + Roundup Ready®[c]/CryIAc + CP4 EPSPS	Performance, milk composition	No differences: milk yield, body condition score, dry matter intake, intake of cottonseeds, and milk composition for fat, protein, lactose, nonfat solids, and urea	Castillo *et al.* (2001b).
Soybeans: raw	Glyphosate-tolerant/CP4 EPSPS	Performance, digestibility, milk composition	No differences: milk yield, composition, somatic cell count, dry matter intake, feed efficiency, dry matter digestibility, nitrogen balance, ruminal volatile fatty acids. Slight advantage for cows fed transgenic soybeans: fat corrected milk.	Hammond *et al.* (1996).
Soybean meal	Glyphosate tolerant/CP4 EPSPS	Detection of transgenic DNA	No transgenic DNA detected: (fragments ≥ 180 base pairs)	Phipps *et al.* (2002).
Soybean meal	Glyphosate tolerant/CP4 EPSPS	Detection of transgenic DNA	No transgenic DNA detected: blood, leukocytes, milk. Transgenic DNA detected in soybean meal.	Klotz and Einspanier (1998).
Beef cattle: corn silage, corn grain				
Corn grain	StarLink[b] *Bt* + glufosinate-tolerant/Cry9C + PAT	Performance carcass characteristics	No differences: daily gain, feed intake, feed efficiency, carcass yield grade, quality grade.	Kerley *et al.* (2001).
Corn silage and grain	Roundup Ready®[c] glyphosate-tolerant	Performance, carcass characteristics	No differences: daily gain, intake, overall feed efficiency, carcass characteristics.	Petty *et al.* (2001b).

(*continues*)

Table 8 (*continued*)

Sheep: sugar beets, fodder beets, beet pulp				
Sugar beets, fodder beets, beet pulp	Roundup Ready®[c] glyphosate-tolerant/CP4 EPSPS	Digestibility	No differences: digestibilities of energy and main nutrients.	Hvelplund and Weisbjerg (2001).
Catfish: soybean meal				
Soybean meal	Glyphosate-tolerant/CP4 EPSPS	Performance, carcass composition	No differences: feed efficiency, livability, fillet content of moisture, fat, protein, ash. Advantage for fish fed one of the transgenic soy meals: feed intake and corresponding gain and final weight.	Hammond *et al.* (1996).
Composition: soybeans, soybean coproducts				
Soybeans, soybean coproducts	Glyphosate-tolerant/CP4 EPSPS	Composition	No biologically relevant differences: proximates, amino acids, fatty acids, trypsin inhibitor, urease activity, lectin activity, isflavones, phytate, raffinose saccharides.	Padgette *et al.* (1996).

[a]Cry, crystalline protein typically derived from *Bacillus thuringiensis* (*Bt*); EPSPS, 5-enolpyruvylshikimate-3-phosphate synthase; mzEPSPS, modified glyphosate tolerant form of maize EPSPS; PAT, phosphinothricin acetyltransferase.
[b]StarLink is a trademark of Aventis CropSciences.
[c]Roundup Ready is a registered trademark of Monsanto Technology LLC.
[d]Bollgard is a registered trademark of Monsanto Technology LLC.

et al., 2001; Gaines *et al.*, 2001a,b; Hvelplund and Weisbjerg, 2001; Kerley *et al.*, 2001; Petty *et al.*, 2001b; Stanisiewski *et al.*, 2001; Taylor *et al.*, 2001a,b). This conclusion may be expected because Padgett *et al.* (1996) reported finding no biologically important differerences for proximate analyses, amino acids, fatty acids, trypsin inhibitor, urease activity, lectin activity, isflavones, phytate, and raffinose saccharides of glyphosate-tolerant and conventional parent varieties of soybeans and their respective coproducts.

Plants with Modified Nutritive Value

For feed crops that are commonly fed to farm animals, several new biotechnology-derived crops with enhanced nutritive value are available; these include canola with the phytase enzyme for degrading plant phytate, soybeans with high levels of oleic acid, and canola with high levels of laurate. Feeds in this first generation of biotechnology-derived plants offer to animal agriculture a glimpse of the future crops that can be developed to improve health and performance of farm animals, healthfulness of their products for humans, and the sustainability of animal agricultural systems. For example, feed crops that contain phytase allow farm animals such as swine and poultry to utilize a larger percentage of phytate-bound phosphorus in crops, thus reducing the overall amount of phosphorus in animal waste for disposal. Experimental varieties of corn low in phytate were reported to increase substantially plant phosphorus utilization by pigs and to reduce by 17.5 to 23.4% the supplemental dicalcium phosphate added to diets with no associated differences in pig growth rates, feed efficiency, or serum osteocalcin—an indicator of bone turnover (Frank *et al.*, 2001; Klunzinger *et al.*, 2001).

Feed crops with modified fatty acid profiles may be beneficial for enhancing the quality of products from farm animals. By modifying the fatty acid profiles of diets for animals such as swine, poultry, sheep, and beef and dairy cattle, it is possible to improve shelf life for milk, meat, and eggs, to produce products such as butter that differ in melting points, to decrease oxidative rancidity for animal fats, and to increase content of conjugated linoleic acid in animal products. Biotechnology and traditional crop breeding techniques are being employed to develop soybeans, corn, and other crops with modified fatty acid profiles and total oil contents (O'Quinn *et al.*, 2000; Owens and Soderlund 2000), and most of the available crops have been developed by means of traditional techniques. Biotechnology-derived soybeans with high oleic acid content are available commercially and were fed as soybean meal in diets to weanling swine from 11.4 to 20.4 kg of body weight. Overall, pig growth rates, feed intake, and feed efficiency did not differ for pigs fed meal from soybeans high in oleic acid (processed at about 80–85°C), control genetic counterpart varieties, and commercial sources (Loughmiller *et al.*, 1998). Similarly, O'Quinn *et al.* (2000) reported that digestibilities of nutrients including amino acids, gross energy, crude protein, and dry matter were similar when young pigs were fed diets containing varieties of corn high in oil and high in both oil and lysine. For biotechnology-derived crops evaluated to date, bioavailability of nutrients has been comparable to bioavailability for conventional varieties.

Additional Implications of Biotechnology-Derived Plants for Feed and Food Safety

Pest-protected and herbicide-tolerant crops likely have additional indirect benefits for safety of feed and ultimately for food chains. For example, crops that are resistant to pests offer the opportunity to reduce reliance on chemical-based pesticides. Reducing the use of chemical pesticides may decrease pesticidal residues in feed, food, and water sources, and further may limit exposure of farm animals and humans to these compounds. Cited advantages for adoption of herbicide-tolerant varieties of crops include overall reductions in the number of herbicide applications for crops and for some events, the ability to use herbicides with fewer potential impacts for the environment. Reduced exposure of farm animals to chemical pesticides has the potential to reduce residues from these substances in animal products namely milk, meat, and eggs.

Some authors argue that reduction of pesticide use attributed to *Bt* corn is tenuous at best because few growers report using foliar-applied pesticides for control of European corn borer (Obrycki *et al.*, 2001). However, others define measurable reductions in pesticide treatments for *Bt* corn and *Bt* cotton in the United Sates; adoption of *Bt* corn in the United States was credited with a reduction in hectare treatments with foliar-applied pesticides of 0.9 and 0.4 million ha in 1998 and 1999, respectively (Carpenter and Gianessi, 2001). In the United States, cotton is a crop for which pest losses can be high; thus pesticide use typically is important for maintaining high yields and good quality. Decreases in pesticide hectare treatments attributed to the adoption of biotechnology-derived *Bt* cotton in the United States are significant; reductions were 3.6 and 6.1 million ha treatments for 1998 and 1999, respectively (Carpenter and Gianessi, 2001; see Table 9). Aggregate decreases in pesticide hectare treatments attributed to adoption of herbicide-tolerant cotton (Roundup Ready® and Bromoxynil-tolerant varieties) and Roundup Ready® soybeans were 0.73 and 6.5 million ha, respectively for 1998 and 0.4 and 7.7 million ha for 1999 (Carpenter and Gianessi, 2001; see Table 9). Indirectly, these advantages may have implications for the safety of the feed and food chains and water supplies.

Table 9

Aggregate Impacts of Decreases in Pesticide Hectare Treatments Attributed to the Planting of Biotechnology-Derived Varieties in the United States (vs Planting No Biotechnology-Derived Varieties)[a]

	Hectares treated ($\times 10^6$)	
Biotechnology-derived crop	1998	1999
Bt corn[b]	0.81	0.40
Bt cotton	3.6	6.1
Herbicide-tolerant cotton[c]	0.73	0.40
Roundup Ready® soybeans[d]	6.5	7.7

[a]An application hectare is defined as the number of different active ingredients applied per hectare times the number of repeat applications. Thus, each of the following examples represent two applications: (1) one spraying to simultaneously apply two active ingredients and (2) two separate sprayings of a single ingredient.
[b]Insecticide reductions represented as number of hectares treated.
[c]Includes varieties tolerant of oxynil and glyphosate herbicide.
[d]Roundup Ready® is a registered trademark of Monsanto Technology LLC.
Source: Carpenter and Gianessi (2001).

In addition, biotechnology-derived crops may provide somewhat greater safety to farm animals and to humans because these crops deter development of fungi and resulting secondary metabolites, mycotoxins. Mycotoxins are produced by opportunistic fungi that can infect crops such as corn, wheat, barley, rice meal, legumes, and sorghum grain, and production of mycotoxins is influenced by environmental factors such as temperature, humidity, drought stress, and rainfall during the preharvest and harvest periods (Orriss, 1997; U.S. FDA, 2001b). Other factors that may predispose plants to infection by mycotoxin-producing fungi include injury caused by insects and other plant pathogens (Munkvold and Desjardins, 1997).

Nearly 500 mycotoxins have been identified, but their total impact on health for farm animals and humans has not been fully characterized. However, biological activity for several important mycotoxins has been studied in animal models, and several of these are important for animal agriculture. Mycotoxins such as aflatoxin B1, ochratoxin A, and fumonisin B1 may be carcinogenic; zearalenone and I and J zearalenols may possess estrogenic activity; ochratoxins and citrinin may be nephrotoxic, and immunosuppressive mycotoxins may include aflatoxin B1, ochratoxin A, and T-2 toxin (Orriss, 1997). Different species of farm animals are more susceptible to the effects of specific mycotoxins, likely owing to species differences in metabolism of the toxins. Meat, milk, eggs, and visceral organs from farm animals fed mycotoxin-infected feeds can contain detectable levels of mycotoxins or their metabolites (Orriss, 1997). Although levels for these toxins found in animal food products are considerably lower than levels ingested by farm animals, the presence of carcinogenic mycotoxin residues in animal products may pose a concern for human health (Orriss, 1997). Organizations such as the U.S. FDA and FAO/WHO define maximum daily intakes of a variety of mycotoxins for humans and farm animals.

Fumonisins are mycotoxins produced commonly on corn by the fungi *Fusarium* spp., and Munkvold and Desjardins (1997) and U.S. FDA (2001a,b) have reviewed health implications of fumonisins for animals. Fumonisins have been associated with leukoencephalomalacia in horses, donkeys, mules, and rabbits; pulmonary edema in swine; liver damage in all animals studied; kidney lesions; and heart failure in swine (Munkvold and Desjardins, 1997; U.S. FDA, 2001a,b). Chronic feeding of fumonisins reportedly produced liver and kidney cancer in experimental animals and may be implicated in the high rates of esophageal cancer in human populations recorded in regions of South Africa and China (Munkvold and Desjardins, 1997; U.S. FDA, 2001b). As a result of concerns for health and food safety, the FDA defined recommended maximum levels of fumonisins in corn products for humans (2–4 ppm) and farm animals (5–100 ppm) (U.S. FDA, 2001a–c).

However, *Bt* corn may be less susceptible to infection by *Fusaria*; Munkvold *et al.* 1997, 1999) reported lowest levels of *Fusarium* fungi and fumonisin mycotoxin for corn grain from hybrids that express *Bt* proteins in kernels. When corn plants were infested manually with neonatal larvae of European corn borers, mean total fumonisin concentrations for *Bt* varieties were 78 to 87% lower than levels for conventional genetic counterpart varieties (Munkvold *et al.*, 1999). Additionally, after reviewing current and future strategies to reduce fumonisin levels in corn, Munkvold and Desjardins (1997) concluded that agricultural biotechnology appears to be the "most attractive" alternative. Thus, biotechnology-derived crops may enhance food safety for farm animals and ultimately for humans, through reduced levels of undesirable mycotoxins.

COMPOSITION OF MEAT, MILK, AND EGGS FROM FARM ANIMALS CONSUMING BIOTECHNOLOGY-DERIVED CROPS

Summary

The milk, meat, and eggs from farm animals fed the commercially available biotechnology-derived crops and their conventional counterparts are compositionally comparable and cannot be differentiated. Thus, products produced by farm animals fed biotechnology-derived crops are as wholesome, safe, and nutritious as similar products produced by animals fed conventional crops.

Introduction

The composition of diets for farm animals can have impacts on the composition of milk, meat, and eggs. For example, the body composition of growing pigs fed diets with higher protein-to-carbohydrate ratios tends to have a higher percentage of lean tissue versus fat. In fact, animal agriculture is expecting to enhance the nutritive value of milk, meat, and eggs as a result of the use of new biotechnology-derived crops with modified composition. However for the biotechnology-derived crops that are available today and are used frequently as animal feeds, no differences in animal products are expected for farm animals consuming conventional and these biotechnology-derived varieties.

The pest-protected and herbicide-tolerant crops now available essentially are compositionally comparable to their conventional genetic counterparts except with respect to the introduced traits. In addition, no biological activity for farm animal species has been identified for the novel proteins introduced into currently available biotechnology-derived crops, and the introduced DNA in these crops is similar to other dietary sources of DNA. Further, the introduced DNA in biotechnology-derived crops represents a relatively insignificant percentage of the total DNA consumed daily by farm animals. Moreover, the United Nations Food and Agriculture Organisation, the World Health Organisation, and the U.S. Food and Drug Administration have stated that the consumption of DNA from all sources, including introduced DNA in biotechnology-derived crops, presents no health or safety concerns (FAO/WHO, 1991; U.S. FDA, 1992). These conclusions are based on the long history of safety associated with the consumption of DNA by farm animals and humans. Consequently when these biotechnology-derived crops are fed to farm animals, no impact on the composition of animal-produced products, is expected, and results from several studies are available for commercial pest-protected and herbicide-tolerant crops.

Pest-Protected Plants: *Bt*

Studies to investigate the composition of milk, meat, and eggs from farm animals fed pest-protected *Bt* crops have been completed for broiler chickens, laying hens, beef cattle, and lactating dairy cattle fed *Bt* corn, *Bt* cottonseeds, and *Bt* soybean meal. Several reports have focused on consumer-related quality parameters for products from these animals (Faust and Miller, 1997; Brake and Vlachos, 1998; Daenicke *et al.*, 1999; Mayer and Rutzmoser, 1999; Folmer *et al.*, 2000b; Hendrix *et al.*, 2000; Weber *et al.*, 2000; Anonymous, 2001; Barrière *et al.*, 2001; Castillo *et al.*, 2001a,b; Kan *et al.*, 2001; Kerley *et al.*, 2001; Petty *et al.*, 2001a; Taylor *et al.*, 2001a; Weber and Richert, 2001). Findings for these studies (Table 7) indicate no biologically relevant differences in composition of products from farm animals fed *Bt* crops.

When broiler chickens, beef cattle, and swine were fed varieties of *Bt* corn and their conventional control varieties, no biologically relevant differences for percentages of different retail cuts and the composition of these cuts were identified (Brake and Vlachos, 1998; Daenicke *et al.*, 1999; Hendrix *et al.*, 2000; Weber *et al.*, 2000; Kerley *et al.*, 2001; Petty *et al.*, 2001a; Taylor *et al.*, 2001a; Weber and Richert, 2001). Kan *et al.* (2001), who fed *Bt* and conventional soybean meal in rations to broiler chickens, reported no differences in carcass weights, carcass yields, and composition. Milk produced by dairy cows fed *Bt* corn (silage, grain, whole-plant green chopped material) in diets did not differ for number of somatic cells and concentrations of milk fat, protein, lactose, total solids, nonfat solids, urea, and individual fatty acids (Faust and Miller, 1997; Mayer and Rutzmoser, 1999; Folmer *et al*, 2000b). Number of somatic cells in milk is an indirect measure of health of the udder and presence of infectious organisms in the udder. Similar findings of no differences were reported for milk from dairy cows fed *Bt* and conventional cottonseeds (Castillo *et al.*, 2001a,b), and Barrière *et al.* (2001) reported finding no differences for several cheese-making measures of milk from cows fed conventional control and *Bt* varieties of corn.

HERBICIDE-TOLERANT PLANTS

Assessments of the composition of tissues and products from animals fed herbicide-tolerant crops have included broiler chickens, laying hens, dairy cattle, beef cattle, and catfish. Studies used diets containing glufosinate-tolerant corn (Anonymons, 2001; Kerley *et al.*, 2001) and glyphosate-tolerant corn (Donkin *et al.*, 2000; Petty *et al.*, 2001b; Stanisiewski *et al.*, 2001; Taylor *et al.*, 2001a,b), cottonseeds (Castillo *et al.*, 2001b), and soybeans and meal (Cromwell *et al.*, 2001; Hammond *et al.*, 1996; Klotz and Einspanier, 1998; Ash *et al.*, 2000; Khumnirdpetch *et al.*, 2001). As outlined in Table 8, no biologically relevant differences were detected for percentages of different retail cuts and the composition of these cuts when broiler chickens (Taylor *et al.*,

2001a,b), swine (Stanisiewski *et al.*, 2001), and beef cattle (Petty *et al.*, 2001b) were fed glyphosate-tolerant corn and when beef cattle were fed glufosinate-tolerant corn (Kerley *et al.*, 2001). Similarly, Hammond *et al.* (1996) and Cromwell *et al.* (2001) reported finding no differences in carcass-related measures when broiler chickens and catfish (Hammond *et al.*, 1996) consumed soybean meal derived from glyphosate-tolerant soybeans. When glyphosate-tolerant corn (Donkin *et al.*, 2000), cottonseeds (Castillo *et al.*, 2001b), raw soybeans (Hammond *et al.*, 1996), and respective control varieties of these crops were fed to lactating dairy cows, number of somatic cells and concentrations of milk fat, protein, lactose, nonfat solids, and urea did not differ.

DETECTING PLANT SOURCE PROTEINS AND DNA IN ANIMAL PRODUCTS

Summary

Despite detection of fragments of endogenous plant chloroplast-specific genes in some samples from farm animals, current information suggests that these fragments (regardless of source) do not pose health risks when consumed. No intact or immunologically reactive fragments of transgenic plant proteins or fragments of transgenic plant source DNA have been detected in samples of milk, meat, eggs, skin, duodenum, leukocytes, lymphocytes, blood, organ tissue, duodenal fluid, and excrement from animals fed biotechnology-derived crops.

Detection of Transgenic Plant Proteins

Researchers have investigated whether introduced proteins from biotechnology-derived plants can be detected in tissues and fluids from farm animals fed these crops. In fact, it is difficult to track absorbed amino acids from digested dietary protein (endogenous and introduced) in meat, milk, and eggs because these proteins are broken down relatively rapidly into amino acids, ammonia, and carbon skeletons, ultimately undergoing transamination, urea production, or protein synthesis. However, to investigate the possibility for detecting these novel plant proteins, samples of animal tissues and products from farm animals fed biotechnology-derived and control varieties were evaluated for the presence of introduced *Bt* proteins. These effects were studied for *Bt* corn in dairy cattle (Faust and Miller, 1997; Faust, 2000) and swine (Weber and Richert, 2001); for glufosinate-tolerant corn in broiler chickens (Anonymous, 2001); and glyphosate-tolerant soybeans in broilers (Ash *et al.*, 2000) and swine (Cromwell *et al.*, 2001).

Milk samples from dairy cows fed fresh chopped, whole-plant corn from one genetic control and two different *Bt* hybrids were evaluated by means of a CryIAb sandwich immunoassay (Faust and Miller, 1997; Faust, 2000). The positive control used for this study consisted of duplicate milk samples spiked with purified *Bt* proteins. Transgenic proteins were detected in all spiked samples, but no *Bt* plant proteins were detected in normal (unspiked) milk samples collected during this study (Faust and Miller, 1997; Faust, 2000). The Japanese government commissioned a study for broiler chickens fed StarLink corn (*Bt* and glufosinate tolerant) and found no biotechnology-derived proteins in samples of blood, liver, and muscle from these birds (Anonymous, 2001). Weber and Richert (2001) reported that a competitive immunoassay was used for samples of loin muscle tissue from pigs fed *Bt* corn; no intact or immunologically reactive fragments of CryIAb protein were detected. Further, biotechnology-derived proteins were not detected in samples of whole egg, egg white, liver, and fecal samples from laying hens fed glyphosate-tolerant soybean meal (Ash *et al.*, 2000) and samples of loin tissue from swine fed glyphosate-tolerant corn (Cromwell *et al.*, 2001). It is important to note that sandwich immunoassays are reliable for detecting intact proteins, whereas competitive immunoassays can detect intact protein and immunologically reactive fragments of the target protein. Additional details for these studies are provided in Tables 7 and 8.

Because the novel proteins in biotechnology-derived crops have no known biological function in farm animal species, it is expected that these proteins and other dietary proteins largely are broken down during digestion into peptides and amino acids. In fact, stability of transgenic proteins during digestion is evaluated during the regulatory process for biotechnology-derived crops. Results in Table 10 indicate that transgenic proteins for biotechnology-derived crops are unstable when exposed to *in vitro* conditions that simulate the gastric environment, indicating that these proteins are broken down relatively quickly in the gastrointestinal tract. Unique proteins in one *Bt* corn variety are relatively more stable to digestion when tested in simulated conditions, however registration for this product has been voluntarily cancelled in the U.S. (U.S. EPA, 2001). Results from several studies (Faust and Miller, 1997; Ash *et al.*, 2000; Faust, 2000; Anonymous, 2001; Cromwell *et al.*, 2001; Weber and Richert, 2001) corroborate results from the regulatory process in that no introduced proteins were identified in products and tissues from farm animals fed biotechnology-derived crops.

Table 10

Stability of Novel Proteins in Biotechnology-Derived Plants to Digestion in Simulated Gastric Fluid

Protein[a]	Stability (s)
CryIA	30
CryIIA	<15
CryIIIA	<15
CP4 EPSPS	<15
mzEPSPS	<15
GOX	<15
GUS	<15
NPTII	<10

[a]Cry, crystalline protein typically derived from *Bacillus thuringiensis*; EPSPS, 5-enolpyruvylshikimate-3-phosphate synthase; mzEPSPS, modified glyphosate-tolerant form of maize EPSPS; NPTII, neomycin phosphotransferase II; GUS, β-D-glucuronidase; GOX, glyphosate oxidase.

Source: AGBIOS Inc. (2001), by permission.

Detection of Plant Source DNA

The gastrointestinal tracts of farm animals and humans are exposed daily to exogenous DNA; consequently the U.S. FDA and World Health Organisation have concluded that consumption of DNA presents no health concerns (FAO/WHO, 1991; U.S. FDA, 1992). Additionally, introduced DNA in biotechnology-derived plants is compositionally comparable to endogenous DNA, and during digestion, ingested DNA is hydrolyzed to nucleotides or smaller constituents relatively rapidly. Consequently, it is extremely difficult to identify large fragments of consumed DNA in tissues and fluids from farm animals. As a further assessment of the safety for biotechnology-derived crops, several studies have been completed to evaluate tissues and/or products for the presence of plant source DNA when dairy cows (Klotz and Einspanier, 1998; Faust, 2000; Einspanier *et al.*, 2001; Phipps *et al.*, 2001, 2002), growing bulls (Einspanier *et al.*, 2001), pigs (Weber and Richert, 2001), broiler chickens (Anonymous, 2001; Einspanier *et al.*, 2001; Khumnirdpetch *et al.*, 2001), and laying hens (Einspanier *et al.*, 2001) are fed diets containing biotechnology-derived crops.

Samples of milk from the study reported by Faust and Miller (1997) also were evaluated for the presence of plant source proteins by means of the polymerase chain reaction (PCR). Plant gene fragments studied were the two

Bt genes found in *Bt* corn varieties fed (*Bt*11 and *Bt*176 constructs), maize zein gene, which is an endogenous multicopy gene in corn, and the maize amylose-extender gene, a single-copy endogenous corn gene. Two endogenous bovine genes served as controls, the bovine actin gene (multicopy) and the bovine interleukin 2 receptor gamma gene (single copy). No fragments of the *Bt* *cry*1Ab genes were found in any milk samples collected during the original study (Faust, 2000). Further, no fragments of any of the plant genes tested were detected, and as expected, the two bovine genes that served as positive controls were detected in these milk samples (Faust, 2000). Other researchers were unable to detect transgenic DNA in milk from cows fed *Bt*-corn containing diets (Phipps *et al.*, 2001) and in milk from those fed soybean meal from herbicide-tolerant soybeans (Phipps *et.al.*, 2002). Weber and Richert (2001) reported similar findings for samples of loin tissue from pigs fed *Bt* and conventional corn by using PCR, followed by Southern blot analysis for evaluating samples for fragments ($\sim$ 200 bp) of the *cry*IAb and *shrunken-2* (endogenous single copy) genes from corn. The endogenous porcine pre-prolactin gene served as the positive control. Limit of detection for this study was 1 to 2.5 pg of target DNA per microgram of input DNA. No *Bt* DNA fragments and no fragments of the *shrunken-2* gene were detected in these samples of loin tissue; fragments of the porcine pre-prolactin were detected in all samples (Weber and Richert, 2001). Khumnirdpetch *et al.* (2001) reported finding no transgenic DNA in samples of meat, skin, duodenum, and liver from broilers fed diets containing glyphosate-tolerant soybean meal Further, no transgenic DNA was found in samples from broiler chickens fed StarLink corn (Anonymous, 2001).

Einspanier *et al.* (2001) used PCR and Light Cycler "real-time" PCR to evaluate samples of tissues and fluids from dairy cows, growing bulls, and laying hens that had been fed *Bt* corn. No fragments of the *cry*IAb transgene were detected in any samples collected from animals in this study (Einspanier *et al.*, 2001). Small fragments of the coding region for a highly abundant chloroplast-specific gene ($tRNA_{leu}$) were detected in duodenal fluid and lymphocytes from dairy cows and in muscle, liver, spleen, and kidney tissue from hens (Einspanier *et al.*, 2001). When dairy cattle were fed diets containing a variety of glyphosate-tolerant soybeans, no transgenic DNA was detected in samples of milk, blood samples, and leukocytes. However, these researchers detected fragments of an endogenous chloroplast-specific gene, chloroplast ribulose-1,5-bisphosphate carboxylase/oxygenase, in blood and leukocyte samples, but not in milk (Klotz and Einspanier, 1998). Also, see Tables 7 and 8 for details of experiments to detect transgenic DNA and proteins in animal products.

As noted earlier, several researchers have been unable to detect fragments from plant-based genes (endogenous) in samples from farm animals (Faust, 2000; Anonymons, 2001; Weber and Richert, 2001); others, however, reported finding fragments of chloroplast-specific genes in some animal tissues and fluids (Klotz and Einspanier, 1998; and Einspanier *et al.*, 2001). Some have theorized that the chloroplast-specific fragments investigated by Klotz and Einspanier (1998) and Einspanier *et al.* (2001) are more readily detectable in animal tissues because these genes are present in abundance in plant tissues (Aumaitre *et al.*, 2002).

It is important to emphasize that although fragments of chloroplast-specific genes have been detected in some samples from farm animals (Klotz and Einspanier, 1998; Einspanier *et al.*, 2001), no laboratory has reported finding fragments of biotechnology-derived plant genes in similarly collected samples when animals were fed diets containing biotechnology-derived crops (Klotz and Einspanier, 1998; Faust, 2000; Anonymons, 2001; Khumnirdpetch *et al.*, 2001; Phipps *et al.*, 2001; Weber and Richert, 2001; Phipps *et al.*, 2002). The detection of fragments (~10% of the size of the original gene) of an abundant, endogenous plant gene in samples collected from animals indicates the likelihood that similarly sized fragments of any consumed DNA are present, also. However, given the small fraction of consumed DNA that consists of transgenic DNA when biotechnology-derived crops are consumed by farm animals (~0.00024% of the total DNA ingested: Beever and Kemp, 2000), some conclude that fragments of less abundant genes such as transgenic DNA are not likely to be detected by currently available methodologies in tissues and fluid samples from animals (Beever and Kemp, 2000; Glenn, 2001). However, three summary points that likely are more important are as follows.

1. Farm animals and humans have a long history of safety associated with the consumption of DNA, consequently, the consumption of DNA from all sources including introduced DNA in biotechnology-derived crops is stated to present no health or safety concerns (FAO/WHO, 1991; U.S. FDA, 1992),
2. When gene fragments from ingested DNA have been detected in animal tissues or fluids, their presence has never been associated with any deleterious effects for animals or with any disruptions of normal animal gene function (as reviewed by Beever and Kemp, 2000)
3. Despite a long history of daily consumption of endogenous plant DNA, no plant gene (or gene fragment) has ever been detected in the genome of animals or humans (as reviewed by Beever and Kemp, 2000).

FUTURE DIRECTIONS

SUMMARY

Future biotechnology-derived crops offer to animal agriculture the potential for improved environmental stewardship, production of animal products with enhanced nutritive value and safety for consumers, and improved animal health and efficiency. These new "designer" crops are expected to produce novel proteins, unique fatty acid profiles, more balanced amino acid profiles, novel carbohydrates, enhanced contents of favorable phytonutrients, reduced levels of antinutritional factors, and compounds with immunological and pharmacological properties. Successful development of new biotechnology-derived crops for use by animal agriculture likely will benefit from development of comprehensive databases for endogenous levels of potential nutritional and antinutritional compounds in plants and the plant tissues in which they are present. Widespread adoption by animal agriculture of new "designer" crops will require evidence of cost-effectiveness for specific crops, rapid and robust methods for on-farm use to verify hybrids/varieties associated with specific traits (first step in "identity preservation"), traceable marketing channels for "identity-preserved" crops, and for some crops, markets for the specialty animal products that are produced.

Discussion

New biotechnology-derived crops to enhance environmental stewardship are expected to include an expanded array of crops that resist target pests, resist common plant viruses, tolerate environmentally friendly herbicides, better assimilate nitrogen and other soil nutrients, tolerate environmental stresses (e.g., temperature extremes, grazing pressures), and have improved bioavailability for plant nutrients such as phosphorus. Direct benefits to animal agriculture for the application of designer crops of these types are likely to be less important than indirect benefits. Indirectly, this future array of biotechnology-derived crops may be expected to enhance the capability for animal agriculture to develop more holistic and sustainable animal-cropping systems. Some of these designer crops will permit livestock producers to improve the efficiency by which farm animals utilize nutrients in the crops, thus reducing levels of excess nutrients in animal waste requiring disposal. Canola with the phytase enzyme, developed by means of agricultural biotechnology, is one early example of this type of designer crop. Others will allow producers to improve the quality and productivity of marginal and fragile soils. Crops that resist pests, resist viruses, and tolerate new generation herbicides can improve the capability of producers to reduce the exposure of farm animals to chemical pesticides and herbicides.

From the perspective of those who raise farm animals, one of the most exciting areas for biotechnology-derived crops is for crops that enhance the quality and nutritive value of animal products. Currently, the majority of animal agriculture is involved in the production of commodity animal products. Although there is very little opportunity for producers of commodities, to differentiate their products for specialty and niche markets, consumers are becoming increasingly aware of health benefits that they can gain by consuming specific nutrients such as omega-3 fatty acids, conjugated linoleic acid (CLA), monounsaturated fatty acids, proteins with high biological value, antioxidants, calcium, and selenium, just to name a few. Milk, meat, and eggs are good sources for many of these nutrients; however, levels for several of these nutrients and other nutritionally beneficial components in animal products can be enhanced by changing the nutrient composition of feedstuffs and diets fed to farm animals. For example, it has been reported that dietary CLA has anticancerous (Ip, 1994; Ip *et al.*, 1994a, Belury *et al.*, 1996) and antiatherogenic properties (Lee *et al.*, 1994). The primary natural dietary sources of CLA are in tissues and milk from ruminants such as sheep, dairy cattle, and beef cattle, where CLA is produced by action of rumen bacteria on the fatty acid linoleic acid. Designer crops can be developed to provide higher levels of linoleic acid, which has been shown to increase levels of CLA in products from ruminant farm animals (Owens and Soderlund, 2000). Fatty acid profiles for meat and eggs, and to some extent, milk can be modified by incorporating into diets for farm animals crops such as soybeans high in oleic acid, as well as other biotechnology-derived oilseeds with more favorable fatty acid composition. Modified fatty acid profiles for animal products can help to improve the nutritive value for these products and enhance flavor and shelf life. Biotechnology-derived designer crops have the potential to help animal agriculture to move from a commodity-based marketing system to one of producing more healthful products for specialty demands of target consumer groups.

Also important to those who raise farm animals is the prospect of improving the health and efficiency of animals by feeding designer crops with unique composition and bioavailability for nutrients. These biotechnology-derived crops may include crops with high oil content (O' Quinn *et al.*, 2000), high protein content, improved amino acid profiles (Falco *et al.*, 1995; O' Quinn *et al.*, 2000), modified starch availability and content, enriched phytonutrient or antioxidant content, reduced levels of antinutrients, enhanced bioavailability for minerals, improved fiber digestibility, and even crops with pharmacological properties. Individual species of farm animals are expected to benefit differently from the broad array of potential biotechnology-derived crops (Dado, 1999; Baumel *et al.*, 1999)

For several potential crops, differing conditions on farms will help to determine whether specific crops offer cost-effective benefits. For example, Dado (1999) projected that dairy farms with lower current productivity per cow would benefit most from a designer corn developed to have high protein content; whereas farms with greater current efficiency would capture more benefit from corn with higher lysine content. As estimated by Baumel *et al.*

(1999), swine farms would derive greatest value for designer corn with high protein and varieties with an enlarged germ; poultry farms would benefit greatly from these two varieties and from corn with high starch digestibility. This potential of farm-specific benefits for individual varieties of "designer" crops will necessitate the development of specialized marketing channels for many biotechnology-derived crops with enhanced value.

Numerous crops contain antinutritional compounds such as soybeans that include endogenous levels of trypsin inhibitors and other legumes that include tannins and cyanogens. Since these compounds reduce growth rates for some species, crops such as soybeans typically require processing to inactivate a portion of these antinutritional compounds. It is possible through biotechnology to produce crops with significantly lower levels of these undesirable compounds. Potential biotechnology-derived crops with pharmacological properties offer the possibility to improve animal health and health for those who consume animal products and may include edible and injectable vaccines from plants. Comprehensive reviews of future biotechnology-derived crops with expected benefits for farm animals are provided by Hartnell (2000), Owens and Soderlund (2000), and Sauber (2000).

As new crops are developed for animal agriculture, it will be critical for animal scientists to cooperate with agronomists, plant scientists, and regulators to identify important nutrients, antinutrients, and other compounds in plants. In fact, NAS (2000) has recommended increased research for developing a comprehensive database of endogenous levels of potential nutritional and antinutritional compounds in plants and the plant tissues in which they are present. Mammalian species differ in their response to plant-produced compounds, including toxicants, enzymes, and nutrients; thus it is vital for animal scientists to actively participate in developing databases to further characterize endogenous levels of potential nutritional and antinutritional compounds in plants. These databases can play a foundational role for developing new biotechnology-derived crops to benefit farm animals and to improve the quality, safety, and nutritive value of animal products.

REFERENCES

AGBIOS Inc. (2001). *The safety of GM feeds*. Agriculture & Biotechnology Strategies (Canada), Merrickville, ON. Available http://64.26.172.90/agbios/gmfeeds.php

Ash, J. A., Scheideler, E., and Novak, C. L. (2000). The fate of genetically modified protein from Roundup Ready® soybeans in the laying hen. *Poultr. Sci.*, **79**(suppl. 1), **26**; abstr. 111.

Anonymous. (2001). No traces of modified DNA in poultry fed GM corn. *Nature*, **409**, 657.

Aulrich, K., Halle, I., and Flachowsky, G. (1998). Inhaltsstoffe und Verdaulichkeit von Maiskörnen Sorte Cesar und der gentechnisch Veränderten Bt-hybride bei Legenhennen, in

Verband deutscher landwirtschaftlicher unterschungs- und forschungsanstalten Reichs Kongress Berichte, Giessen, Germany, pp. 465–468.

Aumaitre, A., Aulrich, K., Chesson, A., Flachowsky, G., and Piva, G. (2002). New feeds from genetically modified plants: Substantial equivalence, nutritional equivalence, digestibility, and safety for animals and the food chain. *Livest. Prod. Sci.* **74**, 223–238.

Barrière, Y., Vérité, R., Brunschwig, P., Surault, F., and Emile, J. C. (2001). Feeding value of corn silage estimated in sheep and dairy cows is not altered by genetic incorporation of Bt176 resistance to *Ostrinia nubialis*. *J. Dairy Sci.*, **84**, 1863–1871.

Baumel, C. P., Yu, T., Hardy, C., Johnson, L. A., McVey, M. J., and Sell, J. L. (1999). GM corn has impact on feed consumption. *Feedstuffs*, 71, **16**, 22–23, 26–28.

Beever, D. E., and Kemp, C. F. (2000). Safety issues associated with the DNA in animal feed derived from genetically modified crops. A review of scientific and regulatory procedures. *Nutr. Abstr. Rev.*, **70**, 175–182.

Belury, M. A., Bird, C., Nickels, K. P., and Wu, B. (1996). Inhibition of mouse skin tumor promotion by dietary conjugated linoleate. *Nutr. Cancer*, **26**, 149–157.

Böhme, H., and Aulrich, K. (1999). Inhaltsstoffe und Verdaulichkeit von transgenen Zuckerrben bzw. Maiskörnern im Vergleich zu den isogenen Sorten beim Schwein, in *111th Verband deutscher landwirtschaftlicher unterschungs und forschungsanstalten Reichs Kongress Berichte*, Halle/Saale, pp. 289–292.

Brake, J., and Vlachos, D. (1998). Evaluation of event 176 "Bt" corn in broiler chickens. *Poultr. Sci.*, **77**, 648–653.

Carpenter, J. E., and Gianessi, L. P. (2001). *Agricultural Biotechnology: Updated Benefits Estimates*. National Center for Food and Agricultural Policy, Washington, DC.

Castillo, A. R., Gallardo, M. R., Maciel, M., Giordano, J. M., Conti, G. A., Gaggiotti, M. C. Quaino, O., Giani, C., and Hartnell, G. F. (2001a). Effect of feeding dairy cows with either BollGard®, BollGard® II, Roundup Ready® or control cottonseeds on feed intake, milk yield, and milk composition. *J. Dairy Sci.*, **84**(suppl. 1), abstr. 1712.

Castillo, A. R., Gallardo, M. R., Maciel, M., Giordano, J. M., Conti, G. A., Gaggiotti, M. C., Quaino, O., Giani, C., and Hartnell, G. F. (2001b). Effect of feeding dairy cows with cottonseeds containing BollGard® and Roundup Ready® genes or control nontransgenic cottonseeds on feed intake, milk yield, and milk composition. *J. Dairy Sci.*, **84**(suppl. 1), abstr. 1713.

Clark, J. H., and Ipharraguerre, I. R. (2001). Livestock performance: Feeding biotech crops. *J. Dairy Sci.*, **84**(E suppl.), E9–E18. Available http://www.adsa.org/jds/papers/2001 jds_es9.pdf

CAST. (1999). *Animal Agriculture and Global Food Supply*. Council for Agricultural Science and Technology, Ames, IA.

CAST. (2001). *Evaluation of the U.S. regulatory process for crops developed through biotechnology*. Issue paper no. 19. Council for Agricultural Science and Technology Ames, IA.

Cromwell, G. L., Lindemann, M. D., Randolph, J. H., Stanisiewski, E. P., and Hartnell, G. F. (2001). Soybean meal from Roundup Ready® or conventional soybeans in diets for growing-finishing pigs. *J. Anim. Sci.*, **79**(suppl. 1), Abstr. 1318.

Dado, R. G. (1999). Nutritional benefits of specialty corn grain hybrids in dairy diets. *J. Dairy Sci.*, **82**(suppl. 2), 208.

Daenicke, R., Gadeken, D., and Aulrich, K. (1999). Einsatz von Silomais herkömmlicher Sorten und der gentechnisch veränderten Bt Hybriden in der Rinderfütterung – Mastrinder, in *Maiskolloquium 12*, Wittenberg, Germany, pp. 40–42.

De Vries, J., and Wackernagel, W. (1998). Detection of *npt* II (kanamycin resistance) genes in genomes of transgenic plants by marker-rescue transformation. *Mol. Gen. Genet.*, **257**, 606–613.

Donkin, S. S., Velez, J. C., Stanisiewski, E. P., and Hartnell, G. F. (2000). Effect of feeding Roundup Ready corn silage and grain on feed intake, milk production, and milk composition in lactating dairy cattle. *J. Dairy Sci.*, **83**(suppl. 1), 273, abstr. 1144.

Dukes' Physiology of Domestic Animals, 11th ed. (1993). M. J. Swenson and W. O. Reece, eds, Cornell University Press, Ithaca, NY.

Einspanier, R., Klotz, A., Kraft, J., Aulrich, K., Poser, R., Schwägele, R., Jahreis, G., and Flachowsky, G. (2001). The fate of forage plant DNA in farm animals: A collaborative case-study investigating cattle and chicken fed recombinant plant material. *Eur. Food Res. Technol.*, **212**, 129–134.

Falco, S. C., Guide, T. Guida, T., Locke, M., Mauvais, J., Sanders, C., Ward, R. T., and Webber, P. 1995. Transgenic canola and soybean seeds with increased lysine. *Bio/technology*. **13**, 577–582

FAO/WHO. 1991. Strategies for assessing the safety of foods processed by biotechnology. Report of a joint FAO/WHO consultation. World Health Organisation, Geneva.

Faust, M. A. (1999). Research update on Bt corn silage. In *Four-State Applied Nutrition and Management Conference, MWPS-4SD5*, MidWest Plans Service, Ames, IA, pp. 158–164.

Faust, M. A. (2000). Livestock products: Composition and detection of transgenic DNA/proteins, in *Selected Proceedings from the Agricultural Biotechnology in the Global Marketplace Symposium*. American Society of Animal Science, Savoy, IL.

Faust, M., and Miller, L. (1997). Study finds no Bt in milk. *IC-478, Fall Special Livestock Edition*, Iowa State University, Ames, pp. 6–7.

Faust, M. A., and Spangler, S. (2000). Nutritive value of silages from MON810 Bt and non-Bt near-isogenic corn hybrids. *J. Dairy Sci.*, **83**, 1184, P301.

Folmer, J. D., Grant, R. J., Milton, C. T., and Beck, J. F. (2002). Utilization of *Bt* corn residues by grazing beef steers and *Bt* corn silage and grain by growing beef cattle and lactating dairy cows. *J. Anim. Sci.*, **80**, 1352–1361.

Frank, J. W., Allee, G. L., and Sauber, T. E. (2001). Comparison of apparent ileal amino acid digestibility values of high oil (HOC), high oil/high oleic acid (HOHOC), and low phytate (LP) corn diets fed to finishing pigs. *J. Anim. Sci.*, **79**(suppl. 1), 319, abstr. 1319.

Gaines, A. M., Allee, G. L., and Ratliff, B. W. (2001a). Nutritional evaluation of Bt (MON810) and Roundup Ready® corn compared with commercial hybrids in broilers. *Poult. Sci.*, 80(suppl. 1), Abstr. 214.

Gaines, A. M., Allee, G. L., and Ratliff, B. W. (2001b). Swine digestible energy evaluations of Bt (MON810) and Roundup Ready® corn compared with commercial varieties. *J. Anim. Sci.*, **79**(suppl. 1), abstr. 453.

Gebhard, F., and Smalla, K. (1998). Transformation of *Acinetobacter* sp. BD413 by transgenic sugar beet DNA. *App. Environ. Microbiol.* **64**, 1550–1554.

Gebhard, F., and Smalla, K. (1999). Monitoring field releases of genetically modified sugar beets: Persistence of transgenic plant DNA in soil and horizontal gene transfer. *FEMS Microbiol. Ecol.*, **28**(3), 261–272.

Glenn, K. (2001). Is DNA or protein from feed detected in livestock products? *J. Anim. Sci.*, **79**(suppl. 1), abstr. 230.

Guyton, A. C. (1976). *Textbook of Medical Physiology*. 5th ed. Saunders, Philadelphia, p. 885.

Halle, I., Aulrich, K., and Flachowsky, G. (1998). Einsatz von Maiskörnen der sorte Cesar und des gentechnisch veränderten Bt-Hybriden in der Broilermast, in *Proc. 5 Tagung, Schweine und Geflügelernährung*, Wittenberg, Germany, pp. 265–267.

Hammond, B. G., Vicini, J. L., Hartnell, G. F., Naylor, M. W., Knight, C. D., Robinson, E. H., Fuchs, R. L., Padgette, S. R. (1996). The feeding value of soybeand fed to rats, chickens, catfish, and dairy cattle is not altered by genetic incorporation of glyphosate tolerance. *J. Nutr.* **126**, 717–727.

Hartnell, G. F. (2000). Benefits of biotech crops for livestock feed, in *Proceedings of the 2000 Cornell Nutrition Conference* Cornell Universities, Ithaca, NY, pp. 46–56.

Hendrix, K. S., Petty, A. T., and Lofgren, D. L. (2000). Feeding value of whole plant silage and crop residues from Bt or normal corns. *J. Anim. Sci.*, **78**(suppl. 1), 273, abstr. 1146.

Hvelplund, T., and Weisbjerg, M. R. (2001). Comparison of nutrient digestibility between Roundup Ready® beets and pulp derived from Roundup Ready® beets and conventional beets and pulps. *J. Anim. Sci.*, **79**(suppl. 1), abstr. 1732.

Ip, C. (1994). *Conjugated linoleic acid in cancer prevention research: A report of current status and issues*. National Livestock and Meat Board, Chicago.

Ip, C., Singh, M., Thompson, H. J., and Scimeca, J. A. (1994). Conjugated linoleic acid suppresses mammary carcinogenesis and proliferative activity of the mammary gland in the rat. *Cancer Res.*, **54**, 1212–1215.

Kan, C. A., Versteegh, H. A. J. Uijttenboogaart, T. G., Reimert, H. G. M., and Hartnell, G. F. (2001). Comparison of broiler performance and carcass characteristics when fed Bt, parental control or commercial varieties of dehulled soybean meal. *Poult. Sci.*, **80**(suppl. 1), abstr. 841.

Kerley, M. S., Felton, E. E. D., Lehmkuhler, J. W., and Shillito, R. (2001). Bt corn that is genetically modified to prevent insect damage is equal to conventional corn in feeding value for beef cattle. *J. Anim. Sci.*, **79**(Suppl. 1), 98, abstr. 301.

Khumnirdpetch, V., Intarachote, U., Treemanee, S., Tragoonroong, S., and Thummabood, S. (2001). Detection of GMOs in the broiler that utilized genetically modified soybean meals as a feed ingredient. Abstract of the Plant and Animal Genome Conference. Available http://www.intl-pag.org/pag/9/abstracts/P08_38.html.

Klotz, A., and Einspanier, R. (1998). Nachweis von "Novel-Feed" im Tier? Beeinträchtigung des Verbrauchers von Fleish oder Milch ist nicht zu erwarten. *Mais*, **3**, 109–111.

Klunzinger, M. W., Roberson, K. D., Hill, G. M., Rozeboom, D. W., and Link, J. E. (2001). Effects of low-phytic-acid corn on growth performance, bone strength, and serum osteocalcin concentration in growing–finishing pigs. *J. Anim. Sci.*, **79**(suppl. 1) 319. abstr. 1320.

Lee, K. N., Kritchevsky, D., and Pariza, M. W. (1994). Conjugated linoleic acid and atherosclerosis in rabbits. *Atherosclerosis*, **108**, 19–25.

Loughmiller, J. A., Nelssen, J. L., Goodband, R. D., Tokach, M. D., Lohrman, T. T., De La Llata, M., O'Quinn, P. R., Woodworth, J. C., Moser, S., and Grinstead, G. S. (1998). Influence of soybean meal variety and processing temperature on the growth performance of pigs from 25 to 45 lb. *Swine Day Report*, Kansas State University, Manhattan, pp. 111–115.

MacKenzie, D. J. (2000). *International Comparison of Regulatory Frameworks for Food Products of Biotechnology*. Canadian Biotechnology Advisory Committee, Ottawa, Ontario.

Mayer, J., and Rutzmoser. K. (1999). Einsatz von silomais herkömmlicher Sorten und der gentechnisch veränderten Bt-hybriden in der Rinderfütterung: Bei Milchkühen. *Maiskolloquium 12*, Wittenberg, Germany, pp. 36–39.

Maynard, L. A., and Loosli, J. K. (1979). *Animal Nutrition*. McGraw-Hill, New York, pp. 48–63, 115–153.

McAllan, A. B. (1982). The fate of nucleic acids in ruminants, in *Proceedings of the Nutrition Society*, vol. 41, pp. 309– 317.

Merck & Co., Inc. (1998). *The Merck Veterinary Manual*, 8 ed, S. E. Aiello, ed. Merck & Co., Inc., and Merial Limited, Whitehouse Station, N. J., pp. 2076–2091.

Mirales, Jr., A., Kim, S., Thompson, R., and Amundsen, B. (2000). GMO (Bt) corn is similar in composition and nutrient availability to broilers as non-GMO corn. *Poult. Sci.*, **79**(suppl. 1), 65–66, abstr. 285.

Munkvold, G. P., and Desjardins, A. E. (1997). Fumonisins in maize: Can we reduce their occurrence? *Plant Dis.* **81**, 556–565.

Munkvold, G. P., Hellmich, R. L., and Showers, W. B. (1997). Reduced *Fusarium* ear rot and symptomless infection in kernels of maize genetically engineered for European corn borer resistance. *Phytopathology*, **87**, 1071.

Munkvold, G. P., Hellmich, R. L. and Rice, L. G. (1999). Comparison of fumonisin concentrations in kernels of transgenic Bt hybrids and nontransgenic hybrids. *Plant Dis.*, **83**, 130.

NAS. (2000). *Genetically modified pest-protected plants: Science and regulation.* National Academy of Sciences, Washington, DC. Available http://www.nap.edu/openbook/0309069300/html/1.html

Nielsen, K. M., Bones, A. M. Smalla, K., and van Elsas, J. D. (1998). Horizontal gene transfer from transgenic plants to terrestrial bacteria – A rare event? *FEMS Microbiol. Rev.*, **22**, 79–103.

Obrycki, J. J., Losely, J. E., Taylor, O. R., and Jesse, L. C. H. (2001). Transgenic insecticidal corn: Beyond insecticidal toxicity to ecological complexity. *BioScience*, **51**, 353.

O' Quinn, P. R., Nelssen, J. L., Goodband, R. D., Knabe, D. A., Woodworth, J. C., Tokach, M. D., and Lohrmann, T. T. (2000). Nutritional value of genetically improved high-lysine, high-oil corn for young pigs. *J. Anim. Sci.* **78**, 2144–2149.

Orriss, G. D. (1997). Animal diseases of public health importance. *Emerging Infect. Dis.*, **3**, 497–502. Available http://www.cdc.gov/ncidod/eid/vol3no4/adobe/orriss.pdf

Owens, F., and Soderlund, S. (2000). Specialty grains for ruminants, in *Proceedings of the 61st Minnesota Nutrition Conference & Minnesota Soybean Research and Promotion Council Technical Symposium.* University of Minnestota Extension, St. Paul, pp. 98–113.

Padgett, S. R., Taylor, N. B., Nida, D. L., Bailey, M. R., MacDonald, J., Holden, L. R., and Fuchs, R. L. (1996). The composition of glyphosate-tolerant soybean seeds is equivalent to that of conventional soybeans. *J. Nutr.*, **126**, 702–716.

Parsons, C. M., Zhang, Y., Johnson, M. L., and Araba, M. (1996). Nutritional evaluations of soybean meals varying in oligosaccharide content. *Poultr. Sci.*, **75**(suppl 1), 39.

Petty, A. T., Hendrix, K. S., Stanisiewski, E. P., and Hartnell, G. F. (2001a). Feeding value of Bt corn grain compared with its parental hybrid when fed in beef cattle finishing diets. *J. Anim. Sci.*, **79**(suppl. 1). **101**, abstr. 320.

Petty, A. T., Hendrix, K. S., Stanisiewski, E. P., and Hartnell, G. F. (2001b). Performance of beef cattle fed Roundup Ready® corn harvested as whole plant silage and grain. *J. Anim. Sci.*, **79**(suppl. 1), 101, abstr. 321.

Phipps, R. H., Beever, D. E., and Humphries, D. J. (2002). Detection of transgenic DNA in milk from cows receiving herbicide tolerant (CP4EPSPS) soybean meal. *Livest. Prod. Sci.* **74**, 269–273.

Phipps, R. H., Beever, D. E., and Tingey, A. P. (2001). Detection of transgenic DNA in bovine milk: Results for cows receiving a TMR containing maize grain modified for insect protection (MON810). *J. Dairy Sci.*, **84**(suppl. 1), abstr. 476.

Piva, G., Morlacchini, M., Pietri, A., Piva, A., and Casadei, G. (2001a). Performance of weaned piglets fed insect-protected (MON810) or near isogenic corn. *J. Anim. Sci.*, **79**(suppl. 1), abstr. 441.

Piva, G., Morlacchini, M., Pietri, A., Rossi, F., and Prandini, A. (2001b). Growth performance of broilers fed insect-protected (MON810) or near-isogenic control corn. *Poult. Sci.*, **80**(suppl. 1), abstr. 1324.

Reuter, T., Aulrich, K., Berk, A., and Flachowsky, G. (2001). Nutritional evaluation of Bt-corn in pigs. *J. Anim. Sci.*, **79**(suppl. 1), abstr. 1073.

Russell, J. R., Farnham, D. E., Berryman, R. K., Hersom, M. J. Puch, A., and Barrett, K. (2000a). Nutritive value of the crop residues from Bt-corn hybrids and their effects on performance of grazing beef cows, *2000 Beef Research Report.* Iowa State University, Ames. A.S. Leaflet R1723.

Russell, J. R., Hersom, M. J., Pugh, A., Barrett, K., and Farnham, D. (2000b). Effects of grazing crop residues from Bt-corn hybrids on the performance of gestating beef cows. *J. Anim. Sci.*, **78**(suppl. 2), 79, abstr. 244.

Russell, J. R., Hersom, M. J., Haan, M. M., Kruse, M., Dixon, P. M., Morrical, D. G., and Farnham, D. E. (2001a). Effects of grazing crop residues of Bt-corn hybrids on performance of pregnant beef cows, *2001 Beef Research Report*. Iowa State University, Ames. A.S. Leaflet R1745.

Russell, J. R., Hersom, M. J., Haan, M. M., Kruse, M., and Morrical, D. G. (2001b). Effects of grazing crop residues from Bt-corn hybrids on pregnant beef cows. *J. Anim. Sci.*, **79** (suppl. 2), 98, abstr. 300.

Sauber, T. E. (2000). Performance of soybean meals produced from genetically enhanced soybeans, in *Proceedings of the 61st Minnesota Nutrition Conference & Minnesota Soybean Research and Promotion Council Technical Symposium*. University of Minnesota Extension, St. Paul, pp. 44–51.

Sidhu, R. S., Hammond, B. G., Fuchs, R. L., Mutz, J. N., Holden, L. R. George, B., and Olson, T. (2000). Glyphosate-tolerant corn: The composition and feeding value of grain from glyphosate-tolerant corn is equivalent to that of conventional corn (*Zea mays L.*). *J. Agric. Food Chem.*, **48**, 2305.

Stanisiewski, E. P., Hartnell, G. F., and Cook, D. R. (2001). Comparison of swine performance when fed diets containing Roundup Ready® corn (GA21), parental line or conventional corn. *J. Anim. Sci.*, **79**(suppl. 1), abstr. 1322.

Taylor, M. L., Hartnell, G. F., Nemeth, M. A., George, B., and Astwood, J. D. (2001a). Comparison of broiler performance when fed diets containing YieldGard® corn, YieldGard® and Roundup Ready® corn, parental lines, or commercial corn. *Poult. Sci.*, **80**(suppl. 1), abstr. 1321.

Taylor, M. L., Hartnell, G. F., Nemeth, M. A., George, B., and Astwood, J. D. (2001b). Comparison of broiler performance when fed diets containing Roundup Ready® corn event NK603, parental line, or commercial corn. *Poult. Sci.*, **80**(suppl. 1), abstr. 1323.

Tsume, Y., Taki, Y., Sakane, T., Nadai, T., Sezaki, H., Watabe, K., Kohno, T., and Yamashita, S. (1996). Quantitative evaluation of the gastrointestinal absorption of protein into blood and lymph circulation. *Biol. Pharm. Bull.*, **19**, 1332–1337.

USDA. (2001). *Acreage*. National Agricultural Statistics Service. U.S. Department of Agriculture, Washington, DC, June. Available http://usda.mannlib.cornell.edu/reports/nassr/field/pcp-bba/acrg0601.txt.

U.S. EPA. (2001). *Bacillus thuringiensis* subspecies *tolworthi* Cry9C protein and the genetic material necessary for its production in corn. Biopesticide Fact Sheet No. 006466. Available http://www.epa.gov/pesticides/biopesticides/factsheets/fs006466t.htm. Accessed May 13, 2002.

U.S. Food and Drug Administration (FDA). (1992). Statement of policy: Foods derived from new plant varieties; Notice. *Fed. Regis.*, **57**, 22984–23005.

U.S. FDA. (2001a). *Background Paper in Support of Fumonisin Levels in Animal Feed: Executive Summary of This Scientific Support Document*. Center for Food Safety and Applied Nutrition, Center for Veterinary Medicine. Available http://www.cfsan.fda.gov/~dms/fumonbg4.html

U.S. FDA. (2001b). *Background Paper in Support of Fumonisin Levels in Corn and Corn Products Intended for Human Consumption*. Center for Food Safety and Applied Nutrition, Center for Veterinary Medicine. Available http://www.cfsan.fda.gov/~dms/fumonbg3.html.

U.S. FDA. (2001c). *Guidance for Industry Fumonisin Levels in Human Foods and Animal Feeds*. Available http://www.cfsan.fda.gov/~dms/fumongu2.html.

U.S. FDA. (2001d). List of completed consultations on bioengineered foods. Center for Food Safety and Applied Nutrition, Office of Food Additive Safety. July. Available http://www.cfsan.fda.gov/~lrd/biocon.html.

Weber, T. E., and Richert, B. T. (2001). Grower–finisher growth performance and carcass characteristics including attempts to detect transgenic plant DNA and protein in muscle from pigs fed genetically modified "*Bt*" corn. *J. Anim. Sci.*, **79**(suppl. 1), **67**, abstr. 162.
Weber, T. E., Richert, B. T., Kendall, D. C., Bowers, K. A. and Herr, C. T. (2000). Grower–finisher performance and carcass characteristics of pigs fed genetically modified "*Bt*" corn. Purdue University, West Lafayette, IN, 2000 Swine Day Report. Available http://www.ansc.purdue.edu/swine/swineday/sday00/psd07–2000.html.

Chapter 7

Preclinical Immunotoxicology Assessment of Cytokine Therapeutics

Robert V. House
Immunotoxicology Laboratory
Covance Laboratories Inc.
Madison, Wisconsin

An increased understanding of the biology of health and disease has led to the ability to modulate endogenous humoral factors for treating disease and restoring health. One potential target of such modulation is the group of factors collectively known as cytokines. Cytokines may be modulated by addition (introduction of exogenous cytokines, either natural or artificial) or by subtraction (by either removal, inactivation, or prevention of release). This brief review describes advances in the development of cytokine modulators (both biological and nonbiological) as well as their use in human therapeutics. In addition, the various types of toxicity associated with cytokine therapeutics, as well as methodologies and approaches for available for assessing the multiplicity of potential toxicities to the immune system, are reviewed.

Biotechnology and Safety Assessment, 3rd edition

INTRODUCTION

For most of medical history, therapeutic intervention has relied on exogenous factors. These treatments have run the gamut from magical incantations to animal- and plant-derived substances and finally to rational drug design including combinatorial chemistry and recombinant DNA technology. With increased understanding of the biology of health and disease, as well as the tools that were developed to gain this understanding, the possibility of utilizing the body's own endogenous factors to treat disease and restore health is beginning to be realized. One potential target of this therapy is the group of factors collectively known as cytokines. This chapter describes cytokines and their use in human therapeutics, providing as well an overview of the multitude of experimental and clinical approaches that have been designed to tap into the body's own healing power. In addition, this chapter discusses the various types of toxicity that have been associated with cytokine therapeutics, and describes methodologies and approaches for assessing the multiplicity of potential toxicities to the immune system.

CYTOKINES AND THEIR ROLE IN HEALTH AND DISEASE

Cytokines are highly bioactive proteins that are usually secreted, although also existing as membrane-associated molecules. They are generally not produced constitutively, but rather in response to cellular activation (Fraser *et al.*, 1993). Cytokine production is highly regulated in both paracrine and autocrine fashions at the transcriptional and post-transcriptional levels (Dendorfer, 1996; Lunney, 1998).

CHARACTERISTICS OF CYTOKINES

Perhaps the most important biological feature of cytokines is that they are both highly pleiotropic and redundant in function. That is, most of the cytokines described to date have multiple actions, and these actions often overlap with the actions of other cytokines. In addition, cytokines often work via cascading mechanisms, allowing them to interact with each other both synergistically and antagonistically. These features are critical from a therapeutic consideration, since the ultimate consequences of manipulating any single cytokine must be evaluated in the context of this overall network of factors. In most cases, cytokines act primarily at a local level and are rapidly cleared from the circulation; these features also are important from a clinical standpoint.

Cytokine Families

For convenience, cytokines may be classified in several groups. In general, the term *interleukins* (IL) refers to cytokines exerting their primary effect on cells of the immune system; IL-1 and IL-6, which mediate a tremendous number of effects besides immunomodulation, are notable exceptions to this broad categorization. *Colony-stimulating factors* (CSF) are important as mediators of hematopoiesis but also exert effects on the immune system. *Tumor necrosis factors* (TNF) regulate a range of biological effects including immunity, inflammation, and cellular death. *Interferons* (IFN) were originally named based on their ability to hinder viral replication, and thus function as important mediators of host defense; however, they also affect various other processes. Finally, we may consolidate the miscellaneous cytokines that do not fit quite so neatly into the scheme owing to their multiplicity of action. Not included in this scheme are the various *peptide growth factors* such as epidermal growth factor and insulin-like growth factor, whose functions appear to be primarily related to growth and tissue repair and homeostasis but may function as cytokines in certain circumstances, and *chemokines*, small polypeptides that are important not only in immunity and inflammation but in other regulatory processes as well.

Cytokine Receptors

Cytokines interact with specific receptors, which are grouped into the hemopoietin superfamily, the TNF family, the immunoglobulin superfamily, and the tyrosine kinase family. These receptors are generally membrane-bound molecules, although soluble receptors have also been described. Many cytokine receptors share a number of similar characteristics, including a subunit structure and an association with signal transducing elements within the cell (Williams and Giroir, 1995). Moreover, several of the receptor subunits are shared between various cytokine receptors, possibly contributing to cytokine pleiotropy and redundancy. Alterations in cytokine receptor structure or function have been implicated in human disease states (Touw *et al.*, 2000).

Type 1/Type 2 Cytokine Responses

A major advance in understanding the role of cytokines in disease, and particularly in diseases incorporating an immune component in their etiology, was the realization that T-helper lymphocytes could be categorized into at least multiple subsets based on their particular pattern of cytokine production.

Originally described *in vitro* using T cell clones, these subsets were designated T-helper 1 and T-helper 2 (subsequent to this, the nomenclature TH1/TH2, T1/T2, and type 1/type 2 has also been used). In general, TH1 cells predominantly produce interferon-gamma (IFN-γ), IL-2, and TNF, whereas TH2 cells primarily make IL-4, IL-5, IL-10, and IL-13; these patterns appear to be essentially the same in humans. IL-2 acts as an autocrine growth factor for TH1 and TH2 cells, although TH1 cells are much more sensitive to this activity. IFN-γ acts to suppress the growth of TH2 cells. Conversely, IL-4 acts as the autocrine growth factor for TH2 cells, while IL-10 suppresses the growth and function of TH1 cells. Various other cytokines are produced in common by these subsets, albeit at relatively lower levels. It should be noted that these distinctions are somewhat fluid, since clones have been isolated in both human and laboratory animal cells that display characteristics of both profiles. Moreover, CD8 (cytotoxic/suppressor) T cells also produce cytokines, as do macrophages and B cells (Pistoia, 1997), thus complicating the elucidation of a characteristic cytokine response. However, the general pattern just described for T-helper cells appears to hold true in most cases.

The induction of a TH1-type response vs a TH2-type response (i.e., a state characterized by the preponderance of either TH1- or TH2-type cytokine production) depends on a variety of factors including the type of antigen-presenting cell, the type and amount of antigen, and factors in the cellular milieu such as other cytokines and hormones (Romagnani, 2000). This particular pattern of functions is now recognized to form the basis for differential immune responses as well as certain pathologies. For example, in normal immune responses TH1 cytokines tend to favor cytolytic (cell-mediated) reactions and the activation of macrophages, while TH2 cytokines provide T-cell help for the production of various antibodies. On the other hand, TH1 responses are usually associated with autoimmune diseases (e.g., multiple sclerosis, rheumatoid arthritis), whereas TH2 responses include atopic conditions, reduced response to infectious challenge, and even successful pregnancy (Del Prete, 1998). In current and future therapies involving cytokines, it will be of vital importance to understand this interlocking network of cytokines in greater detail, possibly allowing for the direct manipulation of immune responses (Powrie and Coffman, 1993; Hetzel and Lamb, 1994).

Cytokine Polymorphism

It has become increasingly obvious that the genes for the production of cytokines and their receptors exhibit a fair degree of polymorphism within the human genome. This polymorphism, at least for some genes, may have selective benefits (Allen, 1999). Conversely, cytokine polymorphism has also been associated with a variety of human disease states (Csaszar and Abel, 2001). Examples of diseases in which cytokine polymorphisms have been implicated include pancreatitis (Powell *et al.*, 2001), susceptibility to leukemia/lymphoma (Tsukasaki *et al.*, 2001), systemic lupus erythematosus (Schotte *et al.*, 2001), psoriasis (Craven *et al.*, 2001), and Alzheimer's disease (McCusker *et al.*, 2001).

MODULATION OF CYTOKINE ACTIVITY AS A THERAPEUTIC MODALITY

For the various reasons already described, it is clear that modulation of cytokine activity (by increasing or decreasing the amount of certain molecules) represents a powerful addition to the medical armamentarium. As evidence of this, Table 1 illustrates a sampling of cytokine therapies that were either marketed or in development/testing in mid-2001.

Given the redundant and pleiotropic nature of cytokines themselves, as well as their role as regulatory molecules for various physiological processes, it becomes apparent that developing therapeutics based on modifying the function of cytokines is far from straightforward. Potentially complicating the altered regulation of cytokine biology are the intrinsic toxicity of the molecules at superphysiological concentrations, the disruption of normal processes in their absence, the unintended consequences of modulated function, the induction of an anticytokine response due to modified chemical structure, and other issues (Kulmatycki and Jamali, 2001; Slifka and Whitton, 2000). In the sections that follow, we will consider a variety of human diseases that may be amenable, to one degree or another, to modulation of cytokine function.

Therapeutic Immunomodulation

Ironically, although cytokines are usually associated with the immune system, direct modulation of the immune response has not been their principal therapeutic application to date. Immunological defects severe enough to warrant modulation of cytokine levels often must be addressed at a level more fundamental than simple infusion of cytokines.

Treatment of Neoplastic Disease

One of the most thoroughly investigated therapeutic uses of cytokines has been in the treatment of neoplastic disease. The intrinsically complex nature

Table 1
Cytokine Drugs: 2001

Cytokine name	Drug name	Indication	Producer
Erythropoietin α	Epogen (Epoetin alfa)	Stimulation of erythrocyte production	Amgen
Granulocyte colony-stimulating factor (G-CSF)	Neulasta (Pefilagristim)	Stimulation of neutrophil production	Amgen
Glycosylated G-CSF	Neutrogen (Lenograstim)	Stimulation of neutrophil production	Chugai
Mutated G-CSF	Nartograstim	Stimulation of neutrophil production	Kyowa Hakko Kogyo
Granulocyte/ macrophage colony-stimulating factor (GM-CSF)	Leukine (Sargramostim)	Stimulation of leukocyte production	Immunex
Stem cell factor	Stemgen (Ancestim)	Stem cell transplantation	Amgen
Interleukin 11	Neumega (Oprelvekin)	Stimulation of platelet production	Wyeth
Interleukin 2	Proleukin (Aldesleukin)	Melanoma and renal cell carcinoma	Chiron
Consensus interferon	Infergen (interferon alfacon-1)	Hepatitis C	Amgen
Interferon-α	Intron A (interferon alfa-2b)	Hairy cell leukemia and melanoma	Schering Plough
Interferon-β	Avonex (interferon beta-1a)	Multiple sclerosis	Biogen
Interferon-γ	Actimmune (interferon gamma-1b)	Chronic granulomatous disease	InterMune Pharmaceuticals
Anti-cytokine agents			
TNF receptor fusion protein	Enbrel (Etanercept)	Arthritis	Immunex
IL-4 receptor	Nuvance	Asthma	Immunex
IL-1 receptor antagonist	Kineret (Anakinra)	Inflammation	Amgen
Chimeric anti-TNF monoclonal antibody	Remicade (Infliximab)	Crohn's disease and rheumatoid arthritis	Centocor
Anti-IL-2-receptor monoclonal antibody	Simulect (Basiliximab)	Organ transplantation	Novartis

of neoplastic disease, suggests that the use of cytokines for treatment is equally complex. Therapeutic options include disruption of autocrine growth, enhancement of natural immune resistance, the use of cytokines in combination with other drugs, and direct tumor cytotoxicity by cytokines.

It is recognized that many tumors, particularly neoplasias of the hematopoietic system, may involve the autocrine growth of cytokine-responsive or cytokine-producing cells. In such cases, selective reduction in the levels of certain growth factors could theoretically be beneficial (although clinical trials employing extracorporeal removal of cytokine have yet to demonstrate clinical improvements). Conversely, deficiencies in cytokine production may impair the function of cellular immune components required for resistance to tumors, such as cytotoxic T lymphocytes or natural killer cells (Romagnani, 1994; Wagstaff *et al.*, 1995). In such cases, the administration of carefully selected cytokines might boost a suboptimal immune response.

An intriguing concept that has been explored is the combined administration of cytokines with cytotoxic drugs. Experimental data suggest that cytokines may result in enhanced susceptibility to cytotoxic drugs under certain circumstances. Certain cytokines may also exert a direct cytotoxic effect on tumor cells, either as single agents or in combination (Kreuser *et al.*, 1992). Finally, a relatively successful use of cytokines in oncology has been the repopulation of bone marrow following transplant subsequent to chemotherapy or radiation.

Transplantation

Another successful therapeutic use of cytokines has been as an invaluable adjunct to transplantation. Ironically, successful treatment may take the form of either cytokine suppression or cytokine addition. In the former situation, cytokines are highly active participants in the rejection process responsible for the failure of organ transplantation and may contribute to engraftment failure by a variety of mechanisms. Successful long-term engraftment has been made possible by new drugs that selectively inhibit cytokine production without completely eliminating host defense.

Conversely, the availability of recombinant cytokines (both natural and modified) has led to great advances in bone marrow transplantation and, more recently, stem cell transplantation (Brugger *et al.*, 2000). Cytokine therapy may be used in a variety of ways for bone marrow transplantation. For example, colony-stimulating factors and other hematopoietic cytokines (particularly in combination) can be administered to enhance repopulation of bone marrow following chemotherapy or radiation treatment, lessening potential morbidity from infectious disease or coagulation dysfunction.

Alternatively, cytokines may be administered prior to therapy to enhance the success of autologous transplantation, particularly of stem cells. This expansion of stem cells may be carried out *in vivo* or *ex vivo*.

Diagnosis and Treatment of Cardiovascular Disease

An excellent example of the increasing relevance of cytokines in medical diagnosis and treatment, outside the original realm of immunology is their role in cardiovascular disease. Proinflammatory cytokines such as IL-1 and TNF have been implicated in the pathogenesis of atherosclerosis, acute ischemic events associated with myocardial infarction, myocardiopathies, and the pathogenesis of congestive heart failure (Dinarello and Pomerantz, 2001; Long, 2001). Modulation of cytokine activity has been proposed as a therapeutic option for various cardiovascular pathologies (Baumgarten *et al.*, 2000; Deswal *et al.*, 2001).

Treatment of Infectious Disease

Resistance to infectious organisms is a primary function of the immune system and thus involves a plethora of cytokines and chemokines (Banyer *et al.*, 2000). However, the complex and redundant mechanisms of natural and acquired host resistance probably are not amenable to simplistic addition or deletion of individual cytokines. Improvements in the control of infectious disease will more likely involve cytokines as a component of the therapeutic regimen.

One promising avenue is the use of cytokines as adjuvants. Although not as powerful as some of the best experimental adjuvants, cytokines have been shown to be at least equal if not superior to adjuvants currently approved for human use (Heath and Playfair, 1992). Much work remains to be done to optimize this approach; some initial attempts have included using combinations of cytokines, increasing exposure time, and physical association of cytokines with antigens. Some cytokines being developed as adjuvants include IL-2 (Gursel and Gregoriadis, 1998), IL-12 (Gherardi et al., 2001), and granulocyte/macrophage colony-stimulating factor (GM-CSF) (Mellstedt and Osterborg, 1999).

Another use of cytokines in treatment and prevention of infectious disease is their use to correct defects in immune function, thus enhancing host resistance to infections that are common in immunocompromised hosts. This approach usually entails the use of colony-stimulating factors to restore more marrow depletion, although other cytokines such as IL-1, IL-6, and IFN-γ display potential benefits (Roilides and Pizzo, 1993). Other potential therapeutic modalities include the manipulation of cytokines to direct the immune system to generate either a type 1 or type 2 response (Golding *et al.*, 1994) or to selectively recruit immunocytes (Nelson, 2001).

CYTOKINE MODULATION PARADIGMS

Addition of Exogenous Cytokines

Native Sequence Cytokines: Cell Derived

Many of the earliest known cytokines were discovered in human or animal cell cultures, and elucidation of their *in vitro* and *in vivo* effects utilized culture supernatants of stimulated primary lymphoid or myeloid cells. These cultures contain a diversity of cytokines that are produced naturally in reaction to cellular activation. Although these native sequence molecules obviously produce desired biological effects, they have not been pursued as intensively as recombinant cytokines for a number of reasons. First, mammalian cell culture does not routinely produce cytokines as economically as bacterial fermentations, and mammalian cells are technically more difficult to work with than bacteria or other biotechnology workhorses such as insect cells. A more important consideration may be the undefined nature of the producer cells, particularly primary cell cultures. Cell culture supernatants may contain a multitude of cytokines and other bioactive molecules, often of an undefined nature, making it difficult or even impossible to determine the activity of individual factors. This ambiguity is highly undesirable for routine drug discovery efforts.

On the other hand, a teleological assessment would suggest that just such an assortment of cytokines is perhaps the best application of cytokines. Utilizing this simplistic logic, a mix of natural cytokines might represent an extension of the body's own healing process. One such preparation that has demonstrated clinical efficacy is Leukocyte Interleukin Inj., trade name Multikine. Multikine is a serum-free cytokine mixture containing TNF-α, TNF-β, IFN-γ, RANTES, IL-8, the macrophage-inhibiting proteins MIP-1α, and MIP-1β, IL-1, IL-2, and IL-6. The preparation is derived from human buffy-coat mononuclear cells and enhances the function of natural killer cells and cytotoxic T lymphocytes (Chirigos *et al.*, 1995). Injection of Multikine into humans with head and neck cancer was found to result in tumor regression, with infiltration of the tumors with lymphocytes.

Native Sequence Cytokines: Recombinant

The majority of discovery and preclinical testing of exogenous cytokines have utilized native sequences produced by means of recombinant DNA technology. The literature on the results of these studies is therefore extensive, and a recounting of the results of these studies is well beyond the scope of this chapter. The interested reader is directed to a number of excellent reviews (e.g., Cardi *et al.*, 1996). Although native sequence cytokines would be an intuitive best choice for replacement or augmentation therapies, the toxicity and propensity toward inducing neutralizing antibodies may limit their therapeutic applicability.

Artificial Cytokines

Seeking to broaden the therapeutic index of cytokines, some investigators have found that rational protein drug design can result in cytokines with enhanced biological activity and reduced toxicity. The various categories of rationale cytokine drug design are described next.

Muteins Muteins (mutated proteins) are molecules that are produced beginning with the native sequence of a known cytokine gene, which is then modified (often employing site-directed mutagenesis). The goal of this modification may be to enhance bioactivity, to reduce toxicity (by increasing the therapeutic index), or even to produce new activities. A number of cytokine muteins have been constructed and are under evaluation for the treatment of various conditions. Examples include IL-4 muteins (Duschl *et al.*, 1995; Srivannaboon *et al.*, 2001) and IL-5 mutein (Devos *et al.*, 1995) for the treatment of allergic diseases, IL-2 muteins for antitumor activity (Shanafelt *et al.*, 2000), TNF muteins for antitumor activity (Takagi *et al.*, 1998), GM-CSF mutein for increasing circulating neutrophils (Eliason *et al.*, 2000), and IL-13 muteins for allergic reactions (Oshima and Puri, 2001).

Mimetics Cytokine mimetics are similar to muteins in that their construction involves the rearrangement of genetic elements. However, many cytokine mimetics are designed rationally, rather than relying solely on off-the-rack sequences. As with muteins, the goal of producing mimetics is usually to reduce toxicity and to "improve" cytokine function. Mimetics have been described for TNF as a treatment for mycobacterial infections (Briscoe *et al.*, 2000), for IL-4 (Domingues *et al.*, 1999), and for IL-2 (Eckenberg et al., 2000). In mid-2001, testing for efficacy of these molecules in modification of the immune response was at an early stage.

Synthetic Cytokines The term "synthetic cytokine" has obvious overlap with the two preceding types of cytokine molecule, since none of these agents are produced in nature per se. In the present context, however, we will use this terminology as an umbrella covering a variety of genetic modifications intended to decrease toxicity, increase efficacy, or enhance such features as biological and thermodynamic stability (Bishop *et al.*, 2001).

One type of synthetic cytokine, a combination of IL-6 with the IL-6 receptor (IL-6R), was produced by Peters and associates (1998). This molecule, designated "Hyper-IL-6," shows more activity toward certain IL-6-responsive cells (e.g., hematopoietic cells, hepatocytes) than does IL-6 alone.

Another example of this approach is the molecule Synthokine SC-55494, a potent IL-3 receptor antagonist. Employing oligonucleotide-directed mutagenesis on a synthetic, natural sequence IL-3 DNA, followed by iterative combination, produced several thousand IL-3 mutants, which were subsequently screened for biological activity. The results of this screening identified Synthokine, a molecule with 48 amino acid sequence changes relative to natural IL-3. Synthokine was found to produce a 10- to 20-fold increase in hematopoietic activity, although its potential for induction of inflammatory responses was only about twice that of natural sequence IL-3. Subsequent primate studies (Farese *et al.*, 1996) demonstrated that administration of Synthokine, either alone or in combination with a granulocyte colony-stimulating factor (G-CSF), significantly enhanced recovery from both neutropenia and thrombocytopenia following radiation-induced bone marrow aplasia. These results suggest that Synthokine and related molecules may demonstrate significant promise in the support of myelosuppressed patients.

Although not designated as a designer molecule in the literature, an interesting example of protein engineering has been described by Anderson *et al.* (1997). These investigators worked with IL-12, a cytokine that has been somewhat problematic to produce by means of recombinant DNA technology because of its heterodimeric structure and because distinct subunits of the molecule are produced by separate genes on different chromosomes. This problem has been overcome by constructing a single-chain fusion protein (Flexi-12), which retains the biological activity of the natural cytokine molecule but is considerably easier to construct.

Another subclass of synthetic cytokine is the chimeric cytokine, also termed harlequin cytokines. These molecules are produced using structural components of two or more cytokines. Grazi and Ferrero (1997) described a series of such molecules combining structural elements of G-CSF and IL-6. Although some of these chimeras exhibited the activity of only one parent molecule (and some displayed no bioactivity), at least one had properties of both molecules. Another group of cytokine chimeras comprises the various synthetic hematopoietic factors such as myelopoietins (MPOs), a family of engineered molecules that bind and activate the IL-3 and G-CSF receptors, and promegapoietin (PMP), which stimulates the human IL-3 and c-mpl receptors (Dempke *et al.*, 2000).

Fusion Proteins A different form of cytokine is the fusion protein. The concept behind fusion proteins is to connect the functional portion of a cytokine (i.e., the receptor-binding moiety) with another molecule; this second molecule can have biological activity of its own, or it may be there to assist the function of the cytokine molecule. In the case of fusion proteins consisting of two different cytokines, the resulting molecule is termed a hybrid cytokine or hybrikine. One of the first examples of a hybrikine was PIXY321 (Pixykine), which consists of the active domains of IL-3 and GM-CSF, coupled with a flexible amino acid linker sequence. By incorporating elements of both IL-3 and GM-CSF (both of which are potent colony-stimulating factors), PIXY321 was found to be more active than either of its constituent cytokines, either alone or in combination. Subsequent clinical trials with PIXY321 have shown it to be a useful adjunct in autologous bone marrow transplantation, among other indications (Vose *et al.*, 1997). Other cytokine/cytokine fusion proteins are now being described, such as CH925, a fusion protein that exhibits erythropoietin-like activity and comprises the recombinant human interleukins rhIL-6 and rhIL-2 (Zhao *et al.*, 1996).

Cytokines may also be combined with other molecules. For example, construction of cytokine–antibody hybrids produces fusion proteins retaining the properties of both parents (i.e., antigen binding and cellular activation). Such constructs display promise as potent immunostimulants (Penichet *et al.*, 1997) or as treatments for conditions such as septic shock (Abraham *et al.*, 1997). A different, but related, immunostimulant property can be obtained by fusing cytokines directly to antigens, as demonstrated by Kim *et al.* (1997) and Dela Cruz *et al.* (2000).

Cytokine Inhibition

Chemical (Nonbiological) Agents Inhibiting Cytokine Production/Action

A variety of chemical (i.e., nonbiological) drugs have been found to either specifically or nonspecifically suppress cytokine production. These drugs act via myriad mechanisms, including alteration in cytokine gene transcription or mRNA translation, inhibition of cytokine release, or inhibition of cytokine processing (Henderson and Blake, 1992). To date, the primary clinical use of these agents has been as adjuncts to organ and tissue transplantation, where they have provided great advances in clinical success. Increasingly, these agents are being evaluated for treatment of other immune-related diseases. Some of these agents are discussed in the subsections that follow.

Glucocorticoids These well-known and powerful immunosuppressive drugs enjoy wide clinical use for a variety of indications in which inappropriate immune reactions contribute to disease etiology. These drugs act at a variety of molecular targets; germane to the present discussion is their role in modulating cytokine production. Glucocorticoids affect cytokine production via a number of mechanisms including transcriptional repression, post-transcriptional alterations, induction of the suppressive cytokine TGF-β, antagonisms of transcription factors, and cooperation with transcription factors (Almawi *et al.*, 1996). Glucocorticoids also alter the expression of cytokine receptors. These drugs have highly potent anti-inflammatory and immunosuppressive effects, hence are associated with numerous side effects and with toxicity.

Cyclosporin (SandImmune) Probably the most thoroughly studied of the immunosuppressive drugs that specifically target cytokine production, cyclosporin is a fungal metabolite with several unique features, including a novel amino acid in its composition. Cyclosporin acts primary at the level of the T lymphocytes, effectively and reversibly inhibiting the action of these cells by binding to cyclophilin, which is a member of a class of molecules generically known as immunophilins. The cyclosporin–cyclophilin complex targets calcineurin, which is then unable to interact with downstream transcription factors such as NF-AT; this ultimately prevents the transcription of cytokine genes. In addition, cyclosporin may block the activation of c-Jun amino terminal Kinases (JNK) and p38 signaling pathways (Matsuda and Koyasu, 2000; see also subsection entitled p38 MAP Kinase Inhibitors).

Tacrolimus (FK506) Another immunosuppressive fungal metabolite, FK506, like cyclosporin, seems to exert its principal effects by altering the expression of cytokine genes, in particular the gene for IL-2. It shares little structural similarity to cyclosporin, although it has a similar mechanism of action (Morris, 1994). Like cyclosporin, FK506 acts by binding to a cytosolic immunophilin; unlike cyclosporin, for FK506 this immunophilin is termed FK506 binding protein (FKBP). Inhibition of IL-2 gene transcription then occurs in a molecular manner similar to that following treatment with cyclosporin (Thomson *et al.*, 1995).

Rapamycin (Rapamune, Sirolimus) Rapamycin shares structural similarities, including identical binding domains, with FK506; like FK506, it also binds to the cyclophilin FKBP. However, unlike its molecular cousin, rapamycin appears to have only limited (or no) effect on cytokine production, but rather appears to affect cell cycle progression in late G1 phase by inhibiting growth factor signal transduction pathways (Sehgal, 1995).

Although rapamycin does not appear to affect cytokine production, there is some evidence that it may decrease the stability of cytokine RNA within the cell. Work on elucidating the precise mechanism of this drug is under way.

Leflunomide (HWA 486) This synthetic immunomodulatory drug (an isoxazole derivative) has been demonstrated to be effective in animal models of autoimmunity and transplantation rejection. Early reports on its mechanism of action were conflicting, although it was recognized early that the drug operates via different mechanisms than the macrolides described earlier. In particular, early studies provided conflicting data regarding its ability to suppress cytokine production or the expression of cytokine receptors (Bartlett *et al.*, 1994). More recently, Cao *et al.* (1996) have demonstrated that leflunomide does in fact work via modulation of cytokine function, enhancing the production of the immunosuppressive cytokine TGF-β. It is clear that more work is required to identify the exact mechanism of action of this drug.

Thalidomide After languishing for many years owing to its etiology in a number of birth defects following its use in pregnant humans, thalidomide has demonstrated potential as an inhibitor of cytokine production and thus may be enjoying a comeback (Jacobson, 2000). For a number of years it has been known to be a potent suppressor of TNF production (Zwingenberger and Wendt, 1995) and has also been demonstrated to suppress production of other cytokines. The mechanism of cytokine suppression by thalidomide is unknown, although it may act as an immunomodulator via selective gene regulation.

Pentoxifylline This methylxanthine drug has been shown to reproducibly inhibit the production of TNF, and IL-12 (Moller *et al.*, 1997) and to variably affect the production of other cytokines such as IL-1, IL-6, IL-8 and IL-10 (D'Hellencourt *et al.*, 1996). Pentoxifylline has promise in treating conditions associated with release of inflammatory cytokines, such as toxic shock (Krakauer and Stiles, 1999).

Pentamidine This antiprotozoal drug is used primarily to treat *Pneumocystis carinii* pneumonia, as well as certain inflammatory conditions. In addition to its antiprotozoal activity, pentamidine inhibits the production of IL-1 via a post-translational event, possibly altering the cleavage of the precursor form of the cytokine (Rosenthal *et al.*, 1991). More recently, it has been shown to inhibit the production of various inflammatory chemokines (Van Wauwe *et al.*, 1996).

p38 MAP Kinase Inhibitors Members of the mitogen-activated protein (MAP) kinase family include p38, JNK, and extracellular response kinases (ERK). All these molecules play a role in mediating signals associated with cytokines, as well as other factors such as stress (Rincon *et al.*, 2000). Thus, modulation of this pathway is seen as a potential therapeutic option for various immune and inflammatory conditions including septic shock, cardiovascular conditions, and arthritis (Lee *et al.*, 1999; Salituro *et al.*, 1999). Although p38 inhibitors can be structurally diverse, these pyridinylimidazole agents appear to work by competitive binding in the ATP pocket (Fijen et al., 2001; Lee *et al.*, 2000). In addition, at least some of these agents also appear to work by destabilizing specific cytokine mRNA. The mechanism for this action is not understood (Wang *et al.*, 1999).

Miscellaneous Small Molecules A number of peptide and nonpeptide small molecules are under development as specific modulators of the cytokine response, both as inhibitors and as agonists (Boger *et al.*, 2001). A particularly important feature of some of these molecules is their apparent ability to selectively modulate the production of T1/T2 cytokines, thus allowing an exquisite control of the immune/inflammatory response that was not available earlier (Mollison *et al.*, 1999; Tamaoki *et al.*, 2000). Small-molecule agents exert their activities via a number of different mechanisms (Table 2), and a comprehensive review of them is not warranted here. From a toxicological standpoint, however, it should be noted that all these agents would require case-by-case assessment for immunotoxicology.

Biological Agents Inhibiting Cytokine Production/Action

Inhibitory Cytokines To date, only one endogenous cytokine antagonist has been identified, namely, the interleukin 1 receptor antagonist (IL-1RA).

Table 2

Miscellaneous small molecules with cytokine-modulatory activity

Agent/action	Ref.
Phosphodiesterase isozyme inhibitors	Yoshimura *et al.* (1997)
Metalloproteinase inhibitors	Gallea-Robache *et al.* (1997)
PKC isozyme-selective inhibitor (Rottlerin)	Kontny *et al.* (2000)
NF-κ B inhibitor (SP100030)	Gerlag *et al.* (2000)

IL-1RA exists in two forms, secreted and intracellular. It is a member of the IL-1 cytokine family (including IL-α and IL-β) according to amino acid sequence, receptor binding avidity, and gene structure and location (Arend, 1995). It seems to function as a pure receptor antagonist and does not exhibit any discernible agonist activity (e.g., internalization of receptor complexes, etc.) Experimental evidence suggest that IL-1RA plays a role in various disease states such as rheumatoid arthritis (RA), sepsis, and diabetes (Arend, 1995).

Cytokine Receptors Although most cytokines appear to exert their biological activity via interaction with specific, cell-surface-bound receptors, a number of cytokine receptors are known to be present in a soluble form in the circulation of normal healthy individuals. These soluble receptors may bind their cognate antigen with the same affinity as the surface receptors, and various physiological functions have been postulated for them. One possible function is to serve as a chaperone molecule, protecting the cytokines as they are ferried through the circulation. Another possible function, and one that serves as the basis for their possible use as therapeutics, is as a binding molecule for free cytokines in the circulation, thus serving to control their bioavailability. Most of the work in this area has involved the IL-2 receptor (Morris and Waldmann, 2000), although a number of other soluble receptors have been investigated (Fernandez-Botran, 1999).

A new concept in cytokine receptor biology that may eventually yield therapeutic options is the concept of decoy receptors (Mantovani *et al.*, 2001). In normal physiology, these soluble receptors, which are incapable of signal transduction, apparently act as molecular sinks, thus limiting the level of circulating cytokines. Several of these molecules are in the very early stages of investigation as anti-cytokine therapy (Attur *et al.*, 2000; D'Amico *et al.*, 2000).

Anti-cytokine Antibodies Like soluble cytokine receptors, endogenous anti-cytokine antibodies have been demonstrated in the circulation of normal healthy individuals. Also like the soluble receptors, these antibodies are thought to represent a mechanism for controlling the bioactivity of cytokines and to prevent them from inducing inappropriate responses. These molecules have been invaluable research reagents in the quest to understand the role of cytokines in myriad biological processes, and they are increasingly being investigated as therapeutics. A number of anti-cytokine antibodies (particularly anti-IL-4 and anti-IL-5 antibodies) have been evaluated with some success in the treatment of allergic disease. A variety of other conditions have been treated with anti-cytokine antibodies with varying degrees of success (Arend, 1995).

Fusion Toxins/Immunotoxins/Chimera Toxins A slightly different use of anti-cytokine biotherapeutics targets the cytokine-producing cell, rather than the cytokine itself. As discussed elsewhere, the overproduction of cytokines, or at least an overly vigorous response by cytokine-receptor-bearing cells (e.g., lymphocytes), can lead to a number of pathological conditions such as autoimmunity or neoplasia. The greatest degree of specificity would be to develop a "magic bullet," a molecule that is both highly specific and toxic.

A number of highly potent toxins are found in nature, and three of these in particular have been used in the construction of immunotoxins: the bacterial products diphtheria toxin and *Pseudomonas* exotoxin, and ricin, a plant product. These molecules share several structural features, including a binding domain allowing them to attach to cells, a component that allows the molecule to cross the target cell membrane, and an enzymatically active domain (the toxophore) that poisons the cell (Strom *et al.*, 1991; Kreitman *et al.*, 1992). Since the binding domain allows the toxins to attach to many different cell types, it is necessary to remove this binding domain and replace it with a portion of the cytokine or growth factor of interest. The resulting hybrid molecule is specific only for cells bearing a particular receptor; the molecular components mediating translocation and intoxication remain functional. Two of the earliest successes with this approach were the toxins DAB_{486}-IL-2 and DAB_{389}-IL-2, which contained fragmentary portions of the IL-2 molecule (Strom *et al.*, 1991). Clinical trials and experimental studies with these immunotoxins have found them to be relatively safe and effective for a wide variety of conditions including mycosis fungoides (Kuzel *et al.*, 1993), rheumatoid arthritis (Schrohenloher *et al.*, 1996), and malignancies that express IL-2 receptor (Nichols *et al.*, 1997). A number of other fusion toxins have been constructed (Table 3). Most of these novel toxins are still experimental, so their actual utility as therapeutics remains to be established.

Delivery of Cytokine Therapeutics

As mentioned earlier, cytokines are proteins or small polypeptide molecules. Since these molecules could not withstand the digestive environment, oral administration has been impractical, and administration has been primarily by intravenous injection. Although arguably the most convenient method clinically, this approach is associated with a number of serious disadvantages, including the necessity of bolus administration (or at least short-term infusion) and the resultant toxicities, as well as the need to administer high doses of cytokines to yield sufficient locally effective concentrations and a resultant lack of targeting.

Table 3
Cytokine Fusion Toxins

Toxin	Cytokine	Ref.
Diphtheria toxin	IL-3	Liger *et al.* (1997)
	IL-6	Chadwick *et al.* (1993a)
	IL-15	vanderSpek *et al.* (1995)
	G-CSF	Chadwick *et al.* (1993b)
	GM-CSF	Hotchkiss *et al.* (1999)
Ricin	IL-1	Frankel *et al.* (1996)
	GM-CSF	Burbage *et al.* (1997)
Pseudomonas exotoxin	IL-4	Puri (1999)
	IL-13	Husain *et al.* (2001)

A variety of advanced methodologies are under development for a more efficient and clinically manageable delivery system for therapeutic cytokines. Although many of these techniques are in the developmental phase, they show exciting promise as future treatment options. Table 4 illustrates some of the methods under development in this area.

Cytokine Gene Transfer

One type of gene transfer is designed specifically for treating neoplastic disease. In this approach, cytokine genes are inserted directly into tumor cells, which subsequently produce cytokines. Cytokines exhibit a number of properties that make them viable candidates for treatment of neoplasia, including immunostimulatory activity, the ability to enhance tumor antigen presentation or antigenicity, and direct cytotoxic activity. Although the toxicity associated with systemic cytokine treatment is a major drawback, a more localized administration might lessen this toxicity. A number of techniques are being developed to specifically target cytokine genes, such as targeted delivery to the nervous system using engineered herpesviruses (Poliani *et al.* 2001).

Table 4
Methodologies for Cytokine Delivery

Delivery system	Example	Ref.
Bacterial colonization	Delivery of IL-2 and IL-6 to mucosa	Steidler *et al.* (1998)
Ultrasound-mediated transdermal delivery	Delivery of erythropoietin and interferon-γ demonstrated	Mitragotri *et al.* (1995)
Biodegradable polymers	Delivery of IL-12 to tumors	Maheshwari *et al.* (2000)
Microencapsulated cytokines	Systemic administration of CSF-1	Maysinger *et al.* (1996)
Vitamin B_{12}-mediated oral delivery	Delivery of orally active EPO and G-CSF	Russell-Jones *et al.* (1995)
Subcutaneous infusion	Stimulation of megakaryocytopoiesis by IL-11	Leonard *et al.* (1996)
Lipid–DNA complexes	Delivery of IL-2 and IL-12 to tumors	Dow *et al.* (1999)
Poly(ethylene glycol) hybrid	Increased plasma half-life of TNF	Tsutsumi *et al.* (1995)

In practice, tumor cells from a patient would be removed, specific cytokine genes would be inserted, and these cells would then be reinfused. The theory behind this technique is that the tumor cells will secrete cytokines *in vivo*; when these cells are encountered by immune effector cells, the secreted cytokines will enhance the antitumor effect and the tumor cells will be eliminated. Another potential problem is that gene transfer may induce growth autonomy in the tumor cells, actually exacerbating the problem (Gansbacher *et al.*, 1993). Clearly, a number of issues remain to be addressed before this technique can be implemented in human therapy.

As further illustration of the increasingly fuzzy distinctions between the various novel cytokine constructions, Arteaga *et al.* (2001) have recently described the design of vectors that express GM-CSF DNA with different localization signals. This approach allowed the researchers to target and express the message for this cytokine in various subcellular-compartments, including the cytoplasm, endoplasmic reticulum, and nucleus.

CLINICAL TOXICITY ASSOCIATED WITH MODULATION OF CYTOKINE ACTIVITY

A common unintended consequence of cytokine treatment, although not a toxicity per se, is the development of anti-cytokine antibodies. This may seem surprising intuitively, since cytokines are endogenous molecules. However, anti-cytokine antibodies have been demonstrated in the sera of normal healthy individuals (Antonelli, 1994) and are now thought to constitute part of a complex regulatory system for controlling the activities of endogenous cytokines. Thus, it should not be surprising that administration of superphysiological concentrations of these proteins should induce the production of specific antibodies. Moreover, as mentioned elsewhere regarding animal studies, even recombinant proteins may not duplicate the natural product, with the results that novel antigens are created and the antibodies subsequently developed.

Anti-cytokine antibodies may be either binding (i.e., directed against epitopes without functional consequences) or neutralizing antibodies (those that abrogate the biological activity of the molecule). Naturally, of the two, neutralizing antibodies represent the greater concern because their presence could effectively negate the therapeutic benefit of treatment. Owing to the relatively small number of studies evaluating these molecules, most of the neutralizing antibody information has been gained from clinical trials of interferons (Antonelli, 1994), interleukin 2 (Allegretta *et al.*, 1986), and hematopoietic growth factors such as GM-CSF (Wadhwa *et al.*, 1999). It is obvious that minimizing the development of anti-cytokine antibodies will be an ongoing challenge to clinicians.

Aside from the issue of neutralizing antibody formation, administration of almost all exogenous cytokines tested to date has been associated with a range of toxicities of varying severity (e.g., Goldman, 1995; Car *et al.*, 1999). Some representative examples of such toxicity following clinical administration are given in Table 5, which lists some of the more common toxicities associated with three different classes of therapeutic cytokines; the list of symptoms is representative of such adverse side effects and is by no means a comprehensive catalog.

One toxic side effect that appears to be common to a number of different cytokines (particularly IL-2 and GM-CSF) is the so-called vascular leak or capillary leak syndrome. This syndrome is characterized by symptoms including peripheral and pulmonary edema, hypoxia, occasional ascites, and perhaps hypotension, and sometimes respiratory failure (Bruton and Koeller, 1994; Vial and Descotes 1995). Although the syndrome is suspected to be of immunological origin, the exact mechanism of toxicity remains to be elucidated (Cohen *et al.*, 1992).

Table 5
Clinical Toxicities Associated with Therapeutic Cytokine Administration

Cytokine(s)	Reported toxicities	Ref.
Interleukin-2	Hypotension, weight gain, capillary leak syndrome, renal dysfunction, gastrointestinal disturbances, neurological symptoms	Bruton and Koeller, (1994)
	Neuropsychiatric disturbances	Lerner *et al.* (1999)
Hematopoietic factors	Bone pain, autoimmune disturbances, vascular leak syndrome, arterial thromboses, cutaneous reactions	Vial and Descotes (1995)
Interferons	"Flu-like syndrome," neurological toxicity (fatigue, behavioral changes), hypo/hypertension, myalgia, autoimmunity (rare), tachycardia, hematotoxicity	
Neuropsychiatric disturbances	Endocrine disorders	Vial and Descotes (1994)
	Cutaneous reactions	Lerner *et al.* (1999)
		Vial *et al.* (2000)
		Prussick (1996)

IMMUNOTOXICOLCOGY ASSESSMENT OF CYTOKINE THERAPEUTICS

Overview

Preclinical toxicology is an integral and vital component of all pharmaceutical and biotechnology drug development. Toxicological evaluation estimates the potential of a drug candidate to adversely affect various organ systems at up to superphysiological doses, thus establishing a margin of safety prior to the initiation of clinical trials in humans (Ryffel, 1996). Although the preceding sections of this chapter have established the multiple actions (and therefore the multiple potential targets of toxicity) of cytokine-associated therapeutic moieties, the balance of the chapter focuses on the assessment of the potential of these agents to adversely affect the structure and function of the immune system (i.e., immunotoxicity).

The immune system comprises a complex collection of interrelated cellular and molecular components that protect the organism from foreign materials (primarily infectious agents). Simplistically, the essential functions of the immune system are to recognize, eliminate, and remember that which is "nonself." It is obvious that the immune system is essentially for survival. At least two features of this system present a challenge when one is assessing the effect of potential toxicants, however. First, the vertebrate immune system is composed of dynamic cellular components that are constantly undergoing adaptation, differentiation, and proliferation. The second key feature is the regulation of the immune system by intricate feedback mechanisms that can be modulated in a positive or negative fashion.

Regulatory Considerations in Immunotoxicology Assessment

The panel (tier) of validated assays and experimental approaches used in most immunotoxicology study designs resulted from the seminal work of Luster *et al.* (1988). The tier approach consists of a structured panel of assays designed to evaluate the structural and functional integrity of the immune response of an experimental animal (in the original work, rodents) following xenobiotic exposure. Initial assessment (screening) consists of a limited panel of assays designed to provide an overview of potential immune deficits. If significant alteration in any of the parameters were observed, a more comprehensive panel of assays could be employed to determine the mechanism of immune dysfunction.

Although the tier approach served as the foundation of all subsequent approaches for assessing immunotoxicity of chemicals and drugs, later studies by Luster and his collaborators indicated that a judiciously chosen combination of tests was often sufficient to immunotoxicity and that the entire panel was, in effect, superfluous for screening purposes. Based on this, most regulatory mandates for testing new molecular entities (NME) echo the tier approach in their design and requirements.

U.S. Food and Drug Administration Center for Drug Evaluation and Research (CDER)

The CDER is responsible for ensuring the safety of human pharmaceuticals produced primarily by synthetic or semisynthetic means (i.e., not biologicals). The immunotoxicology guidance document entitled "Immunotoxicology Evaluation of Investigational New Drugs," issued in draft format (U.S. FDA, 2001), details guidance for evaluation of potential immunotoxicity that would pertain to nonbiological cytokine therapeutic agents including small-molecule agonists and inhibitors, metabolic inhibitors (e.g., p38 inhibitors), and immunomodulatory drugs with cytokine-modulating activity (e.g., cyclosporin). The document addresses potential immunotoxicity as immunosuppression, antigenicity, hypersensitivity, autoimmunity, and adverse immunostimulation.

Unintended immunosuppression is particularly relevant in the present context because an increasingly precise comprehension of the immune response is resulting in a variety of NME designed specifically to suppress immunity. Although unintended drug-induced immunosuppression is relatively uncommon, it is of sufficient concern to warrant consideration. Routine toxicology tests that may indicate immunosuppression include hematology (total counts and differentials), serum immunoglobulin levels, and descriptive histopathological examination of thymus, spleen, and draining lymph nodes. If results from these tests suggest immunosuppression, follow-up (level I) tests could include assessment of leukocyte subsets by flow cytometric analysis, as well as the antibody-forming cell assay. Evidence of immune modulation in level I tests might prompt follow-up (level II) tests, which could be designed to measure host resistance (infectious organisms or transplantable tumors) and to determine the specific mechanistic of immunomodulation.

A second area of concern to CDER is antigenicity, which for nobiological drugs refers primarily to the induction of (drug allergy) reactions—any of the types I to IV hypersensitivity reactions. Related to this concern is pseudoallergy, an adverse drug reaction whose mechanisms of action are not well understood. Although pseudoallergy is not necessarily immune mediated, it may be difficult to differentiate from true IgE-mediated hypersensitivity reactions.

Autoimmunity is the condition in which exposure to a drug is associated with the induction of an immune response against the patient's own (self) antigens; one classical example is drug-induced lupus. The mechanisms for drug-related autoimmunity are very poorly understood, and there are unfortunately no standard methods for prediction of this adverse effect. Closely related to drug-induced autoimmunity is unintended immunostimulation. As with immunosuppression, NMEs (such as most of the nonbiologicals discussed in this chapter) are being designed to augment or restore a suboptimal immune response (e.g., repopulation of bone marrow). Unintended immunostimulation may be subtle and difficult to identify; moreover, its identification may be confounded by overlap with a drug's intended mechanism of action.

U.S. Food and Drug Administration Center for Biologics Evaluation and Research (CBER)

The CBER arm of the FDA regulates the safety of human pharmaceuticals that are produced as biological agents, rather than synthesized. This includes blood and blood products, vaccines, and biotechnology products such as cytokines, therapeutic proteins, and monoclonal antibodies. In mid-2001 CBER did not have a specific Guidance Document for immunotoxicology testing. Rather, testing for potential immunotoxicity was to be evaluated on a case-by-case basis. In many cases, testing of drugs regulated by CBER (i.e., biologicals) would be covered by the ICH S6 document, discussed shortly.

Case-by-case assessment of biological products generally emphasizes the importance of testing an NME in an appropriate species (i.e., one in which the agent is biologically active), determination of potential immunogenicity, the interpretation of multiple-dose toxicity study results, and the availability of specific and sensitive assays for determining pharmacodynamic properties of the agent (Griffiths and Lumley, 1998; Dempster, 2000).

U.S. Food and Drug Administration, Center for Devices and Radiological Health (CDRH)

Traditionally, immunotoxicology assessment has evaluated the effect on the immune system of repeated exposure to test materials (drugs, chemicals, etc.). Although the precise regimen of exposure varies, the repeated-dose nature of the exposure generally has not. However, it has been recognized that medical devices designed to remain in contact with the body (essentially a permanent exposure regimen) may enhance the opportunity for an agent to modulate the immune response (Anderson and Langone, 1999). In the context of cytokine therapeutics, this property would be relevant for the alternative methods of cytokine delivery described earlier.

In 1999 CDRH, the branch of the FDA responsible for ensuring the safety of medical devices and for preventing unnecessary exposure to radiation from electronic products, published a guidance document entitled "Immunotoxicity Testing Guidance" (U.S. FDA, 1999). This document is intended to be used in conjunction with CDRH's Memorandum G95-1, which is an FDA-modified version of the International Standard ISO-10993 document "Biological Evaluation of Medical Devices-Part 1: Evaluation and Testing." The immunotoxicology guidance document describes provides a flowchart for determining whether immunotoxicology testing may be needed to support the safety of a medical device. There are also several tables summarizing potential immunotoxic effects based on body contact and contact duration. This includes type of body contact such as surface (skin, mucosal, or compromised surface), external communicating (blood path, tissue/bone/dentin), or implanted (tissue/bone, blood, body fluids). Testing approaches or methods based on immunotoxic effect are also described (hypersensitivity, inflammation, immunosuppression, immunostimulation, and autoimmunity). Finally, a list of models that can be employed in evaluating immune responses (cross-referenced by immune parameter assessed and end point evaluated) is included.

International Conference on Harmonisation of Technical Requirement for Registration of Pharmaceuticals for Human Use (ICH)

Comprising representatives from Europe, the United States, and Japan, the ICH was instituted to provide an opportunity for the pharmaceutical industry and regulatory agencies to discuss requirements for product registration in the three member regions. ICH has published a document with direct relevance to immunotoxicology assessment, "Preclinical Safety Evaluation of Biotechnology-Derived Pharmaceuticals," commonly referred to as the "S6" document (ICH, 1997). Pertinent sections of the document include 3.6 (Immunogenicity) and 4.5 (Immunotoxicity Studies).

Section 3.6 of the S6 document recognizes the following: many biotechnology-derived drugs are potentially immunogenic; detection of antibodies should not be sole criterion for termination or modification of safety testing, and antibody formation in animals is not necessarily predictive for humans. Section 4.5 of S6 recognizes that certain drug candidates may be intended to alter the immune response; that testing strategies may require screening studies, followed by mechanistic studies; and that the standard, so-called tier approach is not recommended for biotechnology-derived pharmaceuticals. This latter point is critical, since it places emphasis on science-based, case-by-case assessment of each molecule based on its intended use and mechanism of action.

Committee for Proprietary Medicinal Products (CPMP)

The CPMP was created by the European Commission to provide technical and scientific support for ICH activities. In July 2000 the CPMP published document CPMP/SWP/1042/99, "Note for Guidance on Repeated Dose Toxicity," which includes guidance on immunotoxicology testing on agents intended for regulatory submission in Europe (CPMP, 2000). The document states that all new medicinal products should be screened for immunotoxic potential in at least one repeated-dose toxicity study, preferably in a study of 28 days duration (although 14-day or 3-month studies may also be acceptable). This guidance document would be applicable for assessment of many of the nonbiological cytokine-modulating drugs discussed in this chapter, although it would not be appropriate for assessment of biologicals such as cytokine receptors or recombinant cytokine molecules (or their protein derivatives).

The initial screening phase consists of both nonfunctional and functional assessment of immune system integrity in rats or mice; all these end points, including the functional assays, are readily incorporated into standard toxicology assessment paradigms. Tests to be performed include hematology, lymphoid organ weights (thymus, spleen, draining and distant lymph nodes), lymphoid organ microscopy (thymus, spleen, draining and distant lymph nodes, Peyer's patches), bone marrow cellularity, natural killer cell activity, and distribution of lymphocyte subsets. These latter two functional assays may be performed by using cells derived from animals during a standard repeated-dose toxicity test and thus do not require the addition of experimental groups, unlike the response to a T-cell-dependent antigen, which requires immunization. Alterations in these parameters may signal the need for additional testing.

Extended studies to define mechanisms of any potential immunotoxicity include functional tests and would be included based on results of initial screening studies, again on a case-by-case basis depending on the nature of the observations from the screening phases. Suggested functional assays include delayed-type hypersensitivity response, lymphocyte proliferation following stimulation with mitogens or specific antigens, macrophage function, primary response to T-dependent antigen antibody, and *in vivo* models of host resistance to infectious agents or tumor cells.

Japanese Immunotoxicology Regulations

The standard tier concept of immunotoxicity assessment is not utilized in Japan (Maki, 1997). Requirements for New Drug Applications in Japan include antigenicity studies, skin sensitization studies, and skin photosensitization studies. Several indirect immunotoxicity tests, used by some investigators, include histopathological changes in lymphoid tissues, weight changes in lymphoid organs, bone marrow effects, and changes in hematology or blood chemistry. In general, "direct" immunotoxicity (i.e., specific for immune function per se) tests are not routinely performed in Japan, where the perceptions are prevalent that the criteria for defining immunotoxicity not clearly defined and that the testing methods are not fully standardized or validated.

Techniques and Approaches Used for Preclinical Immunotoxicology Assessment

The variety of methods and experimental approaches for evaluating the immunotoxicology of drugs and chemicals is varied; a comprehensive overview is beyond the scope of the chapter but may be obtained elsewhere (Burleson *et al.*, 1995; Dean *et al.*, 2001; House and Thomas, 2001). The sections that follow highlight some of the techniques most commonly encountered in regulatory immunotoxicology guidance documents.

Model Species

During the initial development and validation studies that formed the genesis of the discipline of immunotoxicology, the mouse was almost exclusively the model of choice. Consequently, the bulk of historical data for both drugs and chemicals is based on mouse-derived data. Although the mouse was adequate for the development of the field, this species is a relatively uncommon model for toxicology and safety assessment studies. More recently, the rat has gained prominence as a model for immunotoxicology testing and has been shown to yield results comparable to those from mice (Ladics *et al.*, 1995, 1998). For evaluation of many of the nonbiological cytokine modulators, particularly ones with mechanisms of action directed to highly conserved pathways (e.g., p38 inhibitors), rodents should serve as an adequate model system.

Rodents are less ideally suited for immunotoxicology evaluation with biotechnology-derived molecules such as peptides, proteins, or other macromolecules. At the risk of fatal oversimplification, the primary difficulties in preclinical and clinical evaluation of biological cytokine modulators come down to two problems. The first problem is that traditional animal toxicology models may not reproducibly predict the ultimate human toxicity of recombinant molecules, particularly cytokines. This may be best understood by considering the basic biology of the cytokines themselves. Three essential features are of particular importance.

1. Certain cytokines exhibit a high degree of species specificity. Unlike most nonbiological therapeutics, cytokines mediate their normal physiological (and apparently at least some of their deleterious) effects via their interaction with highly specific receptor molecules. This receptor interaction is also the basis for the strict species specificity of certain cytokines. Thus, administration of such molecules to an inappropriate species might be expected to result in difficulties in interpretation.
2. Although many recombinant proteins are highly homologous to their endogenous counterparts, minor biochemical discrepancies (e.g., base pair substitution, alternative forms of glycosylation) may result in altered biological function or, more likely, antigenic stimulation leading to the induction of neutralizing antibodies. Although the effect is not as common when cytokines are evaluated in homologous species (e.g., mouse/mouse), it has been found that when human sequence cytokines are injected into laboratory animals, neutralization of the protein's bioactivity will result within a few weeks. Consequently, effects that may become apparent in humans after chronic administration will effectively be masked in animal models.

3. Whereas structure and function have been relatively well conserved in most mammalian cytokines, certain of these molecules display divergent function between species. In such cases, the clinical sequelae of human cytokine therapy might not be predictable even from appropriately designed animal studies.

The second problem, namely, that the clinical response to *in vivo* cytokine administration is not necessarily predictable by *in vitro* data, is likewise a complex issue. This is best examined in light of cytokine biology.

1. Cytokines are pleiotropic and redundant in function. An example of this is provided by tumor necrosis factor, which exhibits potent cytotoxicity for isolated (i.e., *in vitro*) tumor cells and induces the regression of solid tumor following *in situ* injection. However, TNF's broad diversity of physiological actions (including its central role as a mediator of shock) precludes its use as a systemic therapeutic. Cytokine redundancy may account for certain toxicities by inappropriately modulating the effects of other cytokines.

2. Cytokines appear to function principally at a local level. Although cytokines are now known to mediate the functions of the immune, nervous, and endocrine systems (the so-called neuroimmunoendocrine axis), most of these interactions probably take place at discrete sites. Indeed, the systemic (circulating) levels of almost all cytokines are extremely low, and exogenously administered cytokines are rapidly cleared from the circulation.

Despite these potential difficulties, a number of approaches are being taken to facilitate the assessment of cytokine toxicity. One approach that is showing great promise in general toxicology as well as in immunotoxicology is the use of genetically modified rodents as model systems. Homologous recombination technology can be used to breed, so-called knockout mice, mice that lack particular genes of interest, or, conversely, transgenic mice, which constitutively express human sequence proteins or overexpress native sequence proteins (Lovik, 1997; Rudolph and Mohler, 1999). Vis-à-vis cytokines, a number of models have been described for elucidating the function of the endogenous molecule in various physiological processes (Taverne, 1993; Dighe *et al.*, 1995). More to the point of immunotoxicological assessment, however, are models for evaluating the effects of cytokine modulation itself on the test system (Bessis *et al.*, 1998). Another approach is the construction of transgenic animals expressing human cytokine receptors (Kyoizumi *et al.* 1993). Although none of these options are ideal, they represent major improvements over the use of standard "wild-type" rodent species.

For biological therapeutic agents, the primate is generally recognized to be the most appropriate species for both efficacy and safety studies. Primates have immune systems that more closely resemble those of humans and thus respond in a more relevant manner to the intended pharmacological activities of the cytokine agents. Also important is the lower likelihood of inducing an anti-cytokine immune response, a situation that can confound pharmacokinetic data and even result in the inactivation of the therapeutic agent due to the induction of neutralizing antibodies. Even then, it may be impractical if not impossible to conduct early safety assessment in nonhuman primates, for example when the relevant test material does not exist in sufficient quantities, or when appropriate test systems are not available in primates. In such cases, rodent homologues of the test material may be evaluated in rodents. However, this is an expensive and time-intensive approach and is not often a viable alternative.

Methods for Descriptive Assessment of Immune Structure

Clinical and Anatomic Pathology

A considerable amount of information on the structural status of the immune system can be obtained from standard toxicology parameters or from end points easily incorporated into standard testing paradigms. Most of these end points are included in regulatory immunotoxicology documents and include histopathology of immune system tissues/organs (spleen, thymus, bone marrow, Peyer's patches, various lymph nodes) and lymphoid organs weights (particularly spleen and lymph node). In many cases, a skilled pathologist can detect alterations in the structure of immune tissues suggestive of immune dysfunction. If evidence is observe on standard hematoxylin/eonin-stained tissues, a variety of more detailed methodologies are available to assess the nature of the structure changes, such as immunohistochemistry and other special stains. For details on available end points, the interested reader is directed to reviews by Harleman (2000) and by Kuper *et al.* (2000).

Flow Cytometry

Flow cytometry is a powerful tool that uses the high specificity available from monoclonal antibodies to assess both the structure and function of the immune system. The early tier testing validation work by Luster *el al.* (1988, 1992) demonstrated that surface marker analysis by flow cytometry was one of the most sensitive indicators of immunotoxicity. For this reason, and because this analysis can be performed with cells derived from animals in standard toxicology study designs, flow cytometric analysis is recommended as an immunotoxicology end point by several regulatory agencies. However, a workshop sponsored by the International Life Sciences Institute Immunotoxicology Technical Committee (ILSIITC, 2001) determined that flow cytometry, by itself, does not provide sufficient information to warrant its use as the sole determinant of immunotoxicity at present. Nevertheless, the inclusion of a carefully designed panel of surface markers for analysis, as well as foresight in considering the multitude of variables that might confound such analysis, can provide much useful information when these markers are used as an adjunct to other structural and functional determinations, particularly in studies using nonhuman primates as a model. Investigators have made progress in developing standard ranges of leukocyte subsets for this purpose (Verdier *et al.* 1995).

Methods for Descriptive Assessment of Immune Function

Although much useful information on the state of the immune system may be obtained from structural analysis of the tissues, the great redundancy in function of the immune system precludes the absolute determination of immune dysfunction by means of this criterion alone. As recognized by the various regulatory agencies and good science, evidence of immune system alteration gained from standard toxicology end points should be followed up by assessment of the ability of the patent immune system to respond appropriately to exogenous stimuli. Whereas the various regulatory guidance documents differ slightly in the particulars, most of them include suggestions for evaluating humoral-mediated immune function, cell-mediated immune function, and some parameter of natural or innate host defense. In the sections that follow we will briefly consider representative of these assays.

Assessment of Humoral-Mediated Immunity

Within days of an *in vivo* injection of an antigen, antibody molecules of the IgM class are produced and released into the systemic circulation. One of the original "tier" assays in immunotoxicology is the antibody-forming-cell assay (AFC), alternatively referred to as the plaque-forming-cell (PFC) assay, which measures the production of specific antibody through enumeration of antibody-producing cells in the spleen following immunization with sheep red blood cells (SRBC), a T-dependent antigen. The AFC response to SRBC is often considered to be a measure of B-cell function only; however, it is also useful for evaluating the proper overall function of the immune system (Luster *et al*, 1988, 1992), since the induction of this response requires complex interaction among and regulation by antigen-presenting cells, T cells, and various molecules including complement and cytokines. The primary antibody response is currently measured by using either a PFC assay (Wilson *et al.*, 1999) or an enzyme-linked immunossay (ELISA) method (Temple *et al.*, 1993). The AFC assay is a standard immunotoxicology assay and is appropriate for evaluation of nonbiological agents in rodent studies. For studies in nonhuman primates, various regimens of immunization may be employed using agents such as tetanus toxoid, followed by quantitation of specific antibody production by immunoassay (e.g., ELISA). At present, there are no "standard" paradigms for primate studies, with individual investigators often utilizing their own assay systems.

Assessment of Cell-Mediated Immune Function

Although measures of humoral-mediated immunity such as the AFC assay are able to detect alterations in the function of regulatory T cells (albeit indirectly), other assays are designed to directly measure both regulatory and effector function of this arm of the specific immune response. In mice, two of the most commonly used assays are the induction of cytotoxic T lymphocytes and the induction of delayed-type hypersensitivity (DTH) (Luster *et al.*, 1988, 1992). Both these assays are highly predictive of immunomodulation produced by chemical agents. Less commonly used is assessment of lymphocyte blastogenic response, which is technically easier to perform but is not considered to be as relevant from a functional standpoint. For nonhuman primate studies, DTH is commonly used as a measure of cell-mediated immunity (Bleavins and de la Iglesia, 1995).

Innate Host Resistance

Natural killer cell activity is measured *in vitro* by culturing single-cell suspensions of lymphoid cells with a tumor cell line known to be sensitive to NK-mediated cytotoxicity. The target cells are radiolabeled prior to the assay; thus, any cells that have been lysed will release their radioactivity into the culture medium, when it can subsequently be quantitated. Modifications of the microculture procedure described by Reynolds and Herberman (1981) is the standard approach for immunotoxicity assessment in rodents, and similar methods have been developed for primates (Rappocciolo *et al.*, 1992). Another, slightly less common parameter for measuring innate host resistance (also known as natural immunity) is macrophage function. A number of other assays are available, including neutrophil function and complement activation, but these are not commonly employed in immunotoxicology assessment at any level except mechanistic evaluations.

CONCLUSIONS

The discipline of immunotoxicology, while still relatively young, has made remarkable advances in increasing the safety level of new pharmaceutical agents. Many questions remain, however, perhaps owing to the complex nature of cytokines or to our infinitesimal knowledge of how they work. For example, when does suppression of a normal immune response become pathological? What is a superphysiological response, and how does it differ from hyperpharmacology? What degree of immunostimulation is beneficial, and what degree is pathological? Related to this, does immunostimulation predispose to autoimmunity? These, and as yet many unanticipated questions, remain to be answered by immunotoxicology.

REFERENCES

Abraham, E., Glauser, M. P., Butler, T., Garbino, J., Gelmont, D., Laterre, P. F., Kudsk, K., Bruining, H. A., Otto, C., Tobin, E., Zwingelstein, C., Lesslauer, W., and Leighton, A. (1997). p55 tumor necrosis factor receptor fusion protein in the treatment of patients with severe sepsis and septic shock. A randomized controlled multicenter trial. *JAMA*, **277**, 1531–1538.

Allegretta, M., Atkins, M. B., Dempsey, R. A., Bradley, E. C., Konrad, M. W., Childs, A., Wolfe, S. N., and Mier, J. W. (1986). The development of anti-interleukin-2 antibodies in patients treated with recombinant human interleukin-2 (IL-2). *J. Clin. Immunol.*, **6**, 481–489.

Allen, R. D. (1999). Polymorphism of the human TNF-alpha promoter—Random variation or functional diversity? *Mol. Immunol.*, **36**, 1017–1027.

Almawi, W., Beyhum, H., Rahme, A. A., and Rieder, M. J. (1996). Regulation of cytokine and cytokine receptor expression by glucocorticoids. *J. Leukocyte Biol.*, **60**, 563–572.

Anderson, J. M., and Langone, J. J. (1999). Issues and perspective on the biocompatibility and immunotoxicity evaluation of implanted controlled release systems. *J. Controller Release*, **57**, 107–113.

Anderson, R., Macdonald, I., Corbett, T., Hacking, G., Lowdell, M. W., and Prentice, H. G. (1997). Construction and biological characterization of an interleukin-12 fusion protein (Flexi-12): Delivery to acute myeloid leukemic blasts using adeno-associated virus. *Hum. Gene Ther.*, **8**, 1125–1135.

Antonelli, G. (1994). Development of neutralizing and binding antibodies to interferon (IFN) in patients undergoing IFN therapy. *Antiviral Res.*, **24**, 235–244.

Arend, W. P. (1995). Inhibiting the effects of cytokines in human diseases. *Adv. Inter. Med.*, **40**, 365–394.

Arteaga, H. J., Mohamed, A. J., Christensson, B., Gahrton, G., Smith, C. I., and Dilber, M. S. (2001). Expression and release of stable and active forms of murine granulocyte–macrophage colony-stimulating factor (mgm-csf) targeted to different subcellular compartments. *Cytokine*, **14**, 136–142.

Attur, M. G., Dave, M., Cipolletta, C., Kang, P., Goldring, M. B., Patel, I. R., Abramson, S. B., and Amin, A. R. (2000). Reversal of autocrine and paracrine effects of interleukin 1 (IL-1) in human arthritis by type II IL-1 decoy receptor. Potential for pharmacological intervention. *J. Biol. Chem.*, **275**, 40307–40315.

Banyer, J. L., Hamilton, N. H., Ramshaw, I. A., and Ramsay, A. J. (2000). Cytokines in innate and adaptive immunity. *Rev. Immunogenet*, **2**, 359–373.

Bartlett, R. R., Campion, G., Musikic, P., and Schleyerbach, R. (1994). Leflunomide: A novel immunomodulating drug, in *Nonsteroidal Anti-Inflammatory Drugs: Mechanisms and Clinical Uses*, A. J. Lewis and D. E. Furst, eds., pp. 349–366. Dekker, New York.

Baumgarten, G., Knuefermann, P., and Mann, D. L. (2000). Cytokines as emerging targets in the treatment of heart failure. *Trends Cardiovasc. Med.*, **10**, 216–223.

Bessis, N., Chiocchia, G., Kollias, G., Minty, A., Fournier, C., Fradelizi, D., and Boissier, M. C. (1998). Modulation of proinflammatory cytokine production in tumour necrosis factor-alpha (TNF-alpha)-transgenic mice by treatment with cells engineered to secrete IL-4, IL-10 or IL-13. *Clin. Exp. Immunol.*, **111**, 391–396.

Bishop, B., Koay, D. C., Sartorelli, A. C., and Regan, L. (2001). Reengineering granulocyte-colony stimulating factor (G-CSF) for enhanced stability. *J. Biol. Chem.* **276**, 33465–33470.

Bleavins, M. R., and de la Iglesia, F. A. (1995). Cynomolgus monkeys (*Macaca fascicularis*) in preclinical immune function safety testing: Development of a delayed-type hypersensitivity procedure. *Toxicology*, **95**, 103–112.

Boger, D. L., and Goldberg, J. (2001). Cytokine receptor dimerization and activation: Prospects for small molecule agonists. *Bioorg. Med. Chem.*, **9**, 557–562.

Briscoe, H., Roach, D. R., Meadows, N., Rathjen, D., and Britton, W. J. (2000). A novel tumor necrosis factor (TNF) mimetic peptide prevents recrudescence of *Mycobacterium bovis* bacillus Calmette–Guerin (BCG) infection in CD4+ T cell-depleted mice. *J. Leukocyte Biol.*, **68**, 538–544.

Brugger, W., Scheding, S., Ziegler, B., Buhring, H. J., and Kanz, L. (2000). Ex vivo manipulation of hematopoietic stem and progenitor cells. *Semin. Hematol.*, **37**(1 suppl. 2), 42–49.

Bruton, J. K., and Koeller, J. M. (1994). Recombinant interleukin-2. *Pharmacotherapy*, **14**, 635–656.

Burbage, C., Tagge, E. P., Harris, B., Hall, P., Fu, T., Willingham, M. C., and Frankel, A. E. (1997). Ricin fusion toxin targeted to the human granulocyte-macrophage colony stimulating factor receptor is selectively toxic to acute myeloid leukemia cells. *Leuk. Res.*, **21**, 681–690.

Burleson, G., Dean, J. and Munson, A. (1995). *Methods in Immunotoxicology*. Wiley-Liss, New York.

Cao, W. W., Kao, P. N., Aoki, Y., Xu, J. C., Shorthouse, R. A., and Morris, R. E. (1996). A novel mechanism of action of the immunomodulatory drug, leflunomide: Augmentation of the immunosuppressive cytokine, TGF-beta 1 and suppression of the immunostimulatory cytokine, IL-2. *Transplant. Proc.*, **28**, 3079–3080.

Car, B. D., Eng, V. M., Lipman, J. M., and Anderson, T. D. (1999). The toxicology of interleukin-12: A review. *Toxicol. Pathol.*, **27**, 58–63.

Cardi, G., Ciardelli, T. L., and Ernstoff, M. S. (1996). Therapeutic applications of cytokines for immunostimulation and immunosuppression: An update. *Prog. Drug Res.*, **47**, 211–250.

Chadwick, D. E., Jean, L. F., Jamal, N., Messner, H. A., Murphy, J. R., and Minden, M. D. (1993a). Differential sensitivity of human myeloma cell lines and normal bone marrow colony forming cells to a recombinant diphtheria toxin-interleukin 6 fusion protein. *Br. J. Haematol.*, **85**, 25–36.

Chadwick, D. E., Williams, D. P., Niho, Y., Murphy, J. R., and Minden, M. D. (1993b). Cytotoxicity of a recombinant diphtheria toxin-granulocyte colony-stimulating factor fusion protein on human leukemic blast cells. *Leuk. Lymphoma*, **11**, 249–262.

Chirigos, M. A., Talor, E., Sidwell, R. W., Burger, R. A., and Warrent, R. P. (1995). Leukocyte interleukin inj. (LI) augmentation of natural killer cells and cytolytic T-lymphocytes. *Immunopharmacol. Immunotoxicol.*, **17**, 247–264.

Cohen, R. B., Siegal, J. P., Puri, R. K., and Pluznik, D. H. (1992). The immunotoxicology of cytokines, in *Clinical Immunotoxicology*, D. S. Newcombe, N. R. Rose, and J. C. Bloom, eds., pp. 93–108. Raven Press, New York.

CPMP (2000). Note for Guidance on Repeated Dose Toxicity. Committee for Proprietary Medicinal Products. Document available at www.eudra.org/humandocs/PDFs/SWP/104299en.pdf.

Craven, N. M., Jackson, C. W., Kirby, B., Perrey, C., Pravica, V., Hutchinson, I. V., and Griffiths, C. E. (2001). Cytokine gene polymorphisms in psoriasis. *Br. J. Dermatol.*, **144**, 849–853.

Csaszar, A., and Abel, T. (2001). Receptor polymorphisms and diseases. *Eur. J. Pharmacol.*, **414**, 9–22.

D'Amico, G., Frascaroli, G., Bianchi, G., Transidico, P., Doni, A., Vecchi, A., Sozzani, S., Allavena, P., and Mantovani, A. (2000). Uncoupling of inflammatory chemokine receptors by IL-10: Generation of functional decoys. *Nat. Immunol.*, **1**, 387–391.

Dean, J. H., House, R. V., and Luster, M. I. (2001). Immunotoxicology: Effects of, and response to, drugs and chemicals, in *Principles and Methods of Toxicology*, 4th ed., A. W. Hayes, ed., pp. 1415–1450. Taylor & Francis, London.

Del Prete, G. (1998). The concept of type-1 and type-2 helper T cells and their cytokines in humans. *Int. Rev. Immunol.*, **16**, 427–455.

Dela Cruz, J. S., Trinh, K. R., Morrison, S. L., and Penichet, M. L. (2000). Recombinant anti-human HER2/neu IgG3-(GM-CSF) fusion protein retains antigen specificity and cytokine function and demonstrates antitumor activity. *J. Immunol.*, **165**, 5112–5121.

Dempke, W., Von Poblozki, A., Grothey, A., and Schmoll, H. J. (2000). Human hematopoietic growth factors: Old lessons and new perspectives. *Anticancer Res.*, **20**, 5155–5164.

Dempster, A. M. (2000). Nonclinical safety evaluation of biotechnologically derived pharmaceuticals. *Biotechnol. Annu. Rev.*, **5**, 221–258.

Dendorfer, U. (1996). Molecular biology of cytokines. *Artific. Organs*, **20**, 437–444.

Deswal, A., Misra, A., and Bozkurt, B. (2001). The role of anti-cytokine therapy in the failing heart. *Heart Failure Rev.* **6**, 143–151.

Devos, R., Plaetinck, G., Cornelis, S., Guisez, Y., Van der Heyden, J., and Tavernier, J. (1995). Interleukin-5 and its receptor: A drug target for eosinophilia associated with chronic allergic disease. *J. Leukocyte Biol.*, **57**, 813–819.

D'Hellencourt, C. L., Diaw, L., Cornillet, P., and Guenounou, M. (1996). Differential regulation of TNF-alpha, IL-1 beta, IL-6, IL-8, TNF beta and IL-10 by pentoxifylline. *Int. J. Immunopharmacol.*, **18**, 739–748.

Dighe, A. S., Campbell, D., Hsieh, C. S., Clarke, S., Greaves, D. R., Gordon, S., Murphy., K. M., and Schreiber, R. D. (1995). Tissue-specific targeting of cytokine unresponsiveness in transgenic mice. *Immunity*, **3**, 657–666.

Dinarello, C. A., and Pomerantz, B. J. (2001). Proinflammatory cytokines in heart disease. *Blood Purif.*, **19**, 314–321.

Domingues, H., Cregut, D., Sebald, W., Oschkinat, H., and Serrano, L. (1999). Rational design of a GCN4-derived mimetic of interleukin-4. *Nat. Struct. Biol.*, **6**, 652–656.

Dow, S. W., Elmslie, R. E., Fradkin, L. G., Liggitt, D. H., Heath, T. D., Willson, A. P., and Potter, T. A. (1999). Intravenous cytokine gene delivery by lipid–DNA complexes controls the growth of established lung metastases. *Hum. Gene Ther.*, **10**, 2961–2972.

Duschl, A., Muller, T., and Sebald, W. (1995). Antagonistic mutant proteins of interleukin-4. *Behring Inst. Mitt.*, **96**, 87–94.

Eckenberg, R., Moreau, J. L., Melnyk, O., and Theze, J. (2000). IL-2R beta agonist P1-30 acts in synergy with IL-2, IL-4, IL-9, and IL-15: Biological and molecular effects. *J. Immunol.*, **165**, 4312–4318.

Eliason, J. F., Greway, A., Tare, N., Inoue, T., Bowen, S., Dar, M., Yamasaki. M., Okabe, M., and Horii, I. (2000). Extended activity in cynomolgus monkeys of a granulocyte colony-stimulating factor mutein conjugated with high molecular weight polyethylene glycol. *Stem Cells*, **18**, 40–45.

Farese, A. M., Herodin, F., McKearn, J. P., Baum, C., Burton, E., and MacVittie, T. J. (1996). Acceleration of hemotopoietic reconstitution with a synthetic cytokine (SC-55494) after radiation-induced bone marrow aplasia. *Blood*, **87**, 581–591.

Fernandez-Botran, R. (1999). Soluble cytokine receptors: Basic immunology and clinical applications. *Crit. Rev. Clin. Lab. Sci.*, **36**, 165–224.

Fijen, J. W., Zijlstra, J. G., De Boer, P., Spanjersberg, R., Cohen Tervaert, J. W., Van Der Werf, T. S., Ligtenberg, J. J., and Tulleken, J. E. (2001). Suppression of the clinical and cytokine response to endotoxin by RWJ-67657, a p38 mitogen-activated protein-kinase inhibitor, in healthy human volunteers. *Clin. Exp. Immunol.*, **124**, 16–20.

Frankel, A. E., Burbage, C., Fu, T., Tagge, E., Chandler, J., and Willingham, M. (1996). Characterization of a ricin fusion toxin targeted to the interleukin-2 receptor. *Protein Eng.*, **9**, 913–919.

Fraser, J. D., Straus, D., and Weiss, A. (1993). Signal transduction events leading to T-cell lymphokine gene expression. *Immunol. Today*, **14**, 357–362.

Gallea-Robache, S., Morand, V., Millet, S., Bruneau, J. M., Bhatnagar, N., Chouaid, S., and Roman-Roman, S. (1997). A metalloproteinase inhibitor blocks the shedding of soluble cytokine receptors and processing of transmembrane cytokine precursors in human monocytic cells. *Cytokine*, **9**, 340–346.

Gansbacher, B., Rosenthal, F. M., and Zier, K. (1993). Retroviral vector-mediated cytokine-gene transfer into tumor cells. *Cancer Invest.*, **11**, 345–354.

Gerlag, D. M., Ransone, L., Tak, P. P., Han, Z., Palanki, M., Barbosa, M. S., Boyle, D., Manning, A. M., and Firestein, G. S. (2000). The effect of a T cell-specific NF-kappa B inhibitor on in vitro cytokine production and collagen-induced arthritis. *J. Immunol.*, **165**, 1652–1658.

Gherardi, M. M., Ramirez, J. C., and Esteban, M. (2001). Towards a new generation of vaccines: The cytokine IL-12 as an adjuvant to enhance cellular immune responses to pathogens during prime-booster vaccination regimens. *Histol. Histopathol.*, **16**, 655–667.

Golding, B., Zaitseva, M., and Golding, H. (1994). The potential for recruiting immune responses toward type 1 or type 2 T cell helps. *Am. J. Trop. Med. Hyg.*, **50**, 33–40.

Goldman, S. J. (1995). Preclinical biology of interleukin 11: A multifunctional hematopoietic cytokine with potent thrombopoietic activity. *Stem Cells*, **13**, 462–471.

Grazi Cusi, M., and Ferrero, D. (1997). Harlequin granulocyte-colony stimulating factor interleukin 6 molecules with bifunctional and antagonistic activities. *Immunotechnology*, **3**, 61–69.

Griffiths, S. A., and Lumley, C. E. (1998). Non-clinical safety studies for biotechnologically derived pharmaceuticals: Conclusions from an international workshop. *Hum. Ext. Toxicol.*, **17**, 63–83.

Gursel, M., and Gregoriadis, G. (1998). The immunological co-adjuvant action of liposomal interleukin-2: The role of mode of localisation of the cytokine and antigen in the vesicles. *J. Drug Target.*, **5**, 93–98.

Harleman, J. H. (2000). Approaches to the identification and recording of findings in the lymphoreticular organs indicative for immunotoxicity in regulatory type toxicity studies. *Toxicology*, **142**, 213–219.

Heath, A. W., and Playfair, J. H. L. (1992). Cytokines as immunological adjuvants. *Vaccine*, **10**, 427–434.

Henderson, B., and Blake, S. (1992). Therapeutic potential of cytokine manipulation. *Trends Pharmacol. Sci.*, **13**, 145–152.

Hetzel, C., and Lamb, J. R. (1994). $CD4^+$ T cell-targeted immunomodulation and the therapy of allergic disease. *Clin. Immunol. Immunopathol.*, **73**, 1–10.

Hotchkiss, C. E., Hall, P. D., Cline, J. M., Willingham, M. C., Kreitman, R. J., Gardin, J., Latimer, A., Ramage, J., Feely, T., De Latte, S., Tagge, E. P., and Frankel, A. E. (1999). Toxicology and pharmacokinetics of DTGM, a fusion toxin consisting of a truncated diphtheria toxin (DT388) linked to human granulocyte–macrophage colony-stimulating factor, in cynomolgus monkeys. *Toxicol. Appl. Pharmacol.*, **158**, 152–160.

House, R. V., and Thomas, P. T. (2001). Immunotoxicology: Fundamentals of preclinical assessment, in *CRC Handbook of Toxicology*, 2nd ed., M. Derelanko and M. Hollinger, eds., pp. 401–438. CRC Press, Boca Raton, FL.

Husain, S. R., Joshi, B. H., and Puri, R. K. (2001). Interleukin-13 receptor as a unique target for anti-glioblastoma therapy. *Int J. Cancer*, **92**, 168–175.

ICH (1997). Preclinical Safety Evaluation of Biotechnology-Derived Pharmaceuticals. International Conference on Harmonisation of Technical Requirement for Registration of Pharmaceuticals for Human Use. Document available at http://www.ifpma.org/pdfifpma/s6.pdf

International Life Sciences Institute, Immunotoxicology Technical Committee. (2001). Application of flow cytometry to immunotoxicity testing: Summary of a workshop. *Toxicology*, **163**, 39–48.

Jacobson, J. M. (2000). Thalidomide: A remarkable comeback. *Expert Opin. Pharmacother.*, **1**, 849–863.

Kim, T. S., DeKruyff, R. H., Rupper, R., Maecker, H. T., Levy, S., and Umetsu, D. T. (1997). An ovalbumin-IL-12 fusion protein is more effective than ovalbumin plus free recombinant IL-12 in inducing a T helper cell type 1–dominated immune response and inhibiting antgen-specific IgE production. *J. Immunol.*, **158**, 4137–4144.

Kontny, E., Kurowska, M., Szczepanska, K., and Maslinski, W. (2000). Rottlerin, a PKC isozyme-selective inhibitor, affects signaling events and cytokine production in human monocytes. *J. Leukocyte Biol.*, **67**, 249–258.

Krakauer, T., and Stiles, B. G. (1999). Pentoxifylline inhibits superantigen-induced toxic shock and cytokine release. *Clin. Diagn. Lab. Immunol.*, **6**, 594–598.

Kreitman, R. J., FitzGerald, D., and Pastan, I. (1992). Targeting growth factor receptors with fusion toxins. *Int. J. Immunopharmcol.*, **14**, 465–472.

Kreuser, E.-D., Wadler, S., and Thiel, E. (1992). Interactions between cytokines and cytotoxic drugs: Putative molecular mechanisms in experimental hematology and oncology. *Semin. Oncol.*, **19**, 1–7.

Kulmatycki, K. M., and Jamali, F. (2001). Therapeutic relevance of altered cytokine expression. *Cytokine*, **14**, 1–10.

Kuper. C. F., Harleman, J. H., Richter-Reichelm, H. B., and Vos. J. G. (2000). Histopathologic approaches to detect changes indicative of immunotoxicity. *Toxicol. Pathol.*, **28**, 454–466.

Kuzel, T. M., Rosen, S. T., Gordon, L. I., Winter, J., Samuelson, E., Kaul, K., Roenigk, H. H., Nylen, P., and Woodworth, T. (1993). Phase I trial of the diphtheria toxin/interleukin-2 fusion protein DAB486IL-2: Efficacy in mycosis fungoides and other non-Hodgkins lymphomas. *Leuk. Lymphoma*, **11**, 369–377.

Kyoizumi, S., Murray, L. J., and Namikawa R. (1993). Preclinical analysis of cytokine therapy in the SCID-hu mouse. *Blood*, **81**, 1479–1488.

Ladics, G. S., Smith, C., Heaps, K., Elliott, G. S., Slone, T. W., and Loveless, S. E. (1995). Possible incorporation of an immunotoxicology functional assay for assessing humoral immunity for hazard identification purposes in rats on standard toxicology study. *Toxicology*, **96**, 225–238.

Ladics, G. S., Smith, C., Elliott, G. S., Slone, T. W., and Loveless, S. E. (1998). Further evaluation of the incorporation of an immunotoxicological functional assay for assessing humoral immunity for hazard identification purposes in rats in a standard toxicology study. *Toxicology*, **126**, 137–152.

Lee, J. C., Kassis, S., Kumar, S., Badger, A., and Adams, J. L. (1999). p38 mitogen-activated protein kinase inhibitors-mechanisms and therapeutic potentials. *Pharmacol. Ther.*, **82**, 389–397.

Lee, J. C., Kumar, S., Griswold, D. E., Underwood, D. C., Votta, B. J., and Adams J. L. (2000). Inhibition of p38 MAP kinase as a therapeutic strategy. *Immunopharmacology*, **47**, 185–201.

Leonard, J. P., Neben, T. Y., Kozitza, M. K., Quinto, C. M., and Goldman, S. J. (1996). Constant subcutaneous infusion of rhIL-11 in mice: Efficient delivery enhances biological activity. *Exp. Hematol.*, **24**, 270–276.

Lerner, D. M., Stoudemire, A., and Rosenstein, D. L. (1999). Neuropsychiatric toxicity associated with cytokine therapies. *Psychosomatics*, **40**, 428–435.

Liger, D., vander Spek, J. C., Gaillard, C., Cansier, C., Murphy, J. R., Leboulch, P., and Gillet, D. (1997). Characterization and receptor specific toxicity of two diphtheria toxin–related interleukin-3 fusion proteins DAB389-mIL3 and DAB389-(Gly4Ser)2-mIL3. *FEBS Lett.*, **406**, 157–161.

Long, C. S. (2001). The role of interleukin-1 in the failing heart. *Heart Failure Rev.*, **6**, 81–94.

Lovik, M. (1997). Mutant and transgenic mice in immunotoxicology: An introduction. *Toxicology*, **119**, 65–76.

Lunney, J. K. (1998). Cytokines orchestrating the immune response. *Rev. Sci. Tech. Off. Int. Epizsot.*, **17**, 84–94.

Luster, M. I., Munson, A. E., Thomas, P. T., Holsapple, M. P., Fenters, J. D., White, K. L., Jr., Lauer, L. D., Germolec, D. R, Rosenthal, G. J., and Dean, J. H. (1988). Development of a testing battery to assess chemical-induced immunotoxicity: National Toxicology Program's guidelines for immunotoxicity evaluation in mice. *Fundam. Appl. Toxicol.*, **10**, 2–19.

Luster, M. I., Portier, C., Pait, D. G., White, K. L., Jr., Gennings, C., Munson, A. E., and Rosenthal, G. J. (1992). Risk assessment in immunotoxicology. I. Sensitivity and predictability of immune tests. *Fundam. Appl. Toxicol.*, **18**, 200–210.

Maheshwari, A., Mahato, R. I., McGregor, J., Han, So, Samlowski, W. E., Park, J. S., and Kim, S. W. (2000). Soluble biodegradable polymer-based cytokine gene delivery for cancer treatment. *Mol. Ther.*, **2**, 121–130.

Maki, E. (1997). The practice of preclinical immunotoxicity testing in Japan. *Drug. Inf. J.*, **31**, 1325–1329.

Mantovani, A., Locati, M., Vecchi, A., Sozzani, S., and Allavena, P. (2001). Decoy receptors: A strategy to regulate inflammatory cytokines and chemokines. *Trends Immunol.*, **22**, 328–336.

Matsuda, S., and Koyasu, S. (2000). Mechanisms of action of cyclosporine. *Immunopharmacology*, **47**, 119–125.

Maysinger, D., Berezovskaya, O., and Fedoroff, S. (1996). The hematopoietic cytokine colony stimulating factor 1 is also a growth factor in the CNS II. Microencapsulated CSF-1 and LM-10 cells as delivery systems. *Exp. Neurol.*, **14**, 47–56.

McCusker, S. M., Curran, M. D., Dynan, K. B., McCullagh, C. D., Urquhart, D. D., Middleton, D., Patterson, C. C., McIlroy, S. P., and Passmore, A. P. (2001). Association between polymorphism in regulatory region of gene encoding tumour necrosis factor alpha and risk of Alzheimer's disease and vascular dementia: A case–control study. *Lancet*, **357**, 436–439.

Mellstedt, H., and Osterborg, A. (1999). Active idiotype vaccination in multiple myeloma. GM-CSF may be an important adjuvant cytokine. *Pathol. Biol.* (Paris), **47**, 211–215.

Mitragotri, S., Blankschtein, D., and Langer, R. (1995). Ultrasound-mediated transdermal delivery of proteins. *Science*, **269**, 850–853.

Moller, D. R., Wysocka, M., Greenlee, B. M., Ma, X., Wahl, L., Flockhart, D. A., Trinchieri, G., and Karp, C. L. (1997). Inhibition of human IL-12 production by pentoxifylline. *Immunology*, **91**, 197–203.

Mollison, K. W., Fey, T. A., Gauvin, D. M., Kolano, R. M., Sheets, M. P., Smith, M. L., Pong, M., Nikolaidis, N. M., Lane, B. C., Trevillyan, J. M., Cannon, J., Marsh, K., Carter, G. W., Or, Y. S., Chen, Y. W., Hsieh, G. C., and Luly, J. R. (1999). A macrolactam inhibitor of T helper type 1 and T helper type 2 cytokine biosynthesis for topical treatment of inflammatory skin diseases. *J. Invest. Dermatol.*, **112**, 729–738.

Morris, J. C., and Waldmann, T. A. (2000). Advances in interleukin 2 receptor targeted treatment. *Ann. Rheum. Dis.*, **59**, i109–i114.

Morris, R. (1994). Modes of action of FK506, cyclosporin A, and rapamycin. *Transplant. Proc.*, **26**, 3272–3275.

Nelson S. (2001). Novel nonantibiotic therapies for pneumonia: Cytokines and host defense. *Chest*, **119**, 419S–425S.

Nichols, J., Foss, F., Kuzel, T. M., LeMaistre, C. F., Platanias, L., Ratain, M. J., Rook, A., Saleh, M., and Schwartz, G. (1997). Interleukin-2 fusion protein: An investigational therapy for interleukin-2 receptor expressing malignancies. *Eur. J. Cancer*, **33**, S34–S36.

Oshima, Y., and Puri, R. K. (2001). Characterization of a powerful high affinity antagonist that inhibits biological activities of human interleukin-13. *J. Biol. Chem.*, **276**, 15185–15191.

Penichet, M. L., Harvill, E. T., and Morrison, S. L. (1997). Antibody-IL-2 fusion proteins: A novel strategy for immune protection. *Hum. Antibodies*, **8**, 106–118.

Peters, M., Blinn, G., Solem, F., Fischer, M., Meyer zum Buschenfelde, K. H., and Rose-John, S. (1998). In vivo and in vitro activities of the gp130-stimulating designer cytokine Hyper-IL-6. *J. Immunol.*, **161**, 3575–3581.

Pistoia, V. (1997). Production of cytokines by human B cells in health and disease. *Immunol. Today*, **18**, 343–350.

Poliani, P. L., Brok, H., Furlan, R., Ruffini, F., Bergami, A., Desina, G., Marconi, P. C., Rovaris, M., Uccelli, A., Glorioso, J. C., Penna, G., Adorini, L., Comi, G., 't Hart, B., and Martino, G. (2001). Delivery to the central nervous system of a nonreplicative herpes simplex type 1 vector engineered with the interleukin 4 gene protects rhesus monkeys from hyperacute autoimmune encephalomyelitis. *Hum. Gene Ther.*, **12**, 905–920.

Powell, J. J., Fearon, K. C., Siriwardena, A. K., and Ross, J. A. (2001). Evidence against a role for polymorphisms at tumor necrosis factor, interleukin-1 and interleukin-1 receptor antagonist gene loci in the regulation of disease severity in acute pancreatitis. *Surgery*, **129**, 633–640.

Powrie, F., and Coffman, R. L. (1993). Cytokine regulation of T-cell function: Potential for therapeutic intervention. *Immunol. Today*, **14**, 270–274.

Prussick, R. (1996). Adverse cutaneous reactions to chemotherapeutic agents and cytokine therapy. *Semin. Cutan. Med. Surg.*, **5**, 267–276.

Puri, R. K. (1999). Development of a recombinant interleukin-4-*Pseudomonas* exotoxin for therapy of glioblastoma. *Toxicol. Pathol.* **27**, 53–57.

Rappocciolo, G., Allan, J. S., Eichberg, J. W., and Chanh, T. C. (1992). A comparative study of natural killer cell activity, lymphoproliferation, and cell phenotypes in nonhuman primates. *Vet. Pathol.*, **29**, 53–59.

Reynolds, C. W., and Herberman, R. B. (1981). *In vitro* augmentation of rat natural killer (NK) cell activity. *J. Immunol.*, **126**, 1581–1585.

Rincon, M., Flavell, R. A., and Davis, R. A. (2000). The JNK and p38 Map kinase signaling pathways in T cell-mediated immune responses. *Free Radicals Biol. Med.* **28**, 1328–1337.

Roilides, E., and Pizzo, P. A. (1993). Biologicals and hematopoietic cytokines in prevention or treatment of infections in immunocompromised hosts. *Hematol./Oncol. Clin. N. Am.*, **7**, 841–864.

Romagnani, S. (1994). Lymphokine production by human T cells in disease states. *Annu. Rev. Immunol.*, **12**, 227–257.

Romagnani S. (2000). T-cell subsets (Th1 versus Th2). *Ann. Allergy Asthma Immunol.*, **85**, 9–18.

Rosenthal, G. J., Corsini, E., Craig, W. A., Comment, C. E., and Luster, M. I. (1991). Pentamidine: An inhibitor of interleukin-1 that acts via a post-translational event. *Toxicol. Appl. Pharmacol.*, **107**, 555–561.

Rudolph, U., and Mohler, H. (1999). Genetically modified animals in pharmacological research: Future trends. *Eur. J. Pharmacol.*, **375**, 327–337.

Russell-Jones, G. J., Westwood, S. W., and Habberfield, A. D. (1995). Vitamin B12 mediated oral delivery systems for granulocyte-colony stimulating factor and erythropoietin. *Bioconjug. Chem.*, **6**, 459–465.

Ryffel, B. (1996). Unanticipated human toxicology of recombinant proteins. *Arch. Toxicol. Suppl.*, **18**, 333–341.

Salituro, F. G., Germann, U. A., Wilson, K. P., Bemis, G. W., Fox, T., and Su, M. S. (1999). Inhibitors of p38 MAP kinase: Therapeutic intervention in cytokine-mediated diseases. *Curr. Med. Chem.*, **6**, 807–823.

Schotte, H., Schluter, B., Rust, S., Assmann, G., Domschke, W., and Gaubitz, M. (2001). Interleukin-6 promoter polymorphism (–174 G/C) in Caucasian German patients with systemic lupus erythematosus. *Rheumatology* (Oxford), **40**, 393–400.

Schrohenloher, R. E., Koopman, W. J., Woodworth, T. G., and Moreland, L. W. (1996). Suppression of in vitro IgM rheumatoid factor production by diphtheria toxin interleukin 2 recombinant fusion protein (DAB 486IL-2) in patients with refractory rheumatoid arthritis. *J. Rheumatol.*, **23**, 1845–1848.

Sehgal, S. N. (1995). Rapamune (Sirolimus, Rapamycin): An overview and mechanism of action. *Ther. Drug Monit.*, **17**, 660–665.

Shanafelt, A. B.. Lin, Y, Shanafelt M. C., Forte C. P., Dubois-Stringfellow, N., Carter, C., Gibbons, J. A., Cheng, S. L., Delaria, K. A., Fleischer, R., Greve, J. M., Gundel, R., Harris, K., Kelly, R., Koh, B., Li, Y., Lantz. L., Mak, P., Neyer, L., Plym, M. J., Roczniak, S., Serban, D., Thrift, J., Tsuchiyama, L., Wetzel, M., Wong, M., and Zolotorev, A. (2000). A T-cell-selective interleukin 2 mutein exhibits potent antitumor activity and is well tolerated in vivo. *Nat. Biotechnol.*, **18**, 1197–1202.

Slifka, M. K., and Whitton, J. L. (2000). Clinical implications of dysregulated cytokine production. *J. Mol. Med.*, **78**, 74–80.

Srivannaboon, K., Shanafelt, A. B., Todisco, E., Forte, C. P., Behm, F. G., Raimondi, S. C., Pui, C. H., and Campana, D. (2001). Interleukin-4 variant (BAY 36-1677) selectively induces apoptosis in acute lymphoblastic leukemia cells. *Blood*, **97**, 752–758.

Steidler, L., Robinson, K., Chamberlain, L., Schofield, K. M., Remaut, E., Le Page, R. W., and Wells, J. M. (1998). Mucosal delivery of murine interleukin-2 (IL-2) and IL-6 by recombinant strains of *Lactococcus lactis* coexpressing antigen and cytokine. *Infect. Immun.*, **66**, 3183–3189.

Strom, T. B., Anderson, P. L., Rubin-Kelley, V. E., Williams, D. P., Kiyokawa, T., and Murphy, J. R. (1991). Immunotoxins and cytokine fusion proteins. *Ann. N. Y. Acad. Sci.*, **636**, 233–250.

Takagi, K., Fukushima, Y., Kanou, J., Honda, T., Tomita, K., Takano, M., and Soma, G. (1998). A long-term survivor case of malignant mesothelioma treated by recombinant tumor necrosis factor-SAM2 (TNF-alpha mutein) and 5-fluorouracil (5-FU): A new therapeutic approach based on host–tumor relationship. *Anticancer Res.*, **18**, 4591–4600.

Tamaoki, J., Kondo, M., Sakai, N., Aoshiba, K., Tagaya, E., Nakata, J., Isono, K., and Nagai, A. (2000). Effect of Suplatast Tosilate, a Th2 cytokine inhibitor, on steroid-dependent asthma: A double-blind randomised study. Tokyo Joshi-Idai Asthma Research Group. *Lancet*, **356**, 273–278.

Taverne, J. (1993). Transgenic mice in the study of cytokine function. *Int. J. Exp. Pathol.*, **74**, 525–546.

Temple, L., *et al.* (1993). Comparison of ELISA and plaque-forming cell assays for measuring the humoral immune response to SRBC in rats and mice treated with benzo[*a*]pyrene or cyclophosphamide. *Fundam. Appl. Toxicol.*, **21**, 412–419.

Thomson, A. W., Bonham, C. A., and Zeevi, A. (1995). Mode of action of Tacrolimus (FK506): Molecular and cellular mechanisms. *Ther. Drug Monitor.*, **17**, 584–591.

Touw, I. P., De Koning, J. P., Ward, A. C., and Hermans, M. H. (2000). Signaling mechanisms of cytokine receptors and their perturbances in disease. *Mol. Cell. Endocrinol.*, **160**, 1–9.

Tsukasaki, K., Miller, C. W., Kubota, T., Takeuchi, S., Fujimoto, T., Ikeda, S., Tomonaga, M., and Koeffler, H. P. (2001). Tumor necrosis factor alpha polymorphism associated with increased susceptibility to development of adult T-cell leukemia/lymphoma in human T-lymphotropic virus type 1 carriers. *Cancer Res.*, **61**, 3770–3774.

Tsutsami, Y., Kihira, T., Tsunoda, S., Kanamori, T., Nakagawa, S., and Mayumi, T. (1995). Molecular design of hybrid tumour necrosis factor alpha with polyethylene glycol increases its anti-tumour potency. *Br. J. Cancer*, **71**, 963–968.

U.S. FDA. (1999). Immunotoxicity Testing Guidance. U.S. Food and Drug Administration, document available at http://www.fda.gov/cdrh/ost/ostggp/immunotox.pdf

U.S. FDA., (2001). Immunotoxicology Evaluation of Investigational New Drugs. U.S. Food and Drug Administration document available at http://www.fda.gov/cder/guidance/3010dft.htm

Van Wauwe, J., Aerts, F., Van Genechten, H., Blcokx, H., Deleersnijder, W., and Walter, H. (1996). The inhibitory effect of pentamidine on the production of chemotactic cytokines by in vitro stimulated human blood cells. *Inflamm. Res.*, **45**, 357–363.

Vander Spek, J. C., Sutherland, J., Sampson, E., and Murphy, J. R. (1995). Genetic construction and characterization of the diphtheria toxin–related interleukin-15 fusion protein DAB389 sIL-15. *Protein Eng.*, **8**, 1317–1321.

Verdier, F., Aujoulat, M., Condevaux, F., and Descotes, J. (1995). Determination of lymphocyte subsets and cytokine levels in cynomolgus monkeys. *Toxicology*, **105**, 81–90.

Vial, T., and Descotes, J. (1994). Clinical toxicity of the interferons. *Drug Saf.*, **10**, 115–150.

Vial, T., and Descotes, J. (1995). Clinical toxicity of cytokines used as haemopoietic growth factors. *Drug Saf.*, **13**, 371–406.

Vial, T., Choquet-Kastylevsky, G., Liautard, C., and Descotes, J. (2000). Endocrine and neurological effects of the therapeutic cytokines. *Toxicology*, **142**, 161–172.

Vose, J. M., Pandite, A. N., Beveridge, R. A., Geller, R. B., Schuster, M. W., Anderson, J. E., LeMaistre, C. F., Ahmed, T., Granena, A., Keating, A., Ranada, J. M. F., Stiff, P. J., Tabbara, I., *et al.* (1997). Granulocyte–macrophage colony-stimulating factor/interleukin-3 fusion protein versus granulocyte–macrophage colony-stimulating factor after autologous bone marrow transplantation for non-Hodgkins lymphoma: Results of a randomized double-blind trial. *J. Clin. Oncol.*, **15**, 1617–1623.

Wadhwa, M., Hjelm Skog, A.-L., Bird, C., Ragnhammar, P., Lilljefors, M., Gaines-Das, R., Mellstedt, H., and Thorpe, R. (1999). Immunogenicity of granulocyte–macrophage colony-stimulating factor (GM-CSF) products in patients undergoing combination therapy with GM-CSF. *Clin. Cancer Res.*, **5**, 1353–1361.

Wagstaff, J., Baars, J. W., Wolbink, G.-J., Hoekman, K., Eerenberg-Belmer, A. J. M., and Hack, C. E. (1995). Renal cell carcinoma and interleukin-2: A review. *Eur. J. Cancer*, **31A**, 401–408.

Wang, S. W., Pawlowski, J., Wathen, S. T., Kinney, S. D., Lichenstein, H. S., and Manthey, C. L. (1999). Cytokine mRNA decay is accelerated by an inhibitor of p38-mitogen-activated protein kinase. *Inflamm. Res.*, **48**, 533–538.

Williams, G., and Giroir, B. P. (1995). Regulation of cytokine gene expression: Tumor necrosis factor, interleukin-1, and the emerging biology of cytokine receptors. *New Horiz.*, **3**, 276–287.

Wilson, S. D., Munson, A. E., and Meade, B. J. (1999). Assessment of the functional integrity of the humoral immune response: The plaque-forming cell assay and the enzyme-linked immunosorbent assay, *Methods,* **19**, 3–7.

Yoshimura, T., Kurita, C., Nagao, T., Usami, E., Nakao, T., Watanabe, S., Kobayashi, J., Yamazaki, F., Tanaka, H., and Nagai, H. (1997). Effects of cAMP–phosphodiesterase isozyme inhibitor on cytokine production by lipopolysaccharide-stimulated human peripheral blood mononuclear cells. *Gen. Pharmcol.*, **29**, 633–638.

Zhao, C., Tang, P., Mao, N., Zhang, S., Fan, E., Dong, B., Li, Q., and Du, D. (1996). Erythropoietin-like activity in vivo of the fusion protein rhIL-6/IL-2 (CH925). *Exp. Hematol.*, **24**, 54–58.

Zwingenberger, K., and Wendt, S. (1995). Immunomodulation by thalidomide: Systematic review of the literature and of unpublished observations. *J. Inflamm.*, **46**, 177–211.

Chapter 8

Ecological Assessment of Crops Derived through Biotechnology

Thomas E. Nickson
Ecological Technology Center
Monsanto Company
St. Louis, Missouri

Michael J. McKee
Ecological Technology Center
Monsanto Company
St. Louis, Missouri

Risk assessment is an important tool that has been developed to help individuals make informed decisions about the impacts from their human-based activities on people and society. Risk assessment is relatively new as a formal, science-based discipline, though. Today, science-based risk assessment is a respected and recognized discipline that is fundamental to decision making in many industries and governmental activities. In 1998 the U.S. Environmental Protection Agency (EPA) published final guidelines for ecological risk assessment, laying out the key principles and processes that we have applied to genetically modified crop plants and traits. This chapter describes an approach to the building of an ecological risk assessment framework for genetically modified plants merging the concepts developed by the National Research Council and EPA with the experience present in industry, recognizing the

Biotechnology and Safety Assessment, 3rd edition

principles outlined elsewhere on ecological risk assessment specific to biotechnology. The chapter presents the framework, key terminology, and several examples of how the principles of ecological risk assessment can be applied to plants derived through biotechnology.

> Everyone is a true risk "expert" in the original sense of the word; we have all been trained by practice and experience in the management of risk. Everyone has a valid contribution to make to a discussion of the subject.
>
> J. Adams, 1995

INTRODUCTION

In broad terms, risk is "the probability that an outcome will occur times the consequence, or level of impact, should the outcome occur" (Kammen and Hassenzahl, 1999). The literal interpretation of "risk" allows for consequences to be judged good or bad. People today generally think of risk in a negative context rather than the chance that something good might happen. Most people, thinking in terms of risk, would associate risk assessment with an evaluation of the potentially harmful consequences of an activity. As such, risk assessment is an important tool purposely developed to help individuals make informed decisions about the impacts from their human-based activities on people and society.

Although risk assessment is relatively new as a formal, science-based discipline, as the quotation from Adams (1995) indicates, it is common in everyday life. In 1983 an attempt to formalize the risk assessment process occurred when the National Research Council (NRC) of the National Academy of Sciences (NAS) convened a committee of experts to outline a process and establish nomenclature (NRC, 1993). Today, science-based risk assessment is a respected and recognized discipline that is fundamental to decision making in many industries and governmental activities. At its core, risk assessment is a tool for making objective decisions by collecting what is known, learning what is needed, and weighing the costs of mitigating risk or reducing uncertainty against the benefits. Several scientific societies have risk assessment as fundamental to their discipline: for example, the Society for Risk Analysis (SRA) and the Society for Environmental Toxicology and Chemistry (SETAC).

Recognizing that environmental aspects of risk could be addressed in a focused manner, the NRC committee just mentioned (NRC, 1993) proposed a conceptual framework for ecological risk assessment. More recently, the U.S. Environmental Protection Agency published final guidelines for ecological risk assessment (U.S. EPA, 1998). Further to these efforts, risk assessment methods and approaches have been developed specifically to evaluate ecological risks (see Suter, 1993, for example).

The development and deployment of crops derived through the techniques of modern biotechnology, also known as genetically modified plants (GMPs), have prompted concerns about potential risks. Focusing on the environmental aspects of GMPs, individual scientists, scientific committees, and international organizations have reviewed and debated the potential ecological risks associated with GMPs (Rissler and Mellon, 1996; NRC, 1989; OECD, 1992) and proposed various risk assessment approaches. One scientific committee concluded that the process of modification does not constitute a new risk (NRC, 1989). Rather, the risk assessment should focus on the nature of the modified plant and the environment into which it will be released (NAS, 1987; NRC, 1989). Others, who view genetic modification as "obviously different and more powerful than traditional breeding" (Rissler and Melon, 1996), call for increased regulation and rigorous risk assessment prior to commercialization. This chapter describes an approach to the building of an ecological risk assessment framework for GMPs merging the concepts developed by the federal government (NRC, 1993; U.S. EPA, 1998) with the experience present in industry, recognizing the principles outlined in Rissler and Melon (1996), NRC (1989), OECD (1992), and other publications on ecological risk assessment (Dale *et al.*, 1993; Kjellsen, G., 1997; Kjær *et al.*, 1999) specific to biotechnology.

TERMINOLOGY

Terms and phrases purposefully developed to communicate the scientific ideas and concepts have been described in excellent reviews of the broader subject of ecological risk assessment (U.S. EPA, 1998; Suter, 1993). Selected terms have been included here (Table 1) to serve as a platform to bring clarity and structure readers' understanding of ecological risk assessment principles. Except for "risk management," the following terms were abstracted verbatim from the EPA's finalized ecological risk assessment guidelines. They are reproduced in Table 1 to provide a basis for a clear understanding of the EPA's ecological risk assessment framework and its application to GMPs.

Table 1

Key Terms and Their Definitions

Analysis: a process that examines the two primary components of risk, exposure and effects, and their relationships between each other and ecosystem characteristics. The objective is to provide the ingredients necessary for determining or predicting ecological responses to stressors under exposure conditions of interest.

Assessment end point: an explicit expression of the environmental value that is to be protected, operationally defined by an ecological entity and its attributes.

Comparative risk assessment: a process that generally uses a professional judgment approach to evaluate the relative magnitude of effects and set priorities among a wide range of environmental problems.

Conceptual model: a written description and visual representation of predicted relationships between ecological entities and the stressor to which they may be exposed.

Problem formulation: a process for generating and evaluating preliminary hypotheses about why ecological effects have occurred, or may occur, from human activity.

Prospective risk assessment: an evaluation of the future risks of a stressor(s) not yet released into the environment or of future conditions resulting from an existing stressor(s).

Retrospective risk assessment: an evaluation of the causal linkages between observed ecological effects and stressor(s) in the environment.

Risk characterization: a phase of ecological risk assessment that integrates the exposure and stressor response profiles to evaluate the likelihood of adverse ecological effects associated with exposure to a stressor.

Risk management: "the process of deciding what actions to take in response to a risk" (Suter, 1993).

Source: U.S. EPA (1998), except as noted.

ECOLOGICAL RISK ASSESSMENT PRINCIPLES

Ecological risk assessment, as defined by the EPA, is "a process that evaluates the likelihood that adverse ecological effects may occur or are occurring as a result of exposure to one or more stressors" (U.S. EPA, 1998). Anyone embarking on this process faces the challenge of defining "adverse ecological effects" in terms of hypotheses that are testable within realistic temporal and spatial constraints. In addition, imperfect knowledge of complex ecological systems limits the certainty with which we can evaluate effects. A well-designed ecological risk assessment consists of information and data organized such to be usable in a way that will both increase the predictability of effects broadly and reduce uncertainty.

We have identified four fundamental principles upon which ecological risk assessment is grounded. Firstly, ecological risk assessment is *inclusive*, drawing information from all available sources. Knowledge from different sources including research, monitoring programs, expert opinion, and individual beliefs is integrated and considered. The second principle is that ecological risk assessment is *systematic*. As already noted, the EPA has provided detailed guidelines relating to the agency's process, spelling out how and when the steps occur. Third, ecological risk assessment is *iterative*. New information is integrated into the process at the appropriate point(s). Conclusions and assumptions are reexamined relative to this new information. Finally, ecological risk assessment is *science based*. The process relies on science and the use of the scientific method to collect and review information. Because ecological risk assessment is both inclusive and science-based, information that is not grounded in objective science is ultimately included in the risk decision. As such, it is possible to have scientifically based conclusion of low risk, but a risk decision of unacceptable risk owing to other factors (e.g. morality, values, aesthetics, etc.). Each principle is developed in more detail in this chapter.

The ecological risk assessment model developed by the EPA prescribes a *systematic* approach comprising distinct steps that occur in phases, logically sequenced. The framework, depicted in Figure 1, provides structure for thinking about ecological risk assessment for potential stressors such as GMPs. These steps are problem formulation, analysis, risk characterization, and risk management.

During problem formulation, investigators gather information about the stressor and the environment and assess it for relevance to the scientific characterization of risk. Hypotheses about possible effects and mechanisms of exposure are generated as a part of problem formulation. Two essential products of the first phase are the selection of assessment end points and the selection of measures. These two extremely important concepts are discussed in more detail later. Since ecological risk assessments are inclusive, problem formulation incorporates information that is direct (actual and estimated) and indirect (inferred) in the development of the analysis plan. Lines of evidence developed during problem formulation will be used to link data quality and adequacy in arriving at an assessment of the relevance and confidence level of measures used in the risk assessment.

The two steps in the EPA process that follow problem formulation are called analysis and risk characterization. During analysis, exposure and effects are characterized. Analysis also involves acquisition of data and the integration of new information as needed. Furthermore, the methods used for data collection are evaluated for their relevance and possible shortcomings. During risk characterization, information on exposure and effects is integrated to estimate the likelihood, magnitude, and nature of an adverse effect. Risk assessors working for the EPA use this process to provide a characterization of risk to risk managers, who make environmental

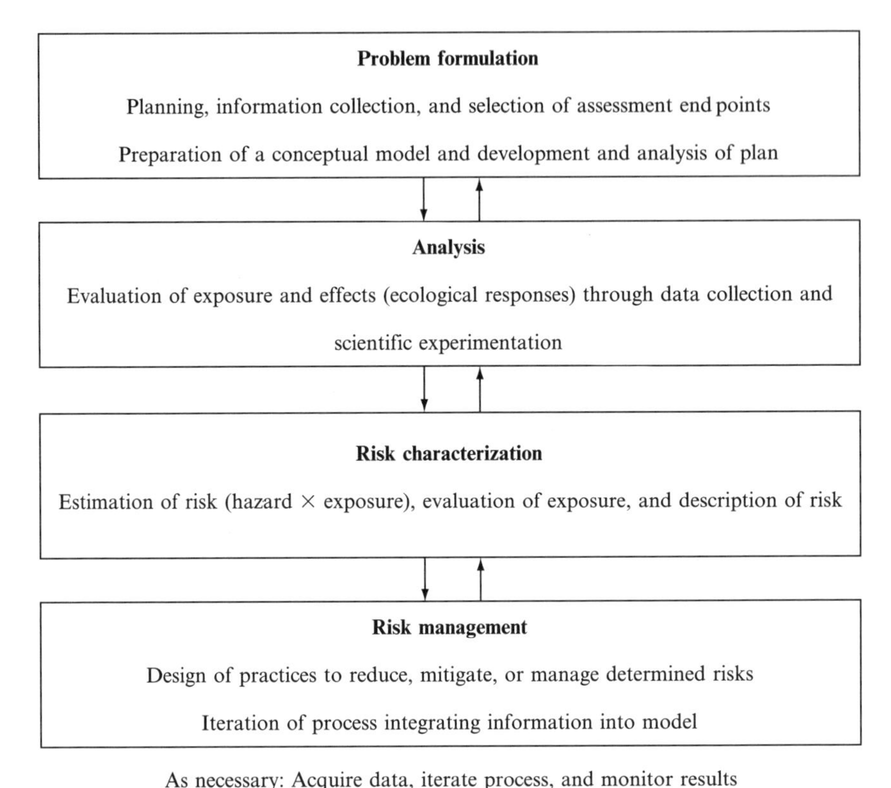

Figure 1 Ecological risk assessment framework [Adapted from U.S. EPA (1998).]

decisions. The risk characterization is based first on an estimation of risk that provides the foundation for the subsequent description of risk (Figure 1). Risk managers use this description of risk to determine whether the risk is acceptable or, if it is not, what risk management is necessary. A characterization of risk could require that actions be designed to reduce, mitigate, or manage risks.

Another important principle of ecological risk assessment is that it is *iterative*. Depending on the results of a particular analysis or a new interpretation of existing information, a specific step or the entire ecological risk assessment process may need to be revisited. As stated earlier, information comes from the available sources regardless of why it was collected. New data can be intentionally collected because of a regulatory requirement or can result from scientific inquiry. For example, a regulatory requirement for risk management may prescribe ecological monitoring of a specific end point (e.g., development of resistance to a pesticide in a target insect species). In this case, the information obtained from monitoring is necessarily integrated into the risk assessment that is iterated, and the risk decision is reevaluated (confirmed or modified) based on the monitoring results. The reevaluation of insect-resistant crops (cotton and corn protected by *Bt*, *Bacillus thuringiensis*) based on information collected from insect resistance management (IRM) is an example of iteration of an ecological risk assessment using information required to be collected (see EPA website for summaries of Scientific Advisory Panel discussions on IRM). Likewise, publications containing scientific findings relevant to a stressor may educe an iteration of a risk assessment and reevaluation of a risk decision.

Observations or new experimental data can unintentionally trigger an iteration of the risk assessment and a reevaluation of a risk decision, or a specific phase in the risk assessment process. A result obtained in a particular assay during analysis could impact the conclusion or validity of another experiment. In this situation, the information is integrated and the analysis phase may need to be iterated. The well-publicized demonstration of an effect of pollen from a line of *Bt* corn on monarch butterfly larvae in a laboratory at Cornell (Losey *et al.*, 1999), is an example of new data obtained through scientific inquiry. While the effect was predicted based on the biology of the insect and known toxicity of the Cry1Ab protein toward lepidopterans, the information initiated an iteration of the risk assessment for all *Bt* (Cry1Ab) corn products toward monarch larvae in the form of a data call-in (DCI), a tool commonly used by the EPA.

The fourth fundamental principle of ecological risk assessment is that it is a *science-based* endeavor. As noted earlier, science is not the sole criterion in risk decisions. It is, however, a key tool for evaluating the relevance and quality of information used in the risk assessment. The use of the scientific method and scientific principles ensures that information is treated objectively and systematically. A valuable science-based approach to ecological risk assessment is comparative ecological risk assessment using knowledge of the system along with expert judgment to evaluate effects relative to some known entity. Those GMP-based crops that are used within food production systems, lend themselves to comparative risk assessment and description of risk relative to familiar agricultural systems. Figure 2 depicts the comparative framework useful for ecological risk assessment of GMPs.

As noted earlier, assessment end points and measures are critically important to ensuring the quality of the scientific basis of risk assessment. An assessment end point is "an explicit expression of the environmental value that is to be protected" (U.S. EPA, 1998) (Table 1). In addition, assessment end points must be "operationally defined by an ecological entity and its attributes" (U.S. EPA, 1998) and as such will be reflective of measurable

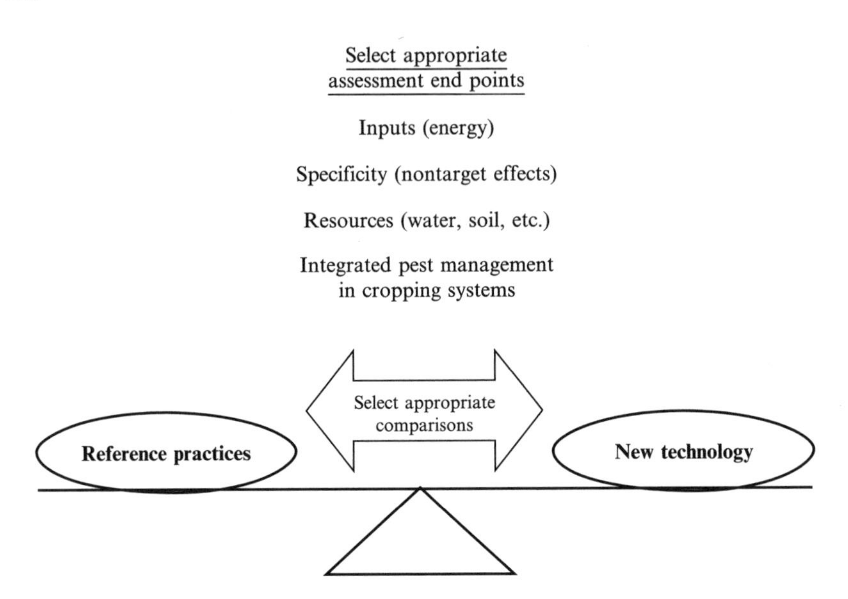

Figure 2 Comparative ecological risk assessment for GMPs.

quantities rather than environmental goals. The criteria for selecting assessment end points are ecological relevance, susceptibility to stressors, and relevance to societal values. For example, effects on a beneficial insect species could be an appropriate assessment end point for a GMP with a pesticidal trait (e.g., *Bt*) because the species is ecologically relevant and susceptible to the stressor. Since various laboratory and field-based methods can be used to measure effects on insects, such assessment end points are practicable. Conversely, weediness is ecologically significant, but presents operational challenges as an assessment end point because of the lack of consensus, science-based criteria. Examples of appropriate assessment end points for the ecological risk assessment of agriculturally based GMPs are listed in Figure 2.

Just as important as the selection of assessment end points is the need to have measures that are clearly relevant and interpretable. The federal government (U.S. EPA, 1998) has described three categories of measures: measures of effects, measures of exposure, and measures of ecosystem characteristics. Direct measures are typically used, but where complicities are present, indirect measures (e.g., surrogates) may be chosen as long as the potential shortcomings are recognized. As noted in the foregoing example of beneficial insects, laboratory and field tests can directly measure effects and exposure. In an agricultural field, quantification of pesticide load is an example of a measure of an ecosystem characteristic.

Finally, science-based ecological risk assessments are *inclusive*; drawing from all available information to reduce the uncertainty in the risk decision. Information can be obtained from published literature, experimental systems, and research programs, as well as from expert judgment and predictive models. There are costs and benefits associated with using the various types of information (Zimmer, 1997). The science-based information in the ecological risk assessments for GMPs that have been approved has primarily focused on expert judgment, published literature, and experimentally generated data. While probabilistic models have great utility in risk assessment of chemical stressors, they have had specific use for GMPs. Simulation models for invasion and gene flow from GMPs have been described and discussed (Karieva and Manasse, 1990; Giddings, 1999), but their limited reliability in predicting effects in complex ecological systems has prevented their broad acceptance. Improved models will most likely be developed and applied to GMPs in the future.

Risk assessment principles do not prescribe who should collect risk assessment data. As with the agricultural chemical products, the EPA requires that some of the data be collected under Good Laboratory Practices (GLP) (U.S. EPA, 1989). Under GLP guidelines, study conduct, including data recording, reporting, and retention, is held to legal standards and subject to inspection and independent review. These standards hold legally enforceable consequences of noncompliance and are more rigorous than any voluntary standards for data integrity. Because information across diverse biological disciplines is needed for a complete ecological risk assessment, independent laboratories with specialized expertise are frequently contracted to collect data. This ecological risk assessment for GMPs as currently practiced utilizes independent as well as industry laboratories to gather data, some of which must be collected under GLP. Some, however, believe that greater independence in the conduct of risk assessment is needed, particularly to address public acceptance (Stewart, 2001). Implementing some of the proposed changes would require marked changes to the regulatory review process and management of confidentiality, as well as significant increases in funding risk assessment work in the public sector.

ECOLOGICAL RISK ASSESSMENT PRINCIPLES FOR GENETICALLY MODIFIED PLANTS

The sections that follow describe a risk assessment model and application of the principles just outlined. A general approach was developed because of the potential offered by biotechnology to introduce diverse traits with desirable characteristics. Much experience has been obtained with the first traits that afford protection against pest insects, such as *Bt* proteins, provide tolerance to environmentally superior herbicides, and protect plants from economically important viral diseases. However, the near future of biotechnology-derived crops will see traits that improve food quality, such as increases in vitamin or essential amino acid content, and plant-produced pharmaceuticals. Ecological risk assessment for the myriad of potential products will demand a flexible yet robust approach.

Model Overview

Evaluation of potential ecological risk for these new products readily segregates into two areas: the plant and the trait (Figure 3). The plant risk assessment evolves along a path created by asking questions such as, Is the plant biologically and ecologically similar to other nonmodified varieties? Or, if there is potential for gene flow to crops or plants in adjacent areas, what might be the consequence? As such, the trait risk assessment focuses on the ability to confer a selective advantage to a plant. In addition, traits are assessed for their potential toxicity to nontarget organisms under field use conditions.

Prior to initiation of the risk assessment, it is important to collect and consider the available background information focusing on that which will guide the assessment (U.S. EPA, 1998). Key concepts are identification of the

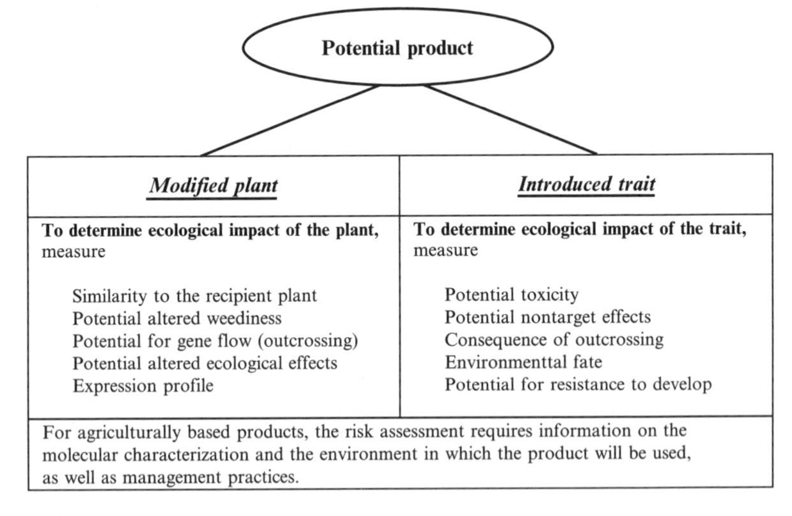

Figure 3 Ecological risk assessment model for GMPs.

risk agent (or stressor, to use the EPA term), potentially exposed or affected organisms, and the specific assessment end points to consider (Table 2). In the case of the GMP, the entire plant can be viewed as the risk agent. For in-field assessments of the GMP, it is important to determine whether the GMP is similar to traditionally grown commercial crops and whether the product has increased potential to persist into rotational crops or otherwise disrupt the agroecosystem. Off-field assessments must consider the potential of the GMP to move from the field either as the parent plant or through pollen-mediated gene flow. Assessment end points reflect these potential effects and provide a basis for developing measures of effects in the risk analysis phase.

Potential effects of the trait as the risk agent can also be evaluated in-field and off-field. For in-field assessments, it is important to define what organisms are of concern and how effects on these organisms are to be judged. For example, herbivorous nontarget insects feeding on green plant tissue will be tolerated only as long as the damage to the crop is too low to affect agronomic performance. Pollinators typically do not adversely affect the crop, and any change in abundance of these organisms due to the trait would be of interest. Beneficial arthropods, such as parasitoid and predatory insects, are important in integrated pest management strategies for many crops. Therefore, the impact of the trait on biological control of pests relative to existing methodologies should be considered. Soil organisms can change substantially in agroecosystems as a result of many activities including tillage and fertilization. Because of this, the focus of the risk assessment for traits in roots or in decaying vegetation should be on macroelements of the soil system that will assure the preservation of the overall function of the soil as evidenced by microbial biomass and nutrient cycling. Off-field assessments focus on direct exposure of nontarget organisms to the trait, either in plants germinated from seed deposited off-field or through plant parts, such as pollen, which can be windblown or distributed by insects.

Tiered Assessment

Some have suggested that evaluation of ecological risk for GMPs should follow a process of tiers according to which progression to higher tiers is required for those GMPs or traits with a higher potential for ecological risk (Rissler and Melon; 1996; Kjellsson, 1997; and Kjaer *et al.*, 1999). This conceptual information can easily be summarized in three tiers (Table 3). Tier 1 evaluations are based on currently available data, whereas tiers 2 and 3 involve collection of data at different levels of complexity. No unconfined release of the trait (growth with no isolation mechanisms) is allowed until the GMP has been thoroughly evaluated.

Table 2

Examples of Assessment End Points for Organisms Potentially Affected by Genetically Modified Plants or the Introduced Traits

Risk agent (i.e. stressor)	Specific agent	In-field/ off-field	Potentially affected taxa/ group	Assessment end point
Modified plant	Whole plant	In-field	Crop	Agronomic characteristics
	Whole plant	In-field	Rotational crop	No change in persistence
	Whole plant (parent)	Off-field	Adjacent vegetation	Change in fitness (weediness, competition, etc.)
	Whole plant (hybrid/ progeny of parent plant)	Off-field	Adjacent vegetation and animal populations	Change in fitness Altered community structure Altered gene pool Effects on nontarget animal populations
Introduced trait	Trait in green tissue (parental)	In-field	Herbivores	Change in species abundance, trophic transfer of trait, and damage to crop
	Trait in nectar	In-field	Butterflies/bees	Change in species abundance
	Trait in pollen	In-field	Herbivores, predatory insects	Change in capacity to control pest populations
	Trait in grain	In-field	Granivorous animals	Change in species abundance, trophic transfer of trait, and damage to crop
	Trait in roots	In-field	Soil organisms	Change in species abundance or ecosystem function
	Trait in decaying vegetation	In-field	Soil organisms	Change in species abundance or ecosystem function
	Trait in pollen (wind or insect transport)	Off-field	Herbivores, predatory insects	Change in species abundance

Table 3

Tiered Progression for Evaluating Ecological Risk of GMP and Trait Should Be Evaluated for Each Geographic Region for Both In-Field and Off-Field Assessments

Tier	Characteristics of tier	Key ecological risk criteria	
		Modified plant	Introduced trait
Tier 1	Evaluate available information	Is plant familiar to existing varieties?	Is the trait common or is there sufficient information to judge safety?
	Tier occurs prior to any environmental release of the GMP	If yes, then minimal concern; due to history of safe use	If yes, then minimal risk is indicated
		If no or insufficient data, then go to tier 2	If no or insufficient data, then tier 2
Tier 2	Conduct laboratory or small field studies	Is plant familiar to existing varieties?	Are there potential adverse ecological effects from exposure to the trait?
	To prevent release of genetic material, all studies are confined	If yes, then minimal risk	In no, then minimal risk is indicated
		If no, then proceed to tier 3	If yes, proceed to tier 3
		If plant is judged not familiar, do wild relatives exist in area?	
		If no, then no risk of gene flow	
		If yes, repeat tier 2 with hybrids	
Tier 3	Evaluate GMP in large-scale field trials	Is there any increased pest potential for the GMP or hybrids?	Are there potential adverse ecological effects from exposure to the trait?
	To prevent release of genetic material, all studies are confined	If no, then minimal risk	If no, then minimal risk
		If yes, then consider risk management options	If yes, then consider risk management options

Analysis of the GMP focuses initially on similarity to conventional varieties. In some cases this can be established based on existing data (tier 1), but other situations may call for the collection of data (tier 2). In either case, a finding that the GMP is similar in ecological characteristics to conventional varieties leads to a minimum risk conclusion based on a history of safe use. Progression to tier 3 for GMPs occurs when information is obtained indicating that the modified plant is ecologically different in some way from the conventional varieties. Tier 3 evaluation investigates the pest potential of the plant. Also, if the GMP is found to be ecologically different from conventional varieties and if wild relatives are known to occur in the geography under analysis, hybrids through tier 2 and possibly tier 3, depending on results, should be evaluated. This tiered analysis culminates in a risk decision and implementation of risk management practices, if necessary.

Analysis of the introduced trait in tier 1 focuses on the plausibility that an effect could occur in potentially exposed organisms based on current information on mechanism of action, expression levels, and other factors. Tier 2 analysis focuses on effects and exposure components of the trait to evaluate potential for adverse effects under field conditions. If the potential for effects in the field cannot be eliminated in tier 2, more focused semifield and field studies should be conducted to reveal whether effects will be observed under actual use conditions. If effects are observed, their magnitude and intensity should be characterized.

Measures of Effects Within Tiers

Evaluation of potential to affect the assessment end points in Table 2 requires the availability of measures that can be readily quantified and easily related. Examples of measures of effects for some of these assessment end points are shown in Table 4. The initial phase of the risk assessment for GMPs focuses on similarity to conventional crops. A broad spectrum of studies can be used to assess this; however, several key studies serve to provide a base for the assessment. Measures of effect that are particularly useful are percentage of germination and altered dormancy, time to flowering, and other measures of growth habit, yield, and volunteering potential. These agronomic characteristics, along with a measure of population replacement capacity, are used to assess similarity to currently used conventional varieties. At tier 3, measures of effects become more complex. Measures of community structure can be expressed as richness or diversity, while measures of interspecific competition, persistence, and invasiveness can be less precise. As mentioned earlier, the evaluation of hybrids will follow a similar path if compatible species are known to exist and if the GMP is shown to be different from conventional varieties.

Measures of effect for toxicological assessments are well defined. Laboratory studies can be designed to predict concentrations of the protein (in the

Table 4

Examples of Measurement End Points at Different Tiers of the Risk Assessment for Genetically Modified Plants or Introduced Traits

Risk agent	Assessment end point	Tier	Measurement end points	Basis for risk characterization
Modified plant	Change in fitness	Tier 1	Background characteristics of crops	Expert judgment
		Tier 2	Agronomic measures (% germination, yield, etc.)	Similarity to traditional varieties
			Population replacement	Net replacement rate (Rissler and Melon, 1996)
		Tier 3	Interspecific competition	Occurrence of weed
			Persistence	Potential disruption of ecosystem
			Invasiveness	
			Genetic biodiversity (allele persistence)	
			Community structure	
			Plant–plant interactions	
Introduced trait	In-field decreases in nontarget organisms	Tier 1	Literature data review	Plausibility of effect
		Tier 2	Hazard assessment by direct ingestion of protein or crop)	Hazard quotient or probability of effect
			In-field realistic worst-case exposure analysis	
		Tier 3	Field study (actual use conditions)	Comparison to current agronomic conditions

case of *Bt*) that result in a 50% effect on the organism investigated. Alternatively, the focus of the investigation may be to identify a concentration that produces no measurable effect (the no-observed-effect-concentration, or NOEC). Comparisons of these measures of effect to concentrations expected to occur in the environment provide an excellent tool for evaluating potential for harm. If the margin of safety predicted between the levels of exposure and the levels associated with biological effects is insufficient, progression to tier 3 would be required. Measures of effect for tier 3 studies are dependent on the design of the study and the questions to be answered. For example, if effects of *Bt* pollen on monarch butterflies are to be investigated in the field, it will be important to know what measures are to be used (number of monarchs per acre, number of monarchs per acre of corn, number of monarchs within 10 m of corn field). Similarly, the measure of exposure needs to be in units that can be related to the potentially exposed organisms, such as pollen per square centimeter of milkweed.

Risk Characterization

Characterization of risk for the GMP revolves principally around two features: familiarity and pest/ecological disruption potential. Tiers 1 and 2 ask questions related to the degree of similarity to conventional varieties. This concept of familiarity is well developed (NRC, 1989; Tedje *et al.*, 1989; Hokanson *et al.*, 1999) and can be characterized by evaluating each measure of effect using the "line-of-evidence" approach (U.S. EPA, 1998). The concept of familiarity was developed jointly by groups of scientists in the United States (NRC, 1989; Tejde *et al.*, 1989). As originally proposed by the NRC (1989), one could arrive at a conclusion of "familiarity" assuming the availability of sufficient information on the biology of the plant, the introduced trait, and the receiving environment. The concept of familiarity was predicated on two important points: first, that the process of developing a GMP was not fundamentally different from traditional breeding, with transgenes behaving like any other genes in the crop genome, and second, that plant breeders have extensive experience and a long history of safety in introducing new varieties of crops into agriculture. Familiarity does not mean that a GMP is "safe," but rather that enough information is available to allow analysts to characterize the risk. To conclude familiarity, one needs knowledge of biology and experience with the crop, its phenotype, and the environmental impact of the introduced DNA (NRC, 1989). When the GMP is judged not familiar to existing varieties, more complex studies are conducted and the information is also integrated into a line-of-evidence decision process. A good example of a parameter that fell outside the range of what would be familiar is the incidence of mycotoxins detected on *Bt* orn (Cry1Ab). In this case, the effect indicated that the risk was reduced (a positive effect) and more complex studies were not required, since the effect was associated with insect protection and not with a direct effect on the fungal source of the mycotoxin (Munkvold, 1999).

In the United States, the Department of Agriculture's Animal and Plant Health Inspection Service has gone further to create a list of familiar traits (Hokenson *et al.*, 1999) that could be used in risk assessment. However, risk characterization of traits, such as those of *Bt* proteins, can follow traditional risk quotient methodologies, whereby the potential concentration in the field is compared with the concentration known not to affect the organisms. However, as is the case for *Bt* proteins, tier 1 assessment may lead to a conclusion of minimal risk for many nontarget organisms because of high selectivity of toxic action (U.S. EPA, 2001). Characterization of risk for more advanced tier 3 testing may require more advanced methods such as probabilistic analysis (ECOFRAM, 2000) or line-of-evidence approaches.

Risk Management

The overall process of risk assessment culminates with a risk decision. Risk can be judged to be minimal, acceptable, acceptable with risk management, or not acceptable. A minimal risk conclusion indicates that based on the information collected, the assessment end point is not likely to be affected by introduction of the new product. An acceptable risk decision indicates that some risk has been identified, but in comparison to other currently used technologies or to anticipated benefits, the risk is judged to be acceptable. Mitigation may be proposed to lessen the risk for a given product or action, whereby the risk is judged to be acceptable. For example, the EPA requires that populations of target lepidopteran insects be monitored for the development of resistance. To date, resistance has not been detected in the field; but if it were, mitigation measures beyond the requirement to limit the planting of a *Bt* crop on a limited number of acres would be instituted. Last, the risk may be judged to be unacceptable relative to benefits and potential mitigation options.

Future Research Needs

The data used in ecological risk assessments are never complete because the variety of circumstances that can exist in natural ecosystems is almost infinite. It can always be argued that additional information could increase the accuracy and reduce uncertainty. Therefore, an assessment of the adequacy of the knowledge base must be made in the context of the assessment end points (Giesy *et al.*, 1999). Many of the assessment end points discussed in this chapter are related to in-field analysis. However, there is a large degree of uncertainty around impact of current agronomic practices on specific ecosystem components, such as soil microbes and beneficial arthropods, especially relative to detailed site-specific impacts of fertilizers and pesticides. Also, impacts of landscape changes in cropping patterns are not well understood. This does not mean that there are high-risk situations existing in today's agriculture, only that the agriculture landscape is different from a "natural" landscape and that this information is important in characterizing risk of new biotechnology products.

As noted earlier, another area of inherent uncertainty is related to the designation of whether a plant is "weedy." A number of proposals have been suggested, but all are limited in their ability to predict with certainty whether a plant will or will not become a weed. In essence, weediness is a highly personal characterization based in aesthetics; it introduces uncertainty, and the involvement of science is limited. Similarly, the challenges in predicting ecologically complex effects such as invasiveness with acceptable predictability have also been described (Parker and Kareiva, 1996). Many regulatory agencies utilize expert judgment in the form of scientific advisory groups to address uncertainty.

CONCLUDING COMMENTS

Given that food production is fundamental to human survival, the benefit of agriculture is self-evident, and the risk associated with new crop varieties has been tacitly understood to be acceptable until recent times. As such, risk assessment has been an embedded, less formal process in the selection of new crop genotypes, and virtually lacking in public scrutiny. This experience with agricultural practices affords a basis for "familiarity," which is valuable to establish a comparative framework for assessing risk from GMPs. The relevant risk hypothesis in this case is, What system presents lower risk?

The process of ecological risk assessment has improved greatly over the past two decades. It is a process that is iterative, adapting to new information and knowledge, and relying on knowledge from all available sources. Unlike plants derived from traditional breeding, GMPs are required to undergo a prospective evaluation of their ecological risks. Based on scientific principles and following the ecological risk assessment framework that has been developed in the United States (NRC, 1993; U.S. EPA, 1998) a risk assessment of a GMP can be designed and conducted that is consistent with a precautionary approach and relevant to the system in which the product will be employed.

The current "debate-based" approach to risk evaluation in the public is neither constructive nor progressive. Several scientifically grounded models, based on well-developed principles of ecological risk assessment, are available. Using dialog-based approach that encourages viewpoints from the breadth of stakeholders and is procedurally operational will lead to further improvements in the models.

REFERENCES

Adams, J. (1995). *Risk*. UCL Press, London, p. 1.

Ammann, K., Jacot, Y, Simonsen, V., and Kjellson, G. (1999). Note Section 4. Monitoring Methods, in The role of modelling in risk assessment for the release of genetically engineered plants, in *Methods for Risk Assessment of Transgenic Plants*, Vol. III, *Ecological Risks and Prospects of Transgenic Plants, Where Do We Go from Here? A Dialogue between Biotech Industry and Science*, K. Ammann, Y. Jacot, V. Simonsen, and G. Kjellsson, eds., pp. 87–122. Birkhauser Verlag, Basel.

Dale, P. J., Irwin, J. A., and Scheffler, J. A. (1993). The experimental and commercial release of transgenic crop plants. *Plant Breeding*, **111**, 1–22.

ECOFRAM (2000). Probabilistic models and methodologies: Advancing the ecological risk assessment process in the EPA Office of Pesticide Program. Final report for the Science Advisory Panel meeting, March 13–16, 2001. SAP Report. 2001–06.

Giddings, G. D. (1999). The role of modelling in risk assessment for the release of genetically engineered plants, in *Methods for Risk Assessment of Transgenic Plants*, Vol. III. *Ecological Risks and Prospects of Transgenic Plants, Where Do We Go from Here? A Dialogue between Biotech Industry and Science*. K. Ammann, Y. Jacot, V. Simonsen, and G. Kjellsson, eds., pp. 31–41. Birkhauser Verlag, Basel.

Giesy, J., Dobson, S., and Solomon, K. (2000). Ecotoxicological Risk Assessment for Roundup® Herbicide. *Reviews of Environmental Contamination and Toxicology*, **167**, 35-120.

Hokanson, K., Heron, D., Gupta, S., Koehler, S., Roseland, C., Shantharam, S., Turner, J., White, J., Schechtman, M., McCammon, S., and Bech, R. (1999). The Concept of Familiarity and Pest Resistant Plants. Presented at "Workshop on Ecological Effects of Pest Resistance Genes in Managed Ecosystems". Traynor, P. L. and Westwood, J. H. editors. *Information System for Biotechnology*, pp. 15–19.

Kammen, D. M., and Hazzenzahl, D. M. (1999). *Should We Risk It? Exploring Environmental, Health, and Technological Problem Solving*. Princeton University Press, Princeton, NJ.

Kareiva, P., and Manasse, R. (1990). Using models to integrate data from field trials and estimate risks of gene escape and gene spread, in *Biological Monitoring of Genetically Engineered Plants and Microbes*, D. R. MacKenzie, and S. R. Henry, eds., pp. 31–42. Agricultural Research Institute, Bethesda, MD.

Kjær, C., Damgaard, C., Kjellson, G., Strandberg, B., and Strandberg, M. (1999). Ecological Risk Assessment of Genetically Modified Higher Plants (GMHP). NERI Technical Report, 303. National Environmental Research Institute, Silkeborg, Denmark.

Kjellsson, G. (1997). In *Methods for Risk Assessment of Transgenic Plants*, Vol. II, *Pollination, Gene-Transfer and Population Impacts*, G. Kjellson, V. Simonsen, and K. Ammann, eds., pp. 221–236. Birkhauser Verlag, Basel.

Munkvold, G., Hellmich, R., and Rice, L., (1999). Comparison of fumonisin concentrations in kernels of transgenic Bt maize hybrids and nontransgenic hybrids. *Plant Dis.*, **83**(2): 130–138.

NAS. (1987). *Introduction of Recombinant DNA-Engineered Organisms into the Environment: Key Issues*. Washington, DC: National Academy Press.

NRC. (1989). *Field Testing Genetically Modified Organisms: Framework for Decisions*. National Research Council Committee on Scientific Evaluation of the Introduction of Genetically

Modified Microorganisms and Plants into the Environment. National Academy Press, Washington, DC.

NRC. (1993). *Issues in Risk Assessment*. National Research Council Committee on Risk Assessment Methodology Board on Environmental Studies and Toxicology Commission on Life Sciences. National Academy Press, Washington, DC.

OECD. (1992). Safety considerations for biotechnology. Organisation for Economic Co-operation and Development Paris, 50 pp. http://www.oecd.org

OECD. (2000). Report of the Working Group on Harmonization of Regulatory Oversight in Biotechnology. Organisation for Economic Co-operation and Development, Paris, 65 pp. http://www.oecd.org

Parker, I. M., and Kareiva, P. (1996). Assessing the risks of invasion for genetically engineered plants: Acceptable evidence and reasonable doubt. *Biol. Conserv.*, **78**, 193–203.

Rissler, J., and Melon, M. (1996). *The Ecological Risks of Engineered Crops*. MIT Press, Cambridge, MA.

Stewart, C. N. (2001). GM crop data—Agronomy and ecology in tandem. Commentary. *Nat. Biotechnol.*, **19**, 3.

Suter, G., (1933). *Ecological Risk Assessment*. G. Suter editor and principle author. Lewis Publishers, Chelsea, Michigan.

Tiedje, J., Colwell, R., Grossman, Y., Hodson, R., Lenski, R., Mack, R., and Regal, P. (1989). The planned introduction of genetically modified organisms: Ecological considerations and recommendations. *Ecology* **70** (2): 298–315.

U.S. EPA (1998). Guidelines for Ecological Risk Assessment. EPA/630/R-95/002F, April 1998 Final. U.S. Environmental Protection Agency, Washington, DC.

U.S. EPA. (1989). Federal Insecticide, Fungicide and Rodenticide Act (FIFRA); Good Laboratory Practice Standards, Final Rule. 40 CFR Part 160. *Fed. Regis.*, August 17.

U.S. EPA (2001). Biopesticides Registration Action Document, revised risks and benefits section: *Bacillus thuringiensis* Plant-Pesticides, July 16, 2001. http://www.epa.gov/oppbppd1/biopesticides/otherdocs/bt_reassess/1-Overview.pdf

Zimmer, R. D. (1997). Environmental risk assessments and the need to cost-effectively reduce uncertainty, in *Biotechnology in the Sustainable Environment*, Sayler *et al.*, eds, pp. 215–221, Plenum Press, New York.

Chapter 9

Ribozyme Technology and Drug Development

Yan Lavrovsky and Arun K. Roy
University of Texas Health Science Center at San Antonio
San Antonio, Texas

This chapter reviews the state of knowledge of the oligonucleotide-based therapeutic approaches, with special emphasis on ribozyme technology. Ribozymes can be considered to be an extension of the antisense oligonucleotides with the additional advantage of catalytic activity. In addition to the naturally occurring ribozymes, a number of modified oligonucleotide structures with endonuclease activity have been used for selective degradation of RNA sequences responsible for pathogenesis. The presently available list of ribozymes includes group I and group II introns, ribonuclease P, hammerhead ribozyme, hairpin ribozyme, VS (Varkud satellite) ribozyme, and HDV (hepatitis delta virus) ribozyme. Many of these ribozymes have been used for treatment of viral diseases and also against neoplastic, cardiovascular, and genetic disorders. Effective use of the antisense oligonucleotides and ribozymes for therapeutic purposes is dependent on optimization of the targeted delivery, pharmacokinetics, and drug metabolism and also protection against adverse cytokine response. Although these are challenging problems, rapid advances in this field hold promise for a greater

Biotechnology and Safety Assessment, 3rd edition

utilization of these oligonucleotide-based therapeutic agents as "designer drugs."

NONENZYMATIC ANTISENSE OLIGONUCLEOTIDES AS INHIBITORS OF SPECIFIC GENE EXPRESSION

Initial findings of sequence-specific inhibition of Rous sarcoma virus by *antisense oligonucleotides* (Zamecnik and Stephenson, 1978) has led to the development of the oligonucleotide-based technology for inhibition of specific gene expression. A number of comprehensive reviews deal with various aspects of this subject concerning antisense therapy, mechanisms of action, and phosphorothioate oligonucleotides, their pharmacokinetics and toxicological characteristics (Crooke, 1998, 2000; Culman, 2000; Gelband *et al.*, 2000; Smith *et al.*, 2000; Stein, 2000). Examined in a chronological manner, these articles reveal a progressive shift of the research emphasis from *in vitro* studies, to animal experimements to clinical trials. Some of the success stories include the following:

Bcl-2 antisense oligonucleotide (C 3139) in patients with B-cell non-Hodgkin's lymphoma

Application of c-*myb* antisense to a group of patients with chronic myelocytic leukemia

Phase 3 trial of ISIS 3521 protein kinase C-α antisense for treatment of lymphoma; phase 2 trial of ISIS 2503 H-*ras* antisense in patients with sarcoma, mesothelioma, melanoma, and tumors of colon and pancreas;

Phase 2 trial of ISIS 5232/c-*raf* kinase antisense in patients with colon, renal, and ovarian tumors (Crooke, 2000)

Attempts to regulate cardiovascular function with anti-vasopressin and anti-oxytocin oligonucleotides (Culman, 2000)

Use of antisense oligonucleotides to regulate systemic hypertension by inhibiting the renin–angiotensin system (antisense oligos against angiotensin II type 1 receptor), targeting angiotensin-converting enzyme, and antisense knockdown of the β_1-adrenergic receptors (Gelband *et al.*, 2000)

Clinical applications of antisense oligonucleotides targeted to the intercellular adhesion molecule 1 and cytomegalovirus immediate early mRNAs (Stein, 2000). Although in complete, this list helps to highlight the growing potential for therapeutic application of oligonucleotides.

In addition to the foregoing clinical applications, there is an extensive body of literature describing the experimental use of oligonucleotides both *in vivo* and *in vitro*. McCarthy *et al.* (2000) described very intriguing animal experiments dealing with behavioral effects of neonatal exposure of antisense oligonucleotides. Estrogen receptor antisense oligos attenuated the masculinizing effects of neonatal testosterone treatment on female sexual behavior in adults, but had no effect on control animals. Antisense oligos to glutamic acid decarboxylase (the rate-limiting enzyme in γ-aminobutyric acid synthesis) had the predicted effect of reducing masculinization induced by exogenous androgen but also unexpectedly reduced female sexual behavior in normal females. Steroid receptor coactivator (SRC-1) antisense oligos reduced the masculinizing effects of exogenous androgen on female sexual behavior but had no influence on male sexual behavior. The authors speculated that this set of results may be due to specific interactions of SRC-1 with estrogen receptor versus androgen-receptor-mediated events and suggested that antisense oligos might be used to selectively influence such complex interactions.

Specificity, side effects, safety, and toxicity are important concerns because many adverse clinical manifestations related to antisense therapy (e.g., complement activation, coagulation prolongation, thrombocytopenia, liver and kidney toxicity) have been reported (Crooke 2000).

Nonspecific effects of antisense oligonucleotides on the molecular level are mediated by RNase H (known as "irregular cleavage"). A number of laboratories are working on elimination of such irregular cleavage. One of the interesting approaches that has emerged is to induce RNase P-mediated cleavage in order to replace RNase H-dependent cleavage (Stein, 2000).

Peptide Nucleic Acids (PNAS)

Peptide nucleic acids (PNAs) are nucleic acid analogues in which the phosphodiester backbone has been replaced with a polyamide backbone made up of repeating *N*-(2-aminoethyl) glycine units (Nielsen *et al.*, 1991). PNAs are extremely stable in PNA–DNA duplex and PNA_2–DNA triplex. One consequence of the polyamide backbone of PNA is that the structure is highly resistant to digestion by either nuclease or protease. In contrast to the PNA itself, hybridized DNA or RNA targets remains sensitive to degradation by a variety of enzymes. Because PNAs are not substrates for RNase H or other RNases, the major antisense mechanism of PNAs is through steric interference. Thus, the greater stability of the PNA–RNA duplexes and the resistance of PNA agonist enzymatic degradation make them very promising antisense agents (Dean, 2000).

TRIPLEX-FORMING OLIGONUCLEOTIDES

Basic rules and canonical motifs for intermolecular triplex formation on double-helical DNA are well characterized (Lavrovsky *et al.*, 1997; Praseuth *et al.*, 1999; Fox, 2000, Kochetkova and Shannon, 2000). Optimal target sequences must harbor consecutive purines on the same strand, since only purine bases are able to establish two Hoogsteen-type hydrogen bonds in the major groove of DNA. Significant progress has been made toward the goal of increasing intermolecular triplex stability while preserving specificity. Some examples include the synthesis of intercalation agents with chemical or photochemical activation properties. A panel of drugs including anthraquinones, benzopyridoindole derivatives, and benzopyridoquinoxaline reportedly specifically stabilized triple helices. Protection of the triplex-forming oligonucleotides against cellular nucleases has been achieved by changing the anomeric configuration of the glycosidic bond from β to α (Praseuth *et al.*, 1999).

A very significant problem in triplex-mediated gene expression regulation is the formation of self-associated-structures that compete with triplex formation, especially in the case of guanine- or cytosine-rich oligonucleotides. Praseuth *et al.* (1999) have suggested several ways to impair the formation of self-associated-structures. These include designing a short hairpin structure to be introduced at one end of the oligonucleotide, additing a oligonucleotide "helper," to form a short duplex at either the 3′ or the 5′ end of the oligonucleotide, and designing clamp or circular oligonucleotides that are less prone to self-association.

Although a role of *intramolecular triplexes* (or H-DNA) has not been elucidated, growing evidence favors their implication in chromatin structure and condensation phenomena. Several examples show that potential H-forming DNA sequences are also critical for gene expression. Agazie *et al.* (1996) reported that antibodies specific for H-DNA can inhibit global transcription and replication processes on isolated nuclei or in permeabilized cells. Much information is expected from identification and cloning of triplex- or H-specific proteins.

Triplex-forming oligonucleotides are being used in several fields of molecular and cellular biology, such as plasmid DNA purification on affinity columns based on triple helix formation; digestion of long DNA fragments, including chromosomes, at specific sites; and site-directed mutagenesis using photochemically cross-linked oligonucleotides. In contrast to antisense oligonucleotides, which are being tested in many clinical trials, triplex-forming oligonucleotides, though studied for more than a decade, have yet to find broad clinical application.

The triplex approach to regulating gene expression *in vivo* has two major limitations. First, conditions of low pH are necessary for formation of the

C+*GC triplet; and second, these structures are often less stable than their duplex counterparts. Fox (2000) outlines the strategies that have been employed to overcome these limitations. The pH problem is addressed by considering the various DNA base analogues that have been used to recognize GC base pairs in a pH-independent fashion. Triplex stability can also be increased by using novel base analogues, by making backbone modifications, and by using triplex-specific binding ligands (Fox, 2000).

One of the approaches is to use *peptide nucleic acids*, mentioned earlier. Because triplex PNA-DNA and PNA-RNA structures are extremely stable, they have been used to interfere with activity of DNA- or RNA-associated enzymes (Dean, 2000). Additionally, because PNAs bind tightly to cellular homopurine DNA targets, they can induce DNA repair pathways within the cell. The PNA-DNA complex can be interpreted by the cell as a DNA lesion in need of repair. One possible outcome of this is the production of site-specific mutants at the site of PNA binding. The frequency of mutagenesis observed with the PNAs was 8×10^{-4}, whereas the background mutagenesis frequency is 9×10^{-5} (Dean, 2000). This technique has the potential to repair the single point mutations important in certain genetic diseases. Although most of the work on PNA applications has focused on antisense strategies, several groups have instead studied the ability to PNAs to turn in gene expression. When PNA triplex structures are formed on one strand of the DNA duplex, the opposing DNA strand is displaced to form a D loop. When sufficiently large, this D loop resembles a transcriptional bubble or initiation–elongation loop (Dean, 2000). Dean's laboratory has utilized fluorescently labeled PNA clamps to label plasmids, which allows investigators to follow the plasmids' nuclear localization. By microinjecting PNA-DNA complexes into the cytoplasm of cells, they were able to follow the movement of plasmids into the nuclei of individual cells in real time. The authors demonstrated that plasmids enter nuclei in a sequence-specific manner and use the same pathways as do proteins containing nuclear localization signal (NLS). However, unlike the case of NLS-mediated protein nuclear import, additional proteins were required for plasmid nuclear localization (Dean, 2000).

RNA-BASED ENZYMES

RNA Folding; Ribozyme Structure and Function

Early studies on transfer RNA carried out in 1960s and 1970s provided the foundation for more recent work on the folding mechanisms of catalytic RNAs (Fresco *et al.*, 1966; Kim *et al.*, 1973).

Because conformational changes in RNA folding in real time are complete in less than a second, biochemical probing of RNA structure cannot be the method of choice. Woodson (2000) reviews such new methods of monitoring the folding reactions as fluorescence resonance energy transfer and x-ray footprinting of RNA. The author proposed a new model for RNA folding called kinetic partitioning mechanism. According to this model, a fraction of the RNA molecule population folds directly and rapidly to the native structure, and the remainder of the population becomes trapped in misfolded intermediates. Transitions from the intermediates to the native state are slow, because at least some interactions must be broken before the correct structure can be formed. The model states that the rates of slow folding processes are determined by the relative stabilities of the trapped intermediates and the native conformation. Supporting the model's hypothesis are findings of increased folding rates obtained by adding destabilizing denaturants and decreased rates obtained when stabilizing compounds, such as higher concentrations of Mg^{2+}, are used. Folding and refolding of RNAs *in vivo* also depends on interactions with RNA chaperones, such as RNA binding proteins and ribosomes (Woodson, 2000).

RNA catalysis was first described by teams who the discoverd the group I intron (Cech *et al.*, 1981) and RNase P (Guerrier-Takada *et al.*, 1983). Ribozymes, catalytic RNA molecules, exist in a range of distinct categories of naturally occurring catalytic RNA. These include a number of small ribozymes involved in replication of viroid genomes, such as hammerhead and hairpin ribozymes (Haseloff and Gerlach, 1988), group I introns (Cech *et al.*, 1981), the RNA component of RNase P (Guerrier-Takada *et al.*, 1983), and hepatitis delta virus ribozyme (Branch and Robertson, 1991). In nature, ribozymes catalyze a large number of reactions, resulting in sequence-specific RNA processing. Although the importance of ribozymes in early evolution is now acknowledged, there has been a considerable effort to study the structure and function of natural ribozymes and convert them into tools for manipulation of gene expression. This led to a significant increase in the number of synthetic catalytic nucleic acids over the last several years.

The general mechanism of ribozymes is similar to that of many protein ribonucleases in which a 2′-oxygen nucleophile attacks the adjacent phosphate in the RNA backbone, resulting in cleavage products with 2′,3′-cyclic phosphate and 5′-hydroxyl termini. Ribozyme as, opposite to protein ribonucleases, cleave only at a specific location based on Watson–Crick interactions. Alignment of the cleavage site within the catalytic core is mediated by tertiary interactions.

RNA packing into stable tertiary structures was thought to be difficult because of the limited number of bases available for tertiary contacts, the high density of negative charges, and the flexibility of the phosphate backbone

(Doherty and Doudna, 2000). Catalytical RNA molecules, however, are capable of accelerating acyl and phosphoryl transfer reactions by 10^5- to 10^{11}-fold (Herschlag and Cech, 1990; Hegg and Fedor, 1995; Hertel *et al.*, 1997; Zhang and Cech, 1997).

These rates were mostly monitored under multiple-turnover conditions when the substrate RNA is in excess of the ribozyme and the ribozyme can catalyze the cleavage of several substrate molecules. However, in the case of long mRNA substrates, the ribozyme is in excess. Cleavage rates under these single-turnover conditions are several orders of magnitude lower than the rates obtained for short substrates.

Divalent metal ions, particularly Mg^{2+}, play an important role in RNA catalysis. A metal ion coordinated to a hydroxide might activate a hydroxyl or water nucleophile by deprotonation, or a divalent ion might directly coordinate the nucleophilic oxygen, exposing the oxygen to deprotonation by hydroxide ions. Metal ions might also stabilize the intermediate structure by donating positive charge. Additionally, Doherty and Doudna (2000) proposed that since divalent cations are required for the folding of many RNAs, and RNA structures often contain specific binding sites for divalent ions, ribozymes may carry out metal-ion-assisted catalysis that is similar to the action of several protein enzymes that catalyze phosphate chemistry.

Ribozymes of Different Types and Their Effects on Gene Expression Regulation *In Vitro*

Group I and Group II Introns

Introns are noncoding sequences that interrupt the coding sequences of most eukaryotic genes. They have to be spliced after transcription to allow expression of functional RNA molecules. In the case of group I introns, excision is mediated by the autocatalytic activity of the intronic sequences themselves. Cech *et al.* (1981) first reported naturally occurring self-splicing activity of *Tetrahymena thermophila*. Since then several hundred examples of group I introns have been characterized in plant and fungal mitochondria, bacteriophage, eubacteria, and chloroplast tRNA. Although they are all different in size and sequence, group I introns preserved an intriguing phylogenetical conservatism with respect to their reaction mechanism and secondary structure (Palmer *et al.*, 2000). The enzymatic activity of group I introns involves a two-step transesterification. In the first step a nucleophilic attack of the guanosine cofactor 3′-hydroxyl group on the 5′-splice site results in the formation of a free 3′ hydroxyl on the 5′ exon. Then this free hydroxyl group makes a nucleophilic attack on the 3′-splice site, releasing the intron as a circular molecule and leaving a ligated exon (Cech *et al.*, 1992). Group I intron folding and cleavage require the presence of divalent cations, such as Mg^{2+} (Heilman-Miller *et al.*, 2001).

With demonstration that group I introns can trans-splice RNA both *in vitro* and in cultured cells, where the intron to be spliced out is part of a different RNA molecule sharing complementarity with the ribozyme, these introns have emerged as potential regulators of gene expression (Phylactou, 2000). The main advantage of this system is the repair of genetically defective messages. Theoretically, group I intron ribozymes can be designed to repair any defective mRNA. They may be used against viral diseases by changing part of their vital RNA information with another one, which will stop protein production. Trans-splicing group I intron ribozymes can be used to repair genetic mutations, which are involved in many forms of cancer. Mutations in tumor suppressor genes are among potential targets for a ribozyme-mediated repair of those transcripts.

Application of group I intron ribozymes in potential treatment of sickle cell anemia and myotonic dystrophy is discussed separately.

Group II introns are large catalytic RNA molecules that act as mobile genetic elements. They were initially identified in the organelle genomes of lower eukaryotes and plants, and it has been suggested that they are the progenitors of nuclear spliceosomal introns (Sharp, 1991). Zimmerly *et al.* (1995) have also suggested that, because of their organization and mobility mechanism, group II introns may be progenitors of telomerase.

Group II introns catalyze their own self-splicing from precursor RNA molecules. The excised intron RNAs can react with other molecules, catalyzing their own insertion into RNA or DNA (Cousineau *et al.*, 2000). In the latter case, reserve splicing into duplex DNA occurs with the help of an intron-encoded protein (Yang *et al.*, 1996). Group II intron ribozymes catalyze cleavage of RNA and DNA molecules with a high degree of sequence specificity and cleavage site fidelity (Su *et al.*, 2001).

Su *et al.* (2001) have demonstrated that the cleavage site of a group II intron ribozyme can be tuned at will by manipulating the thermodynamic stability and structure of the exon binding site/intron binding site pairing. Instead of recognition of specific nucleotides, the ribozyme detects a structure at the junction between single- and double-stranded residues on the bound substrate. The authors suggested that these results are consistent with the lack of phylogenetic conservation in ribozyme and substrate sequences near group II intron target sites.

The mechanism for group II intron cleavage site fidelity, proposed by Su *et al.* (2001), allows a design of ribozymes capable of cleaving virtually any target. This, in combination with a report by Guo *et al.* (2000) showing that mobile group II introns can be designed to insert themselves into the DNA of plasmids transfected into human cell lines, makes group II introns particularly appealing agents for gene therapy.

Ribonuclease P (RNase P)

RNase P is involved in maintaining the functionality of the translation machinery by processing the 5′ termini of tRNA precursors during their maturation. RNA cleavage is mediated by nucleophilic attack on the phosphodiester bond, leaving one 5′-phosphate and one 3′-hydroxyl terminus at the cleavage site. Similar to many other catalytic RNAs, there is an absolute requirement for divalent metal ions. Bacterial RNase P is comprised of an RNA domain of approximately 400 nucleotides and also contain a protein component having a mass of 14 kDa. The RNA component of RNase P is a ribozyme *in vitro*; however, there is a requirement for protein component for catalytic activity *in vivo* (Guerrier-Takada *et al.*, 1983). A detailed analysis of RNase P protein structure by Stams *et al.* (1998) revealed an arrangement of three α helices surrounding a central β sheet, three putative RNA binding regions, and a metal ion binding pocket, suggesting a role for RNase in metal-mediated RNA binding. Stams *et al.* (1998) have used the very close similarity between structures of RNase P protein and several universally conserved proteins of the translation machinery as a basis for suggesting that an ancient RNase P protein may have mediated the conversion from an RNA to a ribonucleoprotein world.

Fang *et al.* (2001), who showned that the *Bacillus subtilis* RNase P contains two RNA and two protein subunits, have suggested a more versatile role for proteins in ribonucleoprotein complexes.

In bacteria, RNase P cleaves not only pre-tRNAs but also other small RNAs (Komine *et al.*, 1994), precursors to 4.55 ribosomal RNA (Bothwell *et al.*, 1976), and, at least, one known, the polycistronic *his* operon, mRNA (Alifano *et al.*, 1994).

Whereas bacterial RNase P contains only one protein subunit, there are nine protein subunits characterized in yeast (Schon, 1999; Xiao *et al.*, 2001). Human RNase P contains an RNA subunit, H1 RNA, and at least seven protein subunits that copurify with enzymatic activity. Interestingly, two of them are scleroderma autoimmune antigens (Eder *et al.*, 1997). The protein complement is highly variable between cellular compartments, with the number of subunits varying between one in yeast mitochondria and nine in the nucleus. Additionally, yeast RNase P shares eight protein subunits with another closely related ribonucleoprotein, RNase MRP. Localized in the nucleolus, RNase MRP is thought to mediate pre-rRNA maturation. It is still unclear how many (if any) of eight shared protein subunits are responsible for subnuclear localization of both enzymes. Unlike most other catalytic

RNAs, RNase P does not bind its substrate via base-pairing interactions. The exact cleavage site is determined by the large number of different substrates integrated in a cell.

Hammerhead Ribozyme

The hammerhead ribozyme model is based on the satellite RNA strand of tobacco ringspot virus. The ribozyme naturally acts in *cis* during replication by the rolling circle mechanism. Uhlenbeck (1987) developed the three-stem trans-acting hammerhead ribozyme, in which stems I and III are base-pairing flanking sequences and stem II is a highly conserved 22-nucleotide catalytic domain. The mRNA substrate and the hammerhead ribozyme hybridize through stems I and III, exposing a recognition sequence on the target RNA such as the GUC nucleotide triplet. The cleavage of the RNA substrate by the hammerhead ribozyme occurs via a transesterification reaction that generates 5′-hydroxyl and 2′-3′-cyclic phosphate termini. Divalent cations, particularly Mg^{2+}, are required not only for catalysis and enhancement of the structural stability of the substrate RNA, but also for conformational folding of the ribozyme itself (Hammann *et al.*, 2001).

The hammerhead ribozyme is one of the smallest and best studied among catalytic RNAs. Structures, catalytic motifs, physicochemical kinetics, target selection, design, and *in vitro* activities have been well characterized (Lavrovsky *et al.*, 1997; Amarzguioui and Prydz, 1998; Stage-Zimmermann and Uhlenbeck, 1998; Scott, 1999; Sun *et al.*, 2000; Suzumura *et al.*, 2000).

Komatsu *et al.* (2000b) deserve special attention, for having since they designed a hammerhead ribozyme employing a pseudo-half-knot motif, a ribozyme that is activated by short oligonucleotides. A hammerhead ribozyme with stem II replaced with a non-self-complementary loop was regulated by the presence of a short oligonucleotide complementary to the 5′-side sequence of loop II. Komatsu *et al.* (2000a) speculate that these allosteric ribozymes may be applicable to the regulation of other functional RNAs with hairpin loops.

Since hammerhead ribozymes have been extensively studied, it is not surprising that they are being tested for ability to combat many infections, as well as genetic and neoplastic diseases. We will discuss *in vivo* applications of hammerhead ribozyme separately.

Hairpin Ribozyme

The hairpin ribozyme is another example of a small catalytic RNA. Similar to the hammerhead ribozyme, the hairpin ribozyme cleaves the phosphodiester bond of substrate RNA to generate 5′-hydroxyl and 2′,3′-cyclic phosphate termini. Also, similar to the hammerhead ribozyme, the hairpin catalytic motif was first discovered in the tobacco ringspot virus satellite RNA (Buzayan *et al.*, 1986). However, the hammerhead ribozyme was found in the positive strand, whereas the hairpin motif was discovered in the negative strand of the virus satellite RNA. There are also several principal differences, first of all, structural. Molecular modeling, cross-linking, fluorescence resonance energy transfer, crystallography, and nuclear magnetic resonance studies of folding and catalysis by the hairpin ribozyme (Lilley, 1999; Fedor, 2000; Komatsu *et al.*, 2000a; Pinard *et al.*, 2001; Rupert and Ferre-D'Amare, 2001) have revealed precise interactions between two principal domains of the ribozyme and an RNA substrate in catalysis facilitation. Additionally, catalysis by the hairpin ribozyme is completely independent of divalent cations and relatively independent of pH, making it unique among catalytic RNAs. Also, in contrast to the hammerhead ribozyme, which favors cleavage over ligation by nearly 200-fold (Hertel *et al.*, 1998), the hairpin ribozyme favors ligation by nearly 10-fold (Fedor, 2000). We will also discuss *in vivo* applications of hairpin ribozymes separately.

The Vankud Satellite (VS) Ribozyme

The VS ribozyme is another type of nucleolytic ribozymes generating 5′-hydroxyl and 2′,3′-cyclic phosphate termini as a result of attack of the 2′ oxygen in transesterification reaction. In the nature, *cis* cleaving the VS ribozyme occurs in the 881nt RNA found in the mitochondria of *Neurospora*, which is transcribed from the Varkud satellite DNA (Kennell *et al.*, 1995). Guo and Collins (1995) have shown that the VS ribozyme can act in *trans* on a hairpin–loop substrate and proposed that tertiary interactions mediate substrate recognition.

Lafontaine *et al.* (2001) used fluorescence resonance energy transfer to study the structure, folding, and activity of the VS ribozyme. They found the ribozyme has a core comprising five helices organized by means of two three-way junctions. One of the junctions, so-called 2–3–6, undergoes two-stage ion-dependent folding into a stable conformation, and there is good correlation between changes in activity and alterations in the folding of junction 2–3–6. Lafontaine *et al.* (2001) proposed that the 2–3–6 junction organizes important aspects of the structure of the VS ribozyme to facilitate productive association with the substrate. This finding is further support for interaction between the VS ribozyme and its substrate occurring mainly through tertiary interactions.

Hepatitis Delta Virus (HDV) Ribozyme

So far, the HDV ribozyme is the only example of a self-cleaving RNA discovered in an animal virus. A single-stranded circular RNA of 1700 nucleotides, HDV is considered to be a satellite virus. It requires the presence of the hepatitis B surface antigen for superinfection. This superinfection by HDV leads to acute hepatitis and eventually liver cirrhosis in many patients.

The HDV ribozyme is involved in a second step of virus replication. Since correct therapeutic regimes to treat the HDV infection are mostly nonspecific, the ribozyme is being considered as a potential target for treatment of HDV infection. Its structure, cleavage kinetics, intracellular activity, and role in replication have been extensively studied (Been and Wickham, 1997; Chen *et al.*, 1997; Shih and Been, 2001, Wadkins *et al.*, 2001). Given the nature of this ribozyme and the pathogenic potential of HDV, it is not surprising that, most of the research efforts to study the HDV ribozyme are aimed of combating replication of the virus. There is a considerable interest in the suppression of ribozyme activities by ligands, including new antibiotics (Jenne *et al.*, 2001).

CATALYTIC DNAs

It is widely accepted now that catalytic RNAs have an advantage over antisense DNA oligonucleotides in suppressing gene expression. This advantage is due to a combinatory effect of antisense RNA interactions and catalytic reaction between a ribozyme and an mRNA molecule. The biggest technical problem with catalytic RNAs is their fragility in biological fluids. Because of this problem, ribozymes can be delivered only through transfection–transgenic expression. An ideal oligonucleotide-based molecule would combine antisense and catalytic properties of a ribozyme with stability of a DNA in biological fluids. That kind of molecule has not yet been identified in biological systems. However, *in vitro* selection designed to specifically derive DNA sequences with RNA-cleaving activities led to the discovery of several Mg^{2+}-dependent DNA enzymes (Santoro and Joyce, 1997; Li and Breaker, 1999). One of the best-characterized molecules from this class is the 10–23 DNA enzyme, named from its origin as the 23rd clone of the 10th cycle of *in vitro* selection (Santoro and Joyce, 1997). It contains a conserved catalytic domain flanked by variable binding domains and has structural similarity to the hammerhead ribozyme.

With the potential to bind any RNA sequence and cleave purine–pyrimidine junctions, the 10–23 DNA enzyme has unprecedented target site flexibility. The AUG start codon of any gene could be used as a target for 10–23 DNA enzyme. Santoro and Joyce (1997) used the start codon sequences from different HIV genes to show that the kinetic efficiency of catalysis mediated by the 10–23 DNA enzyme varies from one sequence to another. These kinetics were found to correlate with thermodynamic stability of the DNA enzyme–substrate heteroduplex. In experiments with different GC content and substrate binding domain length ranging between 4/4 and 13/13, the maximum efficiency under physiological reactions conditions was found with an arm length of between 8 and 9 base pairs (Santoro and Joyce, 1998).

Sun *et al.* (2000) have found that in the case of c-*myc* translation initiation region, cleavage can be enhanced by asymmetric arm length truncation. They observed the maximum cleavage efficiency when they used the 10–23 DNA enzyme with a binding arm length ratio (5′ to 3′ bp) of 6:10. Since its discovery, the 10–23 DNA enzyme has been successfully used to suppress expression of many genes *in vitro* and in cell lines (Sun *et al.*, 2000). Attempts to use them *in vivo* are discussed next.

THERAPEUTIC APPLICATIONS OF CATALYTIC OLIGONUCLEOTIDES

Application as Antiviral Agents

An RNA genome of retroviruses in general and HIV-1 in particular has been an appealing target for catalytic RNA applications for the past decade (Sarver *et al.*, 1990; Yu *et al.*, 1995; Ramezani *et al.*, 1997; Wang *et al.*, 1998). Combined use of protease and reverse transcriptase inhibitors ("cocktails") has achieved significant progress in the treatment of HIV infection. Because of the high cost of the drugs, their side effects, and the high mutation frequency of the HIV genome, however, additional therapeutic approaches must be developed. A ribozyme capable of sequence-specifically cleaving HIV RNA may potentially be useful addition to the "cocktail."

Ribozymes can be designed to cleave HIV RNA at multiple target sites, overcoming the problem of high mutation frequency of the HIV genome (Chen *et al.*, 1992).

In the first clinical trial involving patients infected with HIV-1, Wong-Staal *et al.* (1998) used T-cell transduction of a retroviral vector expressing hairpin ribozymes targeted to the *Pol* gene and 5′ long terminal repeat of HIV-1 region. Since then, a number of different catalytic RNA and DNA molecules have been tested in pre- and clinical trials (Rossi, 2000; Sun *et al.*, 2000). Gene Shears Pty Ltd is conducting two clinical trials with hammerhead ribozymes targeted to the *Tat* gene and the RNA packaging sequence of HIV (de Feyter and Li, 2000).

Shahi *et al.* (2001) have reported successful cleavage by hammerhead ribozymes of a double-stranded RNA-containing virus. With the *S1* gene of reovirus targeting these ribozymes, remarkable protection was shown against the pathological effects of reovirus.

Treatment of Neoplastic Diseases

Ribozymes have been shown to successfully cleave mRNA of many transcription factors and oncogenes, leading to phenotypical changes in cell cultures and transgenic animals (Lavrovsky *et al.*, 1997; Amarzguioui and Prydz, 1998; Stage-Zimmermann and Uhlenbeck, 1998; Scott, 1999; Sun *et al.*, 2000; Suzumura *et al.*, 2000). However, cancer in a clinic is a complex phenotype resulting from changes in the expression of many genes, as well as cell and tissue reorganization. Therefore, clinical expectations for catalytic nuclei acids capable of cleaving certain RNAs are far below those in the case of viral infections. Nevertheless, Sioud and Leirdal (2000) have reported a very significant reduction of glioma solid tumor after a single injection of vascular endothelial growth factor–specific ribozyme into the tumor. They have also shown that this effect was due to inhibition of angiogenesis. Lui *et al.* (2001) have reported that a hammerhead ribozyme, driven by U6 promoter and targeted to c-*neu* mRNA at the tyrosine kinase domain, inhibited growth of ovarian cancer. A report by Mistry *et al.* (2001) on an anti-stathmin ribozyme is very interesting because stathmin is overexpressed in all human cancer and is an attractive molecular target for anticancer interventions. More comprehensive reviews on design and testing of ribozymes for cancer gene therapy can be found in Sioud (1999), James (1999), Turner (2000), and Norris *et al.* (2000).

Treatment of Cardiovascular Diseases

Cardiovascular diseases are multisystemic disorders, and from the theoretical point of view, it is difficult to stop their progression or even reverse their pathology by application of catalytic nucleic acids to suppress expression of a certain gene. That is, perhaps, the reason for the failure to test ribozymes in models of cardiovascular diseases until relatively recently. In one of the earlier reports, Morishita *et al.* (1998) used DNA-RNA hybrid hammerhead ribozyme to inhibit production of apolipoprotein A, a high concentration of which was implicated in pathogenesis of atherosclerosis. In a later report, Enjoji *et al.* (2000) used adenoviral vector to target hammerhead ribozyme to apolipoprotein B mRNA in the liver of a dyslipidemic mouse that is deficient in apolipoprotein B mRNA editing enzyme and overexpresses human apolipoprotein B. The level of apolipoprotein B mRNA decreased approximately 80% after infection with adenovirus. Additionally, there was a significant reduction of levels of cholesterol and triglycerides in plasma from infected mice.

One of the most serious problems of modern cardiovascular surgery is neointimal hyperplasia after angioplasty, leading to restenosis of the arteries that had been operated on. Although pathogenesis of this hyperplasia is not well understood, several genes have been implicated in this process. Genes encoding for transforming growth factor β and platelet-derived growth factor are among them. Su *et al.* (2000) have successfully used ribozymes to inhibit transforming growth factor β production, and Yamamoto *et al.* (2000) have additionally shown that ribozyme-mediated inhibition of transforming growth factor β production leads to suppression of neointimal formation after vascular injury in animal models. Hu *et al.* (2001) have shown that DNA-RNA chimeric hammerhead ribozyme targeted to platelet-derived growth factor A-chain mRNA inhibited the proliferation of vascular smooth muscle cells.

Since 12-lipoxygenase products of arachidonic acid metabolism have mitogenic effects on vascular smooth muscle cells, Gu *et al.* (2001) tested the hypothesis of whether a decrease in 12-lipoxygenase level will prevent neointimal hyperplasia. The authors used hammerhead ribozyme targeted to rat 12-lipoxygenase and showed that ribozyme-mediated suppression of 12-lipoxygenase inhibited neointimal formation in balloon-injured rat carotid arteries. Perlman *et al.* (2000) have analyzed the effects of disrupting Bcl-2 expression in vascular smooth muscle cells by adenovirus-mediated delivery of a hammerhead ribozyme against bcl-2 mRNA. As expected, infected cells underwent apoptosis, resulting in a reduction of cell number and inhibition of neointimal hyperplasia in balloon-injured carotid arteries.

At this point it seems very unlikely that catalytic nucleic acids will be used alone to treat complex cardiovascular diseases, but they may find service as supplementary tools in certain pathologies, such as restenosis after angioplasty or other disorders, especially those caused by pathological overexpression of a single gene.

Treatment of Genetic Disorders

The ability to induce site-specific cleavage of RNA makes smaller catalytic RNAs such as hammerhead and hairpin ribozymes, very attractive candidates for the treatment of dominant genetic diseases by down-regulating the expression of mutant alleles.

Kilpatrick and Phylactou (1998) proposed an RNA-based therapy for Marfan syndrome. Marfan syndrome is caused by mutations in the gene for fibrillin 1, most of which appear to be acting in a dominant-negative fashion. Reduction in the level of mutant *fibrillin 1* RNA can be beneficial for patients with Marfan syndrome. Authors demonstrated that hammerhead ribozyme specifically and efficiently cleaved mutant *fibrillin 1* mRNA *in vitro* and in cell culture system.

Dawson and Marini (2000) isolated fibroblasts from a patient with osteogenesis imperfecta. These cells contain a mutation in one alpha 1 (I) collagen allele, which causes the skeletal disorder. The authors designed a ribozyme capable of cleaving a mutant RNA substrate, while leaving normal substrate intact. This resulted in a significant reduction in mutant type I collagen protein in cells expressing the ribozyme.

Suzuki *et al.* (2000) demonstrated that a chemically modified (long-acting and resistant to RNase) ribozyme targeted to IgV gene mRNA inhibited anti-DNA production and the formation of immune deposits caused by lupus lymphocytes in a mouse model.

Familial amyloidotic polyneuropathy type 1 is an autosomal dominant inherited disorder with systemic deposition of a mutant *trans*-thyretin. Tanaka *et al.* (2001) reported designing hammerhead and hairpin ribozymes capable of cleaving specifically mutant, but not a wild-type, *trans*-thyretin mRNA *in vitro* and in a cell culture system.

Yen *et al.* (1999) investigated the activity of the 10–23 DNA enzymes targeted to *huntingtin* mRNA. The mutant protein expressed from this transcript is thought to cause Huntington's disease. Two of the tested DNA enzymes demonstrated substantial activity against the huntingtin transcript *in vitro*, and protein in human embryonic kidney-derived 293 cells cotransfected with the Huntington gene.

Shaw *et al.* (2001) have used an allele-specific hammerhead ribozyme for gene therapy of autosomal dominant retinitis pigmentosa in a transgenic porcine model. Perspectives on the use of ribozymes to treat inherited retinal diseases are reviewed by Hauswirth and Lewin (2000) and Shaw *et al.* (2000).

Larger *trans*-splicing RNAs, such as the group I intron ribozyme, can be engineered to replace part of an RNA with a sequence attached to its 3′ end. This approach may be very useful in genetic diseases caused by the presence of a mutant mRNA.

Lan *et al.* (1998) reported *trans*-splicing catalyzed by group I introns to correct sickle cell transcripts in umbilical cord erythrocyte precursors and in nucleated blood cells isolated from sickle cell patients. Sickle cell anemia is the result of an A-to-T mutation at the 5′ end of the β-globin gene. Since γ-globin impedes polymerization of mutant sickle cell β-globin, Lan *et al.* (1998) were able to demonstrate that the group I intron ribozyme can accurately splice fetal γ-globin mRNA onto the mutant γ-globin transcript.

A group I intron ribozyme capable of *trans*-splicing the genetic defect resulting in myotonic dystrophy was reported by Phylactou *et al.* (1998). Myotonic dystrophy is characterized by a CTG trinucleotide repeat expansion in the untranslated region of the myotonic dystrophy protein kinase gene. The investigators corrected the repeat expansion by using a group I intron ribozyme capable of *trans*-splicing a short triplet repeat sequence to a myotonic dystrophy protein kinase message that has the CUG expanded repeat.

All these examples prove that patients suffering from dominant genetic disorders can benefit from ribozyme-based therapies. However, other factors, such as stability, efficiency, and delivery of catalytic nucleic acids, play an important role in designing a precise strategy for the therapy of genetic disorder.

RIBOZYME DELIVERY, PHARMACOKINETICS, AND METABOLISM

There are two principal methods of delivering catalytic nucleic acid. The first utilizes synthetic nucleic acids, which can be unmodified or modified and used with or without carriers. Either way, the situation with respect to delivery, pharmacokinetics, metabolism, toxicity, and immune modulation is very similar to that of antisense or triplet forming oligonucleotides. Obviously, modified or unmodified catalytic DNA enzymes can be delivered by means of this method alone. Extensive information about oligonucleotide stability, biodistribution, pharmacokinetics, and toxicity can be obtained in Crooke (1998, 2000), Henry *et al.* (1999), Sereni *et al.* (1999), Stein (2000), and Sun *et al.* (2000).

However, despite on all the developments and improvements, this method may be limited even in the future to surface organs (such as skin) and to organs (such as the eyes or testes) that are separated by physiological barriers.

Interestingly, oligonucleotides by themselves were suggested for use as delivery vehicles for gene therapy. Dean (2000) has reported use of peptide-PNA conjugates as targeting reagents for plasmids. Critical events at the level of nuclear import, transcription, or mitochondrial import can be targeted by using ligands specific for nuclear localization signal, leucine zipper, or mitochondrial presequencing (Dean, 2000).

Alternatively, the second principal method of delivery utilizes physiological or pathophysiological human and virus integration, recombination, and reverse transcription/transcription systems. This virus-mediated approach is theoretically ideal for the delivery of catalytic RNA. Advantages and disadvantages of multiple virus-mediated delivery systems are well characterized (Hanania *et al.*, 1995; Bramlage *et al.*, 1998; Cotter and Robertson, 1999; Hackett *et al.*, 2000; Somia and Verma, 2000; Spencer 2000).

Retroviral vectors are the most utilized for delivering ribozymes into cells owing to their high transfection efficiency and stable integration into the genome. With the exception of lentiviruses, a major limitation is their inability to infect nondividing cells. When used *ex vivo*, retrovirus can infect physiologically nondividing or slowly dividing cells. An additional advantage of this method is that an individual is not exposed to the retrovirus.

Johnson & Johnson Research Laboratories have initiated two independent phase I clinical trials to test the hypothesis that ribozymes can protect $CD4^+$ T lymphocytes from rapid HIV-1-mediated destruction in an infected individual. Both trials utilize a vector based on Moloney murine leukemia virus containing ribozyme, and each trial uses a separate target cell population: $CD4^+$ peripheral blood lymphocytes and $CD34^+$ stem. The first trial involves identical twins, discordant for infection with HIV. Healthy $CD4^+$ lymphocytes from the infected twin are transfected with a retroviral vector containing the ribozyme gene. These transfected cells are cultured *ex vivo* before transfusion into the bloodstream of the HIV-positive twin. The second clinical trial was designed to ascertain the ability of $CD34^+$ stem cells to differentiate into a variety of lineages that express the ribozyme after removal of the cells, the transfection of the cells *ex vivo*, and their transfusion back into the bloodstream of the HIV-positive individual (Sun *et al.*, 2000). Gene Shears Pty Ltd is using a similar approach, conducting two phase I clinical trials with ribozymes introduced into hematopoietic cells *ex vivo* (de Feyter and Li, 2000).

Adenoviruses are DNA viruses capable of infecting both dividing and nondividing cells without integrating into the host genome. They replicate in the nucleus of the host cells as extrachromosomal elements and express large amounts of transgene. Limiting factors for long-term therapy are an inflammatory response to the virus and the short time of expression. The commonly used human adenovirus serotype 5 can include transgene sequences as long as 10 kilobases. Normal adenoviral transcripts include E1, E2, E3, and E4. Transcripts E1 and E3 are commonly deleted and replaced with transgenes. Deletion of E1 renders the adenovirus defective for replication. This makes the vector incapable of producing infectious viruses in target cells. To generate high titers of viruses, the adenovirus must be introduced into a "packaging" cell line such as 293 or 911 cells, which endogenously express E1 and allow viral production. The E3 transcript is involved in the evasion of host immune responses and is therefore not needed for viral replication. The most commonly employed technique for generating the desired recombinant adenovirus is to introduce a replication-deficient adenovirus and a shuttle vector containing the transgene into a packaging cell line to allow homologous recombination. The packaging cell line provides the E1 gene necessary for generating infectious viral particles. Although this approach is useful, recombination efficiency is low, and the process of screening and purifying plaques is long and tedious.

We have used a novel, simplified system (He *et al.*, 1998) for generating recombinant adenoviruses that contain ribozymes specific for androgen and estrogen receptor (Chen *et al.*, 1998; Lavrovsky *et al.*, 1999). Transformation of *E. coli* strain BJ5183 with adenoviral "backbone plasmid" and a shuttle vector plasmid containing the ribozyme results in homologous recombination via the bacteria's efficient recombination system. After the desired recombinant adenovirus had been identified, it was introduced into the packaging cell line to produce infectious viral particles. Human breast adenocarcinoma MCF-7 cells were seeded into 35-mm plates and infected with different amounts of adenovirus containing either wild-type or mutant estrogen-receptor-specific ribozymes (Ad/v-Rz 2 and Ad/v-Rz 2-M). (Figure 1). Infection efficiency was estimated according to the amount of green fluorescent protein (GFP) produced by the cells. The *GFP* gene is a part of the adenoviral vector and serves as a marker of expression. *In vivo* experiments with these recombinant viruses are under way.

However, despite new developments, including the generation of helper-dependent (gutless) adenoviral vector, use of adenoviruses may be limited in the future.

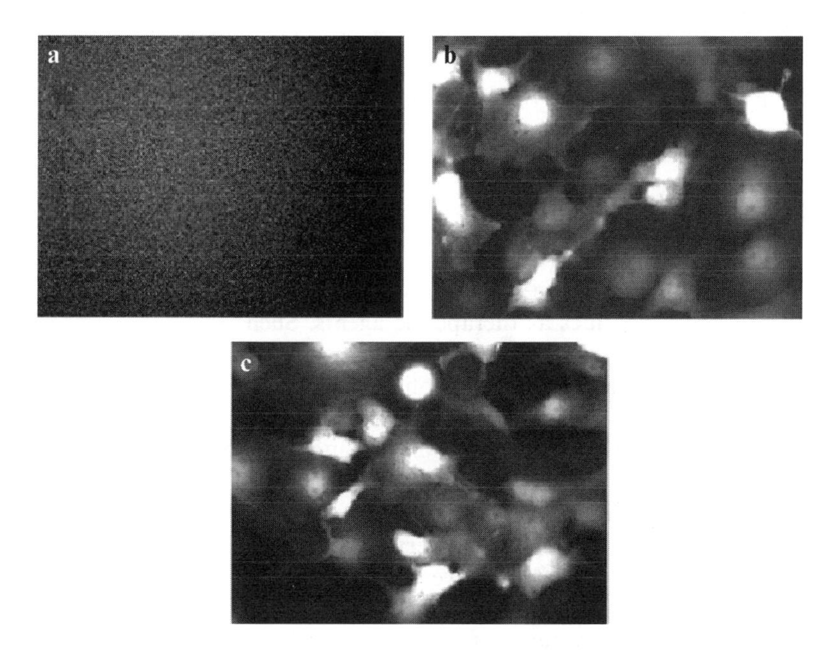

Figure 1 Human breast adenocarcinoma MCF-7 cells infected with adenovirus containing either wild-type or mutant estrogen receptor-specific ribozyme-2: (*a*) control cells—no adenoviral infection; (*b*) Ad/v-Rz 2 infected cells—cells infected with adenovirus containing estrogen-receptor-specific ribozyme-2; (*c*) Ad/v-Rz 2-M infected cells—cells infected with adenovirus containing mutant estrogen-receptor-specific ribozyme-2.

Two other viral vectors are very promising for gene therapy: lentivirus (Naldini and Verma, 2000) and adeno-associated virus (AAV). Human parvovirus AAV-2 has been recognized as a potential vector for human gene therapy (Hermonat and Muzyczka, 1984; Flotte and Carter, 1998; Hauswirth *et al.*, 2000). AAV efficiently, stably and site-specifically integrates into human chromosomal DNA in its wild-type form. It infects a variety of dividing and nondividing cells without pathogenetic effects. AAV requires a helper virus, such as herpesvirus or adenovirus, to replicate, and without that "help," AAV undergoes chromosomal integration and remains latent.

Current AAV vectors rely on the production of the other viral capsid and *Rep* proteins *in trans*, encoded on a separately transfected plasmid DNA, permitting replacement of these genes in the AAV-2 genome with the genes of interest.

The only restriction on AAV vectors is with respect to the size of the insert permitted (4–5 kb), which can be a very serious limitation for gene replacement therapy, especially considering the size of a tissue-specific promoter. But an average hammerhead ribozyme has only 40 to 50 nucleotides, and thus ribozyme-based gene therapy AAV may be an ideal delivery vehicle for small catalytic RNAs. Hauswirth and Lewin (2000) have successfully used AAV to deliver ribozymes to retinal cells. Additionally, AAV has been used to delivery antisense oligonucleotides to combat arterial hypertension (Kimura *et al.*, 2001).

FUTURE PROSPECTS

Notable progress has been made over the past few years toward using catalytic oligonucleotides as therapeutic agents. Such progress is especially significant in the areas of pharmacodynamics, pharmacokinetics, toxicity, stability, and cellular uptake of oligonucleotides. We anticipate that the next wave in the development and application of catalytic oligonucleotides will result from integration of human genome data with new technologies, such as "arrayed" ribozymes (Seetharaman *et al.*, 2001). However, the most progress is expected from developments in delivery, and especially topical (cell- and tissue-specific) delivery, of catalytic oligonucleotides. Despite these advances, targeted delivery of oligonucleotides and ribozymes still needs improvement, and induction of unwarranted cytokine responses after long-term therapy remains a vexing problem. The former problem can be partially overcome with gene-based therapeutic approaches with target-cell-specific promoters. The latter must be examined on a case-by-case basis. All in all, the future prospects of the field look highly promising.

ACKNOWLEDGMENT

Studies in our laboratory are supported by the National Institutes of Health Grants RO1 DK14744 and R37 AG10486.

REFERENCES

Agazie, Y. M., Burkholder, G. D., and Lee, J. S. (1996). Triplex DNA in the nucleus: Direct binding of triplex-specific antibodies and their effect on transcription, replication and cell growth. *Biochem. J.*, **316**, 461–466.

Alifano, P., Rivellini, F., Piscitelli, C., Arraiano, C. M., Bruni, C. B., and Carlomagno, M. S. (1994). Ribonuclease E provides substrates for ribonuclease P-dependent processing of a polycistronic mRNA. *Genes Dev.*, **8**, 3021–3031.

Amarzguioui, M., and Prydz, H. (1998). Hammerhead ribozyme design and application. *Cell Mol. Life Sci.*, **54**, 1175–1202.

Been, M. D., and Wickham, G. S. (1997). Self-cleaving ribozymes of hepatitis delta virus RNA. *Eur. J. Biochem.*, **247**, 741–753.

Bothwell, A. L., Garber, R. L., and Altman, S. (1976). Nucleotide sequence and in vitro processing of a precursor molecule to *Escherichia coli* 4.5 *S* RNA. *J. Biol. Chem.*, **251**, 7709–7716.

Bramlage, B., Luzi, E., and Eckstein, F. (1998). Designing ribozymes for the inhibition of gene expression. *Trends Biotechnol.*, **16**, 434–438.

Branch, A. D., and Robertson, H. D. (1991). Efficient trans cleavage and a common structural motif for the ribozymes of the human hepatitis delta agent. *Proc. Natl. Acad. Sci. USA*, **88**, 10163–10167.

Buzayan, J. M., Hampel, A., and Bruening, G. (1986). Nucleotide sequence and newly formed phosphodiester bond of spontaneously ligated satellite tobacco ringspot virus RNA. *Nucleic Acids Res.*, **14**, 9729–9743.

Cech, T. R., Zaug, A. J., and Grabowski, P. J. (1981). In vitro splicing of the ribosomal RNA precursor of *Tetrahymena*: Involvement of a guanosine nucleotide in the excision of the intervening sequence. *Cell*, **27**, 487–496.

Cech, T. R., Herschlag, D., Piccirilli, J. A., and Pyle, A. M. (1992). RNA catalysis by a group I ribozyme. Developing a model for transition state stabilization. *J. Biol. Chem.*, **267**, 17479–17482.

Chen, C. J., Banerjea, A. C., Harmison, G. G., Haglund, K., and Schubert, M. (1992). Multi-target-ribozyme directed to cleave at up to nine highly conserved HIV-1 env RNA regions inhibits HIV-1 replication–potential effectiveness against most presently sequenced HIV-1 isolates. *Nucleic Acids Res.*, **20**, 4581–4589.

Chen, P. J., Wu, H. L., Wang, C. J., Chia, J. H., and Chen, D. S. (1997). Molecular biology of hepatitis D virus: Research and potential for application. *J. Gastroenterol. Hepatol.*, **12**, S188–S192.

Chen, S., Song, C. S., Lavrovsky, Y., Bi, B., Vellanoweth, R., Chatterjee, B., and Roy, A. K. (1998). Catalytic cleavage of the androgen receptor messenger RNA and functional inhibition of androgen receptor activity by a hammerhead ribozyme. *Mol. Endocrinol.*, **12**, 1558–1566.

Cotter, M. A., and Robertson, E. S. (1999). Molecular genetic analysis of herpesviruses and their potential use as vectors for gene therapy applications. *Curr. Opin. Mol. Ther.*, **1**, 633–644.

Cousineau, B., Lawrence, S., Smith, D., and Belfort, M. (2000). Retrotransposition of a bacterial group II intron. *Nature*, **404**, 1018–1021.

Crooke, S. T. (1998). Antisense therapeutics. *Biotechnol. Genet. Eng Rev.*, **15**, 121–157.
Crooke, S. T. (2000). Progress in antisense technology: The end of the beginning. *Methods Enzymol.*, **313**, 3–45.
Culman, J. (2000). Antisense oligonucleotides in the study of central mechanisms of the cardiovascular regulation. *Exp. Physiol.*, **85**, 757–767.
Dawson, P. A., and Marini, J. C. (2000). Hammerhead ribozymes selectively suppress mutant type I collagen mRNA in osteogenesis imperfecta fibroblasts. *Nucleic Acids Res.*, **28**, 4013–4020.
de Feyter, R., and Li, P. (2000). Technology evaluation: HIV ribozyme gene therapy, Gene Shears Pty Ltd. *Curr. Opin. Mol. Ther.*, **2**, 332–335.
Dean, D. A. (2000). Peptide nucleic acids: Versatile tools for gene therapy strategies. *Adv. Drug Delivery Rev.*, **44**, 81–95.
Doherty, E. A., and Doudna, J. A. (2000). Ribozyme structures and mechanisms. *Annu. Rev. Biochem.*, **69**, 597–615.
Eder, P. S., Kekuda, R., Stolc, V., and Altman, S. (1997). Characterization of two scleroderma autoimmune antigens that copurify with human ribonuclease P. *Proc. Natl. Acad. Sci. USA*, **94**, 1101–1106.
Enjoji, M., Wang, F., Nakamuta, M., Chan, L., and Teng, B. B. (2000). Hammerhead ribozyme as a therapeutic agent for hyperlipidemia: Production of truncated apolipoprotein B and hypolipidemic effects in a dyslipidemic murine model. *Hum. Gene Ther.*, **11**, 2415–2430.
Fang, X. W., Yang, X. J., Littrell, K., Niranjanakumari, S., Thiyagarajan, P., Fierke, C. A., Sosnick, T. R., and Pan, T. (2001). The *Bacillus subtilis* RNase P holoenzyme contains two RNase P RNA and two RNase P protein subunits. *RNA*, **7**, 233–241.
Fedor, M. J. (2000). Structure and function of the hairpin ribozyme. *J. Mol. Biol.*, **297**, 269–291.
Flotte, T. R., and Carter, B. J. (1998). Adeno-associated virus vectors for gene therapy of cystic fibrosis. *Methods Enzymol.*, **292**, 717–732.
Fox, K. R. (2000). Targeting DNA with triplexes. *Curr. Med. Chem.*, **7**, 17–37.
Fresco, J. R., Adams, A., Ascione, R., Henley, D., and Lindahl, T. (1966). Tertiary structure in transfer ribonucleic acids. *Cold Spring Harbor Symp. Quant. Biol.*, **31**, 527–537.
Gelband, C. H., Katovich, M. J., and Raizada, M. K. (2000). Current perspectives on the use of gene therapy for hypertension. *Circ. Res.*, **87**, 1118–1122.
Gu, J. L., Pei, H., Thomas, L., Nadler, J. L., Rossi, J. J., Lanting, L., and Natarajan, R. (2001). Ribozyme-mediated inhibition of rat leukocyte-type 12-lipoxygenase prevents intimal hyperplasia in balloon-injured rat carotid arteries. *Circulation*, **103**, 1446–1452.
Guerrier-Takada, C., Gardiner, K., Marsh, T., Pace, N., and Altman, S. (1983). The RNA moiety of ribonuclease P is the catalytic subunit of the enzyme. *Cell*, **35**, 849–857.
Guo, H. C., and Collins, R. A. (1995). Efficient trans-cleavage of a stem–loop RNA substrate by a ribozyme derived from *Neurospora* VS RNA. *EMBO J.*, **14**, 368–376.
Guo, H., Karberg, M., Long, M., Jones, J. P., Sullenger, B., and Lambowitz, A. M. (2000). Group II introns designed to insert into therapeutically relevant DNA target sites in human cells. *Science*, **289**, 452–457.
Hackett, N. R., Kaminsky, S. M., Sondhi, D., and Crystal, R. G. (2000). Antivector and antitransgene host responses in gene therapy. *Curr. Opin. Mol. Ther.*, **2**, 376–382.
Hammann, C., Cooper, A., and Lilley, D. M. (2001). Thermodynamics of ion-induced RNA folding in the hammerhead ribozyme: An isothermal titration calorimetric study. *Biochemistry*, **40**, 1423–1429.
Hanania, E. G., Kavanagh, J., Hortobagyi, G., Giles, R. E., Champlin, R., and Deisseroth, A. B. (1995). Recent advances in the application of gene therapy to human disease. *Am. J. Med.*, **99**, 537–552.
Haseloff, J., and Gerlach, W. L. (1988). Simple RNA enzymes with new and highly specific endoribonuclease activities. *Nature*, **334**, 585–591.

Hauswirth, W. W., and Lewin, A. S. (2000). Ribozyme uses in retinal gene therapy. *Prog. Retin. Eye Res.*, **19**, 689–710.

Hauswirth, W. W., Lewin, A. S., Zolotukhin, S., and Muzyczka, N. (2000). Production and purification of recombinant adeno-associated virus. *Methods Enzymol.*, **316**, 743–761.

He, T. C., Zhou, S., da Costa, L. T., Yu, J., Kinzler, K. W., and Vogelstein, B. (1998). A simplified system for generating recombinant adenoviruses. *Proc. Natl. Acad. Sci. USA*, **95**, 2509–2514.

Hegg, L. A., and Fedor, M. J. (1995). Kinetics and thermodynamics of intermolecular catalysis by hairpin ribozymes. *Biochemistry*, **34**, 15813–15828.

Heilman-Miller, S. L., Thirumalai, D., and Woodson, S. A. (2001). Role of counterion condensation in folding of the *Tetrahymena* ribozyme. I. Equilibrium stabilization by cations. *J. Mol. Biol.*, **306**, 1157–1166.

Henry, S. P., Templin, M. V., Gillett, N., Rojko, J., and Levin, A. A. (1999). Correlation of toxicity and pharmacokinetic properties of a phosphorothioate oligonucleotide designed to inhibit ICAM-1. *Toxicol. Pathol.*, **27**, 95–100.

Hermonat, P. L., and Muzyczka, N. (1984). Use of adeno-associated virus as a mammalian DNA cloning vector: Transduction of neomycin resistance into mammalian tissue culture cells. *Proc. Natl. Acad. Sci. USA*, **81**, 6466–6470.

Herschlag, D., and Cech, T. R. (1990). Catalysis of RNA cleavage by the *Tetrahymena thermophila* ribozyme. 1. Kinetic description of the reaction of an RNA substrate complementary to the active site. *Biochemistry*, **29**, 10159–10171.

Hertel, K. J., Peracchi, A., Uhlenbeck, O. C., and Herschlag, D. (1997). Use of intrinsic binding energy for catalysis by an RNA enzyme. *Proc. Natl. Acad. Sci. USA* **94**, 8497–8502.

Hertel, K. J., Stage-Zimmermann, T. K., Ammons, G., and Uhlenbeck, O. C. (1998). Thermodynamic dissection of the substrate–ribozyme interaction in the hammerhead ribozyme. *Biochemistry*, **37**, 16983–16988.

Hu, W. Y., Fukuda, N., Nakayama, M., Kishioka, H., and Kanmatsuse, K. (2001). Inhibition of vascular smooth muscle cell proliferation by DNA-RNA chimeric hammerhead ribozyme targeting to rat platelet-derived growth factor A-chain mRNA. *J. Hypertens.*, **19**, 203–212.

James, H. A. (1999). The potential application of ribozymes for the treatment of hematological disorders. *J. Leukocyte Biol.*, **66**, 361–368.

Jenne, A., Hartig, J. S., Piganeau, N., Tauer, A., Samarsky, D. A., Green, M R., Davies, J., and Famulok, M. (2001). Rapid identification and characterization of hammerhead-ribozyme inhibitors using fluorescence-based technology. *Nat. Biotechnol.*, **19**, 56–61.

Kennell, J. C., Saville, B. J., Mohr, S., Kuiper, M. T., Sabourin, J.R., Collins, R. A., and Lambowitz, A. M. (1995). The VS catalytic RNA replicates by reverse transcription as a satellite of a retroplasmid. *Genes Dev.*, **9**, 294–303.

Kilpatrick, M. W., and Phylactou, L. A. (1998). Towards an RNA-based therapy for Marfan syndrome. *Mol. Med. Today*, **4**, 376–381.

Kim, S. H., Quigley, G. J., Suddath, F. L., McPherson, A., Sneden, D., Kim, J. J., Weinzierl, J., and Rich, A. (1973). Three-dimensional structure of yeast phenylalanine transfer RNA: Folding of the polynucleotide chain. *Science*, **179**, 285–288.

Kimura, B., Mohuczy, D., Tang, X., and Phillips, M. I. (2001). Attenuation of hypertension and heart hypertrophy by adeno-associated virus delivering angiotensinogen antisense. *Hypertension*, **37**, 376–380.

Kochetkova, M., and Shannon, M. F. (2000). Triplex-forming oligonucleotides and their use in the analysis of gene transcription. *Methods Mol. Biol.*, **130**, 189–201.

Komatsu, Y., Kanzaki, I., Shirai, M., Kumagai, I., Yamashita, S., and Ohtsuka, E. (2000a). Functional domain-assembly in hairpin ribozymes. *J. Biochem.* (Tokyo), **127**, 531–536.

Komatsu, Y., Yamashita, S., Kazama, N., Nobuoka, K., and Ohtsuka, E. (2000b). Construction of new ribozymes requiring short regulator oligonucleotides as a cofactor. *J. Mol. Biol.*, **299**, 1231–1243.

Komine, Y., Kitabatake, M., Yokogawa, T., Nishikawa, K., and Inokuchi, H. (1994). A tRNA-like structure is present in 10*Sa* RNA, a small stable RNA from *Escherichia coli*. *Proc. Natl. Acad. Sci. USA*, **91**, 9223–9227.

Lafontaine, D. A., Norman, D. G., and Lilley, D. M. (2001). Structure, folding and activity of the VS ribozyme: Importance of the 2–3–6 helical junction. *EMBO J.*, **20**, 1415–1424.

Lan, N., Howrey, R. P., Lee, S. W., Smith, C. A., and Sullenger, B. A. (1998). Ribozyme-mediated repair of sickle beta-globin mRNAs in erythrocyte precursors. *Science*, **280**, 1593–1596.

Lavrovsky, Y., Chen, S., and Roy, A. K. (1997). Therapeutic potential and mechanism of action of oligonucleotides and ribozymes. *Biochem. Mol. Med.*, **62**, 11–22.

Lavrovsky, Y., Tyagi, R. K., Chen, S., Song, C. S., Chatterjee, B., and Roy, A. K. (1999). Ribozyme-mediated cleavage of the estrogen receptor messenger RNA and inhibition of receptor function in target cells. *Mol. Endocrinol.*, **13**, 925–934.

Li, Y., and Breaker, R. R. (1999). Deoxyribozymes: New players in the ancient game of biocatalysis. *Curr. Opin. Struct. Biol.*, **9**, 315–323.

Lilley, D. M. (1999). Folding and catalysis by the hairpin ribozyme. *FEBS Lett.*, **452**, 26–30.

Lui, V. W., He, Y., and Huang, L. (2001). Specific down-regulation of her-2/neu mediated by a chimeric u6 hammerhead ribozyme results in growth inhibition of human ovarian carcinoma. *Mol. Ther.*, **3**, 169–177.

McCarthy, M. M., Auger, A. P., Mong, J. A., Sickel, M. J., and Davis, A. M. (2000). Antisense oligodeoxynucleotides as a tool in developmental neuroendocrinology. *Methods* **22**, 239–248.

Mistry, S. J., Benham, C. J., and Atweh, G. F. (2001). Development of ribozymes that target stathmin, a major regulator of the mitotic spindle. *Antisense Nucleic Acid Drug Dev.*, **11**, 41–49.

Morishita, R., Yamada, S., Yamamoto, K., Tomita, N., Kida, I., Sakurabayashi, I., Kikuchi, A., Kaneda, Y., Lawn, R., Higaki, J., and Ogihara, T. (1998). Novel therapeutic strategy for atherosclerosis: Ribozyme oligonucleotides against apolipoprotein(a) selectively inhibit apolipoprotein(a) but not plasminogen gene expression. *Circulation*, **98**, 1898–1904.

Naldini, L., and Verma, I. M. (2000). Lentiviral vectors. *Adv. Virus Res.*, **55**, 599–609.

Nielsen, P. E., Egholm, M., Berg, R. H., and Buchardt, O. (1991). Sequence-selective recognition of DNA by strand displacement with a thymine-substituted polyamide. *Science*, **254**, 1497–1500.

Norris, J. S., Hoel, B., Voeks, D., Maggouta, F., Dahm, M., Pan, W., and Clawson, G. (2000). Design and testing of ribozymes for cancer gene therapy. *Adv. Exp. Med. Biol.*, **465**, 293–301.

Palmer, J. D., Adams, K. L., Cho, Y., Parkinson, C. L., Qiu, Y. L., and Song, K. (2000). Dynamic evolution of plant mitochondrial genomes: Mobile genes and introns and highly variable mutation rates. *Proc. Natl. Acad. Sci. USA*, **97**, 6960–6966.

Perlman, H., Sata, M., Krasinski, K., Dorai, T., Buttyan, R., and Walsh, K. (2000). Adenovirus-encoded hammerhead ribozyme to Bcl-2 inhibits neointimal hyperplasia and induces vascular smooth muscle cell apoptosis. *Cardiovasc. Res.*, **45**, 570–578.

Phylactou, L. A. (2000). Ribozyme and peptide–nucleic acid-based gene therapy. *Adv. Drug Delivery Rev.*, **44**, 97–108.

Phylactou, L. A., Darrah, C., and Wood, M. J. (1998). Ribozyme-mediated trans-splicing of a trinucleotide repeat. *Nat. Genet.*, **18**, 378–381.

Pinard, R., Lambert, D., Heckman, J. E., Esteban, J. A., Gundlach, C. W., IV, Hampel, K. J., Glick, D., Walter, N. G., Major, F., and Burke, J. M. (2001). The hairpin ribozyme substrate binding domain: A highly constrained d-shaped conformation. *J. Mol. Biol.*, **307**, 51–65.

Praseuth, D., Guieysse, A. L., and Helene, C. (1999). Triple helix formation and the antigene strategy for sequence-specific control of gene expression. *Biochim. Biophys. Acta*, **1489**, 181–206.

Ramezani, A., Ding, S. F., and Joshi, S. (1997). Inhibition of HIV-1 replication by retroviral vectors expressing monomeric and multimeric hammerhead ribozymes. *Gene Ther.*, **4**, 861–867.
Rossi, J. J. (2000). Ribozyme therapy for HIV infection. *Adv. Drug Delivery Rev.*, **44**, 71–78.
Rupert, P. B., and Ferre-D'Amare, A. R. (2001). Crystal structure of a hairpin ribozyme-inhibitor complex with implications for catalysis. *Nature*, **410**, 780–786.
Santoro, S. W., and Joyce, G. F. (1997) A general purpose RNA-cleaving DNA enzyme. *Proc. Natl. Acad. Sci. USA*, **94**, 4262–4266.
Santoro, S. W., and Joyce, G. F. (1998). Mechanism and utility of an RNA-cleaving DNA enzyme. *Biochemistry*, **37**, 13330–13342.
Sarver, N., Cantin, E. M., Chang, P. S., Zaia, J. A., Ladne, P. A., Stephens, D. A., and Rossi, J. J. (1990). Ribozymes as potential anti-HIV-1 therapeutic agents. *Science*, **247**, 1222–1225.
Schon, A. (1999). Ribonuclease P: The diversity of a ubiquitous RNA processing enzyme. *FEMS Microbiol. Rev.*, **23**, 391–406.
Scott, W. G. (1999). Biophysical and biochemical investigations of RNA catalysis in the hammerhead ribozyme. *Q. Rev. Biophys.*, **32**, 241–284.
Seetharaman, S., Zivarts, M., Sudarsan, N., and Breaker, R. R. (2001). Immobilized RNA switches for the analysis of complex chemical and biological mixtures. *Nat. Biotechnol.*, **19**, 336–341.
Sereni, D., Tubiana, R., Lascoux, C., Katlama, C., Taulera, O., Bourque, A., Cohen, A., Dvorchik, B., Martin, R. R., Tournerie, C., Gouyette, A., and Schechter, P. J. (1999). Pharmacokinetics and tolerability of intravenous trecovirsen (GEM 91), an antisense phosphorothioate oligonucleotide, in HIV-positive subjects. *J. Clin. Pharmacol.*, **39**, 47–54.
Shahi, S., Shanmugasundaram, G. K., and Banerjea, A. C. (2001). Ribozymes that cleave reovirus genome segment S1 also protect cells from pathogenesis caused by reovirus infection. *Proc. Natl. Acad. Sci. USA*, **98**, 4101–4106.
Sharp, P. A. (1991). Five easy pieces. *Science*, **254**, 663.
Shaw, L. C., Whalen, P. O., Drenser, K. A., Yan, W., Hauswirth, W. W., and Lewin, A. S. (2000). Ribozymes in treatment of inherited retinal disease. *Methods Enzymol.*, **316**, 761–776.
Shaw, L. C., Skold, A., Wong, F., Petters, R., Hauswirth, W. W., and Lewin, A. S. (2001). An allele-specific hammerhead ribozyme gene therapy for a porcine model of autosomal dominant retinitis pigmentosa. *Mol. Vis.*, **7**, 6–13.
Shih, I. H., and Been, M. D. (2001). Involvement of a cytosine side chain in proton transfer in the rate-determining step of ribozyme self-cleavage. *Proc. Natl. Acad. Sci. USA*, **98**, 1489–1494.
Sioud, M. (1999). Application of preformed hammerhead ribozymes in the gene therapy of cancer (review). *Int. J. Mol. Med.*, **3**, 381–384.
Sioud, M., and Leirdal, M. (2000). Therapeutic RNA and DNA enzymes. *Biochem. Pharmacol.*, **60**, 1023–1026.
Smith, L., Andersen, K. B., Hovgaard, L., and Jaroszewski, J. W. (2000). Rational selection of antisense oligonucleotide sequences. *Eur. J. Pharm. Sci.*, **11**, 191–198.
Somia, N., and Verma, I. M. (2000). Gene therapy: Trials and tribulations. *Nat. Rev. Genet.*, **1**, 91–99.
Spencer, D. M. (2000). Developments in suicide genes for preclinical and clinical applications. *Curr. Opin. Mol. Ther.*, **2**, 433–440.
Stage-Zimmermann, T. K., and Uhlenbeck, O. C. (1998). Hammerhead ribozyme kinetics. *RNA*, **4**, 875–889.
Stams, T., Niranjanakumari, S., Fierke, C. A., and Christianson, D. W. (1998). Ribonuclease P protein structure: Evolutionary origins in the translational apparatus. *Science*, **280**, 752–755.
Stein, C. A. (2000). Is irrelevant cleavage the price of antisense efficacy? *Pharmacol. Ther.*, **85**, 231–236.

Su, J. Z., Fukuda, N., Hu, W. Y., and Kanmatsuse, K. (2000). Ribozyme to human TGF-beta-1 mRNA inhibits the proliferation of human vascular smooth muscle cells. *Biochem. Biophys. Res. Commun.*, **278**, 401–407.

Su, L. J., Qin, P. Z., Michels, W. J., and Pyle, A. M. (2001). Guiding ribozyme cleavage through motif recognition: The mechanism of cleavage site selection by a group II intron ribozyme. *J. Mol. Biol.*, **306**, 655–668.

Sun, L. Q., Cairns, M. J., Saravolac, E. G., Baker, A., and Gerlach, W. L. (2000). Catalytic nucleic acids: From lab to applications. *Pharmacol. Rev.*, **52**, 325–347.

Suzuki, Y., Funato, T., Munakata, Y., Sato, K., Hirabayashi, Y., Ishii, T., Takasawa, N., Ootaka, T., Saito, T., and Sasaki, T. (2000). Chemically modified ribozyme to V gene inhibits anti-DNA production and the formation of immune deposits caused by lupus lymphocytes. *J. Immunol.*, **165**, 5900–5905.

Suzumura, K., Warashina, M., Yoshinari, K., Tanaka, Y., Kuwabara, T., Orita, M., and Taira, K. (2000). Significant change in the structure of a ribozyme upon introduction of a phosphorothioate linkage at P9: NMR reveals a conformational fluctuation in the core region of a hammerhead ribozyme. *FEBS Lett.*, **473**, 106–112.

Tanaka, K., Yamada, T., Ohyagi, Y., Asahara, H., Horiuchi, I., and Kira, J. (2001). Suppression of transthyretin expression by ribozymes: A possible therapy for familial amyloidotic polyneuropathy. *J. Neurol. Sci.*, **183**, 79–84.

Turner, P. C. (2000). Ribozymes. Their design and use in cancer. *Adv. Exp. Med. Biol.*, **465**, 303–318.

Uhlenbeck, O. C. (1987). A small catalytic oligoribonucleotide. *Nature*, **328**, 596–600.

Wadkins, T. S., Shih, I., Perrotta, A. T., and Been, M. D. (2001). A pH-sensitive RNA tertiary interaction affects self-cleavage activity of the HDV ribozymes in the absence of added divalent metal ion. *J. Mol. Biol.*, **305**, 1045–1055.

Wang, L., Witherington, C., King, A., Gerlach, W. L., Carr, A., Penny, R., Cooper, D., Symonds, G., and Sun, L. Q. (1998). Preclinical characterization of an anti-*tat* ribozyme for therapeutic application. *Hum. Gene Ther.*, **9**, 1283–1291.

Wong-Staal, F., Poeschla, E. M., and Looney, D. J. (1998). A controlled, phase 1 clinical trial to evaluate the safety and effects in HIV-1 infected humans of autologous lymphocytes transduced with a ribozyme that cleaves HIV-1 RNA. *Hum. Gene Ther.*, **9**, 2407–2425.

Woodson, S. A. (2000). Recent insights on RNA folding mechanisms from catalytic RNA. *Cell Mol. Life Sci.*, **57**, 796–808.

Xiao, S., Houser-Scott, F., and Engelke, D. R. (2001). Eukaryotic ribonuclease P: increased complexity to cope with the nuclear pre-tRNA pathway. *J. Cell Physiol*, **187**, 11–20.

Yamamoto, K., Morishita, R., Tomita, N., Shimozato, T., Nakagami, H., Kikuchi, A., Aoki, M., Higaki, J., Kaneda, Y., and Ogihara, T. (2000). Ribozyme oligonucleotides against transforming growth factor-beta inhibited neointimal formation after vascular injury in rat model: Potential application of ribozyme strategy to treat cardiovascular disease. *Circulation*, **102**, 1308–1314.

Yang, J., Zimmerly, S., Perlman, P. S., and Lambowitz, A. M. (1996). Efficient integration of an intron RNA into double-stranded DNA by reverse splicing. *Nature*, **381**, 332–335.

Yen, L., Strittmatter, S. M., and Kalb, R. G. (1999). Sequence-specific cleavage of *huntingtin* mRNA by catalytic DNA. *Ann. Neurol.*, **46**, 366–373.

Yu, M., Leavitt, M. C., Maruyama, M., Yamada, O., Young, D., Ho, A. D., and Wong-Staal, F. (1995). Intracellular immunization of human fetal cord blood stem/progenitor cells with a ribozyme against human immunodeficiency virus type 1. *Proc. Natl. Acad. Sci. USA*, **92**, 69w9–703.

Zamecnik, P. C., and Stephenson, M. L. (1978). Inhibition of Rous sarcoma virus replication and cell transformation by a specific oligonucleotide. *Proc. Natl. Acad. Sci. USA*, **75**, 280–284.

Zhang, B., and Cech, T. R. (1997). Peptide bond formation by in vitro selected ribozymes. *Nature*, **390**, 96–100.

Zimmerly, S., Guo, H., Perlman, P. S., and Lambowitz, A. M. (1995). Group II intron mobility occurs by target DNA-primed reverse transcription. *Cell*, **82**, 545–554.

Chapter 10

Biotherapeutics: Current Status and Future Directions

James E. Talmadge
University of Nebraska Medical Center
Omaha, Nebraska

The goal of regulating the body's immune responses, as a therapeutic strategy for neoplastic, infectious, autoimmune, and inflammatory diseases, has begun to be achieved. Optimism about this strategy has fluctuated over the years, but at present, numerous immunoregulatory drugs have been approved. During the last decade, we have observed an explosion in the cloning of immunoregulatory genes and their receptors and the development of novel and exciting therapeutic approaches. These critical advances represent the culmination of efforts with crude and fractionated natural products, supernatants, and cell products. With the advent of molecular techniques, there has been a surfeit of recombinant molecules, including not only cytokines and growth factors, but also monoclonal antibodies and vaccines. Indeed, these strategies have been expanded to include gene therapy, and many of

Biotechnology and Safety Assessment, 3rd edition

the current therapeutic strategies are reliant or focused on gene delivery vectors. Biotherapeutics can be subdivided into recombinant proteins and natural or synthetic products. The latter do not currently share the enthusiasm afforded recombinant proteins, yet they have given us therapeutically important drugs such as bestatin, FK-506, levamisole, and natural products such as bacille Calmette–Guerin (BCG), currently used to treat bladder cancer.

INTRODUCTION

Immunostimulants have been used to treat human disease since William B. Coley treated cancer with experimental mixed bacterial toxins early in the twentieth century (Smyth *et al.*, 2001). These early studies spawned the clinical approval and use of such microbially derived substances as bacille Calmette-Guerin (BCG) (bladder cancer, United States), Krestin, Picibanil, and Lentinan (gastric and other cancers, Japan), and Biostim and Broncho-Vaxom (recurrent infections, Europe). While these "crude" drugs induce numerous immunopharmacological activities, they pose considerable regulatory obstacles owing to impurity, lot-to-lot variability, unreliability, and adverse side effects. Similarly, traditional herbal medicines (Asia) also provide a source of active substances for immunotherapy. Importantly, the purification, characterization, and synthetic production of the active moieties from natural products (bestatin, taxol) and culture supernatants (FK-506, rapamycin, deoxyspergualin and cyclosporin) provide valuable candidates for drug development. The current emphasis is on the use of recombinant proteins (cytokines), although, their focused bioactivity and pharmacologic deficiencies can limit the utility of these drugs. Thus, there remains a potentially important role for classical biological response modifiers (BRMs), with their potential for oral bioavailability and ability to induce multiple cytokines for optimal immune augmentation and hematopoietic restoration.

The overall approach in this chapter is to limit the discussion of biotherapeutics to recombinant, natural, and synthetic drugs currently approved or in the clinic, subdivided by recombinant and synthetic moieties. A brief discussion follows regarding combination and cellular therapy as future prospects. In this chapter, we do not discuss the development and clinical utilization of vaccines, be they genetic or recombinant, or monoclonal antibodies. Chapter 12 addresses monoclonal antibodies, Chapter 14 addresses vaccines, and Chapter 7 discusses the immunotoxicological assessment of cytokines.

RECOMBINANT PROTEINS: APPROVED FOR CLINICAL USE

Therapeutic proteins have emerged as an important class of drugs for the treatment of cancer, immunosuppression, myeloid dysplasia, and infectious disease. However, our limited understanding of their pharmacology and mechanism of action has slowed their development. To facilitate the development of immunoregulatory proteins, information is needed on their pharmacology and mechanism of action (Mihich, 1986; Talmadge and Herberman, 1986). One approach to the development of such biotherapeutics is to identify a clinical hypothesis based on therapeutic surrogate(s) identified during preclinical immunopharmacology (Ellenberg, 1993; Holden, 1993). A surrogate for clinical efficacy may be a phenotypic, biochemical, enzymatic, functional (immunological, molecular, or hematological), or quality-of-life measurement that is believed to be associated with therapeutic activity. Phase I clinical trials can then be designed to identify the optimal immunomodulatory dose (OID) and treatment schedule that maximizes the augmentation of the surrogate end point(s). Subsequent phase II/III trials can then be established to determine whether the changes in the surrogate levels correlate with therapeutic activity. Table I lists the immunologically and hematologically active cytokines that are approved for use in the United States.

In contrast to strategies based on the identification of surrogates for therapeutic efficacy, protocols for recombinant proteins are often identified based on practices that were developed for conventional drugs and may not be advantageous for the demonstration of efficiency by biotechnology-derived drugs. The pharmacological attributes of recombinant biotherapeutics require selective or targeted delivery (Tomlinson, 1991) and strategies to prolong their short half-life, which also limits their biological activity. Pegylated cytokines [i.e., poly(ethylene glycol) modified cytokines] including interferon (IFN) and granutocyle colony-stimulating factor (G-CSF), have demonstrated significant biologic activity, owing in part to their improved pharmacologic profile (Kim *et al.*, 2001; Jen *et al.*, 2001, Van der Auwera *et al.*, 2001). Thus, strategies to limit the pharmacological deficiencies are critical challenges in the development of recombinant biotherapeutics. One additional difficulty in the development of a recombinant biotherapeutic is that in many instances there is little relation between the dose administered and the biological effect. Indeed, in some instances there is a nonlinear dose–response relationship that has been described as "bell shaped" (Talmadge *et al.*, 1987). This lack of a linear dose–response relationship may be due to the drug's nonlinear dispersal throughout the body, a poor ability to enter into a saturatable receptor-mediated transport process, chemical instability, sequence of administration, with other agents or an incorrect time of administration, and/or an

Table 1

Clinically Approved CytoKines

Drug	Corporation	Target	Approval
Actimmune® interferon gamma-1b	InterMune Pharmaceuticals	Management of chronic granulomatous disease	December-90
Actimmune® interferon gamma-1b	InterMune Pharmaceuticals	Osteopetrosis	February-00
Alferon N Injection® interferon alfa-n3 (human leukocyte derived)	Interferon Sciences	Genital warts	October-89
Aranesp (novel erythropoiesis stimulating protein or NESP)	Amgen	Anemia associated with chronic kidney disease (Europe)	June-01
Avonex® interferon beta-1a	Biogen	Relapsing multiple sclerosis	May-96
Betaseron® recombinant interferon beta-1b	Berlex Laboratories	Relapsing, remitting multiple sclerosis	July-93
Enbrel® TNFR:Fc	Immunex	Moderate to severe active juvenile rheumatoid arthritis	May-99
Enbrel® TNFR:Fc	Immunex	Moderate to severe active rheumatoid arthritis	November-98
EPOGEN® Epetin alfa (rEPO)	Amgen	Anemia caused by chemotherapy	April-93
EPOGEN® Epetin alfa (rEPO)	Amgen	Anemia, chronic renal failure, anemia in Retrovir®-treated HIV-infected patients	June-89
EPOGEN® Epetin alfa (rEPO)	Amgen	Chronic renal failure, dialysis	November-99
EPOGEN® Epetin alfa (rEPO)	Amgen	Surgical blood loss	December-96
Infergen® interferon alfacon-1	Amgen	Treatment of chronic hepatitis C viral infection	October-97
Intron® A interferon alfa-2b	Schering-Plough	AIDS-related Kaposi's sarcoma	November-88
Intron® A interferon alfa-2b	Schering-Plough	Follicular lymphoma	November-97
Intron® A interferon alfa-2b	Schering-Plough	Genital warts	June-88
Intron® A interferon alfa-2b	Schering-Plough	Hairy cell leukemia	June-86
Intron® A interferon alfa-2b	Schering-Plough	Hepatitis B	July-92
Intron® A interferon alfa-2b	Schering-Plough	Hepatitis C	February-91
Intron® A interferon alfa-2b	Schering-Plough	Malignant melanoma	December-95

(*continues*)

Table 1 (*continued*)

Leukine® sargramostim (GM-CSF)	Immunex	Allogeneic bone marrow transplantation	November-95
Leukine® sargramostim (GM-CSF)	Immunex	Autologous bone marrow transplantation	March-91
Leukine® sargramostim (GM-CSF)	Immunex	Neutropenia resulting from chemotherapy	September-95
Leukine® sargramostim (GM-CSF)	Immunex	Peripheral blood progenitor cell mobilization	December-95
Neumega®, oprelvekin (rIL-11)	Genetics Institute	Chemotherapy-induced thrombocytopenia	November-97
NEUPOGEN® Filgrastim (rG-CSF)	Amgen	Acute myelogenous leukemia	April-98
NEUPOGEN® Filgrastim (rG-CSF)	Amgen	Autologous or allogeneic bone marrow transplantation	June-94
NEUPOGEN® Filgrastim (rG-CSF)	Amgen	Chemotherapy-induced neutropenia	February-91
NEUPOGEN® Filgrastim (rG-CSF)	Amgen	Chronic severe neutropenia	December-94
NEUPOGEN® Filgrastim (rG-CSF)	Amgen	Peripheral blood progenitor cell transplantation	December-95
Peginterferon alfa-2b (PEG-Intron®)	Schering-Plough	Treatment of chronic hepatitis C in patients not previously treated with interferon alfa who have compensated liver disease and are at least 18 years of age	January-01
Proleukin® aldesleukin (interleukin-2)	Chiron	Metastatic melanoma	January-98
Proleukin® aldesleukin (interleukin-2)	Chiron	Renal cell carcinoma	May-92
Rebetron® ribavirin/interferon alfa-2b	Schering-Plough	Chronic hepatitis C	June-98
Rebetron® ribavirin/interferon alfa-2b	Schering-Plough	Chronic hepatitis C, compensated liver disease	December-99
Roferon® interferon alfa-2a	Hoffman-La Roche	AIDS-related Kaposi's sarcoma	November-88
Regranex®, Becaplermin, rHPDGF-BB	Ortho-McNeil	Diabetic neuropathy, foot ulcers	December-97

(*continues*)

Table 1 (*continued*)

Drug	Corporation	Target	Approval
Roferon® interferon alfa-2a	Hoffman-La Roche	Chronic myelogenous leukemia	November-95
Roferon® interferon alfa-2a	Hoffman-La Roche	Hairy cell leukemia	June-86
Roferon® interferon alfa-2a	Hoffman-La Roche	Hepatitis C	November-96
Stemgen (stem cell factor)	Amgen	Mobilization (Australia, New Zealand, Canada)	1997
Wellferon® interferon-n	Glaxo Wellcome	Treatment of hepatitis C in patients 18 years of age or older without decompensated liver disease	March-99

inappropriate location and response of the target cells. Further, a "bell shaped" dose–response curve may be associated with the tachyphylaxis of receptor expression or a signal transduction mechanism whereby the cells become refractory to subsequent receptor-mediated augmentation such as occurs with chemokines. Because immune reactivity regulation can lead to unwanted physiological events, it is important that administration of recombinant proteins be optimal to ensure the desired biological activity.

INTERFERON ALFA (IFN-α)

The initial, nonrandomized, clinical studies with IFN-α suggested that this protein had therapeutic activity for malignant melanoma, osteosarcoma, and various lymphomas. (Misset *et al.*, 1982). Subsequent randomized trials, however, demonstrated significant therapeutic activity only against less common tumor histiotypes, including hairy cell and chronic myelogenous leukemia (CML) (Misset *et al.*, 1982; Quesada *et al.*, 1984; Golom *et al.*, 1988), and a few types of lymphoma (Quesada *et al.*, 1984), including low-grade non-Hodgkin's lymphoma (O'Connell *et al.*, 1986) and cutaneous T-cell lymphoma (Bunn *et al.*, 1984). Subsequently, the list of responding indications expanded to include malignant melanoma (Kickwood *et al.*, 1996), AIDS and Kaposi's sarcoma (Lane *et al.*, 1988), genital warts, and hepatitis B and C.

It has taken almost three decades to translate the concept of IFN-α as an antiviral agent to its routine utility in clinical oncology and infectious diseases. Despite extensive study, the development of IFN-α is still in its early stages, and such basic parameters as optimal dose and therapeutic schedule remain to be determined (Quesada *et al.*, 1984; Golomb *et al.*, 1988; Pfeffer *et al.*, 1998). The mechanism of activity is also controversial, since IFN-α has been shown to have dose-dependent antitumor activities *in vitro*, and yet is active at low doses for hairy cell leukemia (Quesada *et al.*, 1984; Golomb *et al.*, 1988). Immunomodulation as the mechanism of therapeutic activity with IFN-α is perhaps best supported by its action against hairy cell leukemia. Treatment with IFN-α in this disease is associated with a 90 to 95% response rate; however, this is not fully achieved until the patients have been on the protocol for a year, and it appears that low doses of IFN-α are as active as higher doses (Teichmann *et al.*, 1988). Although the clinical use of IFN-α has been supplanted by other, even more effective drugs, it was approval for use in treating hairy cell leukemia that precipitated expanded clinical studies of IFN-α in treating other disease states.

The initial dose-finding studies determined that a dose of 12×10^6U/M^2 (units/meters^2) of the recombinant forms of rIFN-α, was not tolerable in patients with hairy cell leukemia (Quesada *et al.*, 1984). Subsequently, it was demonstrated that highly purified natural IFN-α at a dose of 2×10^6 U/M^2 was both well tolerated and effective when administered three times per week for 28 days (Black *et al.*, 1993); however, it retained some toxicity, including myelosuppression as well as neurotoxicity and cardiotoxicity. In these studies, a lower dose of 2×10^5 U/M^2 was also administered for 28 days and found to be better tolerated than the standard dose, while inducing equivalent improvements in peripheral neutrophil and platelet counts. In this trial, substantial clinical improvement, primarily in terms of increased platelet and neutrophil counts, was observed within the first 4 to 8 weeks of treatment. This yielded an improved quality of life resulting from a depression in cardiac and neurological toxicity, mitigation of flu-like syndrome, and reduced myelosuppression, need for platelet transfusions, and incidence of bacterial infections. Further, it appears that significant improvements in thrombocytopenia and neutropenia can be rapidly induced in the majority of patients when low and minimally toxic doses of IFN-α are used. However, IFN-α also generates a therapeutic dose–response effect, whereby higher doses of IFN-α will induce a quantitatively greater antileukemic response than is observed with low doses of IFN-α. It appears that once improvements are obtained at 2×10^5 U/M^2 and patients become tolerant to the acute toxicity of IFN-α, the dose can be increased to 2×10^6 U/M^2 to obtain the greater antileukemia effect of the higher dose. In CML, sustained therapeutic responses are found in more than 75% of patients treated (Alimena *et al.*, 1988; Italian Cooperative Study

Group, (1994), and a higher dose (5×10^6 U/M^2) than that required for hairy cell leukemia is needed to achieve optimal therapeutic efficacy. In addition to reducing leukemic cell mass, there is a gradual reduction in the frequency of cells bearing a 9–22 chromosomal translocation. (Guilhot *et al.*, 1997).

The unique cellular and molecular activities of IFN-α complement the mechanisms of action of other therapies (Wadler and Schwartz, 1990). Thus, the combination of IFN-α with other therapies is likely to result in additional clinical efficacy. At present, therapeutic applications for IFN-α are focused on synergistic or additive effects with IFN-γ or interleukin 2 (IL-2). In addition, IFN-α appears to have therapeutic activity via oral administration for at least a limited number of diseases, including oral warts (Friedman-Kien *et al.*, 1988), Sjogren's syndrome (ship *et al.*, 1999), and fibromyalgia (Russell *et al.*, 1999). Responses to orally administered IFN-α appear to be active at low doses and to follow the concept of a "bell-shaped" dose–response curve whereby there is activity at low doses and high doses induce a form of anergy (Tompkins, 1999).

Interferon Gamma (IFN-γ)

Preclinical studies have suggested that recombinant IFN-γ (rIFN-γ) has significant therapeutic activity in animal models of experimental and spontaneous metastasis, which occurs with a reproducible "bell-shaped" dose-response curve (Talmadge *et al.*, 1987). Studies of immune response in normal animals have revealed the same "bell-shaped" dose–response curve for the augmentation of macrophage tumoricidal activity (Talmadge *et al.*, 1987; Black *et al.*, 1990). Thus, optimal therapeutic activity is observed with the same dose and protocol of rIFN-γ but with significantly less therapeutic activity at lower and higher doses. A significant correlation between macrophage augmentation and therapeutic efficacy has been reported (Black *et al.*, 1990), suggesting that immunological augmentation provides one mechanism for the therapeutic activity of rIFN-γ and supports the hypothesis that treatment with the maximum tolerated dose (MTD) of rIFN-γ may not be optimal in an adjuvant setting.

The preclinical hypothesis of a "bell–shaped" dose–response curve for r-IFN-γ has been confirmed in clinical studies on the immunoregulatory effects of rIFN-γ, which defined an OID (Jaffe *et al.*, 1988; Maluish *et al.*, 1988). In general, the OID for rIFN-γ has been found to be between 0.1 and 0.3 mg/M^2 following intravenous or intramuscular injection (Maluish *et al.*, 1988). In contrast, the MTD for rIFN-γ may range from 3 to 10 mg/M^2 depending upon the source of the rIFN-γ and/or the clinical center. The identification of an OID for rIFN-γ in patients with minimal tumor burden has resulted in the development of clinical trials to test the hypothesis that the immunological enhancement induced by rIFN-γ will result in prolongation of the disease-free period and overall survival of patients in an adjuvant setting (Jaffe *et al.*, 1988). However, rIFN-γ was found on an empirical basis to have therapeutic activity in chronic granulomatous disease (CGD) (International CGD Cooperative Study Group, 1991), and it was for this indication that the U.S. Food and Drug Administration (FDA) first approved rIFN-γ. The studies in CGD suggested that the mechanism of therapeutic activity for IFN-γ is associated with enhanced phagocytic oxidase activity and increased superoxide production by neutrophils. However, more recent data suggest that the majority of CGD patients obtain clinical benefit by prolonging IFN-γ therapy and that the mechanism of action may not be due to enhanced neutrophil oxidase activity but, rather, to the correction of a respiratory-burst deficiency in a subset of monocytes (Woodman *et al.*, 1992). It was suggested, as well, that IFN-γ administration may induce nitric oxide (NO) synthetase activity by polymorphic nuclear cells (PMN) in patients with CGD (Ahlin *et al.*, 1999). Following 2 days of IFN-γ administration with 50 or 100 μg of IFN-γ/M^2, a significant increase in PMN-produced NO was observed, revealed by a significant increase in the bactericidal capacity of the PMN (Ahlin *et al.*, 1999). Since these PMN lack the capacity to produce superoxide anions, at least in patients with CGD, it is possible that the increased NO release and *in vitro* bactericidal activity could be instrumental in augmenting host defenses and reducing the morbidity of CGD (Ahlin *et al.*, 1999). Similarly, IFN-γ has been shown to increase PMN expression of FcγRI and to improve Fcγ receptor-mediated phagocytosis, at least in normal subjects (Schiff *et al.*, 1997). Therefore, the mechanism of action critical to using IFN-γ to reduce the frequency of infections in patients with CGD may not be associated with phagocytic oxidase activity and increased superoxide production but, rather, with other mechanisms of granulocyte activity, including NO production. In addition to its licensing for CGD, IFN-γ has been approved for the treatment of rheumatoid arthritis in Germany, and most recently, osteopetrosis in the United States.

INTERLEUKIN 2 (IL-2)

Interleukin 2 is a cytokine that induces T-cell proliferation, augments natural killer (NK) cells, and activates lymphokine-activated killer (LAK) cells. The LAK cells have been cultured with IL-2, *in vitro* for 72 hours or longer and have markedly increased nonspecific cellular cytotoxicity. As such, IL-2 is important to all facets of T- and NK-cell augmentation and proliferation. IL-2 has been approved for use as a single agent for the treatment of renal cell carcinoma, metastatic melanoma, and hepatitis C. It is also administered in conjunction with LAK- or T-cell infiltrating lymphocytes (TILs) in adoptive cellular therapy protocols. TILs are T cells obtained from a tumor and expanded *in vitro* with lower levels of IL-2 than that used with LAK cells and in the presence of tumor antigen(s), with the goal of expanding a population of tumor-specific cytotoxic T cells. However, it has been questioned whether the adoptive transfer of LAK cells is necessary or adds to the clinical efficacy of rIL-2. Indeed, there has been little indication of an improved therapeutic effect of rIL-2 plus LAK cells versus IL-2 alone (West *et al.*, 1987; Rosenberg *et al.*, 1993). When the clinical trials with IL-2 are rigorously examined, neither strategy has impressive (as opposed to significant) therapeutic activity (West *et al.*, 1987; Rosenberg *et al.*, 1993). The overall response rate with rIL-2 is 7 to 14% and is associated with considerable toxicity (Lotze *et al.*, 1986); however, it should be remarked that these responses are durable. In one of the first clinical studies (Heslop *et al.*, 1989), partial responses were observed in 4 out of 31 patients. Interestingly, these partial responders did not correspond to the patients with increased activity of LAK or NK cells. The antitumor effect of both TIL and LAK cells could be either due to a direct effect or secondary to the generation of other cytokine mediators, as suggested by the observation that rIL-2-stimulated lymphocytes produce IFN-γ and tumor necrosis factor (TNF) as well as other cytokines and the possibility that the therapeutic activity of rIL-2 may be synergistic with these cytokines (Heslop *et al.*, 1989). Recently, IL-2 has also been examined as an adjuvant to augment the tumor host response to human immunodeficiency virus (HIV) vaccines (Barouch *et al.*, 1998; Kim *et al.*, 2001).

Many of the IL-2 clinical trials in metastatic renal cell carcinoma, with or without LAK cells, have used an MTD of IL-2. A study by Fefer's laboratory (Thompson *et al.*, 1992) compared maintenance IL-2 therapy at the MTD of 6×10^6 U/M^2/day to 2×10^6 U/M^2/day. These investigators found that it was possible to maintain the patients for a median of 4 days at 6×10^6 U/M^2/day but in the presence of severe hypertension and capillary leak syndrome. In the lower dose protocol, none of the patients experienced severe hypertension or capillary leak syndrome, and the median duration of maintenance IL-2 therapy was 9 days. Further, in the lower dose protocol there was a total response rate of 41% in contrast to the higher dose protocol (with a shorter duration of administration), which had a 22% response rate. The investigators suggest that there may be an improved therapeutic activity associated with a longer maintenance protocol at lower doses.

An IL-2 dose–response study examined the transcriptional regulation of cytokine mRNA levels in the peripheral blood leukocytes (PBL) of cancer patients. The results suggested that doses of rIL-2 as low as 3×10^4 U/day could augment T-cell function and that doses of rIL-2 of 10^5 U/day or more increase not only T-cell, but also macrophage, function (Hladik *et al.*, 1994). The latter was measured as TNF levels and the upregulation of TNF at the higher dose of IL-2 combined with the T-cell production of IFN-γ, which occurs at the lower dose of IL-2 and may be responsible for the toxicity of IL-2 (Mier *et al.*, 1988).

In more recent work, renal cell cancer patients were randomized to receive either a high-dose regimen or one using one-tenth of the dose (72,000 IU/kg/8 h) administered by the same schedule (days 1–5 and 15–19, repeated every 4–6 weeks) (Yang *et al.*, 1994, 1997). An interim report of this trial (Yang *et al.*, 1994) reported similar response rates in the two cohorts: approximately 7% complete responses (CR) and 8% partial responses (PR) in the low-dose group and 3% CR and 17% PR in the high-dose group. However, the toxicity of the low-dose regimen was substantially less than that of the high-dose regimen. A follow-up this study (Yang *et al.*, 1997) also reported similar overall response rates between the low-dose and the high-dose protocols. However, the authors reported that the responses to high-dose IL-2 tended to be more durable. A similar trial comparing high and low doses of IL-2 has been undertaken in HIV-infected patients using push administration, which allows much higher doses (Dazzi *et al.*, 2000). In this trial (Dazzi *et al.*, 2000), the low dose was 1.5×10^6 IU twice a day, in comparison to 7.5×10^6 IU twice a day by subcutaneous injection, both doses given for 5 days, with the cycle being administered every 4 or every 8 weeks. At 6 months, the high-dose recipients had a 95% increase in mean CD4 cells compared with a 19% increase in the low-dose recipients. It was concluded that the high-dose therapy was well tolerated and induced a "dramatic, sustained rise" in $CD4^+$ cells (Davey *et al.*, 1999).

More recently, a study was undertaken to define optimal duration and cycle frequency (Miller *et al.*, 2001). In these studies, patients received IL-2 by continuous infusion, initially starting at 9×10^6 IU/day and decreasing in increments of 1.5 to 3×10^6 IU as needed to manage toxicity (Miller *et al.*, 2001). In this 32-week study, patients were randomized to one of three treatment protocols. In the control protocol, participants received a 5-day IL-2 cycle every 8 weeks for 32 weeks. In the second group, participants received four cycles of longer duration therapy (average 7.7 days), and the last group received a protocol with an increased frequency of 5-day IL-2 cycles resulting in a cycle every 4 weeks. In this study, all three groups experienced a significant increase in their mean CD4 cell counts, with no statistically significant differences between the cohorts. Further, HIV viral loads decreased during the study period in all three cohorts. The conclusion was that the frequency and duration of IL-2 administration, at least by continuous infusion, was not critical and that an IL-2 regimen consisting of 5-day cycles administered no more frequently than every 8 weeks was sufficient to significantly increase CD4 cell counts and decrease viral load (Miller *et al.*, 2001).

Chronic IL-2 administration at low doses (~200,000 IU/M^2/day) has been found to increase CD4$^+$ cell number and the CD4:CD8 ratio in AIDS patients (Kovacs *et al.*, 1995; Jacobson *et al.*, 1996; Smith, 2001). The goal of one such study (Jacobson *et al.*, 1996) was to give asymptomatic HIV-positive individuals IL-2 without promoting viral replication, using an approach patterned after that described by Soiffer *et al.*, (1994), who reported that low doses of IL-2 could be given to cancer patients for periods up to 3 months with minimal toxicity. Their results indicate that extremely low IL-2 doses are nontoxic and are effective in stimulating immune reactivity. Further, low-dose IL-2 in HIV-positive patients not only increased CD4 cell counts, it also reduced lymphocytic apoptosis and specifically increased naïve CD4$^+$ cells (Pandolfi *et al.*, 2000). Further, continuous IL-2 infusion can selectively expand the absolute number of NK cells after 4 to 6 weeks of administration (Fehniger *et al.*, 2000). This occurs at doses of IL-2 that are relatively low but capable of inducing CD34$^+$ hematopoietic progenitor cell differentiation into CD56$^+$CD3$^-$ NK cells. Thus, it is possible that there is selective recruitment of NK cells during low-dose infusion of IL-2 from bone marrow progenitors, resulting in NK cell expansion and potential contribution to therapeutic activity (Fehniger *et al.*, 2000).

RECOMBINANT PROTEINS IN CLINICAL DEVELOPMENT

Numerous recombinant biotherapeutics are in clinical trials (see Table 2). These include not only cytokines and growth factors but also soluble receptors and cytokine and receptor inhibitors and antagonists. The availability of this rapidly growing group of biotherapeutics is due primarily to the evolution in molecular biology, which has made biotherapeutics available in pharmacological amounts, speeding our ability to develop effective immunotherapeutics. In addition to the clinical development of recombinant biotherapeutics, which have been cloned, numerous new indications for licensed recombinant biotherapeutics are also under development. The majority of the indications and biotherapeutics currently in clinical development are found in Table 2. While the intent has been to be all-encompassing, it is probable that relevant therapeutics have been omitted. Equally, given the rapidity with which biotherapeutics are cloned, produced, and entered into clinical trials, no list will remain current. Further, the convergence of cytokine biology and the rapidly expanding sophistication in gene therapy has resulted in the genetic delivery of biotherapeutics. This innovation targets the

Table 2
Cytokines in Clinical Development

Drug	Corporation	Target	Stage
Abarelix	Amgen	Endometriosis	Phase II
Actimmune® interferon gamma-1b	InterMune Pharmaceuticals	Idiopathic pulmonary fibrosis	Phase II/III
Actimmune® interferon gamma-1b	InterMune Pharmaceuticals	Tuberculosis	Phase III
Actimmune® interferon gamma-1b	InterMune Pharmaceuticals	Cystic fibrosis	Phase II
Alferon N Injection® interferon alfa-n3 (human leukocyte derived)	Interferon Sciences	Chronic hepatitis C infections	Phase III
Alferon N Injection® interferon alfa-n3 (human leukocyte derived)	Interferon Sciences	Co-infection (HIV/HCV)	Phase II
Alferon N Injection® interferon alfa-n3 (human leukocyte derived)	Interferon Sciences	Human immunodeficiency virus disease	Phase III
Alferon N Injection® interferon alfa-n3 (human leukocyte derived)	Interferon Sciences	Human papilloma virus	Phase II
Amevive™ recombinant LFA-3/IgG1	Biogen	Psoriasis	Phase III
Anakinra IL-1R antagonist	Amgen	Rheumatoid arthritis	
Avonex® interferon beta-1a	Biogen	Glioma	Phase II
Avonex® interferon beta-1a	Biogen	Idiopathic pulmonary fibrosis	Phase II
Avonex® interferon beta-1a	Biogen	Secondary, progressive multiple sclerosis	Phase III
Avrend CD40L	Immunex	Epithelial solid tumors	Phase I
Avrend CD40L	Immunex	Metastatic renal carcinoma	Phase II
Bataseron® interferon beta-1b	Berlex Laboratories	Secondary, progressive multiple sclerosis	Phase III
ConXn® recombinant human relaxin	Connetics	Scleroderma	Phase II/III
CTLA4Ig	Bristol-Myers Squibb	Immunosuppression	Phase II
Enbrel® TNFR:Fc	Immunex	Psoriasis	Phase II
Enbrel® TNFR:Fc	Immunex	Psoriatic arthritis	Submitted?

(*continues*)

Table 2 (*continued*)

Drug	Corporation	Target	Stage
Flt3L Mobista®	Immunex	Melanoma/renal carcinoma	Phase I/II
GA-EPO gene-activated EPO	Transkaryotic Therapies	Anemia associated with renal disease	Phase III
GA-EPO gene-activated EPO	Transkaryotic Therapies	Anemia of chemotherapy	Phase III
IFN beta-1a	Serono Laboratories	Crohn's disease	Phase II
IL-10 Tenovil	Schering-Plough	Crohn's disease	Phase III
IL-10 Tenovil	Schering-Plough	Acute lung injury	Phase I
IL-10 Tenovil	Schering-Plough	Hepatitis C	Phase I
IL-10 Tenovil	Schering-Plough	Ischemic reperfusion injury	Phase I
IL-10 Tenovil	Schering-Plough	Psoriasis	Phase II
IL-10 Tenovil	Schering-Plough	Rheumatoid arthritis	Phase III
IL-11	Genetics Institute	Crohn's disease	Phase II
IL-1Ra	Amgen	Rheumatoid arthritis	Submitted?
IL-12	Genetics Institute	Renal cell carcinoma, malignant melanoma, hepatitis vaccine adjuvant	
Leridistim	Searle	Mobilization and myelorestoration	Phase II
Leucotropin™ GM-CSF	Cangene	Myeloid reconstitution stem cell transplantation	Phase III
Leukine® or sargramostim (GM-CSF)	Immunex	Immunostimulation in melanoma	Phase III
Leukine® or sargramostim (GM-CSF)	Immunex	Mucositis	Phase III
Leukine® or sargramostim (GM-CSF)	Immunex	Prevention of opportunistic infections	Phase III
Mobista Flt3L	Immunex	Mobilization, DC expansion	Phase II
Mobista Flt3L	Immunex	Prostate carcinoma, non-Hodgkin's lymphoma, melanoma	Phase II
Neumega® oprelvekin	Genetics Institute	Cancer treatment support	Phase II
Nevance A39™ interleukin 4 receptor	Immunex	Asthma	Phase II

(*continues*)

Table 2 (*continued*)

Omniferon™ neutral alpha interferon	Viragen	Hepatitis C	Phase II
Osteoprotegerin (OPG)	Amgen	Cancer	Phase I
Osteoprotegerin (OPG)	Amgen	Osteoporosis	Phase I
Oxsodrol recombinant human superoxide dismutase (rhSOD)	Bio-Technology General	Prevention of asthma	Phase II
Oxsodrol recombinant human superoxide dismutase (rhSOD)	Bio-Technology General	Asthma, bronchopulmonary, dysplasia, and developmental deficits in premature infants	Phase II
Pegasys Peg interferon alfa-2a	Hoffmann-La Roche	Chronic hepatitis C	Phase III
Pegasys Peg interferon alfa-2a	Hoffmann-La Roche	Chronic myelogenous leukemia, melanoma, renal carcinoma	Phase II
Pegasys Peg interferon alfa-2a	Hoffmann-La Roche	Hepatitis B	Phase II
PEG-Intron A® interferon alfa 2b	Enzon/Schering-Plough	Chronic myelogenous leukemia, melanoma	Phase III
Progenipoietin	Searle	Cancer immunotherapy	Phase I
Proleukin® aldesleukin (interleukin 2)	Chiron	Acute myelogenous leukemia	Phase III
Proleukin® aldesleukin (interleukin 2)	Chiron	Leukemia, myelodysplastic syndromes	Phase III
Rebif® recombinant IFN beta-1a	Serono Laboratories	Chronic hepatitis C	Phase II
Rebif® recombinant interferon beta-1a	Serono Laboratories	Early treatment of multiple sclerosis	Phase III
Rebif® recombinant interferon beta-1a	Serono Laboratories	Guillain–Barré syndrome	PhaseII
Recombinant human interleukin-12 (rhIL-12)	Genetics Institute	Hepatitis C	Phase I/II
ReFacto® r factor VIII	Genetics Institute	Hemophilia A	submitted
rhIL-11	Genetics Institute	Psoriasis	Phase II
rhTNF-binding protein 1 (rHuTBP-1)	Serono Laboratories	Cardiac reperfusion injury	Phase II
rhTNF-binding protein 1 (rHuTBP-1)	Serono Laboratories	Crohn's disease	Phase II

(*continues*)

Table 2 (*continued*)

Drug	Corporation	Target	Stage
rhTNF-binding protein 1 (rHuTBP-1)	Serono Laboratories	Rheumatoid arthritis	Phase II
SB251353 cxc chemokine GRO beta	GSB	Mobilization	Phase I/II
SD/01 Peg-G-CSF	Amgen	Chemotherapy-induced neutropenia	Phase III
Soluble TNF-R1	Amgen	Rheumatoid arthritis	Phase II
Stemgen® ancestim	Amgen	Peripheral blood progenitor cell transplantation	Submitted
Thrombopoietin (TPO)	Pharmacia UpJohn	Myelosuppression/ myeloablation	Phase II
Thrombopoietin (TPO)	Genentech	Thrombocytopenia	Phase III
TNF alpha	Boehringer Ingelheim	Advanced melanoma	Phase II
TP10 rTNFR	AVANT Immunotherapies	Transplantation	Phase I/II
TP10 rTNFR	AVANT Immunotherapies	Acute respiratory distress syndrome	Phase II
TP10 rTNFR	AVANT Immunotherapies	Heart attack	Phase I
Veldona® natural IFN-α	Amarillo Biosciences	Fibromyalgia syndrome	Phase II
Veldona® natural IFN-α	Amarillo Biosciences	Sjögren's syndrome	Phase III
Zenapax® Daclizumab	Hoffmann-La Roche	Prevention of graft rejection in pediatric kidney transplant	Phase III
Zovant™ (recombinant human activated protein C)	Eli Lilly	Sepsis	Phase III

transfection of tumor cells not only as an adjuvant approach, but also as a strategy to provide vaccine adjuvants.

While a complete discussion of the recombinant biotherapeutics in development could not be addressed here, two examples of such biotherapeutics are discussed: IL-12, an interleukin with significant activity for T and NK cells, and Flt3 ligand (Flt3L), a cytokine with hematopoietic and dendritic cell (DC) growth factor activity. Other biologics in development include soluble receptors and receptor antagonists for cytokines such as IL-4, IL-1β and TNF, targeting the treatment of autoimmune and inflammatory diseases.

Flt3L is expressed on most $CD34^+$ cells (Rappold *et al.*, 1997) and can increase the peripheral blood leukocyte count (Brasel *et al.*, 1996; Lyman, 1998), NK cell number and activity (Brasel *et al.*, 1996; Shaw *et al.*, 1998), and DC in the peripheral blood (PB) and other tissues (Maraskovsky *et al.*, 1996; Pulendran *et al.*, 1997; Somlo *et al.*, 1997). Flt3L can also expand and mobilize hematopoietic progenitor and stem cells (Brasel *et al.*, 1996; Ashihara *et al.*, 1998; Robinson *et al.*, 2000). Dendritic cells, the "professional" antigen-presenting cells of the immune system, play a major role in immune-based therapies for oncology, infectious disease, autoimmunity, and transplantation. In mice, Flt3L also stimulates the production of NK cells that can directly kill tumor cells (Yu *et al.*, 1998).

Administration of Flt3L, a *fms*–like tyrosine kinase, either before or after tumor injection can inhibit tumor growth *in vivo* (Chen *et al.*, 1997; Lynch *et al.*, 1997; Esche *et al.*, 1998; Lynch, 1998; Braun *et al.*, 1999). Part of this antitumor activity may be a consequence of enhanced NK activity following Flt3L administration (Chen *et al.*, 1997; Esche *et al.*, 1998; Peron *et al.*, 1998; Shaw *et al.*, 1998; Wang *et al.*, 2000). However, transfer of splenic T cells from tumor-free animals, after immunization with Flt3L-expressing tumor cells, protects naïve animals from tumor challenge (Chen *et al.*, 1997; Lynch *et al.*, 1997; Braun *et al.*, 1999). Moreover, mice vaccinated with Flt3L-transduced tumors have greater protection against tumor challenge. Than is conferred by vaccination using GM-CSF-transduced tumors (Braun *et al.*, 1999). Our earlier studies demonstrated a role for T-cell augmentation by Flt3L following vaccination with protein (Pisarev *et al.*, 2000) or genetic vaccines (Parajuli *et al.*, 2001). The mechanism of T-cell augmentation by Flt3L appears to occur by multiple mechanisms. Flt3L administration enhances the $CD11c^{high}$ $CD11b^{low}$ type DC, which leads to an antigen-specific, type 1 T-cell response. (Pulendran *et al.*, 1999b). Flt3L administration preferentially increases the number of DC expressing $CD11c^{high}$ $CD11b^{low}$ (type 1) phenotype, relative to the $CD11c^{high}$ $CD11b^{high}$ (type 2) phenotype, resulting in a significant increase in the frequency of IL-12-secreting cells (Parajuli *et al.*, 2001); this is accompanied by a higher frequency of IFN-γ-producing type 1 cells in the spleens.

Flt3L preferentially induces a type 1 T-cell response (Daro *et al.*, 2000; Pulendran *et al.*, 2000). The adjuvant activity of Flt3L is due, in part, to effects on DC (Lynch, 1998). It has been demonstrated that distinct DC subsets may be responsible for regulating T-cell responses (Daro *et al.*, 2000; Pulendran *et al.*, 2000). Indeed, in mice, DC1 ($CD11c^+$ $CD11b^-$ $CD8a^-$) and DC2 ($CD11c^+$ $CD11b^+$ $CD8a^+$) subsets (Rissoan *et al.*, 1999; Patterson, 2000) induce type 1 and 2 T-cell responses to antigen, respectively (Maldonado-Lopez *et al.*, 1999a,b; Rissoan *et al.*, 1999).

The mechanism of DC-mediated induction of type 1 T cells may be dependent on the preferential production of IL-12, a cytokine critical for the development of type 1 T cells (Cella *et al.*, 2000; Langen Kamp *et al.*, 2000). In contrast, IL-10, which is also produced by DC, inhibits type 1 T-cell activity including T-cell proliferation (Bouloc *et al.*, 2000; Yang *et al.*, 2000). Besides these cytokines, other factors are involved in T-cell polarization and activation (Zou *et al.*, 2000). Flt3L also significantly increases T cells with a memory phenotype (Parajuli *et al.*, 2001).

There are several potential mechanisms for the expansion and stimulation by Flt3L of type 1 T cells. These include Flt3L-induced expansion and stimulation of precursor DC and/or DC1 cells, resultant increased production of IL-12, which stimulates type 1 T cells, increased levels of type 1 cytokines (IFN-γ or IL-2), decreased levels of type 2 cytokines (IL-4 or IL-10), and/or expansion of T-cell precursors or immature, resting T cells. Indeed, the increase in T-cell numbers and type 1 T-cell bias (IL-2 and IFN-γ levels) may contribute to the antitumor activity of Flt3L. Further, Flt3L selectively enhances the number of DC1 cells relative to DC2 cells (Maraskovsky *et al.*, 1996; Pulendran *et al.*, 1997, 1999b; Lyman, 1998; Maldonado-Lopez *et al.*, 1999a). Isolated DC1 also showed a significantly higher mRNA expression for IL-12 (p40) and lower mRNA expression for IL-10, while this profile was reversed in isolated DC2 cells (Maldonado-Lopez *et al.*, 1999a; Pulendran *et al.*, 1999b). The splenocytes of Flt3L-treated mice also had significantly higher IL-12 (p40) and lower IL-10 mRNA levels than control splenocytes. In addition, both the frequency of IL-12-secreting spleen cells (ELISPOT) as well as the level of IL-12 (p40) mRNA were augmented by Flt3L in a dose-dependent fashion. This increase in IL-12 (p40) could contribute to the skewing of the T-cell cytokine profile toward a type 1 phenotype, inasmuch as DC1 cells obtained from IL-12 p40 knockout mice failed to induce an antigen-specific, type 1 response in a wild-type host (Maldonado-Lopez *et al.*, 1999a). Flt3L can increase both DC subsets, or their precursors, including the $CD11c^+$ subset and the $CD11c^-$ IL-3-receptor-positive precursor in humans (Pulendran *et al.*, 2000). Further, freshly sorted $CD11c^+$, but not $CD11c^-$, DC can stimulate $CD4^+$ T cells in an allogeneic mixed lymphocyte response. Two DC subsets were associated with distinct cytokine profiles in $CD4^+$ T cells, with the $CD11c^-$ subset inducing levels of IL-10, a Th2 cytokine (Pulendran *et al.*, 2000). However, it remains unclear whether the differences among DC subsets reflect differentiation (DC1 vs DC2 lineage) (Kusakabe *et al.*, 2000; Traver *et al.*, 2000) or maturation (Ebner *et al.*, 2001). Taken together, these studies suggest that Flt3L-enhanced tumor antigen presentation results in therapeutic responses against disseminated tumor cells and improved survival.

A significantly enhanced number of $CD4^+$ and $CD8^+$ T cells have been observed in mice injected with Flt3L compared with controls (Parajuli *et al.*,

2001). However, the increase in $CD4^+$- and $CD8^+$-positive cells was due to an overall increase in cellularity, not in the frequency of either $CD4^+$ or $CD8^+$ cells. The number of both naïve and effector/memory $CD4^+$ cells was also significantly increased in Flt3L-treated mice (Parajuli *et al.*, 2001). Further, Flt3L enhanced both naïve and effector/memory $CD8^+$ T cells. (Parajuli *et al.*, 2001). These results suggest that Flt3L administration is associated with an expansion of $CD4^+$ and $CD8^+$ T cells *in vivo* and that this occurs predominantly with effector/memory T cells.

In the initial phase I clinical trials, Flt3L was administered to normal volunteers at 10 to 100 μg/kg/day for 14 consecutive days with minimal side effects (Lebsack *et al.*, 1997; Maraskovsky *et al.*, 1997). These studies demonstrated a significant increase in the white blood call count, predominantly occurring in the monocytes. The studies also revealed a significant increase in lineage-negative $CD11c^+$ DC (30-fold) (Maraskovsky *et al.*, 1997). More recent studies in normal volunteers have demonstrated the mobilization of two DC populations, including $CD11c^+CD14^-CD125^{dim}$ (DC1 cells) and $CD11c^-CD14^-CD125^{bright}$ (DC2) (Fay *et al.*, 1999; Pulendran *et al.*, 1999a). The coadministration of Flt3L with G-CSF or granulocyte/macrophage CSF (GM-CSF) to patients revealed similar increases in the number of circulating DC (Gasparetto *et al.*, 1999), as well as significantly improved $CD34^+$ cell mobilization (Morse *et al.*, 2000). Further, the administration of Flt3L is not only associated with an increase in and mobilization of DC, but also results in DC infiltration in a peri-tumor manner and an increase in delayed type hypersensitivity (DTH) responses to recall antigens in metastatic colon cancer patients (Maraskovsky *et al.*, 2000). Consistent with the reported differences in mice (Parajuli *et al.*, 2001), there is also a difference in the DC subsets expanded clinically between G-CSF and Flt3L. In one clinical trial (Pulendran *et al.*, 2000), Flt3L was found to increase both the $CD11c^+$ DC subset (48-fold) and the $CD11c^-$ $CD125^+$ DC subset (13-fold). In contrast, G-CSF also increased the $CD11c^-$ $CD125^+$ subset (approximately seven fold). Thus, Flt3L has the potential to differentially regulate the human immune response. Further, the Flt3L-expanded DC can be pulsed with peptides and induce both a cytotoxic T-lymphocyte (CTL) response and tumor regression (Fong *et al.*, 2001). In one study advanced cancer patients were injected with Flt3L, and the expanded DC were isolated and pulsed with a peptide derived from a carcinoembryonic antigen (CEA). Following immunization with this antigen-pulsed DC, a CTL population was induced that could be stained with peptide major histocompatibility complex (MHC) tetramers, suggesting an expansion of $CD8^+$ T cells that recognize CEA. (Fong *et al.*, 2001). Further, these tetramer-staining cells had a CTL phenotype defined as $CD45RA^+$ $CD27^-CCR7^-$. (Fond *et al.*, 2001). In addition, two of the 12 patients underwent significant tumor regression, one had a mixed response and two had disease stabilization. (Fong *et al.*, 2001). Furthermore, the clinical response was reported to be correlated with the expansion of the tetramer-positive T cells. These results suggest that immunization with Flt3L-expanded DCs, which have been pulsed with peptides, can break tolerance, at least to CEA, which is a self-antigen.

Interleukin 12 is a heterodimeric glycoprotein that binds to a single receptor present on activated T and NK cells. It stimulates the proliferation of activated T (Gately *et al.*, 1991; Bertagnolli *et al.*, 1992; Rodolfo and Colombe, 1999) and NK cells and can also activate NK-cell cytotoxicity (Chan *et al.*, 1991; Robertson *et al.*, 1992). In type 1 lymphocytes, IL-12 upregulates the synthesis of IFN-γ, IL-2, and TNF-α (German *et al.*, 1993; Win *et al.*, 1993). This increase in type 1 T-cell activity can be inhibited by the type 2 cytokine, IL-10. In contrast, with type 2 T cells, IL-12 can reduce the synthesis of IL-4 and IL-10 (D'Andrea *et al.*, 1993; Naume *et al.*, 1993). IL-12 also enhances the myelopoiesis of primitive hematopoietic progenitor cells (Jacobsen *et al.*, 1993) and can mobilize stem cells for subsequent use in transplantation (Jackson *et al.*, 1995). The ability of IL-12 to induce IFN-γ synthesis and to stimulate activated T-cell proliferation has resulted in studies with IL-12 as a vaccine adjuvant.

IL-12 has shown therapeutic activity for murine tumors resulting in tumor regression and reduction of metastases (Wendrzak and Brunda, 1996). In studies comparing the antitumor efficacy of different cytokines, IL-12 has, in general, shown superior antitumor activity (Nastala *et al.*, 1994; Mu *et al.*, 1995; Cavallo *et al.*, 1997; Rappold *et al.*, 1997). Further, IL-12 has demonstrated activity in preventing and inhibiting the growth of primary tumors induced by chemical carcinogens (Nishimure *et al.*, 1995; Noguchi *et al.*, 1995) or by spontaneously arising tumors in transgenic mice expressing the rat HER-2/*neu* oncogene in the mammary gland (Boggio *et al.*, 1998). IL-12 also increases the vaccine response to DC, (Coughlin *et al.*, 1998), peptide in QS21 adjuvant (Noguchi *et al.*, 1995), and recombinant viruses (Rao *et al.*, 1996) used therapeutically in mice bearing micrometastases. Nonetheless, several studies have suggested that the effectiveness of IL-12 against murine tumors is due primarily to inhibition of angiogenesis rather than to antigen-specific tumor immunity (Brunda *et al.*, 1995; Voest *et al.*, 1995; Boggio *et al.*, 1998; Coughlin *et al.*, 1998; Kurzawa *et al.*, 1998; Ogawa *et al.*, 1998).

The first clinical trial of IL-12 in cancer patients showed that a single injection of up to 500 ng/kg could be administered with acceptable toxicity (Kohl *et al.*, 1996). However, a phase II study resulted in severe and even fatal toxicity (Cohen, 1995; Atkins *et al.*, 1997). Subsequent IL-12 immunopharmacology studies in mice (Hendrzak and Brunda, 1995; Orange *et al.*, 1995), and in pilot clinical trials (Cohen, 1995) revealed that the timing and dose of IL-12 administration are critical for effectiveness and reduced toxicity. It was concluded that prophylactic and therapeutic use of IL-12 requires delivery to relevant sites to circumvent toxicity. The potential to desensitize mice to the toxic effects of rIL-12 by preadministration of a low dose of rIL-12 has been tested; however, pretreatment also attenuated rIL-12 therapeutic outcome (Coughlin *et al.*, 1997).

Data from the rodent toxicity studies demonstrated that a single-dose IL-12 exposure 2 weeks before consecutive dosing not only abrogated toxicity but also reduced IL-12-induced IFN-γ production (Leonard, 1997). Similarly, subcutaneously administered recombinant human IL-12 (rhIL-12) as a single dose was well tolerated, with marked effects on immune parameters and some antitumor activity in pretreated melanoma patients (Bajetta *et al.*, 1998). In another study, cancer patients were injected with rIL-12 subcutaneously, revealing the expansion of a subset of peripheral blood $CD8^{+}$ T cells due to upregulation of the IL-12 receptor (Gollub *et al.*, 1998). *In vitro* characterization of these cells revealed expression of CD18 and TcRVa and $IL - 12^{-}$ and IL-2-stimulated production of IFN-γ, proliferation, and enhanced non-MHC-restricted cytolytic activity. The MTD of intravenously injected IL-12 in cohorts of four to six patients with advanced cancer was found to be 500 ng/kg (Atkins *et al.*, 1997). In this protocol, IL-12 was injected intravenously once followed by a 2-week rest period and then by daily administration for 5 days every 3 weeks. Common toxicities observed included fever, chills, fatigue, nausea, vomiting, and headache (Atkins *et al.*, 1997). Fever was observed at doses as low as 3 ng/kg/day 8 to 12 hours after the initial administration and was not fully responsive to nonsteroidal anti-inflammatory drugs (Atkins *et al.*, 1997). Blood chemistry changes included anemia, neutropenia, lymphopenia, hyperglycemia, thrombocytopenia, and hypoalbuminuria (Atkins *et al.*, 1997). The dose-limiting toxicities observed were oral stomatitis and liver function abnormalities, predominantly elevated transaminase levels, which were observed at the 500-ng/kg dose (Atkins *et al.*, 1997). The immune modulatory effects observed included dose-dependent increases in serum IFN-γ levels that were attenuated in the second cycle of administration (Atkins *et al.*, 1997). In contrast, serum neopterin levels rose in a dose-dependent manner regardless of cycle (Atkins *et al.*, 1997). Lymphopenia was also observed at all dose levels with recovery occurring several days after completion of dosing without rebound lymphocytosis (Atkins *et al.*, 1997). A subsequent report extending this study reported that the lymphopenia had a nadir at 10 hours following a single IL-12 injection (Robertson *et al.*, 1999). Further, the lymphopenia involved all major lymphocyte subsets, with NK-cell numbers being the most profoundly affected, and $CD4^{+}$ T cells the least. Further, CD2, LFA-1, and CD56 were transiently upregulated on the NK cells, and NK activity was augmented in these phase I cancer patients. Defective NK-cell cytotoxicity and T-cell mitogenic responses were observed despite preadministration of a single 500-ng/kg dose of IL-12 (Robertson *et al.*, 1999). A phase II trial of human IL-2 in renal cell carcinoma patients (Motzer *et al.*, 2001) reported elevated serum cytokine levels following IL-12 administration at 1250 ng/kg. In these studies, IFN-γ, IL-10, and neopterin were significantly increased and maintained with weekly IL-12 injections. However, only 7% of the patients treated with IL-12 received a partial response (Motzer *et al.*, 2001). In an Italian study with metastatic renal cell carcinoma (Lissoni *et al.*, 1998), serum levels of neopterin, soluble IL-2 receptor (sIL-2R), TNF, IL-2, and IL-6 were examined. In patients given 1250 ng/kg recombinant IL-12 subcutaneously once a week for 3 weeks, there was a significant increase in neopterin, sIL-2R, and TNF but no effect on IL-6 and IL-2 levels (Lissoni *et al.*, 1998). In another dose-escalating study, where IL-12 was administered twice a week for 6 weeks, an MTD was again identified as 500 ng/kg, with dose-limiting toxicities of elevated hepatic transaminases and cytopenia (Gollub *et al.*, 2000). In these studies, IFN-γ, IL-15, and IL-18 were observed to be significantly increased following treatment with IL-12 (Gollub *et al.*, 2000). However, a tachyphylaxis to IFN-γ induction occurred following prolonged IL-12 administration, which was associated not with IL-10 levels or loss of IL-12 bioavailability but with an acquired defect in lymphocyte IFN-γ production in response not only to IL-12, but also to IL-2 and IL-15 (Gollub *et al.*, 2000).

Based on the poor clinical responses, apparent toxicity, and tachyphylaxis to immune modulation, current emphasis is on using IL-12 as a vaccine adjuvant, with genetic vaccines and peptidyl or DC vaccines.

NATURAL BIOLOGICAL RESPONSE MODIFIERS

The use of biological response monitors (BRMs: see Tables 3 and 4) to treat human disease has its origins in the work of William B. Coley, who used bacterial toxins to treat cancer (Nauts, 1975). These early studies resulted in the use of microbially derived substances such as bacille Calmette–Guerin vaccine (BCG) or Picibanil, carbohydrates from plants or fungi, such as Krestin, and Lentinan, and other products such as Biostim and Broncho-Vaxom (Table 3). However, there is considerable lot-to-lot variation in the purity of these compounds. In addition, intravenous injection of these particulate substances can result in pulmonary thrombosis and respiratory distress as well as the potential development of focal or multifocal granulomatous disease following either dermal administration, scarification, or intravenous administration.

The most commonly used microorganism for cancer therapy in the United States is BCG, which has been used systemically for metastatic disease or

Table 3

linically Useful Biologic Response Modifiers

Agent	Chemical nature	Action	Clinical Use
Microbial derived			
BCG (United States and Europe)	Live mycobacteria	Macrophage activator	Bladder cancer
Picibanil (OK432)(Japan)	Extract *Streptomyces pyogenes*	Macrophage activator	Gastric/other cancers
Krestin (PSK)(Japan)	Fungal polysaccharide	Macrophage activator	Gastric/other cancers
Lentinan (Japan)	Fungal polysaccharide	Macrophage activator	Gastric/other cancers
Biostim (Europe)	Extract *Klebsiella pneumoniae*	Macrophage activator	Chronic or recurrent infections
Thymus derived			
Thymostimulin (Europe)	Thymic peptide extract	T-cell stimulant	Cancer & infection
T-activin (Russia)	Thymic peptide extract	T-cell stimulant	Cancer & infection
Thym-Uvocal (Germany)	Thymic peptide extract		Cancer & infection
Chemically defined			
Azathiopine Imuran	Purine antimetabolite	Immunosuppressant	Graft versus host disease, allograft, et al.
Bestatin (Japan)	Dipeptide	Macrophage and T-cell stimulant	Acute myelosis leukemia
Cyclosporin (Sandimmune®)	Cyclic undecapeptide that is a metabolite of soil fungus	Immunosuppressant	Graft versus host disease, allografts, rheumatoid arthritis, psoriasis
Deoxyspergualin	Peptide fermentation product of *Bacillus laterosporus*	Immunosuppressant	Acute renal rejection
Isoprinosine (Europe)	Inosine–salt complex	T-cell stimulant	Infection
Levamisole (United States)	Phenylimidothiazole	T-cell stimulant	Cancer
Mycophenolate mofetil (MMF) Celleept®	2-Morpholine ethyl ester of mycophenolic acid	Immunosuppressant	

(*continues*)

Table 3 (*continued*)

Agent	Chemical nature	Action	Clinical Use
Rapamycin Rapamune® (sirolimus)	Metabolite of *Streptomyces hygroscopicus*	Immunosuppressant	Solid organ transplantation, graft versus host disease
Romurtide (Japan)	18 Lys MDP	Macrophage stimulant	Bone marrow recovery
Tacrolimus, FK505, Prograf®	Macrolide lactone, which is a fermentation product of *Streptomyces*	Immunosuppressant	Eczema, solid organ transplantation, graft versus host disease
Thymopentin TP-5 (Italy and Germany)	Pentapeptide	T-cell stimulant	Rheumatoid arthritis infection and cancer

adjuvant therapy, intralesionally (especially for cutaneous metastatic malignant melanoma), topically for superficial bladder cancer, and in combination with other immune modulators, tumor vaccines, and chemotherapy. When given intravesically, its use has been to treat superficial bladder cancer in residual disease and in the adjuvant setting (Haaff *et al.*, 1986). A well-controlled, randomized study has shown a prolonged disease-free interval and time to progression in patients treated with intradermal and intravesical BCG versus controls (Pinsky *et al.*, 1985). The mechanism by which BCG mediates its antitumor response is not known, but BCG treatment induces granulomatous inflammation in the bladder (Lage *et al.*, 1986) and elevates IL-2 levels in the urine of treated patients (Haaff *et al.*, 1986a), suggesting that an augmented local immune response may be important.

Recent studies have shown that intravesicular installation of BCG in patients with superficial bladder cancer results in a significant increase in IL-1β, IL-2, IL-6, TNF-α, IFN-γ, and macrophage colony-stimulating factor (M-CSF) with a concomitant and significant increase in IL-2 and IFN in the serum of the same patients (Taniguchi *et al.*, 1999). There appears to be a relationship between cytokine production and therapeutic efficacy as evidenced by the demonstration by means of a multivariant logistic analysis that inducibility of IL-2 provided a discriminating parameter yielding a predictive value of 97% for remission in patients receiving BCG treatment for their superficial bladder carcinoma (Kaempfer *et al.*, 1996).

CHEMICALLY DEFINED BIOLOGICAL RESPONSE MODIFIERS

The use of nonspecific immunostimulants has also been extensively studied (Table 3). The microbially derived agents have in common widespread effects on the immune system and side effects akin to infection (fever, malaise, myalgia, etc.). These agents can enhance nonspecific resistance to microbial or neoplastic challenge when administered prior to challenge (immunoprophylactic) but rarely when administered following challenge (immunotherapeutic). This is an important distinction in that the primary objective for the oncologist is the treatment of preexistent metastatic disease. Following a long history of experimental use in many different cancers and diseases, levamisole became the first chemically defined, orally active immunostimulant to be licensed (in the United States) for clinical use (Mutch and Hutson, 1991; Amery and Bruynseels, 1992). It was approved for the treatment of Duke's C colon cancer in combination with 5-fluorouracil (5-FU). This agent promotes T-lymphocyte, macrophage, and neutrophil function. It stimulates T-cell function *in vivo*, particularly in immunodeficient individuals, presumably through the action of its sulfur moiety. A recent dose–response study with levamisole demonstrated a significant increase in the frequency of peripheral blood mononuclear cells expressing the NK-cell Ag CD16 at all dose levels, although lower toxicity was observed at the lower doses of levamisole (Holcombe *et al.*, 2001). The authors suggested that short-term levamisole administration was only minimally immunomodulatory and that chronic administration at low doses may be better tolerated as well as providing levels of immune modulation similar to those observed with higher doses (Holcombe *et al.*, 2001). It is relatively nontoxic (flu-like symptoms, gastrointestinal upset, metallic taste, skin rash, and Antabuse reaction) but can produce an agranulocytosis, particularly in HLA B-27$^+$ patients with rhenmatoid arthritis, in whom its use has been discontinued. It is currently being considered for use in other cancers, with an emphasis on gastrointestinal disease. It is noted that the adjuvant therapeutic activity of levamisole has been questioned in recent years. In one phase III trial comparing 5-FU with leucovorin to 5-FU with levamisole it was found that the 5-FU and levamisole significantly prolonged disease-free survival and overall survival in patients with type III colon cancer who had undergone curative resection relative to adjuvant therapy with levamisole (Porschen *et al.*, 2001).

One of the largest and best studied classes of synthetic agents are the muramyl dipeptides (MDP). The first to be licensed, Romurtide (Japan), induces bone marrow recovery following cancer chemotherapy (Ellouz *et al.*, 1974). Romurtide's mechanism of action is the activation of macrophages to secrete colony-stimulating factors, IL-1, and TNF, resulting in the stimulation of marrow precursors to produce increased numbers of progenitor and mature granulocytes and monocytes. Therefore, the period of granulocytopenia and the risk of secondary infections are reduced allowing more frequent and/or intense chemotherapy. Murabutide (Table 4), an orally active form of MDP that does not induce fevers, is in clinical trials in cancer and infection in France. MTP-PE encapsulated in liposome is also in clinical trials for cancer (in the United States and Europe). The MDPs are also potent adjuvants alone and with oil, and are under consideration for use with HIV vaccines employing various synthetic peptide epitopes.

Table 4

Biological Response Modifiers in Clinical Development

Drug	Corporation	Target	
Acemannan	Carrington	Wound care	Phase III
Ampligen	HemispheRx Biopharma	Chronic fatigue syndrome	Phase III
Ampligen®	HemispheRx Biopharma	Hepatitis	Phase I/II
Ampligen®	HemispheRx Biopharma	Human immunodeficiency virus infection	Phase II
Ampligen®	HemispheRx Biopharma	Renal cancer	Phase II
Aranesp [novel erythropoiesis stimulating protein (NESP)]	Amgen	Anemia of chemotherapy	
Aranesp [novel erythropoiesis stimulating protein (NESP)]	Amgen	Chronic renal failure requiring dialysis	
CellCept® mycophenolate mofetil	Hoffmann-La Roche	Prevention of acute kidney transplant rejection	Phase III
CM 101 polysaccharide	CarboMed	Acute spinal cord injury	Phase I/II
CM 101 polysaccharide	CarboMed	Prostate carcinoma	Phase II
CpG 7909	Coley Pharmaceutical	Allergy, asthma	Phase I/II
CpG 7909	Coley Pharmaceutical	Hepatitis B	Phase I/II
CVT-E002 (natural polysaccharide)	CV Technologies	Upper respiratory tract infections	Phase II

(*continues*)

Table 4 (*continued*)

E5564 (endotoxin antagonist)	Eisai	Sepsis	Phase I
ImmTher® (MDP derivative)	Endorex	Ewing's sarcoma	Phase II
ImmTher® (MDP derivative)	Endorex	Ewing's sarcoma	Phase II
IRX-2 aloe vera extract (PBL supernatant)	Immuno-Rx	Head and neck squamous cell carcinoma	Phase II
MPL® lipid A derivative	Corixa	Vaccine adjuvant	Phase III
MPL® vaccine adjuvant	Corixa	Hepatitis B, herpes, human papilloma virus, malaria, respiratory syncytial virus	Phase III
MTP-PE	Jenner Technologies	Osteogenic sarcoma	Phase III
Multikine (PBL supernatant)	Cel-Sci	Cervical carcinoma, head and neck carcinoma	Phase II
ONO-4007	Oho Pharmaceutical	Vaccine adjuvant	Phase II
Peptide T	Advanced Immunity	Treatment of multifocal leukoencephalopathy (PML), HIV-associated cognitive impairment	Phase I
Tacrolimus	Fujisawa	Eczema	Approved

MDP was discovered based on the isolation of the minimally active substitute for intact BCG in Freund's adjuvant (Woodman *et al.*, 1992). Unfortunately, as with many of the polypeptides, the low molecular weight MDP has a short serum half-life and requires frequent high doses to be active. In addition, agents such as MDP are strongly pyrogenic, presumably due, in part, to their ability to induce IL-1. MDP has been incorporated into multilamellar vesicles (MLV) for higher stability and to facilitate monocytic phagocytosis of the MLV. To further stabilize the incorporation of MDP into MLV, lipophilic analogues of MDP such as MTP-PE have been developed. MTP-PE in liposomes is in an ongoing phase III clinical trial for osteosarcoma (Kleinerman, 1995; Kleinerman *et al.*, 1995). It has been studied in patients with resectable melanoma (Gianan and Kleinerman, 1998) and preclinically has shown protection to the mucosal epithelium from cytoreduction therapy (Killion *et al.*, 1996).

ImmTher is a liposome-encapsulated lipophilic disaccharide tripeptide derived from the MDP family that has the capacity to activate macrophages.

ImmTher has been evaluated in phase I and II trials in large tumors in advanced colorectal cancer patients. It is being evaluated in randomized phase II clinical trials for the treatment of Ewing's sarcoma and osteosarcoma (Worth *et al.*, 1999).

Bestatin (Ubenimex) is a potent inhibitor of aminopeptidase N and aminopeptidase B, (Aoyagi *et al.*, 1976), which was isolated from a culture filtrate of *Streptomyces olivoreticuli* during the search for specific inhibitors of enzymes present on the membrane of eukaryotic cells (Morahan *et al.*, 1989). Inhibitors of aminopeptidase activity are associated with macrophage activation and differentiation. Bestatin has shown significant therapeutic effects in several clinical trials (Urabe *et al.*, 1993). In a multi-institutional study, 101 patients with acute nonlymphocytic leukemia (ANLL) were randomized to receive bestatin or control (Yasumitsu *et al.*, 1990). Patients received 30 mg of bestatin orally after completion of induction and consolidation therapy, and concomitant with maintenance chemotherapy. Remission duration was prolonged in the bestatin group, although this difference did not reach statistical significance. However, overall survival was prolonged in the bestatin group. A confirmatory phase III trial in ANLL was reported which extended the observation to a significant prolongation of remission (Hiraoka *et al.*, 1992). In a multicenter study, bestatin was administered to acute leukemia and chronic myelogenous leukemia patients who did not develop any graft-versus-host disease (GVHD) within 30 days following bone marrow transplantation (BMT) (Goldstein, 1984). Bestatin-treated acute leukemia patients had an increased incidence of chronic low-grade GVHD compared with the control arm, as well as a lower relapse rate.

COMBINATION CHEMOTHERAPY, IMMUNOTHERAPY, AND CELLULAR THERAPY

Because cytokines have unique mechanisms of action, they are ideal candidates for combination therapy with chemotherapeutic agents. However, increased knowledge and consideration of the potential interactions is necessary for optimal clinical use. The use of high-dose chemotherapy (HDT) and stem cell rescue provides the ultimate in cytoreductive therapy and post-transplant immunotherapy. Stem cell transplantation (SCT) provides one of the few statistically supported demonstrations of clinical therapeutic efficacy by T cells based on the survival of patients receiving an allogeneic versus an autologous transplant (Storek and Storb, 2000). Thus, strategies to upregulate T-cell function following autologous SCT provide one focus for cytokine therapy. This is important because the return of immunological function in transplanted patients is slow and is accompanied by depressed numbers of

$CD4^+$ T cells, a low ratio of CD4 to CD8 T cells, and suppressed T-cell responses (Maraninchi *et al.*, 1987). The role of T cells in controlling neoplastic disease has been demonstrated in allotransplanted patients and is described as a graft-versus-tumor (GVT) reaction. A significantly higher risk of relapse is associated with the use of T-cell-depleted bone marrow cells or the clinical use of cyclosporine A (CSA) to prevent GVHD (Weiden *et al.*, 1979; Mitsuyasu *et al.*, 1986; Maraninchi *et al.*, 1987; Horowitz *et al.*, 1990). Similar relapse rates are observed in recipients of non-T-cell-depleted transplants receiving CSA (Peters *et al.*, 1998), suggesting that GVT cells are also removed by T-cell depletion (TCD). Clearly, GVHD can also have unfavorable effects on transplant-related mortality. In first remission, the decreased relapse rate with acute and/or chronic GVHD is more than offset by the increased risk of death from other causes. Consequently, patients with GVHD have a lower risk of treatment failure, but an increased risk of morbidity due to GVHD.

TCD markedly reduces the incidence of severe GVHD as discussed earlier. However, TCD is associated with an increased rate of severe, and often, fatal infections, a higher incidence of graft rejection, and an increased risk of disease recurrence. The increase in infectious complications is associated with the slow recovery of $CD4^+$ and $CD8^+$ T cells that occurs following transplantation and prior to the reconstitution of T-cell immunity, since the initial T-cell recovery that occurs with an unmanipulated stem cell product is associated with the T cells transplanted with the stem cells (Storek and Storb, 1000). Similarly, the increased graft failure following transplantation with a TCD product likely reflects the contribution of infused T cells toward the eradication of the remaining host T cells following the transplant preparative regimen. The increase in the leukemic relapse rate seen following TCD further stresses the importance of T cells in slowing tumor outgrowth. Further, TCD impacts not only tumor relapse, but also the incidence of infections, and it has now become a relatively common practice to undertake a donor leukocyte infusion (DLI) to reduce the incidence of graft loss, disease relapse, and secondary infections (Champlin *et al.*, 1990; Soiffer *et al.*, 1994; Dazzi *et al.*, 2000). However, DLI is also associated with an increased incidence of GVHD, and, thus, alternatives to TCD and DLI such as strategies that can induce antigen-specific tolerance shortly after allo-SCT are appealing insofar as they might prevent GVHD without resulting in a requirement for postgraft immunosuppression.

Adjuvant studies are focused on immunotherapy for transplant patients. The therapeutically intense preparative regimens, commonly referred to as HDT, are administered before transplantation, and a number of cytokine- and/or vaccine-associated protocols are given following transplant with a TCD product or intact product in an attempt to improve immunological function, particularly that directed against tumor cells. One therapeutic strategy is the use of antigens (vaccines) capable of inducing antigen-specific effector T cells. In addition, T cells from the donor may be stimulated *ex vivo*, expanded, and then reinfused. Strategies have also focused on the initiation of CTL, which can reduce the incidence of treatment-related lymphomas or infections such as cytomegalovirus (CMV) (Riddell *et al.*, 1992; Walter *et al.*, 1995).

Another approach to improving survival of cancer patients has been to use immunotherapy following HDT and stem cell transplantation to induce an autologous GVT response. Based on this strategy, studies using rIL-2 alone following BMT have shown an increase in NK-cell phenotype and function (Blaise *et al.*, 1990; Negrier *et al.*, 1991; Hiraoka *et al.*, 1992; Soiffer *et al.*, 1992.) In one such study (Negrier *et al.*, 1991), with 18 evaluable patients, three responses were observed. In another study, rIL-2 was infused following both autologous and allogeneic transplants for a median of 85 days at a dose of $2 \times 10^5 U/M^2$/day (Soiffer *et al.*, 1992). Toxicity was minimal, and the treatment could be undertaken in the outpatient setting via a Hickman catheter. In this study, no patient developed any signs of GVHD, hypertension, or pulmonary capillary leak syndrome. The treatment did not affect the absolute neutrophil count or hemoglobin level, although eosinophilia was observed. Despite the administration of this r IL-2 low dose, significant immunological changes were noted with a 5- to 40-fold increase in the number of NK cells. In addition, there was a significant augmentation of *ex vivo* cytotoxicity against K-562 and colon tumor targets. In a similar study, it was shown that following continuous infusion of rIL-2 in patients receiving autologous bone marrow transplant (AuBMT), the $CD3^+$ and $CD16^+$ cells secreted increased levels of IFN-γ and TNF following *in vitro* culture and there was a significant increase in serum levels of IFN-γ but not TNF (Soiffer *et al.*, 1992). Post-transplantation IL-2 administration has been extended to include the use of IL-2 or IL-2 and G-CSF for the mobilization of stem cells for transplantation (Burns *et al.*, 2000a, b; Toh *et al.*, 2000; Schiller *et al.*, 2001; Sosman *et al.*, 2001). The objective of using IL-2 as a mobilizing agent is to mobilize T cells or to change the population of T cells to ones that may have improved antitumor activity as well as the potential to reduce secondary infections. These studies are too immature to identify therapeutic efficacy, and the only conclusion that can be assessed at present is that mobilization with IL-2 occurs as originally reported in preclinical studies (Talmadge *et al.*, 1987).

Similar post-transplantation strategies with rIFN-α alone have been undertaken with a provisional observation of a reduced risk of relapse and an increase in myelosuppression (Meyers *et al.*, 1987; Klingemann *et al.*, 1991). In an early study of the prophylactic use of rIFN-α following allogeneic BMT, the Seattle group (Meyers *et al.*, 1987) found that adjuvant treatment with IFN-α had no effect on the probability or severity of CMV infections or GVHD in acute lymphocytic leukemia (ALL) patients who were in remission at the time of transplantation. In this large study, there was a significant reduction in the probability of relapse in the rIFN-α recipients ($p \leq 0.004$) compared with transplant patients who did not receive rIFN-α, although survival rates did not differ between the two groups. It was suggested that the administration of rIFN-α following transplantation reduced the risks of relapse but did not affect CMV infection, perhaps because rIFN-α was not initiated until a median of 18 days after transplantation and was not administered chronically. Ratanatharathorn *et al.* (1994) have extended the approach of inducing a GVT reaction to a combination study that utilized both CSA-induced GVHD and IFN-α augmentation of this effect in autologous transplant patients. Twenty-two patients were enrolled, of which 17 were considered evaluable. Thirteen of the patients who received recombinant human IFN-α2a developed GVHD regardless of whether they received CSA, whereas only 2 of the 4 patients who received CSA alone developed detectable GVHD. A patient receiving 1×10^6U/day of rhIFN-α2a concomitant with CSA showed a trend toward increased severity of clinical GVHD in comparison to patients receiving CSA alone ($p \leq 0.06$). The researchers concluded that IFN-α administration can be safely started on day 0 of autologons BMT and can induce autologous GVHD as a single agent with the potential to improve therapy.

In similar studies, Kennedy *et al.* (1994) treated women with advanced breast cancer with combined therapy of CSA for 28 days with 0.025 mg/m^2 of subcutaneous IFN-γ every other day on days 7 to 28 after HDT and autologons BMT. They observed that autologous GVHD developed in 56% of the patients, an incidence comparable to that observed earlier with CSA alone. The severity of GVHD was greater with CSA plus IFN-γ than with CSA alone: 16 patients required corticosteroid therapy for dermatological GVHD. Recently, IFN-γ therapy was administered following DLI to a patient who had received a matched sibling graft to treat first chronic phase CML (Leda *et al.*, 2001).

Further, several studies have used of IL-2 and IFN-α together as a combination immune modulatory approach following HDT in hematopoietic stem cell transplantation (Meehan *et al.*, 1999; Vivances *et al.*, 1999; Vik *et al.*, 2000). At present, the only valid conclusion from these studies is that combination IL-2/IFN-α therapy following HDT and SCT can be undertaken with acceptable toxicity. Note that strategies to induce an autologous GVT, while conceptually interesting, have not matured sufficiently to allow a discussion of their efficacy.

CONCLUSION

In the last 20 years, nonspecific immunostimulation has progressed from initial trials with crude microbial mixtures and extracts to more sophisticated uses with a large collection of targeted immunopharmacologically active compounds (only a few of which are discussed here) having diverse actions on the immune system. A body of immunopharmacological knowledge has evolved which shows substantial divergence from conventional pharmacology, particularly in terms of the relationship of dosing schedules to immunopharmacodynamics. This knowledge is important in evaluating agents and predicting appropriate use. While much remains to be learned, and many new compounds to be extracted and/or cloned, the future of immunotherapy seems bright. A number of the cytokines have been approved, as well as numerous supplemental indications (Gosse and Nelson, 1977), in the United States, Europe, and Asia. However, it is apparent that the combinations of cytokines and biological response modulators will have optimal activity when used as adjuvants with more traditional therapeutic modalities.

REFERENCES

Ahlin, A., Larfars, G., Elinder, G., Palmblad, J., and Gyllenhammar, H. (1999). Gamma interferon treatment of patients with chronic granulomatous disease is associated with augmented production of nitric oxide by polymorphonuclear neutrophils. *Clin. Diagn. Lab. Immunol.*, **6**, 420–424.

Alimena, G., Morra, E., Lazzarino, M., Liberati, A. M., Montefusco, E., Inverardi, D., Bernasconi, P., Mancini, M., Donti, E., and Grignani, F. (1988). Interferon alpha-2b as therapy for Ph'-positive chronic myelogenous leukemia: A study of 82 patients treated with intermittent or daily administration. *Blood*, **72**, 642–647.

Amery, W. K., and Bruynseels, J. P. (1992). Levamisole, the story and the lessons. *Int. J. Immunopharmacol.*, **14**, 481–486.

Aoyagi, T., Suda, H., Nagai, M., Ogawa, K., and Suzuki, J. (1976). Aminopeptidase activities on the surface of mammalian cells. *Biochim. Biophys. Acta*, **452**, 131–143.

Ashihara, E., Shimazaki, C., Sudo, Y., Kikuta, T., Hirai, H., Sumikuma, T., Yamagata, N., Goto, H., Inaba, T., Fujita, N., and Nakagawa, M. (1998). FLT-3 ligand mobilizes hematopoietic primitive and committed progenitor cells into blood in mice. *Eur. J. Haematol.*, **60**, 86–92.

Atkins, M. B., Robertson, M. J., Gordon, M., Lotze, M. T., DeCoste, M., Dubois, J. S., Ritz, J., Sandler, A. B., Edington, H. D., Garzone, P. D., Mier, J. W., Canning, C. M., Battiato, L., Tahara, H., and Sherman, M. L. (1997). Phase I evaluation of intravenous recombinant human interleukin 12 in patients with advanced malignancies. *Clin. Cancer Res.*, **3**, 409–417.

Bajetta, E., Del Vecchio, M., Mortarini, R., Nadeau, R., Rakhit, A., Rimassa, L., Fowst, C., Borri, A., Anichini, A., and Parmiani, G. (1998). Pilot study of subcutaneous recombinant human interleukin 12 in metastatic melanoma. *Clin. Cancer Res.*, **4**, 75–85.

Barouch, D. H., Santra, S., Steenbeke, T. D., Zheng, X. X., Perry, H. C., Davies, M. E., Freed, D. C., Craiu, A., Strom, T. B., Shiver, J. W., and Letvin, N. L. (1998). Augmentation and suppression of immune responses to an HIV-1 DNA vaccine by plasmid cytokine/Ig administration. *J. Immunol.*, **161**, 1875–1882.

Bertagnolli, M. M., Lin, B. Y., Young, D., and Herrmann, S. H. (1992). IL-12 augments antigen-dependent proliferation of activated T lymphocytes. *J Immunol.*, **149**, 3778–3783.

Black, P. L., Phillips, H., Tribble, H. R., Pennington, R. W., Schneider, M., and Talmadge, J. E. (1993). Antitumor response to recombinant murine interferon gamma correlates with enhanced immune function of organ-associated, but not recirculating cytolytic T lymphocytes and macrophages. *Cancer Immunol. Immunother.*, **37**, 299–306.

Blaise, D., Olive, D., Stoppa, A. M., Viens, P., Pourreau, C., Lopez, M., Attal, M., Jasmin, C., Monges, G., Mawas, C., Mannoni, P., Palmer, P., Franks, C., and Phillip, T. (1990). Hematologic and immunologic effects of the systemic administration of recombinant interleukin-2 after autologous bone marrow transplantation. *Blood*, **76**, 1092–1097.

Boggio, K., Nicoletti, G., Di Carlo, E., Cavallo, F., Landuzzi, L., Melani, C., Giovarelli, M., Rossi, I., Nanni, P., De Giovanni, C., Bouchard, P., Wolf, S., Modesti, A., Musiani, P., Lollini, P. L., Colombo, M. P., and Forni, G. (1998). Interleukin 12-mediated prevention of spontaneous mammary adenocarcinomas in two lines of HER-2/*neu* transgenic mice. *J Exp. Med.*, **188**, 589–596.

Bouloc, A., Bagot, M., Delaire, S., Bensussan, A., and Boumsell, L. (2000). Triggering CD101 molecule on human cutaneous dendritic cells inhibits T cell proliferation via IL-10 production. *Eur. J. Immunol.*, **30**, 3132–3139.

Brasel, K., McKenna, H. J., Morrissey, P. J., Charrier, K., Morris, A. E., Lee, C. C., Williams, D. E., and Lyman, S. D. (1996). Hematologic effects of Flt3 ligand in vivo in mice. *Blood*, **88**, 2004–2012.

Braun, S. E., Chen, K., Blazar, B. R., Orchard, P. J., Sledge, G., Robertson, M. J., Broxmeyer, H. E., and Cornetta, K. (1999). Flt3 ligand antitumor activity in a murine breast cancer model: A comparison with granulocyte–macrophage colony-stimulating factor and a potential mechanism of action. *Hum. Gene Ther.*, **10**, 2141–2151.

Brunda, M. J., Luistro, L., Hendrzak, J. A., Fountoulakis, M., Garotta, G., and Gately, M. K. (1995). Role of interferon-gamma in mediating the antitumor efficacy of interleukin-12. *J. Immunother. Emphasis Tumor Immunol.*, **17**, 71–77.

Bunn, P. A., Jr., Foon, K. A., Ihde, D. C., Longo, D. L., Eddy, J., Winkler, C. F., Veach, S. R., Zeffren, J., Sherwin, S., and Oldham, R. (1984). Recombinant leukocyte A interferon: An active agent in advanced cutaneous T-cell lymphomas. *Ann. Intern. Med.*, **101**, 484–487.

Burns, L. J., Weisdorf, D. J., DeFor, T. E., Repka, T. L., Ogle, K. M., Hummer, C., and Miller, J. S. (2000a). Enhancement of the anti-tumor activity of a peripheral blood progenitor cell graft by mobilization with interleukin 2 plus granulocyte colony-stimulating factor in patients with advanced breast cancer. *Exp. Hematol.*, **28**, 96–103.

Burns, L. J., Weisdorf, D. J., DeFor, T. E., Repka, T. L., Ogle, K. M., Hummer, C., and Miller, J. S. (2000b). Enhancement of the anti-tumor activity of a peripheral blood progenitor cell graft by mobilization with interleukin 2 plus granulocyte colony-stimulating factor in patients with advanced breast cancer, erratum. *Exp. Hematol.*, **28**, 352.

Cavallo, F., Signorelli, P., Giovarelli, M., Musiani, P., Modesti, A., Brunda, M. J., Colombo, M. P., and Forni, G. (1997). Antitumor efficacy of adenocarcinoma cells engineered to produce interleukin 12 (IL-12) or other cytokines compared with exogenous IL-12. *J. Natl. Cancer Inst.*, **89**, 1049–1058.

Cella, M., Facchetti, F., Lanzavecchia, A., and Colonna, M. (2000). Plasmacytoid dendritic cells activated by influenza virus and CD40L drive a potent TH1 polarization. *Nat. Immunol.*, **1**, 305–310.

Champlin, R., Ho, W., Gajewski, J., Feig, S., Burnison, M., Holley, G., Greenberg, P., Lee, K., Schmid, I., and Giorgi, J., (1990). Selective depletion of CD8+ T lymphocytes for prevention of graft-versus-host disease after allogeneic bone marrow transplantation. *Blood*, **76**, 418–423.

Chan, S. H., Perussia, B., Gupta, J. W., Kobayashi, M., Pospisil, M., Young, H. A., Wolf, S. F., Young, D., Clark, S. C., and Trinchieri, G. (1991). Induction of interferon gamma production by natural killer cell stimulatory factor: Characterization of the responder cells and synergy with other inducers. *J Exp. Med.*, **173**, 869–879.

Chen, K., Braun, S., Lyman, S., Fan, Y., Traycoff, C. M., Wiebke, E. A., Gaddy, J., Sledge, G., Broxmeyer, H. E., and Cornetta, K. (1997). Antitumor activity and immunotherapeutic properties of Flt3–ligand in a murine breast cancer model. *Cancer Res.*, **57**, 3511–3516.

Cohen, J. (1995). IL-12 deaths: Explanation and a puzzle. *Science*, **270**, 908.

Coughlin, C. M., Wysocka, M., Trinchieri, G., and Lee, W. M. (1997). The effect of interleukin 12 desensitization on the antitumor efficacy of recombinant interleukin 12. *Cancer Res.*, **57** 2460–2467.

Coughlin, C. M., Salhany, K. E., Gee, M. S., LaTemple, D. C., Kotenko, S., Ma, X., Gri, G., Wysocka, M., Kim, J. E., Liu, L., Liao, F., Farber, J. M., Pestka, S., Trinchieri, G., and Lee, W. M. (1998). Tumor cell responses to IFN-gamma affect tumorigenicity and response to IL-12 therapy and antiangiogenesis. *Immunity*, **9**, 25–34.

D'Andrea, A., Aste-Amezaga, M., Valiante, N. M., Ma, X., Kubin, M., and Trinchieri, G. (1993). Interleukin 10 (IL-10) inhibits human lymphocyte interferon gamma production by suppressing natural killer cell stimulatory factor/IL-12 synthesis in accessory cells. *J Exp. Med.*, **178**, 1041–1048.

Daro, E., Pulendran, B., Brasel, K., Teepe, M., Pettit, D., Lynch, D. H., Vremec, D., Robb, L., Shortman, K., McKenna, H. J., Maliszewski, C. R., and Maraskovsky, E. (2000). Polyethylene glycol–Modified GM-CSF Expands CD11b(High)CD11c(high) but not CD11b(low) CD11c(high) murine dendritic cells in vivo: A comparative analysis with Flt3 ligand. *J. Immunol.*, **165**, 49–58.

Davey, R. T., Jr., Chaitt, D. G., Albert, J. M., Piscitelli, S. C., Kovacs, J. A., Walker, R. E., Falloon, J., Polis, M. A., Metcalf, J. A., Masur, H., Dewar, R., Baseler, M., Fyfe, G., Giedlin, M. A., and Lane, H. C. (1999). A randomized trial of high-versus low-dose subcutaneous interleukin-2 outpatient therapy for early human immunodeficiency virus type 1 infection. *J. Infect. Dis.*, **179**, 849–858.

Dazzi, F., Szydlo, R. M., Craddock, C., Cross, N. C., Kaeda, J., Chase, A., Olavarria, E., Van Rhee, F., Kanfer, E., Apperley, J. F., and Goldman, J. M. (2000). Comparison of single-dose and escalating-dose regimens of donor lymphocyte infusion for relapse after allografting for chronic myeloid leukemia. *Blood*, **95**, 67–71.

Ebner, S., Ratzinger, G., Krosbacher, B., Schmuth, M., Weiss, A., Reider, D., Kroczek, R. A., Herold, M., Heufler, C., Fritsch, P., and Romani, N. (2001). Production of IL-12 by human monocyte-derived dendritic cells is optimal when the stimulus is given at the onset of maturation, and is further enhanced by IL-4. *J. Immunol.*, **166**, 633–641.

Ellenberg, S. S. (1993). Surrogate endpoints. *Br. J. Cancer*, **68**, 457–459.

Ellouz, F., Adam, A., Cirobaru, R., and Lederer, E. (1974). Minimal structural requirements for adjuvant activity of bacterial peptidoglycan derivatives. *Biochem. Biophys. Res. Commun.*, **59**, 1317.

Esche, C., Subbotin, V. M., Maliszewski, C., Lotze, M. T., and Shurin, M. R. (1998). Flt3 Ligand administration inhibits tumor growth in murine melanoma and lymphoma. *Cancer Res.*, **58**, 380–383.

Fay, J., Palucka, K., Pulendran, B., *et al.* (1999). In vivo mobilization of dendritic cell precursors in normal volunteers after Flt3–L administration. *Blood*, **94**, 379 (abst.).

Fehniger, T. A., Bluman, E. M., Porter, M. M., Mrozek, E., Cooper, M. A., Van Deusen, J. B., Frankel, S. R., Stock, W., and Caligiuri, M. A. (2000). Potential mechanisms of human natural killer cell expansion in vivo during low-dose IL-2 therapy. *J Clin. Invest.*, **106**, 117–124.

Fong, L., Hou, Y., Rivas, A., Benike, C., Yuen, A., Fisher, G. A., Davis, M. M., and Engleman, E. G. (2001). Altered peptide ligand vaccination with Flt3 ligand expanded dendritic cells for tumor immunotherapy. *Proc. Natl. Acad. Sci. USA*, **98**, 8809–8814.

Friedman-Kien, A. E., Eron, L. J., Conant, M., Growdon, W., Badiak, H., Bradstreet, P. W., Fedorczyk, D., Trout, J. R., and Plasse, T. F. (1988). Natural interferon alfa for treatment of condylomata acuminata. *JAMA*, **259**, 533–538.

Gasparetto, C., Rooney, B., Gasparetto, M., *et al.* (1999). Mobilization of dendritic cells from patients with breast cancer using Flt3-ligand and G-CSF or GM-CSF. *Blood*, 94, 636. (abstr.).

Gately, M. K., Desai, B. B., Wolitzky, A. G., Quinn, P. M., Dwyer, C. M., Podlaski, F. J., Familletti, P. C., Sinigaglia, F., Chizonnite, R., and Gubler, U. (1991). Regulation of human lymphocyte proliferation by a heterodimeric cytokine, IL-12 (cytotoxic lymphocyte maturation factor). *J. Immunol.*, **147**, 874–882.

Germann, T., Gately, M. K., Schoenhaut, D. S., Lohoff, M., Mattner, F., Fischer, S., Jin, S. C., Schmitt, E., and Rude, E. (1993). Interleukin-12/T cell stimulating factor, a cytokine with multiple effects on T helper type 1 (Th1) but not on Th_2 cells. *Eur. J. Immunol.*, **23**, 1762–1770.

Gianan, M. A., and Kleinerman, E. S. (1998). Liposomal muramyl tripeptide (CGP 19835A lipid) therapy for resectable melanoma in patients who were at high risk for relapse: An update. *Cancer Biother. Radiopharm.*, **13**, 363–368.

Goldstein, A. L. (1984). *Thymic Hormones and Lymphokines*. Plenum Press, New York.

Gollob, J. A., Schnipper, C. P., Orsini, E., Murphy, E., Daley, J. F., Lazo, S. B., Frank, D. A., Neuberg, D., and Ritz, J. (1998). Characterization of a novel subset of CD8(+) T cells that expands in patients receiving interleukin-12. *J Clin. Invest*, **102**, 561–575.

Gollob, J. A., Mier, J. W., Veenstra, K., McDermott, D. F., Clancy, D., Clancy, M., and Atkins, M. B. (2000). Phase I trial of twice-weekly intravenous interleukin 12 in patients with metastatic renal cell cancer or malignant melanoma: Ability to maintain IFN-gamma induction is associated with clinical response. *Clin. Cancer Res.*, **6**, 1678–1692.

Golomb, H. M., Fefer, A., Golde, D. W., Ozer, H., Portlock, C., Silber, R., Rappeport, J., Ratain, M. J., Thompson, J., Bonnem, E., Spiegel, R., Tensen, L., Burke, J. S., and Vardiman, J. W. (1988). Report of a multi-institutional study of 193 patients with hairy cell leukemia treated with interferon-alfa2b. *Semin. Oncol.*, **15**, 7–9.

Gosse, M. E., and Nelson, T. F. (1977). Approval times for supplemental indications for recombinant proteins. *Nat. Biotechnol.*, **15**, 130–134.

Guilhot, F., Chastang, C., Michallet, M., Guerci, A., Harousseau, J. L., Maloisel, F., Bouabdallah, R., Guyotat, D., Cheron, N., Nicolini, F., Abgrall, J. F., and Tanzer, J. (1997). Interferon Alfa-2b combined with cytarabine versus interferon alone in chronic myelogenous leukemia. French Chronic Myeloid Leukemia Study Group. *N. Engl. J. Med.*, **337**, 223–229.

Haaff, E. O., Caralona, W. J., and Ratliff, T. L. (1986a). Detection of interleukin-2 in the urine of patients with superficial bladder tumors after treatment with intravesical BCG. *J. Urol.*, **136**, 970.

Haaff, E. O., Dresner, S. M., Ratliff, T. L., and Catalona, W. J. (1986b). Two courses of intravesical bacillus Calmette–Guerin for transitional cell carcinoma of the bladder. *J. Urol.*, **136**, 820.

Hendrzak, J. A., and Brunda, M. J. (1995). Interleukin-12. Biologic activity, therapeutic utility, and role in disease. *Lab Invest.* **72**, 619–637.

Hendrzak, J. A., and Brunda, M. J. (1996). Antitumor and antimetastatic activity of interleukin-12. *Curr. Top. Microbiol. Immunol.*, **213** (Pt 3), 65–83.

Heslop, H. E., Gottlieb, D. J., Bianchi, A. C. M., Meager, A., Prentice, H. G., Mehta, A. B., Hoffbrand, A. V., and Brenner, M. K. (1989). In vivo induction of gamma interferon and tumor necrosis factor by interleukin-2 infusion following intensive chemotherapy or autologous marrow transplantation. *Blood*, **74**, 1374–1380.

Higuchi, C. M., Thompson, J. A., Petersen, F. B., Buckner, C. D., and Fefer, A. (1991). Toxicity and immunomodulatory effects of interleukin-2 after autologous bone marrow transplantation for hematologic malignancies. *Blood*, **77**, 2561–2568.

Hiraoka, A., Shibata, H., and Masaoka, T. (1992). Immunopotentiation with Ubenimex for prevention of leukemia relapse after allogeneic BMT. The Study Group of Ubenimex for BMT. *Transplant. Proc*, **24**, 3047–3048.

Hladik, F., Tratkiewicz, J. A., Tilg, H., Vogel, W., Schwulera, U., Kronke, M., Aulitzky, W. E., and Huber, C. (1994). Biologic activity of low dosage IL-2 treatment in vivo. Molecular assessment of cytokine network interaction. *J. Immunol.*, **153**, 1449–1454.

Holcombe, R. F., Milovanovic, T., Stewart, R. M., and Brodhag, T. M. (2001). Investigating the role of immunomodulation for colon cancer prevention: Results of an in vivo dose escalation trial of Levamisole with immunologic endpoints. *Cancer Detect. Prev.*, **25**, 183–191.

Holden, C. (1993). FDA Okays surrogate markers. *Science*, **259**, 32.

Horowitz, M. M., Gale, R. P., Sondel, P. M., Goldman, J. M., Kersey, J., Kolb, H. J., Rimm, A. A., Ringden, O., Rozman, C., Speck, B., Truitt, R. L., Swaan, F. E., and Bortin, M. M. (1990). Graft-versus leukemia reactions after bone marrow transplantation. *Blood*, **75**, 555–562.

The International Chronic Granulomatous Disease Cooperative Study Group. (1991). A controlled trial of interferon gamma to prevent infection in chronic granulomatous disease. *N. Engl. J. Med.*, **324**, 509–516.

Jackson, J. D., Yan, Y., Brunda, M. J., Kelsey, L. S., and Talmadge, J. E. (1995). IL-12 enhances peripheral hematopoiesis in vivo. *Blood*, **85**, 2371–2376.

Jacobsen, S. E., Veiby, O. P., and Smeland, E. B. (1993). Cytotoxic lymphocyte maturation factor (interleukin 12) Is a synergistic growth factor for hematopoietic stem cells. *J. Exp. Med.*, **178**, 413–418.

Jacobson, E. L., Pilaro, F., and Smith, K. A. (1996). Rational interleukin 2 therapy for HIV positive individuals: Daily low doses enhance immune function without toxicity. *Proc. Natl. Acad. Sci. USA*, **93**, 10405–10410.

Jaffe, H. S., and Herberman, R. B. (1988). Rationale for recombinant human interferon-gamma adjuvant immunotherapy for cancer. *J. Natl. Cancer Inst.*, **80**, 616–618.

Jen, J. F., Glue, P., Ezzet, F., Chung, C., Gupta, S. K., Jacobs, S., and Hajian, G. Population pharmacokinetic analysis of pegylated interferon alfa-2b and interferon alfa-2b in patients with chronic hepatitis C. *Clin. Pharmacol. Ther.*, **69**, 407–421.

Kaempfer, R., Gerez, L., Farbstein, H., Madar, L., Hirschman, O., Nussinovich, R., and Shapiro, A. (1996). Prediction of response to treatment in superficial bladder carcinoma through pattern of interleukin-2 gene expression. *J. Clin. Oncol.*, **14**, 1778–1786.

Kennedy, M. J., Vogelsang, G. B., Jones, R. J., Farmer, E. R., Hess, A. D., Altomonte, V., Huelskamp, A. M., and Davidson, N. E. (1994). Phase I trial of interferon gamma to potentiate cyclosporine-induced graft-versus-host disease in women undergoing autologous bone marrow transplantation for breast cancer. *J. Clin. Oncol.*, **12**, 249–257.

Killion, J. J., Bucana, C. D., Radinsky, R., Dong, Z., O'Reilly, T., Bilbe, G., Tarcsay, L., and Fidler, I. J. (1996). Maintenance of intestinal epithelium structural integrity and mucosal leukocytes during chemotherapy by oral administration of muramyl tripeptide phosphatidylethanolamine. *Cancer Biother. Radiopharm.*, **11**, 363–371.

Kim, J. J., Yang, J., Manson, K. H., and Weiner, D. B. (2001). Modulation of antigen-specific cellular immune responses to DNA vaccination in rhesus macaques through the use of IL-2, IFN-gamma, or IL-4 gene adjuvants. *Vaccine*, **19**, 2496–2505.

Kirkwood, J. M., Strawderman, M. H., Ernstoff, M. S., Smith, T. J., Borden, E. C., and Blum, R. H. (1996). Interferon Alfa-2b adjuvant therapy of high-risk resected cutaneous melanoma: The Eastern Cooperative Oncology Group Trial EST 1684. *J. Clin. Oncol.*, **14**, 7–17.

Kleinerman, E. S. (1995). Biologic therapy for osteosarcoma using liposome-encapsulated muramyl tripeptide. *Hematol. Oncol. Clin. North Am.*, **9**, 927–938.

Kleinerman, E. S., Meyers, P. A., Raymond, A. K., Gano, J. B., Jia, S. F., and Jaffe, N. (1995). Combination therapy with ifosfamide and liposome-encapsulated muramyl tripeptide:

Tolerability, toxicity, and immune stimulation. *J. Immunother. Emphasis. Tumor Immunol.*, **17**, 181–193.

Klingemann, H. G., Grigg, A. P., Wilkie-Boyd, K., Barnett, M. J., Eaves, A. C., Reece, D. E., Shepherd, J. D., and Phillips, G. L. (1991). Treatment with recombinant interferon (alpha-2b) early after bone marrow transplantation in patients at high risk for relapse. *Blood*, **78**, 3306–3311.

Kohl, S., Sigaroudinia, M., Charlebois, E. D., and Jacobson, M. A. (1996). Interleukin-12 administered in vivo decreases human NK cell cytotoxicity and antibody-dependent cellular cytotoxicity to human immunodeficiency virus-infected cells. *J. Infect. Dis.*, **174**, 1105–1108.

Kovacs, J. A., Baseler, M., Dewar, R. J., Vogel, S., Davey, R. T., Jr., Falloon, J., Polis, M. A., Walker, R. E., Stevens, R., and Salzman, N. P. (1995). Increases in CD4 T lymphocytes with intermittent courses of interleukin-2 in patients with human immunodeficiency virus infection. A preliminary study. *N. Engl. J. Med.*, **332**, 567–575.

Kurzawa, H., Wysocka, M., Aruga, E., Chang, A. E., Trinchieri, G., and Lee, W. M. (1998). Recombinant interleukin 12 enhances cellular immune responses to vaccination only after a period of suppression. *Cancer Res.*, **58**, 491–499.

Kusakabe, K., Xin, K. Q., Katoh, H., Sumino, K., Hagiwara, E., Kawamoto, S., Okuda, K., Miyagi, Y., Aoki, I., Nishioka, K., Klinman, D., and Okuda, K. (2000). The timing of GM-CSF expression plasmid administration influences the Th1/Th2 response induced by an HIV-1-specific DNA vaccine. *J. Immunol.*, **164**, 3102–3111.

Lage, J. M., Bauer, W. C., Kelley, D. R., Ratliff, T. L., and Catalona, W. J. (1986). Histological parameters and pitfalls in the interpretation of bladder biopsies in bacillus Calmette–Guerin treatment of superficial bladder cancer. *J. Urol.*, **135**, 916.

Lane, H. C., Feinberg, J., Davey, V., Deyton, L., Baseler, M., Manischewitz, J., Masur, H., Kovacs, J. A., Herpin, B., Walker, R., Metcalf, J. A., Salzman, N., Quinnan, G., and Fauci, A. S. (1988). Anti-retroviral effects of interferon-alpha in AIDS-associated Kaposi's sarcoma. *Lancet*, **2**, 1218–1222.

Langenkamp, A., Messi, M., Lanzavecchia, A., and Sallusto, F. (2000). Kinetics of dendritic cell activation: Impact on priming of TH1, TH2 and nonpolarized T cells. *Nat. Immunol.*, **1**, 311–316.

Lebsack, M. E., McKenna, J. A., Hoek, R., *et al.* (1997). Safety of Flt3 ligand in healthy volunteers. *Blood* 90, 170 (abstr.).

Leda, M., Ladon, D., Pieczonka, A., Boruczkowski, D., Jolkowska, J., Witt, M., and Wachowiak, J. (2001). Donor lymphocyte infusion followed by interferon-alpha plus low dose cyclosporine A for modulation of donor CD3 cell activity with monitoring of minimal residual disease and cellular chimerism in a patient with first hematologic relapse of chronic myelogenous leukemia after allogeneic bone marrow transplantation. *Leuk. Res.*, **25**, 353–357.

Leonard, J. P., Sherman, M. L., Fisher, G. L., Buchanan, L. J., Larsen, G., Atkins, M. B., Sosman, J. A., Dutcher, J. P., Vogelzang, N. J., and Ryan, J. L. (1997). Effects of single-dose interleukin-12 exposure on interleukin-12-associated toxicity and interferon-gamma production. *Blood*, **90**, 2541–2548.

Lissoni, P., Rovelli, F., Giani, L., Fumagalli, L., and Mandala, M. (1998). Immunomodulatory effects of IL-12 in relation to the pineal endocrine function in metastatic cancer patients. *Nat. Immun.*, **16**, 178–184.

Lotze, M. T., Chang, A. E., Seipp, C. A., Simpson, C., Vetto, J. T., and Rosenberg, S. A. (1986). High-dose recombinant interleukin 2 in the treatment of patients with disseminated cancer. Responses, treatment-related morbidity and histologic findings. *JAMA*, **256**, 3117–3124.

Lyman, S. D. (1998). Biologic effects and potential clinical applications of Flt3 ligand. *Curr. Opin. Hematol.*, **5**, 192–196.

Lynch, D. H. (1998). Induction of dendritic cells (DC) by Flt3 ligand (FL) promotes the generation of tumors-specific immune responses in vivo. *Crit. Rev. Immunol.*, **18**, 99–107.

Lynch, D. H., Andreasen, A., Maraskovsky, E., Whitmore, J., Miller, R. E., and Schuh, J. C. L. (1997). Flt3 ligand induces tumor regression and antitumor immune responses in vivo. *Nat. Med.*, **3**, 625–631.

Maldonado-Lopez, R., De Smedt, T., Michel, P., Godfroid, J., Pajak, B., Heirman, C., Thielemans, K., Leo, O., Urbain, J., and Moser, M. (1999a). CD8α+ and CD8α– subclasses of dendritic cells direct the development of distinct T helper cells in vivo. *J. Exp Med.*, **189**, 587–592.

Maldonado-Lopez, R., De Smedt, T., Pajak, B., Heirman, C., Thielemans, K., Leo, O., Urbain, J., Maliszewski, C. R., Moser, M. (1999b). Role of CD8α+ and CD8α– dendritic cells in the induction of primary immune responses in vivo. *J. Leukocyte Biol.*, **66**, 242–246.

Maluish, A. E., Urba, W. J., Longo, D. L. O., Overton, W. R., Coggin, D., Crisp, E. R., Williams, R., Sherwin, S. A., Gordon, K., and Steis, R. G. (1988). The determination of an immunologically active dose of interferon-gamma in patients with melanoma. *J. Clin. Oncol.*, **6**, 434–445.

Maraninchi, D., Gluckman, E., Blaise, D., Guyotat, D., Rio, B., Pico, J., Leblond, V., Michallet, M., Dreyfus, F., and Ifrah, N. (1987). Impact of T-cell depletion on outcome of allogeneic bone marrow transplantation for standard-risk leukaemias. *Lancet*, **2**, 175–178.

Maraskovsky, E., Roux, E., Teepe, M., Braddy, S., Hoek, J., Lebsack, M., and McKenna, H. J. (1997). Flt3 ligand increases peripheral blood dendritic cells in healthy volunteers. Presented at American Society of Hematology meeting. Abstr. 25850.

Maraskovsky, E., Brasel, K., Teepe, M., Roux, E. R., Lyman, S. D., Shortman, K., and McKenna, H. J. (1996). Dramatic increase in the numbers of functionally mature dendritic cells in Flt3 ligand-treated mice: Multiple dendritic cell subpopulations identified. *J. Exp. Med.*, **184**, 1953–1962.

Maraskovsky, E., Daro, E., Roux, E., Teepe, M., Maliszewski, C. R., Hoek, J., Caron, D., Lebsack, M. E., and McKenna, H. J. (2000). In vivo generation of human dendritic cell subsets by Flt3 ligand. *Blood*, **96**, 878–884.

Meehan, K. R., Arun, B., Gehan, E. A., Berberian, B., Sulica, V., Areman, E. M., Mazumder, A., and Lippman, M. E. (1999). Immunotherapy with interleukin-2 and alpha-interferon after IL-2-activated hematopoietic stem cell transplantation for breast cancer. *Bone Marrow Transplant*, **23**, 667–673.

Meyers, J. D., Flournoy, N., Sanders, J. E., McGuffin, R. W., Newton, B. A., Fisher, L. D., Lum, L. G., Appelbaum, F. R., Doney, K., Sullivan, K. M., Storb, R., Buckner, C. D., and Thomas, E. D. (1987). Prophylactic use of human leukocyte interferon after allogeneic marrow transplantation. *Ann. Intern. Med.*, **107**, 809–816.

Mier, J. W., Vachino, G., van der Meer, J. W., Numerof, R. P., Adams, S., Cannon, J. G., Bernheim, H. A., Atkins, M. B., Parkinson, D. R., and Dinarello, C. A. (1988). Induction of circulating tumor necrosis factor (TNF-alpha) as the mechanism for the febrile response to interleukin-2 (IL-2) in cancer patients. *J. Clin. Immunol.*, **8**, 426–432.

Mihich, E. (1986). Future perspectives for biological response modifiers: A viewpoint. *Semin. Oncol.*, **13**, 234–254.

Miller, K. D., Spooner, K., Herpin, B. R., Rock-Kress, D., Metcalf, J. A., Davey, R. T., Jr., Falloon, J., Kovacs, J. A., Polis, M. A., Walker, R. E., Masur, H., and Lane, H. C. (2001). Immunotherapy of HIV-infected patients with intermittent interleukin-2: Effects of cycle frequency and cycle duration on degree of CD4(+) T-lymphocyte expansion. *Clin. Immunol.*, **99**, 30–42.

Misset, J. L., Mathe, G., Gastiaburu, J., Goutner, A., Dorval, T., Gouveia, J., Schwarzenberg, L., Machover, D., Ribaud, P., and de Vassal, F. (1982). Treatment of leukemias and lymphomas by interferons: II. Phase II of the trial treatment of chronic lymphoid leukemia by human interferon alfa. *Biomed. Pharmacother.*, **39**, 112–116.

Mitsuyasu, R. T., Champlin, R. E., Gale, R. P., Ho, W. G., Lenarsky, C., Winston, D., Selch, M., Elashoff, R., Giorgi, J. V., Wells, J., Terasaki, P., Billing, R., and Feig, S. (1986). Treatment of donor bone marrow with monoclonal anti-T-cell antibody and complement for the prevention of graft-versus-host disease. A prospective, randomized, double-blind trial. *Ann. Intern. Med.*, **105**, 20–26.

Morahan, P. S., Edelson, P. J., and Gass, K. (1980). Changes in macrophage ectoenzymes associated with anti-tumor activity. *J. Immunol.*, **125**, 1312–1317.

Morse, M. A., Nair, S., Fernandez-Casal, M., Deng, Y., St Peter, M., Williams, R., Hobeika, A., Mosca, P., Clay, T., Cumming, R. I., Fisher, E., Clavien, P., Proia, A. D., Niedzwiecki, D., Caron, D., and Lyerly, H. K. (2000). Preoperative mobilization of circulating dendritic cells by Flt3 ligand administration to patients with metastatic colon cancer. *J. Clin. Oncol.*, **18**, 3883–3893.

Motzer, R. J., Rakhit, A., Thompson, J. A., Nemunaitis, J., Murphy, B. A., Ellerhorst, J., Schwartz, L. H., Berg, W. J., and Bukowski, R. M. (2001). Randomized multicenter phase II trial of subcutaneous recombinant human interleukin-12 versus interferon-alpha 2a for patients with advanced renal cell carcinoma. *J. Interferon Cytokine Res.* **21**, 257–263.

Mu, J., Zou, J. P., Yamamoto, N., Tsutsui, T., Tai, X. G., Kobayashi, M., Herrmann, S., Fujiwara, H., and Hamaoka, T. (1995). Administration of recombinant interleukin 12 prevents outgrowth of tumor cells metastasizing spontaneously to lung and lymph nodes. *Cancer Res.*, **55**, 4404–4408.

Mutch, R. S., and Hutson, P. R. (1991). Levamisole in the adjuvant treatment of colon cancer. *Clin. Pharm.*, **10(2)**, 95–109.

Nastala, C. L., Edington, H. D., McKinney, T. G., Tahara, H., Nalesnik, M. A., Brunda, M. J., Gately, M. K., Wolf, S. F., Schreiber, R. D., and Storkus, W. J. (1994). Recombinant IL-12 administration induces tumor regression in association with IFN-gamma production. *J. Immunol.*, **153**, 1697–1706.

Naume, B., Johnsen, A. C., Espevik, T., and Sundan, A. (1993). Gene expression and secretion of cytokines and cytokine receptors from highly purified CD56+ natural killer cells stimulated with interleukin-2, interleukin-7 and interleukin-12. *Eur. J. Immunol*, **23**, 1831–1838.

Nauts, H. C. (1975). *The Bibilography of Reports Concerning the Experimental Clinical Use of Coley Toxins.* Cancer Research Institute, New York.

Negrier, S., Ranchere, J. Y., Phillip, I., Merrouche, Y., Biron, P., Blaise, D., Attal, M., Rebattu, P., Clavel, M., Pourreau, C., Palmer, P., Favrot, M., Jasmin, C., Maraninchi, D., and Phillip, T. (1991). Intravenous interleukin-2 just after high dose BCNU and autologous bone marrow transplantation. Report of a multicentric French pilot study. *Bone Marrow Transplant.*, **8**, 259–264.

Nishimura, T., Watanabe, K., Lee, U., Yahata, T., Ando, K., Kimura, M., Hiroyama, Y., Kobayashi, M., Herrmann, S. H., and Habu, S. (1995). Systemic in vivo antitumor activity of interleukin-12 against both transplantable and primary tumor. *Immunol. Lett.*, **48**, 149–152.

Noguchi, Y., Richards, E. C., Chen, Y. T., and Old, L. J. (1995). Influence of interleukin 12 on p53 peptide vaccination against established meth A sarcoma. *Proc. Natl. Acad. Sci. USA*, **92**, 2219–2223.

O'Connell, M. J., Colgan, J. P., Oken, M. M., Ritts, R. E., Jr., Kay, N. E., and Itri, L. M. (1986). Clinical trial of recombinant leukocyte A interferon as initial therapy for favorable histology non-Hodgkin's lymphomas and chronic lymphocytic leukemia. An Eastern Cooperative Oncology Group Pilot Study. *J. Clin. Oncol.*, **4**, 128–136.

Ogawa, M., Yu, W. G., Umehara, K., Iwasaki, M., Wijesuriya, R., Tsujimura, T., Kubo, T., Fujiwara, H., and Hamaoka, T. (1998). Multiple roles of interferon-gamma in the mediation of interleukin 12-induced tumor regression. *Cancer Res.*, **58**, 2426–2432.

Orange, J. S., Salazar-Mather, T. P., Opal, S. M., Spencer, R. L., Miller, A. H., McEwen, B. S., and Biron, C. A. (1995). Mechanism of interleukin 12-mediated toxicities during experimental viral infections: Role of tumor necrosis factor and glucocorticoids. *J. Exp. Med.*, **181**, 901–914.

Pandolfi, F., Pierdominici, M., Marziali, M., Livia, B. M., Antonelli, G., Galati, V., D'Offizi, G., and Aiuti, F. (2000). Low-dose IL-2 reduces lymphocyte apoptosis and increases naive CD4 cells in HIV-1 patients treated with HAART. *Clin. Immunol.*, **94**, 153–159.

Parajuli, P., Mosley, R. L., Pisarev, V., Chavez, J., Ulrich, A., Varney, M., Singh, R. K., and Talmadge, J. E. (2001). Flt3 Ligand and granulocyte-macrophage colony-stimulation factor preferentially expand and stimulate different dendritic cell and T cell subsets. *Exp. Hematol.*, **29**: 1185–1193.

Parajuli, P., Pisarev, V. M., Sublet, J., Steffel, A., Varney, M., Singh, R. K., and Talmadge, J. E. (2001). Immunization with wild-type p53 gene sequences co-administered with Flt3 ligand, induces an antigen-specific type 1 T cell response. *Cancer Res.*, **61**: 8227–8234.

Patterson, S. (2000). Flexibility and cooperation among dendritic cells. *Nat. Immunol.*, **1**, 273–274.

Peron, J. M., Esche, C., Subbotin, V. M., Maliszewski, C., Lotze, M. T., and Shurin, M. R. (1998). FLT3-Ligand Administration Inhibits Liver Metastases: Role of NK Cells. *J. Immunol.*, **161**, 6164–6170.

Peters, C., Minkov, M., Gadner, H., Klingebiel, T., and Niethammer, D. (1998). Proposal for Standard Recommendations for Prophylaxis of Graft-Versus-Host Disease in Children. European Group for Blood and Marrow Transplantation (EBMT) Working Party Paediatric Diseases and the International Study Committee of the BFM Family, Subcommittee on Bone Marrow Transplantation (IBFM-STG). *Bone Marrow Transplant.*, **21** (Suppl. 2), S57–S60.

Pfeffer, L. M., Dinarello, C. A., Herberman, R. B., Williams, B. R., Borden, E. C., Bordens, R., Walter, M. R., Nagabhushan, T. L., Trotta, P. P., and Pestka, S. (1998). Biological properties of recombinant alpha-interferons: 40th anniversary of the discovery of interferons. *Cancer Res.*, **58**, 2489–2499.

Pinsky, C. M., Camacho, F. J., Kerr, D., Geller, N. L., Klein, F. A., Herr, H. A., Whitmore, W. F., and Oettgen, H. F. (1985). Intravesical administration of bacillus Calmette–Guerin in patients with recurrent superficial carcinoma of the urinary bladder: Report of a prospective, randomized trial. *Cancer Treat. Rep.*, **69**, 47.

Pisarev, V. M., Parajuli, P., Mosley, R. L., Sublet, J., Kelsey, L., Sarin, P. S., Zimmerman, D. H., Winship, M. D., and Talmadge, J. E. (2000). Flt3 ligand enhances the immunogenicity of a Gag-based HIV-1 vaccine. *Int. J. Immunopharmacol.*, **22**, 865–876.

Porschen, R., Bermann, A., Loffler, T., Haack, G., G., Rettig, K., Anger, Y., and Strohmeyer, G. (2001). Fluorouracil plus leucovorin as effective adjuvant chemotherapy in curatively resected stage III colon cancer: Results of the trial AdjCCA-01. *J Clin. Oncol.*, **19**, 1787–1794.

Pulendran, B., Lingappa, J., Kennedy, M. K., Smith, J., Teepe, M., Rudensky, A., Maliszewski, C. R., Maraskovsky, E. (1997). Developmental pathways of dendritic cells in vivo: Distinct function, phenotype, and localization of dendritic cell subsets in Flt3 ligand-treated mice. *J. Immunol.*, **159**, 2222–2231.

Pulendran, B., Burkeholder, S., Kraus, E., *et al.* (1999a). Differential mobilization of distinct DC subsets in vivo by Flt3-ligand and G-CSF. *Blood*, **94**, 213a.

Pulendran, B., Smith, J. L., Caspary, G., Brasel, K., Pettit, D., Maraskovsky, E., and Maliszewski, C. R. (1999b). Distinct dendritic cell subsets differentially regulate the class of immune response in vivo. *Proc. Natl. Acad. Sci. USA*, **96**, 1036–1041.

Pulendran, B., Banchereau, J., Burkeholder, S., Kraus, E., Guinet, E., Chalouni, C., Caron, D., Maliszewski, C., Davoust, J., Fay, J., and Palucka, K. (2000). Flt3-ligand and granulocyte colony-stimulating factor mobilize distinct human dendritic cell subsets in vivo. *J. Immunol.*, **165**, 566–572.

Quesada, J. R., Reuben, J., Manning, J. T., Hersh, E. M., and Gutterman, J. U. (1984). Alpha interferon for induction of remission in hairy-cell leukemia. *N. Eng. J. Med.*, **310**, 15–18.

Rakhmilevich, A. L., Janssen, K., Turner, J., Culp, J., and Yang, N. S. (1997). Cytokine gene therapy of cancer using gene gun technology: Superior antitumor activity of interleukin-12. *Hum. Gene Ther.*, **8**, 1303–1311.

Rao, J. B., Chamberlain, R. S., Bronte, V., Carroll, M. W., Irvine, K. R., Moss, B., Rosenberg, S. A., and Restifo, N. P. (1996). IL-12 is an effective adjuvant to recombinant vaccinia virus-based tumor vaccines: Enhancement by simultaneous B7–1 expression. *J. Immunol.*, **156**, 3357–3365.

Rappold, I., Ziegler, B. L., Kohler, I., Marchetto, S., Rosnet, O., Birnbaum, D., Simmons, P. J., Zannettino, A. C., Hill, B., Neu, S., Knapp, W., Alitalo, R., Alitalo, K., Ullrich, A., Kanz, L., and Buhring, H. J. (1997). Functional and phenotypic characterization of cord blood and bone marrow subsets expressing Flt3 (CD135) receptor tyrosine kinase. *Blood*, **90**, 111–125.

Ratanatharathorn, V., Uberti, J., Karanes, C., Lum, L. G., Abella, E., Dan, M. E., Hussein, M., and Sensenbrenner, L. L. (1994). Phase I study of alpha-interferon augmentation of cyclosporine-induced graft versus host disease in recipients of autologous bone marrow transplantation. *Bone Marrow Transplant.*, **13**, 625–630.

Riddell, S. R., Watanabe, K. S., Goodrich, J. M., Li, C. R., Agha, M. E., and Greenberg, P. D. (1992). Restoration of viral immunity in immunodeficient humans by the adoptive transfer of T cell clones. *Science*, **257**, 238–241.

Rissoan, M. C., Soumelis, V., Kadowaki, N., Grouard, G., Briere, F., de Waal, M. R., and Liu, Y. J. (1999). Reciprocal control of T helper cell and dendritic cell differentiation. *Science*, **283**, 1183–1186.

Robertson, M. J., Soiffer, R. J., Wolf, S. F., Manley, T. J., Donahue, C., Young, D., Herrmann, S. H., and Ritz, J. (1992). Response of human natural killer (NK) cells to NK cell stimulatory factor (NKSF): Cytolytic activity and proliferation of NK cells are differentially regulated by NKSF. *J. Exp. Med.*, **175**, 779–788.

Robertson, M. J., Cameron, C., Atkins, M. B., Gordon, M. S., Lotze, M. T., Sherman, M. L., and Ritz, J. (1999). Immunological effects of interleukin 12 administered by bolus intravenous injection to patients with cancer. *Clin. Cancer Res.*, **5**, 9–16.

Robinson, S., Mosley, R. L., Parajuli, P., Pisarev, V., Sublet, J., Ulrich, A., and Talmadge, J. (2000). Comparison of the hematopoietic activity of Flt-3 ligand and granulocyte–macrophage colony-stimulating factor acting alone or in combination. *J. Hematother. Stem Cell Res.*, **9**, 711–720.

Rodolfo, M., and Colombo, M. P. (1999). Interleukin-12 As an adjuvant for cancer immunotherapy. *Methods* **19**, 114–120.

Rosenberg, S. A., Lotze, M. T., Yang, J. C., Topalian, S. L., Chang, A. E., Schwartzentruben, D. J., Aebersold, P., Leitman, S., Linehan, W. M., and Seipp, C. A. (1993). Prospective randomized trial of high-dose interleukin-2 alone or in conjunction with lymphokine-activated killer cells for the treatment of patients with advanced cancer. *J. Natl. Cancer Inst.*, **85**, 622–632.

Russell, I. J., Michalek, J. E., Kang, Y. K., and Richards, A. B. (1999). Reduction of morning stiffness and improvement in physical function in fibromyalgia syndrome patients treated sublingually with low doses of human interferon-alpha. *J. Interferon Cytokine Res.*, **19**, 961–968.

Schiff, D. E., Rae, J., Martin, T. R., Davis, B. H., and Curnutte, J. T. (1997). Increased phagocyte Fc gammaRI expression and improved Fc gamma-receptor-mediated phagocytosis after in vivo recombinant human interferon-gamma treatment of normal human subjects. *Blood*, **90**, 3187–3194.

Schiller, G., Wong, S., Lowe, T., Snead, G., Paquette, R., Sawyers, C., Wolin, M., Kunkel, L., Ting, L., Li, G., and Territo, M. (2001). Transplantation of IL-2-mobilized autologous peripheral blood progenitor cells for adults with acute myelogenous leukemia in first remission. *Leukemia*, **15**, 757–763.

Shaw, S. G., Maung, A. A., Steptoe, R. J., Thomson, A. W., and Vujanovic, N. L. (1998). Expansion of functional NK cells in multiple tissue compartments of mice treated with Flt3-ligand: Implications for anti-cancer and anti-viral therapy. *J. Immunol.*, **161**, 2817–2824.

Ship, J. A., Fox, P. C., Michalek, J. E., Cummins, M. J., and Richards, A. B. (1999). Treatment of primary sjøgren's syndrome with low-dose natural human interferon-alpha administered by the oral mucosal route: A phase II clinical trial. IFN Protocol Study Group. *J. Interferon Cytokine Res.*, **19**, 943–951.

Smith, K. A. (2001). Low-dose daily interleukin-2 immunotherapy: Accelerating immune restoration and expanding HIV-specific T-cell immunity without toxicity. *AIDS*, **15** (suppl. 2), S28–S35.

Soiffer, R. J., Murray, C., Cochran, K., Cameron, C., Wang, E., Schow, P. W., Daley, J. F., and Ritz, J. (1992). Clinical and immunologic effects of prolonged infusion of low-dose recombinant interleukin-2 after autologous and T cell-depleted allogeneic bone marrow transplantation. *Blood*, **79**, 517–526.

Soiffer, R. J., Murray, C., Gonin, R., and Ritz, J. (1994). Effect of low-dose interleukin-2 on disease relapse after T-cell-depleted allogeneic bone marrow transplantation. *Blood*, **84**, 964–971.

Somlo, G., Sniecinski, I., Odom-Maryon, T., Nowicki, B., Chow, W., Hamasaki, V., Leong, L., Margolin, K., Morgan, R., Jr., Raschko, J., Shibata, S., Tetef, M., Molina, A., Berenson, R. J., Forman, S. J., and Doroshow, J. H. (1997). Effect of $CD34^+$ selection and various schedules of stem cell reinfusion and granulocyte colony-stimulating factor priming on hematopoietic recovery after high-dose chemotherapy for breast cancer. *Blood*, **89**, 1521–1528.

Sosman, J. A., Stiff, P., Moss, S. M., Sorokin, P., Martone, B., Bayer, R., van Besien, K., Devine, S., Stock, W., Peace, D., Chen, Y., Long, C., Gustin, D., Viana, M., and Hoffman, R. (2001). Pilot trial of interleukin-2 with granulocyte colony-stimulating factor for the mobilization of progenitor cells in advanced breast cancer patients undergoing high-dose chemotherapy: Expansion of immune effectors within the stem-cell graft and post-stem-cell infusion. *J Clin. Oncol.*, **19**, 634–644.

Storek, J., and Storb, R. (2000). T-cell reconstitution after stem-cell transplantation – By which organ? *Lancet*, **355**, 1843–1844.

Talmadge, J. E., and Herberman, R. B. (1986). The preclinical screening laboratory. Evaluation of immunomodulatory and therapeutic properties of biological response modifiers. *Cancer Treat Res.*, **70**, 171–182.

Talmadge, J. E., Tribble, H. R., Pennington, R. W., Phillips, H., and Wiltrout, R. H. (1987). Immunomodulatory and immunotherapeutic properties of recombinant *Y*-interferon and recombinant tumor necrosis factor in mice. *Cancer Res.*, **47**, 2563–2570.

Talmadge, J. E., Scheider, M., Keller, J., Ruscetti, F., Longo, D., Pennington, R., Bowersox, O., and Tribble, H. (1989). Myelostimulatory activity of recombinant human interleukin-2 in mice. *Blood*, **73**, 1458–1467.

Taniguchi, K., Koga, S., Nishikido, M., Yamashita, S., Sakuragi, T., Kanetake, H., and Saito, Y. (1999). Systemic immune response after intravesical instillation of bacille Calmette–Guerin (BCG) for superficial bladder cancer. *Clin. Exp. Immunol*, **115**, 131–135.

Teichmann, J. V., Sieber, G., Ludwig, W. D., and Ruehl, H. (1988). Modulation of immune functions by long-term treatment with recombinant interferon-alpha 2 in a patient with hairy-cell leukemia. *J. Interferon Res.*, **8**, 15–24.

The Italian Cooperative Study Group on Chronic Myeloid Leukemia (1994). Interferon alpha-2a as compared with conventional chemotherapy for the treatment of chronic myeloid leukemia. *N. Engl. J. Med.*, **330**, 820–825.

Thompson, J. A., Shulman, K. L., Kenyunes, M. C., Lindgren, C. G., Collins, C., Lange, P. H., Bush, W. H., Jr., Benz, L. A., and Fefer, A. (1992). Prolonged continuous intravenous infusion interleukin-2 and lymphokine-activated killer-cell therapy for metastatic renal cell carcinoma. *J. Clin. Oncol.*, **10**, 960–968.

Toh, H. C., McAfee, S. L., Sackstein, R., Multani, P., Cox, B. F., Garcia-Carbonero, R., Colby, C., and Spitzer, T. R. (2000). High-dose cyclophosphamide + carboplatin and interleukin-2 (IL-2) activated autologous stem cell transplantation followed by maintenance IL-2 therapy in metastatic breast carcinoma – A phase II study. *Bone Marrow Transplant.*, **25**, 19–24.

Tomlinson, E. (1991). Site-specific proteins. In *Polypeptide and Protein Drugs: Production, Characterization and Formulation*; R. C. Hider, and D. Barlow, eds. Ellis Horwood, Chichester, U.K.

Tompkins, W. A. (1999). Immunomodulation and therapeutic effects of the oral use of interferon-alpha: Mechanism of action. *J. Interferon Cytokine Res.*, **19**, 817–828.

Traver, D., Akashi, K., Manz, M., Merad, M., Miyamoto, T., Engleman, E. G., and Weissman, I. L. (2000). Development of CD8 alpha-positive dendritic cells from a common myeloid progenitor. *Science*, **290**, 2152–2154.

Urabe, A., Mutoh, Y., Mizoguchi, H., Takaku, F., and Ogawa, N. (1993). Ubenimex in the treatment of acute nonlymphocytic leukemia in adults. *Ann. Hematol.*, **67**, 63–66.

Van der Auwera, P., Platzer, E., Xu, Z. X., Schulz, R., Feugeas, O., Capdeville, R., and Edwards, D. J. (2001). Pharmacodynamics and pharmacokinetics of single doses of subcutaneous pegylated human G-CSF mutant (Ro 25–8315) in healthy volunteers: Comparison with single and multiple daily doses of filgrastim. *Am. J Hematol.*, **66**, 245–251.

Vlk, V., Eckschlager, T., Kavan, P., Kabickova, E., Koutecky, J., Sobota, V., Bubenik, J., and Pospisilova, D. (2000). Clinical ineffectiveness of IL-2 and/or IFN alpha administration after autologous PBSC transplantation in pediatric oncological patients. *Pediatr. Hematol. Oncol.*, **17**, 31–44.

Vivancos, P., Granena, A., Jr., Sarra, J., and Granena, A. (1999). Treatment with interleukin-2 (IL-2) and interferon [IFN(alpha 2b))] after autologous bone marrow or peripheral blood stem cell transplantation in onco-hematological malignancies with a high risk of relapse. *Bone Marrow Transplant*, **23**, 169–172.

Voest, E. E., Kenyon, B. M., O'Reilly, M. S., Truitt, G., D'Amato, R. J., and Folkman, J. (1995). Inhibition of angiogenesis in vivo by interleukin 12. *J. Natl. Cancer Inst.*, **87**, 581–586.

Wadler, S. and Schwartz, E. L. (1990). Antineoplastic activity of the combination of interferon and cytotoxic agents against experimental and human malignancies: A review. *Cancer Res.*, **50**, 3473–3486.

Walter, E. A., Greenberg, P. D., Gilbert, M. J., Finch, R. J., Watanabe, K. S., Thomas, E. D., and Riddell, S. R. (1995). Reconstitution of cellular immunity against cytomegalovirus in recipients of allogeneic bone marrow by transfer of T-cell clones from the donor. *N. Engl. J. Med.*, **333**, 1038–1044.

Wang, A., Braun, S. E., Sonpavde, G., and Cornetta, K. (2000). Antileukemic activity of Flt3 ligand in murine leukemia. *Cancer Res.*, **60**, 1895–1900.

Weiden, P. L., Flournoy, N., Thomas, E. D., Prentice, R., Fefer, A., Buckner, C. D., and Storb, R. (1979). Antileukemic effect of graft-versus-host disease in human recipients of allogeneic-marrow grafts. *N. Eng. J. Med.*, **300**, 1068–1073.

West, W. H., Tauer, K. W., Yannelli, J. R., Marshall, G. D., Orr, D. W., Thurman, G. B., and Oldham, R. K. (1987). Constant-infusion recombinant interleukin-2 in adoptive immunotherapy of advanced cancer. *N. Eng. J. Med.*, **316**, 898–905.

Wood, D. D., Staruch, M. J., Durette, P. L., Melvin, W. V., and Graham, B. K. (1983). Role of interleukin-1 in the adjuvanticity of muramyl dipeptide in vivo. In *Interleukins, Lymphokines and Cytokines*; J. J., Oppenheim, and S. Cohen, Eds. Raven Press: New York, 1983.

Woodman, R. C., Erickson, R. W., Rae, J., Jaffe, H. S., and Curnutte, J. T. (1992). Prolonged recombinant interferon-gamma therapy in chronic granulomatous disease: Evidence against enhanced neutrophil oxidase activity. *Blood*, **79**, 1558–1562.

Worth, L. L., Jia, S. F., An, T., and Kleinerman, E. S. (1999). ImmTher, a lipophilic disaccharide derivative of muramyl dipeptide, up-regulates specific monocyte cytokine genes and activates monocyte-mediated tumoricidal activity. *Cancer Immunol. Immunother.*, **48**, 312–320.

Wu, C. Y., Demeure, C., Kiniwa, M., Gately, M., and Delespesse, G. (1993). IL-12 induces the production of IFN-gamma by neonatal human CD4 T cells. *J. Immunol.*, **151**, 1938–1949.

Yang, J. C., Topalian, S. L., Parkinson, D., Schwartzentruber, D. J., Weber, J. S., Ettinghausen, S. E., White, D. E., Steinberg, S. M., Cole, D. J., and Kim, H. I. (1994). Randomized comparison of high-dose and low-dose intravenous interleukin-2 for the therapy of metastatic renal cell carcinoma: An interim report. *J. Clin. Oncol.*, **12**, 1572–1576.

Yang, J. C., and Rosenberg, S. A. (1997). An ongoing prospective randomized comparison of interleukin-2 regimens for the treatment of metastatic renal cell cancer. *Cancer J. Sci. Am.*, **3** (suppl. 1), S79–S84.

Yang, J. S., Xu, L. Y., Huang, Y. M., Van Der Meide, P. H., Link, H., and Xiao, B. G. (2000). Adherent dendritic cells expressing high levels of interleukin-10 and low levels of interleukin-12 induce antigen-specific tolerance to experimental autoimmune encephalomyelitis. *Immunology*, **101**, 397–403.

Yasumitsu, T., Ohshima, S., Nakano, N., Kotake, Y., and Tominaga, S. (1990). Bestatin in resected lung cancer. A randomized clinical trial. *Acta Oncol.*, **29**, 827.

Yu, H., Fehniger, T. A., Fuchshuber, P., Thiel, K. S., Vivier, E., Carson, W. E., and Caligiuri, M. A. (1998). Flt3 ligand promotes the generation of a distinct CD34(+) human natural killer cell progenitor that responds to interleukin-15. *Blood*, **92**, 3647–3657.

Zou, W., Borvak, J., Marches, F., Wei, S., Galanaud, P., Emilie, D., and Curiel, T. J. (2000). Macrophage-derived dendritic cells have strong Th1-polarizing potential mediated by beta-chemokines rather than IL-12. *J. Immunol.*, **165**, 4388–4396.

Chapter 11

Food Allergy Assessment for Products Derived through Plant Biotechnology

Steve L. Taylor
University of Nebraska-Lincoln
Food Allergy Research and Resource Program
Lincoln, Nebraska

Susan L. Hefle
University of Nebraska-Lincoln
Food Allergy Research and Resource Program
Lincoln, Nebraska

Virtually all food allergens are proteins, but only a few of the many naturally occurring proteins that exist in foods are known to be allergenic. Protein allergens also exist in pollens, mold spores, animal danders, insect venoms, and other biological sources. Since plant biotechnology involves the transfer of DNA that encodes for proteins from one biological source into another, the possibility for the introduction of a novel allergenic protein must be assessed. A gene of interest obtained from a known allergenic source may encode for an allergen from that source, and this possibility must be considered. Even when the gene of interest is obtained from a source with no history of allergenicity, the possible allergenicity of each novel protein should be assessed. This can be done by comparing the amino acid sequence of the novel protein to the sequences of known allergens, by screening for IgE

Biotechnology and Safety Assessment, 3rd edition

binding between the novel protein and serum IgE from individuals with known allergies to the source material or related species, by evaluating the degree of pepsin resistance of the novel protein, and by evaluating the immunogenicity of the novel protein in suitable animal models. The application of these approaches provides a reasonably rigorous assessment of the potential allergenicity of the novel protein(s) and thus the genetically modified foods.

INTRODUCTION TO FOOD ALLERGIES AND ALLERGENS

Some individuals in the population cannot tolerate the ingestion of certain specific foods that are safe and nutritious for the vast majority of the consuming public. These individualistic adverse reactions to foods, or food sensitivities, can occur through a number of different mechanisms and can be categorized as either food allergies or food intolerances (Taylor *et al.*, 2001). Food allergies are the most relevant issue to discuss with respect to agricultural biotechnology.

PREVALENCE, SYMPTOMS, AND MECHANISMS OF FOOD ALLERGIES

True food allergies involve abnormal immunological responses to certain substances in foods (Taylor *et al.*, 2001). Immediate hypersensitivity reactions are mediated by allergen-specific IgE antibodies, and symptoms occur within minutes of ingestion of the offending food. Delayed hypersensitivity reactions are mediated by sensitized T lymphocytes, usually in the intestinal mucosa, and are characterized by symptoms occurring 24 to 72 hours after ingestion of the offending foods.

In the case of IgE-mediated, immediate hypersensitivity reactions to foods (Mekori, 1996), susceptible individuals respond to exposure to specific allergens in foods with the production of allergen-specific IgE antibodies from B lymphocytes. The specific IgE antibodies become attached to the membrane surfaces of mast cells in the tissues and basophils in the blood. This is the process of allergic sensitization and must occur before allergic reactions to this food will become evident. Once an individual has been sensitized to a particular food allergen, exposure to that specific food allergen will elicit an allergic reaction. The ingested allergen will bind to the specific IgE antibodies on the surface of the mast cell and basophil membranes. The cross-linking of two IgE antibodies by the allergen will elicit the release of various mediators of the allergic reaction. Histamine, the leukotrienes, and the prostaglandins are among the dozens of mediators released from mast cells and basophils by this allergen–antibody interaction. These mediators find their way to tissue-based receptors and elicit the various symptoms that are associated with food allergies. The mechanism of IgE-mediated reactions is depicted in Figure 1.

A wide variety of symptoms can be involved in IgE-mediated food allergies (Table 1). Gastrointestinal symptoms such as nausea, vomiting, abdominal cramping, and diarrhea are common manifestations but not necessarily definitive of food allergies. Cutaneous symptoms such as urticaria, angioedema, pruritis, and dermatitis (eczema) are also common and are also more definitive. Respiratory symptoms such as rhinitis and asthma occur less frequently with food allergies, where the allergen is ingested, than they do with pollen allergies, where the allergen is inhaled. However, asthma is a particularly severe manifestation of food allergy (Sampson *et al.*, 1992). The most severe manifestation of food allergy is anaphylactic shock, a systemic disorder that can involve the gastrointestinal tract, the skin, the respiratory tract, and the cardiovascular system. Only a small fraction of food-allergic

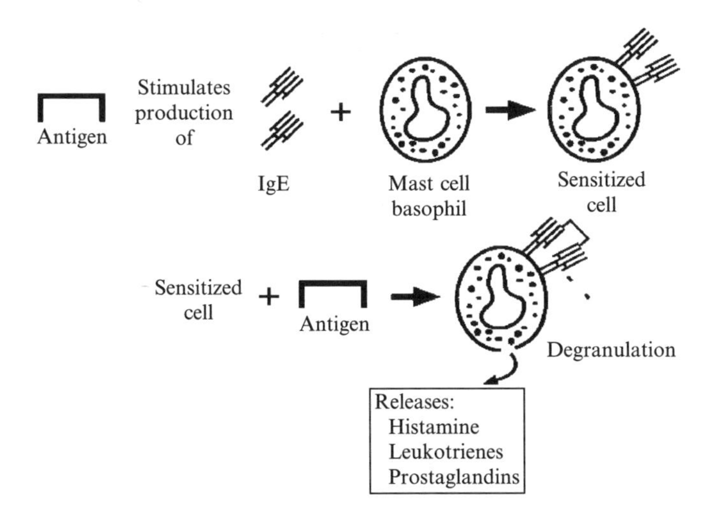

Figure 1 Mechanism of IgE-mediated allergic reaction.

Table 1
Symptoms of IgE-Mediated Food Allergies

Cutaneous	Urticaria Dermatitis or eczema Angioedema Pruritis
Gastrointestinal	Nausea Vomiting Diarrhea Abdominal cramping
Respiratory	Rhinitis Laryngeal edema Asthma
Generalized:	Anaphylactic shock

individuals are so severely affected that they can suffer from anaphylactic shock. However, anaphylactic shock can result in death within minutes if untreated. Although anaphylactic shock is the most severe manifestation of food allergies, the oral allergy syndrome is probably the most frequent, and it is among the mildest manifestations of food allergies. Oral allergy syndrome (OAS) involves symptoms confined to the oropharyngeal region (e.g., urticaria, pruritis, angioedema); systemic reactions are rarely encountered (Ortolani *et al.*, 1988).

Food allergies mediated by IgE affect approximately 2.0 to 2.5% of the population in the United States (Taylor and Hefle, 2001). The prevalence of IgE-mediated food allergies is higher in infants than among adults. The prevalence among infants is estimated at 5 to 8% (Sampson and McCaskill, 1985). On a worldwide basis, eight foods or food groups, cow's milk, eggs, fish, crustacean shellfish (shrimp, crab, lobster, etc.), peanuts, tree nuts (almonds, walnuts, cashews, etc.), soybeans, and wheat, account for perhaps 90% of all IgE-mediated food allergies, other than foods involved in elicitation of OAS (FAO, 1995). More than 160 other foods have been implicated in food allergies on at least rare occasions (Hefle *et al.*, 1996). OAS is most frequently associated with the ingestion of fresh fruits and vegetables (Ortolani *et al.*, 1988). Individuals experiencing OAS usually have been sensitized to airborne pollens and are then reactive to the ingestion of structurally similar proteins that exist in foods (Calkhoven *et al.*, 1987).

IDENTIFICATION AND CHARACTERIZATION OF FOOD ALLERGENS

The vast majority of food allergens are naturally occurring proteins (Bush and Hefle, 1996). Allergenic foods may have from one up to a dozen allergenic proteins, although most allergenic foods contain from one to three major allergens. A major allergen is defined as a protein that elicits a specific IgE antibody response in more than 50% of affected individuals. Although virtually all allergens are proteins, only a small fraction of the many proteins found in nature are known to be allergenic. The major allergens have been identified from many of the commonly allergenic foods and several of the less commonly allergenic foods also. Table 2 lists some of the known food allergens. Of course, many other allergens exist in pollen, mold spores, animal danders, dust mites, insects, and venoms. Many of these allergens have also been identified.

Table 2
Known Food Allergens

Food	Allergen nomenclature	Allergen function
Peanut	*Ara h* 1	Storage protein
	Ara h 2	Storage protein
	Ara h 3	Storage protein
Soybean	*Gly m* 1	Oleosin (??)
Walnut	*Jug r* 1	Storage protein
	Jug r 2	Storage protein
Hazelnut	*Cor a* 1	Storage protein
Brazil nut	*Ber e* 1	Storage protein
Yellow mustard	*Sin a* 1	Storage protein
Oriental mustard	*Bra j* 1	Storage protein
Shrimp	*Pen a* 1	Tropomyosin
Cod	*Gad c* 1	Parvalbumin
Egg	*Gal d* 1	Ovomucoid
	Gal d 2	Ovalbumin
	Gal d 3	Ovotransferrin
	Gal d 4	Lysozyme
Cow's milk		Casein
		β-Lactoglobulin
		α-Lactalbumin

The allergenicity of a particular protein cannot be predicted on the basis of its structure. Examination of the many known allergens reveals no common or shared features. However, allergenic proteins contain specific arrangements of amino acids, called epitopes, which are able to bind to the specific IgE antibodies and elict mediator release from mast cells. Proteins, even from dissimilar species, that contain structurally related proteins with cross-reactive epitopes can elicit reactions in individuals sensitized to one of these proteins. The example of OAS, where individuals sensitized to pollen react to ingestion of certain foods that contain proteins with structurally related epitopes, is illustrative. Allergic individuals are even more likely to react to structurally similar proteins from closely related species. For example, most individuals with allergies to crustacean shellfish are reactive to all species of shrimp, crab, lobster, and crayfish because these crustaceans contain structurally related tropomyosin allergens (Leung *et al.*, 1996). Tropomyosin from muscles of more distantly related species such as beef, pork, and chicken does not elicit reactions (Reese *et al.*, 1999). However, related species do not invariably contain proteins with sufficient similarity to elicit cross-reactions. For example, both peanuts and soybeans are legumes, along with several hundred other foods, yet despite the existence of some cross-reactive individuals, most allergic persons react only to peanuts or only to soybeans, not to both legumes (Herian *et al.*, 1990). Food allergens do tend to fall into certain functional categories (Breiteneder and Ebner, 2000). Many of the families of pathogenesis-related proteins that are of interest to plant biotechnologists contain examples of food allergens (Breiteneder and Ebner, 2000).

Food allergens tend to share other characteristics (Taylor *et al.*, 1987b). Most food allergens are present in comparatively high amounts in the particular food. Food allergens tend to be resistant to digestive proteolysis, although the allergens involved in OAS are exceptions. Allergenic food proteins are relatively heat resistant also, although again the allergens involved in OAS seem to be exceptions. Many food allergens are glycoproteins, although since many proteins in nature are glycoproteins, this is hardly a very distinguishing feature. Food allergens have been described as falling in the mass range of 10,000 to 70,000 Dal (Aas, 1978; Taylor *et al.*, 1987b), but this is an artifact of using detergents to dissociate large polymeric structures in the laboratory procedures used to estimate molecular weight.

NOVEL PROTEINS: JUSTIFICATION FOR ALLERGENICITY ASSESSMENT

In the safety assessment of foods produced through agricultural biotechnology, the assessment of the potential allergenicity of the novel proteins introduced into these foods is one of the key issues. The potential allergenicity of the newly introduced proteins must be assessed in all foods produced through agricultural biotechnology. Because virtually all allergens are proteins, the possibility exists that any novel food protein might be or become an allergen. Since so few of the proteins in nature are allergenic, the probability that a novel protein will be or become a food allergen is rather low. However, this possibility must be considered in each and every case. Unfortunately, no single test is available that is fully predictive of the potential allergenicity of any specific novel protein. Thus, the favored approach is typically a decision tree strategy that employs multiple tests to improve the predictive capability of the overall assessment.

ALLERGY ASSESSMENT IN THE GENE DISCOVERY PHASE

The development of novel foods through agricultural biotechnology involves a three-stage process: gene discovery, line selection, and product advancement to commercialization (Taylor, 2001). The assessment of the allergenicity of genetically modified foods should ideally begin during the initial gene discovery stage. In the gene discovery stage, the scientist develops a product concept and screens and selects genes that might allow the fulfilment of that concept.

A wide variety of factors should be considered during the gene selection phase including the source of the gene and previous consumer exposure to this source material, the history of safe use of the source material in the diet and otherwise, the gene and the product of the gene, any ethical issues that might occur, and any environmental and ecological concerns. Allergy assessment is but one part of this overall consideration. However, the initial allergenicity assessment at the gene discovery phase is extraordinarily important because it can draw attention to concerns and highlight questions that must be effectively addressed later in the safety assessment process if the concept is not abandoned at this point. For example, if the developer selects a gene from a source with a known history of allergy (peanuts, tree nuts, fish, etc.), then assurance must be sought that the gene product is not one of the allergens from that source. However, such issues are not often so directly obvious. If the novel protein belongs to a class of pathogenesis-related proteins that contains known allergens, then the possible allergenicity of the particular novel protein becomes a more relevant concern. For example, chitinase genes might be selected as a means of preventing various fungal diseases common to some crop plants. But, chitinases from some sources are allergens (Breitenender and Ebner, 2000), so the possible cross-reactivity with other chitinases needs to be assessed.

DECISION TREE APPROACH: THE 2001 FAO/WHO DECISION TREE

The first decision tree strategy for the assessment of the potential allergenicity of genetically modified foods was developed in 1996 by a task force of the International Food Biotechnology Council (IFBC) and the Allergy & Immunology Institute of the International Life Sciences Institute (ILSI) (Metcalfe *et al.*, 1996). This decision tree strategy focused upon evaluating the source of the gene, the sequence homology of the newly introduced protein to known allergens, the immunoreactivity of the novel protein with serum IgE from individuals with known allergies to the source of the transferred DNA, and various physicochemical properties of the newly introduced protein such as heat stability and digestive stability. This recommended approach was used to assess for allergenicity the genetically modified foods currently on the market. While the application of this strategy provided reasonable assurance that the newly introduced protein was unlikely to become an allergen, the decision tree approach was subjected to some criticism (Vieths, 1997).

In January 2001, the Food and Agriculture Organisation of the United Nations (FAO) and the World Health Organisation (WHO) convened an expert consultation that developed a modified and more rigorous approach to the assessment of the allergenicity of foods produced through agricultural biotechnology (FAO/WHO, 2001). This new FAO/WHO decision tree (Figure 2) relies on some of the same elements as the IFBC/ILSI decision tree, including the source of the gene, the sequence homology of the newly introduced protein to known allergens, the immunoreactivity of the novel protein with serum IgE from individuals with known allergies to the source of the transferred DNA, and the resistance of the novel protein to digestion with pepsin. Additional criteria that were included in the new decision tree were the immunoreactivity of the novel protein with serum IgE from individuals with known allergies to species that are broadly related to the source of the transferred DNA and the immunogenicity of the novel protein in appropriate animal models.

Although the decision tree strategy is indeed more rigorous than the earlier approach of IFBC/ILSI, it might yield more false positive results, which would complicate regulatory approval of new foods produced through agricultural biotechnology. Certainly, further modification of the new decision tree should be considered as science knowledge allows more accurate predictions regarding the potentially allergenicity of novel proteins in humans. The scientific basis for the various tests advocated in the new FAO/WHO decision tree is provided next.

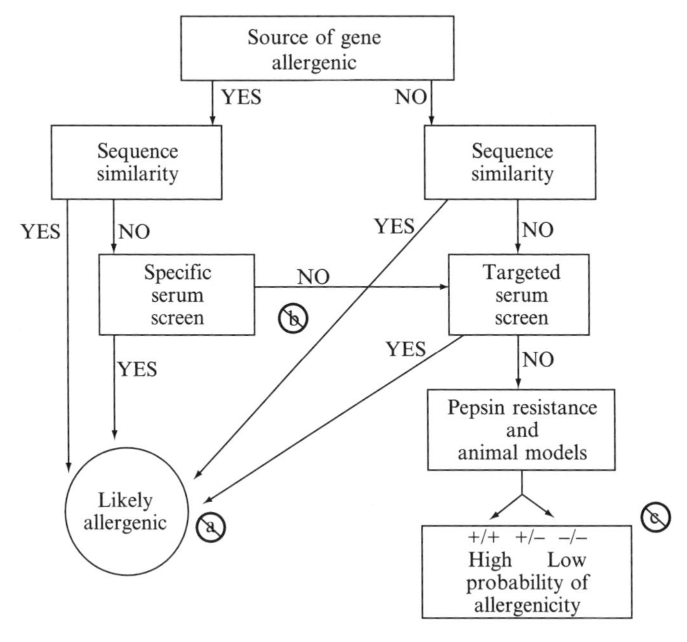

Figure 2 FAO/WHO decision tree strategy for assessment of the potential allergenicity of novel proteins from genetically modified foods. (a) Any positive results obtained from sequence homology comparisons to the sequences of known allergens in existing allergen databases or from serum screening protocols, both conducted in accordance with the guidelines established in Section 6.1 of FAO/WHO (2001) indicate that the expressed protein is likely allergenic. (b) The degree of confidence in negative results obtained in the specific serum screen is enhanced by the examination of larger numbers of individual sera as explained in Section 5.3 of FAO/WHO (2001). Conducting the specific serum screen with small numbers of individual sera when larger numbers of such sera are readily available should be discouraged. When positive results are obtained in both the pepsin resistance and animal model protocols, the expressed protein has a high probability of becoming an allergen. (c) When negative results are obtained in both protocols, the expressed protein is unlikely to become an allergen. When different results are obtained in the pepsin resistance and animal model protocols, the probability of allergenicity is intermediate, although rational explanations may be possible in some situations.

Source of the Gene

Initially, the FAO/WHO decision tree focuses on the source of gene to be introduced into the host organism. If the gene source is known to be allergenic, it must be assumed that this gene encodes for an allergen from that source unless data are generated to disprove that assumption. The source of the gene would be considered to be allergenic if it was associated with food, environmental (e.g. pollen), or occupational allergies. Obviously, the concerns would be greatest when the gene is obtained from a source that is commonly allergenic. As noted earlier, commonly allergenic foods include peanuts, soybeans, tree nuts, and wheat from the plant kingdom and milk, eggs, fish, and crustaceans from the animal kingdom (FAO, 1995). These few foods are thought to account for more than 90% of all food allergies on a worldwide basis (FAO, 1995). More than 160 other foods and food-related substances have been associated with allergic reactions in individuals on at least some occasions (Hefle *et al.*, 1996). All these sources and any sources associated with environmental or occupational allergies would be classified as known allergenic sources under the FAO/WHO decision tree strategy. Obviously, considerable diligence is needed to determine whether the selected source of the gene has any history of allergenicity.

If the gene source is known to be allergenic, the potential allergenicity of the novel gene product can be determined with a reasonable degree of certainty by using the specific serum screening test outlined shortly. Specific serum screening employs blood serum from human subjects known to be allergic to the source of the gene.

If the gene is obtained from a source with no history of allergenicity, then a specific serum screening test is obviously not possible. In many cases in agricultural biotechnology, the gene is obtained from a source with no history of allergenicity.

Sequence Homology

A comparison of the amino acid sequence homology of the novel protein to the amino acid sequences of known allergens is a useful initial approach in the determination of its allergenic potential regardless of the source of the gene (FAO/WHO, 2001; Metcalfe *et al.*, 1996). The amino acid sequences of many food and environmental allergens are known (Metcalfe *et al.*, 1996; Gendel, 1998). If sufficient homology exists, then some potential exists for cross-reaction of the novel protein with the known allergen, and for the provoking of symptoms upon ingestion by individuals with that particular allergy. The criterion for determining significant sequence homology in the FAO/WHO strategy is a match of at least six contiguous, identical amino acids or an overall homology of 35% or greater (FAO/WHO, 2001).

In IgE-mediated food allergy, certain specific proteins are capable of inducing IgE sensitization. However, the immune system does not recognize the entire structure of the allergenic protein but instead responds to smaller sections called allergenic determinants or epitopes. Two types of epitope are known to exist: continuous (a linear sequence of amino acids) and discontinuous (depending on the three-dimensional conformational structure of the protein) (Taylor and Lehrer, 1996). Since food allergens are often stable to heat processing and digestion, it has been hypothesized that linear epitopes may be more important with food allergens than with environmental allergens that are primarily inhaled (Taylor *et al.*, 1987a). However, recent evidence with the major peanut allergen *Ara h* 1 suggests that this hypothesis may not be universally true and that discontinuous epitopes are important in IgE-binding with at least some food allergens (Shin *et al.*, 1998).

In IgE-mediated allergies, a two-phase process of sensitization and elicitation is involved (Mekori, 1996). In the sensitization phase, proteins are processed by proteolysis in an antigen-presenting cell. With food allergies, gut-associated antigen-processing cells are primarily involved. The peptides that result then react with T lymphoctyes to provoke B lymphocytes to switch to production of allergen-specific IgE antibodies. As explained earlier, once the allergen-specific IgE antibodies have become attached to mast cells and basophils, the individual is sensitized. Upon interaction of IgE binding epitopes on the same proteins with the cell-bound IgE antibodies, histamine and the other mediators of the allergic reactions are released from these cells, provoking an allergic reaction. The allergen-specific IgE antibodies can bind to either linear or conformational epitopes depending upon the particular allergen involved. However, T-cell epitopes are most likely to be exclusively linear, and sensitization to a novel protein could not occur without the involvement of T-cell epitopes. Thus, linear epitopes are likely to be quite important in allergic sensitization to novel proteins derived from sources with no history of allergenicity. However, if a novel protein introduced through agricultural biotechnology is identical to an existing allergen or is cross-reactive with an existing allergen, both the linear and conformational epitopes recognized by the cell-bound IgE could be important.

The use of six contiguous, identical amino acids as a match in the FAO/WHO decision tree was predicated upon observations that the minimal IgE-binding epitopes of *Ara h* 1 and *Ara h* 2 involve six contiguous amino acids (Burks *et al.*, 1997; Stanley *et al.*, 1997). However, the minimum peptide length for a T-cell binding epitope is probably eight contiguous amino acids (Metcalfe *et al.*, 1996). Since this approach assesses the entire protein sequence, it is not based on the identity of amino acid sequences just to known T-cell and B-cell binding epitopes of known allergens. Thus, this approach will likely identify matching sequences that are unrelated to the allergenic potential of the novel proteins. While specific serum screening as described shortly would be able to eliminate clinically insignificant matches, this approach may be unwieldy if too many matches are found with the six amino acid criterion. In contrast, the IFBC/ILSI decision tree advocated use of an eight amino acid criterion for sequence homology. Experience with this approach indicates that it does not identify a substantial number of clinically irrelevant matches. However, concerns about the known length of the IgE binding epitopes caused the selection of a more conservative approach in the FAO/WHO decision tree.

The criteria for 35% overall structural homology is intended to identify proteins that share similar functions. Many common plant allergens, such as various proteins related to pathogenesis, fall within a few functional categories (Breiteneder and Ebner, 2000). Novel proteins introduced into foods developed through agricultural biotechnology and falling into functional categories that contain known food allergens are likely to have greater than 35% overall structural homology with these known allergens. In such situations, great caution must be exercised in assessing the potential allergenicity of these particular proteins.

If the sequence homology tests are positive, the conclusion can be reached that the novel protein is likely to be allergenic. A decision can be made to halt commercial development of that particular modified crop as a result. However, confirmation of the results of the sequence homology tests can be sought with specific serum screening. This may be advisable when one of the sequence homology tests appears to yield a false positive result, such as an irrelevant match of six contiguous amino acids.

Specific Serum Screening

When the novel gene is obtained from a known allergenic source or the search for sequence homology identifies a match with a known allergenic source, an assessment of the immunoreactivity of the novel protein with IgE antibodies from the sera of individuals allergic to the source material can be conducted, if desired (FAO/WHO, 2001). However, the structures of all the allergens from all the allergenic sources are not yet known, so specific serum screening is advisable for every gene that is derived from a known allergenic source (FAO/WHO, 2001). In specific serum screening, the novel protein is bound to a solid phase and then blood serum from individuals known to be allergic to the specific allergenic source is used to determine whether the allergen-specific IgE antibodies in the serum will react with epitopes on the novel protein. The availability of sera from well-characterized patients is an important and sometimes challenging issue. Serum would be relatively easy to obtain when the gene is obtained from commonly allergenic sources. But, relevant serum could be very difficult to obtain when the source of the gene was known to be allergenic but only on rare occasions. The quality of the blood serum would also be a source of concern. Sera would be well characterized if the patient had a positive and convincing history of allergy to the gene source, had a positive skin prick test or radioallergosorbent test to an extract of the gene source, and ideally had a positive clinical challenge trial with the source material.

Another concern with specific serum screening is the possibility for clinically insignificant immunoreactivity especially associated with carbohydrate or glycan moieties that exist on many plant proteins. The phenomenon of IgE binding to carbohydrates is well known (Aalberse and Van Ree, 1997). Some investigators believe these cross-reactive carbohydrate determinants are clinically insignificant (Aalberse and Van Ree, 1997). Thus, the possibility of clinically insignificant IgE binding to carbohydrate determinants must be excluded during specific serum screening (FAO/WHO, 2001).

The nature of the protein used in the specific serum screening test should also be a consideration. Novel proteins obtained from recombinant bacterial systems may, in some cases, differ structurally from the same protein as expressed in plant systems. Certainly, it is easier to obtain protein from recombinant bacterial systems, but in some situations it may be preferable to purify the novel protein from the plant source for the purposes of this test and the targeted serum screening test described shortly.

A positive specific screening test would certainly elicit concerns about the possible allergenicity of the novel protein. Of course, such tests could be discounted by additional *in vivo* testing in allergic patients as was initially recommended in the IFBC/ILSI decision tree (Metcalfe *et al.*, 1996). However, obtaining approval from ethics boards for *in vivo* testing (skin prick tests and/or double-blind, placebo-controlled food challenges) with genetically modified foods has proven difficult in some countries. Therefore, *in vivo* testing was not recommended as part of the FAO/WHO decision tree (FAO/WHO, 2001). In most circumstances, a positive result with the specific serum screen will lead to the conclusion that the novel protein is likely allergenic, and commercial development will likely cease.

Targeted Serum Screening

When negative or equivocal results are obtained in the specific serum screening as just described, the FAO/WHO decision tree advocates that targeted serum screening be used to further investigate the novel food (FAO/WHO, 2001). Additionally, targeted serum screening is recommended when the gene source has no history of allergenicity and has no sequence homology to known allergens (FAO/WHO, 2001).

In targeted serum screening, human blood serum is obtained from individuals who are allergic to materials that are broadly related to the source of the gene. The possibility exists that the source material may contain proteins that are cross-reactive with allergens from related sources even if the gene source itself is not known to be allergenic. If the structure of such allergens is known, such cross-reactivity will likely be evident in the sequence homology testing in situations involving allergens with known structure. However, since the structures of all the allergens from all sources are not yet known, targeted serum screening is advisable. The FAO/WHO approach suggests several broad categories for targeted serum screening: monocotyledons, dicotyledons, invertebrates, vertebrates, and molds/yeast/fungi. For example, if the source of a gene was a monocotyledonous plant with no history of allergenicity, targeted serum screening would involve assessing the immunoreactivity of the novel protein with sera from individuals with known allergies to such other monocotyledonous sources as grass pollen. The same caveats regarding the importance of well-characterized sera, exclusion of clinically insignificant IgE binding, and the production of the novel protein for testing also apply to targeted serum screening. If the gene source is bacterial, targeted serum screening is not conducted. Bacterial proteins are rarely allergenic because of the low exposure levels and lack of allergic sensitization to these proteins.

If targeted serum screening is positive, the novel protein is considered likely to be an allergenic substance. If the targeted serum screening is negative, further testing, including resistance to pepsin and immunoreactivity in animal models, is recommended.

Pepsin Resistance and Digestive Fate

Proteolytic stability is recognized as a useful criterion in the assessment of the protein's potential to become a food allergen. Allergenic food proteins must reach the intestinal tract in a form that is sufficiently intact to provoke the immune system. If the protein is rapidly digested, it is unlikely to be allergenic. In simulated gastric and intestinal digestive models, known food allergens exhibited greater proteolytic stability than known nonallergenic food proteins (Astwood *et al.*, 1996). Many of the novel proteins introduced into foods produced through agricultural biotechnology are also rapidly digested in these same model systems (Astwood *et al.*, 1996). Resistance to digestive proteolysis was one of the criteria recommended in the IFBC/ILSI decision tree (Metcalfe *et al.*, 1996).

The FAO/WHO decision tree approach advocates use of resistance to proteolysis with pepsin as a comparative measure of digestive stability for novel proteins introduced into foods through agricultural biotechnology (FAO/WHO, 2001). Obviously, digestive proteolysis in humans is individually variable. Thus, pepsin resistance should not be construed to be predictive of the digestive stability of a novel protein in all humans. However, the comparative resistance to proteolysis with pepsin is likely a reasonable comparative measure in the allergenicity assessment. Novel proteins that are pepsin resistant are more likely to become allergenic than proteins that are rapidly hydrolyzed by pepsin. However, pepsin resistance is probably not a perfect indicator of allergenic potential. Some allergens in fresh fruits and vegetables are known to be sensitive to proteolysis (Moneret-Vautrin *et al.*, 1997). These particular allergens are often ones that are cross-reactive with known pollen allergens (Calkhoven *et al.*, 1987) and would thus likely be discovered in the sequence homology testing. There are probably food proteins that are stable to pepsin and to digestion that are not allergenic, but this possibility has not been studied very thoroughly. The FAO/WHO decision tree provides a specific protocol for the pepsin resistance test (FAO/WHO, 2001). Use of a standardized protocol is important so that comparative data can be obtained.

Animal Models

Well-validated animal models do not exist for prediction of the allergenicity of novel proteins. Several animal models, including the brown Norway rat and several different mouse strains, are undergoing evaluation for use in the allergenicity assessment of novel proteins (Ermel *et al.*, 1997; Knippels *et al.*, 1998, 2000; Li *et al.*, 1999). Despite the lack of a well-validated animal model, the FAO/WHO decision tree approach advocated the use of animal models. This was probably done in an effort to stimulate more research on development of animal models that would be suitable for this purpose. Certainly, further research is needed on the development of animal models to assess and improve their predictive accuracy.

The ideal animal model should incorporate several attributes:

- Sensitization and challenge should occur via the oral route, since the natural digestive and gastrointestinal barriers can be considered only with such an approach.
- Preferably, adjuvants should not be used, since the focus should be on the intrinsic allergenicity of the novel protein.
- The test animal should produce a significant amount of IgE or other T_h2-specific antibody class.
- The test animal should tolerate most food proteins, especially proteins that are known non allergens such as RUBISCO (ribulosebisphosphate carboxytase/oxygenase).
- The test animal should develop allergen-specific IgE antibodies on oral exposure to known food allergens.
- The test should be relatively easy and the results reproducible.

Oral exposure of rodents to food proteins most often results in immunological tolerance (Strobel and Ferguson, 1984; Steinmann and Wottge, 1990). However, recent research has led to the development of several promising animal models. In mice, repeated enteral administration of proteins in combination with adjuvants indicates that allergic sensitization can be achieved with certain protocols (Li *et al.*, 1999, 2000). When specific sensitization protocols are used, allergic sensitization to cow's milk has been achieved in mice via oral administration (Li *et al.*, 1999) and to peanut proteins (Li *et al.*, 2000), with cholera toxin as an adjuvant. Oral feeding of mice with casein or ovalbumin as a constituent of the diet without adjuvant administration resulted in allergic sensitization to casein only (Ito *et al.*, 1997). When proteins were administered intraperitoneally, the characteristic antibody (IgG and IgE) isotype profiles were different for ovalbumin (an allergenic protein) and bovine serum albumin (a weakly allergenic protein) (Dearman *et al.*, 1999). Another promising animal model is the brown Norway rat, which is known to produce high levels of IgE when provoked with antigen. IgE responses were elicited to known food allergens in cow's milk and eggs by daily oral gavage dosing of brown Norway rats (Knippels *et al.*, 1998). However, further validation of the brown Norway rat model using food proteins that are not allergens would be helpful. The beagle dog may also be a useful animal model to consider in further research (Ermel *et al.*, 1997).

In the FAO/WHO decision tree, novel proteins that are resistant to pepsin and able to stimulate IgE responses in suitable animal models are considered to be more likely to become allergens than proteins that are rapidly hydrolyzed by pepsin and fail to stimulate immune responses in these same animal models.

EXAMPLE APPLICATIONS OF ALLERGENICITY ASSESSMENT

The IFBC/ILSI decision tree approach has been applied to products currently available in the marketplace. The FAO/WHO decision tree is so new that there is little experience with its application. It cannot be reliably predicted whether any of the foods that have been approved with consideration of allergenicity by the IFBC/ILSI approach would fail the FAO/WHO strategy. However, there is absolutely no evidence that any of the current products of agricultural biotechnology have resulted in allergic sensitization or allergic reactions among consumers. A brief examination of three examples may be helpful.

Glyphosate-Tolerant Soybean

Glyphosate-tolerant soybeans (GTS) have been modified to be tolerant to the broad-spectrum herbicide, glyphosate. This modification provides an advantage to farmers for weed control in soybean fields. GTS were developed with the insertion of a single gene that encodes for a glyphosate-tolerant 5-enolpyruvylshikimate-3-phosphate synthase from *Agrobacterium* sp. strain CP4 (CP4 EPSPS) (Padgette *et al.*, 1995). With respect to allergenicity assessment, CP4 EPSPS was obtained from a bacterial source with no history of allergenicity. Furthermore, EPSPS is present in many plant, bacteria, and fungi, so related enzymes have been present in the human diet, apparently without consequence, for centuries. Using the eight contiguous amino acid matching specification of the IFBC/ILSI decision tree, no sequence homology is found between CP4 EPSPS and known allergens. Additionally, CP4 EPSPS is rapidly digested by pepsin (Harrison *et al.*, 1996). GTS are compositionally equivalent to conventional soybeans (Padgette *et al.*, 1996), and CP4 EPSPS is expressed at a rather low level in GTS. Therefore, GTS containing CP4 EPSPS are unlikely to contain any novel allergens.

Insect-Resistant Corn

Corn has been modified to be resistant to common insect pests by the introduction of several genes encoding for specific proteins from *Bacillus thuringiensis* (*Bt* proteins). The most successful variety contains the *cry 1Ab* gene that makes the corn tolerant to European corn borers (Sanders *et al.*, 1998). The *Bt* proteins are selectively toxic to certain insect species but are harmless to humans. In terms of allergenicity assessment, the Cry1 Ab protein has no history of allergenicity despite decades of commercial use in agriculture as a pesticidal spray (Sanders *et al.*, 1998). The *cry 1Ab* gene was obtained from a bacterial source with no history of allergenicity. The Cry1Ab protein is rapidly digested (Astwood *et al.*, 1996) by pepsin. Use of the eight contiguous amino acid matching specification of the IFBC/ILSI decision tree failed to reveal any sequence homology between the Cry 1Ab protein and known allergens. Thus, corn containing the *cry 1Ab* gene is unlikely to contain novel allergens.

Corn has also been modified by insertion of a separate *Bt* gene, the *cry9c* gene, to be resistant to the Southern corn borer. This product, marketed as StarLink®, has not been approved for human consumption anywhere in the world but was approved for animal feeding in the United States. Some of this corn was found in products intended for human consumption, causing numerous recalls in the recent past. Like other *Bt* proteins, the Cry9C protein is toxic to certain insects but displays no acute toxicity in mammals. The *cry9c* gene was obtained from a bacterial source with no history of allergenicity. When the IFBC/ILSI criterion of eight contiguous amino acids matching, was applied, the Cry9C protein showed no sequence homology to known allergens. However, the Cry9C protein was stable to pepsin under certain conditions. Thus, questions were raised about its potential allergenicity, and StarLink® was not approved for human consumption. While a few consumers claimed possible allergic reactions to corn products contaminated with StarLink®, no evidence of Cry9C-specific IgE antibodies was found in sera from these individuals. Thus, no proven allergies to corn containing the *cry9c* gene exist. However, this example serves to illustrate the use of the decision tree approach to reach regulatory decisions regarding the possible allergenicity of foods produced through agricultural biotechnology.

Methionine-Enhanced Soybeans

In the early 1990s, Pioneer Hi-Bred International, now a division of Du Pont, developed a high-methionine variety of soybeans. Soybeans are inherently deficient in methionine. Thus, farmers who feed soybean meal to nonruminant farm animals must supplement the animals' diets with methionine. High-methionine soybeans were developed with the insertion of a gene from Brazil nuts that encodes for a high-methionine protein. The high-methionine protein from Brazil nuts was expressed at a reasonably high level in the novel variety of soybeans. In terms of allergenicity assessment, this particular gene was obtained from a known allergenic source. At the time of this development, the allergenicity of Brazil nuts was well established (Arshad *et al.*, 1991), but the identity of the allergens in Brazil nuts was not known. Nordlee *et al.*, (1996) used blood sera from individuals with documented Brazil nut allergy to evaluate the possible allergenicity of the novel soybeans and the purified high-methionine protein. Specific serum screening showed that the gene obtained from the Brazil nut likely encoded for a major Brazil nut allergen (*Ber e* 1). This possibility was confirmed by positive skin prick tests on three of the individuals who were allergic to Brazil nuts (Nordlee *et al.*, 1996). As a result, Pioneer Hi-Bred International decided not to commercialize this variety of soybeans.

CONCLUSIONS

Foods produced through agricultural biotechnology should be assessed for their potential allergenicity. Decision tree strategies do exist for allergenicity assessments. Further research will aid in the implementation of these strategies, particularly with respect to development of an improved database on the amino acid sequences and epitopes of known food allergens, more information on the level of exposure to novel proteins needed to provoke allergic sensitization, and the development of well-validated animal models. With the application of such strategies to foods produced through agricultural biotechnology, consumers and regulators should be confident that the likelihood for allergenicity of these novel foods is quite low.

REFERENCES

Aalberse, R. C., and Van Ree, R. (1997). Crossreactive carbohydrate determinants. *Clin. Rev. Allergy Immunol.*, **15**, 375–387.

Aas, K. (1978). What makes an allergen an allergen? *Allergy*, **33**, 3–14.

Arshad, S. H., Malmberg, E., Krapf, K., and Hide, D. W. (1991). Clinical and immunological characteristics of Brazil nut allergy. *Clin. Exp. Allergy*, **21**, 373–376.

Astwood, J. D., Leach, J. N., and Fuchs, R. L. (1996). Stability of food allergens to digestion in vitro. *Nat. Biotechnol.*, **14**, 1269–1273.

Breiteneder, H., and Ebner, C. (2000). Molecular and biochemical classification of plant-derived food allergens. *J. Allergy Clin. Immunol.*, **106**, 27–36.

Burks, A. W., Shin, D., Cockrell, G., Stanley, J. S., Helm, R. M., and Bannon, G. A. (1997). Mapping and mutational analysis of the IgE-binding epitopes of Ara h 1, a legume vicilin protein and a major allergen in peanut hypersensitivity. *Eur. J. Biochem.*, **245**, 334–339.

Bush, R. K., and Hefle, S. L. (1996). Food allergens. *Crit. Rev. Food Sci. Nutr.*, **36**, S119–S163.

Calkhoven, P. G., Aalbers, M., Koshte, V. L., Pos, O., Oei, H. D., and Aalberse, R. C. (1987). Cross-reactivity among birch pollen, vegetables and fruits as detected by IgE antibodies is due to at least three distinct cross-reactive structures. *Allergy*, **42**, 382–390.

Dearman, R. J., Caddick, H., Basketter, D. A., and Kimber, I. (1999). Divergent antibody isotype responses induced in mice by systemic exposure to proteins: A comparison of ovalbumin with bovine serum albumin. *Food Chem. Toxicol.*, **38**, 351–360.

Ermel, R. W., Kock, M., Griffey, S. M., Reinhart, G. A., and Frick, O. L. (1997). The atopic dog: A model for food allergy. *Lab. Animal Sci.*, **47**, 40–49.

FAO. (1995). Report of the FAO Technical Consultation on Food Allergies, Rome, November 13–14. Food and Agriculture Organisation of the United Nations, Rome.

FAO/WHO. (2001). Evaluation of the Allergenicity of Genetically Modified Foods: Report of a Joint FAO/WHO Expert Consultation. Rome. Food and Agriculture Organisation and World Health Organisation of the United Nations, Rome.

Gendel, S. M. (1998). The use of amino acid sequence alignments to assess potential allergenicity of proteins used in genetically modified foods. *Adv. Food Nutr. Res.*, **42**, 45–62.

Harrison, L. A., Bailey, M. R., Naylor, M. W., Ream, J. E., Hammond, B. G., Nida, D. L., Burnette, B. L., Nickson, T. E., Mitsky, T. A., Taylor, M. L., Fuchs, R. L., and Padgette, S. R. (1996). The expressed protein in glyphosate-tolerant soybean, 5-enolpyruvylshikimate-3-phosphate synthase from *Agrobacterium* sp. strain CP4, is rapidly digested in vitro and is not toxic to acutely gavaged mice. *J. Nutr.*, **126**, 728–740.

Hefle, S. L., Nordlee, J. A., and Taylor, S. L. (1996). Allergenic foods. *Crit. Rev. Food Sci. Nutr.*, **36**, S69–S89.

Herian, A. M., Taylor, S. L., and Bush, R. K. (1990). Identification of soybean allergens by immunoblotting with sera from soy-allergic adults. *Int. Arch. Allergy Appl. Immunol.*, **92**, 193–198.

Ito, K., Inagaki-Ohara, K., Murosaki, S., Nishimura, H., Shimokata, T., Torii, S., Matsuda, T., and Yoshikai, Y. (1997). Murine model of IgE production with a predominant T_h2-response by feeding protein antigen without adjuvants. *Eur. J. Immunol.*, **27**, 3427–3437.

Knippels, L. M. J., Penninks, A. H., Spanhaak, S., and Houben, G. F. (1998). Oral sensitization to food proteins: A brown Norway rat model. *Clin. Exp. Allergy*, **28**, 368–375.

Knippels, L. M. J., van der Kleij, H. P. M., Koppelman, S. J., Houben, G. F., Penninks, A. H., and Felius, A. A. (2000). Comparison of antibody responses to hen's egg and cow's milk proteins in orally sensitized rats and food-allergic patients. *Allergy*, **55**, 251–258.

Leung, P. S. C., Chow, W. K., Duffey, S., Kwan, H. S., Gershwin, M. E., and Chu, K. H. (1996). IgE reactivity against a cross-reactive allergen in Crustacea and Mollusca: Evidence for tropomyosin as the common allergen. *J. Allergy Clin. Immunol.*, **98**, 954–961.

Li, X., Schofield, B. H., Huang, C. K., Kleiner, G. I., and Sampson, H. A. (1999). A murine model of IgE-mediated cow's milk hypersensitivity. *J. Allergy Clin. Immunol.*, **103**, 206–214.

Li, X., Serebrinsky, D., Lee, S. J., Huang, C. K., Bardina, L., Schofield, B. H., Stanley, J. S., Burks, A. W., Bannon, G. A., and Sampson, H. A. (2000). A murine model of peanut anaphylaxis: T- and B-cell responses to a major peanut allergen mimic human responses. *J. Allergy Clin. Immunol.*, **106**, 150–158.

Mekori, Y. A. (1996). Introduction to allergic diseases. *Crit. Rev. Food Sci. Nutr.*, **36**, S1–S18.

Metcalfe, D. D., Astwood, J. D., Townsend, R., Sampson, H. A., Taylor, S. L., and Fuchs, R. L. (1996). Assessment of the allergenic potential of foods derived from genetically engineered crop plants. *Crit. Rev. Food Sci. Nutr.*, **36**, S165–S186.

Moneret-Vautrin, D. A., Kanny, G., Rance, F., and Lemerdy, P. (1997). Les allergens végétaux alimentaires. Allergies associées et réactions croisées. *Rev. Fr. Allergol. Immunol. Clin.*, **37**, 316–324.

Nordlee, J. A., Taylor, S. L., Townsend, J. A., Thomas, L. A., and Bush, R. K. (1996). Identification of a Brazil nut allergen in transgenic soybeans. *N. Engl. J. Med.*, **334**, 688–692.

Ortolani, C., Ipsano, M., Pastorello, E., Bigi, A., and Ansaloni, R. (1988). The oral allergy syndrome. *Ann. Allergy*, **61**, 47–52.

Padgette, S. R., Kolacz, K. H., Delannay, X., Re, D. B., La Vallee, B. J., Tinius, C. N., Rhodes, W. K., Otero, Y. I., Barry, G. F., Eichholtz, D. A., Peschke, V. M., Nida, D. L., Taylor, N. B., and Kishore, G. M. (1995). Development, identification and characterization of glyphosate-tolerant soybean line. *Crop Sci.*, **35**, 1451–1461.

Padgette, S. R., Taylor, N. B., Nida, D. L., Bailey, M. R., MacDonald, J., Holden, L. R., and Fuchs, R. L. (1996). The composition of glyphosate-tolerant soybean seeds is equivalent to that of conventional soybeans. *J. Nutr.*, **126**, 702–716.

Reese, G., Ayuso, R., and Lehrer, S. B. (1999). Tropomyosin: An invertebrate pan-allergen. *Int. Arch. Allergy Immunol.*, **119**, 247–258.

Sampson, H. A., and McCaskill, C. M. (1985). Food hypersensitivity and atopic dermatitis: Evaluation of 113 patients. *J. Pediatr.*, **107**, 669–675.

Sampson, H. A., Mendelson, L., and Rosen, J. (1992). Fatal and near-fatal anaphylactic reactions to foods in children and adolescents. *N. Engl. J. Med.*, **327**, 380–384.

Sanders, P. R., Lee T. C., Groth, M. E., Astwood, J. D., and Fuchs, R. L. (1998). Safety assessment of insect-protected corn, in *Biotechnology and Safety Assessment*, 2nd ed., J. A. Thomas, ed., pp. 241–256. Taylor & Francis, London.

Shin, D. S., Compadre, C. M., Maleki, S. J., Kopper, R. A., Sampson, H., Huang, S. K., Burks, A. W., and Bannon, G. A. (1998). Biochemical and structural analysis of the IgE binding sites on Ara h 1, an abundant and highly allergenic peanut protein. *J. Biol. Chem.*, **273**, 13753–13759.

Stanley, J. S., King, N., Burks, A. W., Huang, S. K., Sampson, H., Cockrell, G., Helm, R. M., West, C. M., and Bannon, G. A. (1997). Identification and mutational analysis of the immunodominant IgE binding epitopes of the major peanut allergen Ara h 2. *Arch. Biochem. Biophys.*, **342**, 244–253.

Steinmann, J., and Wottge, H. U. (1990). Immunogenicity testing of food proteins: In vivo and in vitro trials in rats. *Int. J. Allergy Appl. Immunol.*, **91**, 62–65.

Strobel, S., and Ferguson, A. (1984). Immune responses to fed protein antigens in mice. Systemic tolerance or priming is related to age at which antigen is first encountered. *Pediatr. Res.*, **18**, 588–594.

Taylor, S. L. (2001). Safety assessment of genetically modified foods. *J. Nematol.* **33**, 178–182.

Taylor, S. L., and Hefle, S. L. (2001). Food allergy, in *Present Knowledge in Nutrition*, 8th ed., (R. M. Russell and B. Bowman, eds.), pp. 463-471, ILSI Press, Washington, DC.

Taylor, S. L., and Lehrer, S. B. (1996). Principles and characteristics of food allergens. *Crit. Rev. Food Sci. Nutr.*, **36**, S91–S118.

Taylor, S. L., Lemanske, R. F., Jr., Bush, R. K., and Busse, W. W. (1987a). Chemistry of food allergens, in *Food Allergy*, R. K. Chandra, ed., pp. 21–44. Nutrition Research Education Foundation, St. John's, Newfoundland.

Taylor, S. L., Lemanske, R. F., Jr., Bush, R. K., and Busse, W. W. (1987b). Food allergens: Structure and immunologic properties. *Ann. Allergy*, **59**, 93–99.

Taylor, S. L., Hefle, S. L., and Gauger, B. J. (2001). Food allergies and sensitivities, in *Food Toxicology*, W. Helferich and C. K. Winter, eds., pp. 1–36. CRC Press, Boca Raton, FL.

Vieths, S. (1997). Allergenic potential of genetically modified plant foods—how reliable is the proposed assessment strategy? In *Proceedings of an International Symposium on Novel Foods Regulation in the European Union—Integrity of the Process of Safety Evaluation*, Berlin, November 18–20.

Chapter 12

Biotechnology: Safety Evaluation of Biotherapeutics and Agribiotechnology Products

John A. Thomas
University of Texas Health Science Center
San Antonio, Texas

Unlike the ready acceptance of biotherapeutics by both the public and the medical profession, agribiotechnology's acceptance has not been as rapid. This, however, is changing as more and more genetically modified (GM) foods enter the market worldwide. Safety and regulatory oversight continues, to ensure that these new and novel GM foods are well investigated before being made available to consumers. The future holds great promise for GM foods and feeds, not only for their nutritional and economical importance but also for their potential as vectors for oral vaccines and for generally meeting the caloric needs of people in developing countries.

Since the introduction of several hormones derived from recombinant DNA, such as human insulin and later human growth hormone, there have been many advances in the production of a variety of therapeutic agents. The interferons, blood clot dissolving agents, and immunomodulators represent new therapeutic opportunities. Increasingly, the safety evaluation and preclinical assessment of newer biotherapeutic agents, while still approached from a regulatory standpoint on a case-by-case basis, has undergone signficant improvement, with enhanced predictability as a result.

Biotechnology and Safety Assessment, 3rd edition

INTRODUCTION

HISTORY

Biotechnology has been used for centuries, but only recently has it taken on more molecular complexities in terms of the genetic manipulation of both plant and animal cells. The term *biotechnology* can accommodate somewhat different interpretations, but classically it refers to any technique that uses living organisms (or components of these organisms) to modify or create life-forms, to improve plants or animals, or to develop microorganisms for specific uses. Biotechnology has been used for centuries in making wine, cheese, yogurt, and bread, and in the selective crossbreeding of animals and plants, all leading to an enhancement of specific desirable traits. It is believed that the term "biotechnology" was first coined by a Hungarian, Karl Ereky, toward the end of World War 1. Ereky purportedly used the term to refer to intensive agricultural methods (Thomas, 1998). Unfortunately, the term "genetically modified organisms" (GMOs), coined nearly two decades ago by those seeking to give it a negative connotation, is somewhat misleading with regard to issues of safety or risk pertaining to foods. In retrospect, many other terms could have been adopted, including "selective cross breeding".

Genetic engineering involves the ability to manipulate, modify, or otherwise "engineer" genetic material to produce desired characteristics. This has been the integral part of biotechnology (Liberman *et al.*, 1991). Genetic engineering also encompasses recombinant DNA (rDNA) technology. The era of biotechnology began about three decades ago with the discovery of endonucleases or "restriction enzymes", and their ability to selectively "cut and paste" segments of DNA. Such enzymes act as highly specific chemical scalpels and can be used to obtain specific sequences within genetic material. Thus, restriction enzymes recognize and cut DNA at precise locations. The use of recombinant DNA and similar genetic engineering procedures led to the transfer of genetic material across species barriers. Actually, cell biology has become another overlapping core discipline of modern biology, along with genetics and biochemistry. Of course the sequencing of the human genome will undoubtedly lead to significant advances in other areas of biotechnology and to a better understanding of genetic-related diseases.

Table 1 reveals several important milestones in biotechnology beginning with early advances in molecular biology to present-day DNA-derived products or materials, and even the sequencing of the human genome. Such mileposts include not only important therapeutic proteins, but also environmental and agricultural developments. The first genetically produced biopharmaceutical was insulin, in 1978, followed by the first genetically improved food, the tomato, in 1993.

Table 1

Selected Milestones in Biotechnology

1953	Double-helix structure of DNA discovered
1965	RNA used to break genetic code
1967	Automatic protein synthesizer developed
1967	Genetically engineered food introduced (potato)
1970	Discovery of restriction enzymes (endonucleases)
1972	Splicing of viral DNA
1975	Discovery of monoclonal antibodies (hybridomas)
1976	First commercial company founded to develop recombinant DNA
1977	Expression of gene for human somatostatin
1978	Recombinant DNA used to produce insulin
1978	Mammal-to-mammal gene transplants
1979	Human growth hormone and interferons reproduced
1979	Effect of recombinant bovine somatotropin (rbST) on milk production in dairy cows reported
1980	Transgenic animals produced (mouse to mouse)
1982	Recombinant DNA animal vaccine developed
1982	Production of synthetic growth hormone
1983	Polymerase chain reaction (PCR) technique developed
1989	Bioremediation technology introduced
1990	Clinical trials of human gene therapy begin
1991	Genetically engineered biopesticide created
1993	Genetically engineered tomato produced
1993	Recombinant DNA blood coagulation factors produced
1995	Clinical use of recombinant follicle-stimulating hormone
1996	Cloning of adult mammal
1997	Plasmid DNA vaccines
1998	Further development of xenotransplants
2000	Transgenic plants enhanced with micronutrients
2002	Human genome mapping completed

There are numerous applications of these new biotechnologies. Hormone substitutes, nutritional supplements, and improved crop yields are but a few examples of how this new biotechnology will impact on the quality of life. The biotechnology industry encompasses several segments, including therapeutics, vaccines, diagnostics (e.g., monoclonal antibodies), single-nucleotide polymorphisms (SNPs), agriculture, and the environment. There is the potential to discover many new drug entities for the treatment of various diseases. Agribiotechnology will play an important role in worldwide food production as well as providing products with enhanced nutritional value.

Transgene Technology

Animals

The introduction of transgene technologies involving the transfer of a gene from one species to another species has occurred along with other advances in recombinant DNA methodologies or genetic engineering. The creation of laboratory mouse models for biomedical research has progressed rapidly, with so-called designer mice (i.e., transgenics) being used to study immune deficiencies, cancer, and developmental biology (Goodnow, 1992; Grosveld and Kollias, 1992; Thomas, 1997) (Table 2). Over a decade has elapsed since the development of a transgenic mouse model and its introduction into biomedical research. Transgenic animals are produced by inserting a foreign gene into an embryo, with this foreign gene becoming an integral part of the host animal's genetic materials. Increasingly, there have been improvements in the techniques used for gene introduction into animals. The techniques that have been used include microinjections of DNA into the pronucleus, retrovirus infection, and the insertion of embryonic stem cells, which can be grown *in vitro* from explanted blastocysts. All these techniques have been employed to create chimeric animals. Transgenic animal models are only one type of at least four other classes of models that can be used to study specific disease states (Thomas, 1997).

Table 2
Human Diseases and Pathological Conditions Simulated by Transgenic Mouse Models

AIDS	Inflammatory diseases
Demyelinating diseases	Neonatal hepatitis
Diabetes mellitus	Oncogenesis
Glomerulosclerosis	Osteogenesis imperfecta
Hepatocarcinogenesis	Sickle cell anemia
Hypertension	

Source: Modified from Goodnow (1992).

Plants

Generally speaking, progress has been slower in developing transgene species in the plant cell than in the animal cell. Plant breeding techniques have evolved over the millennium with the purpose of improving crops. Artificially induced mutations and selection have hastened the improvements in major cereal crops. Initially, improvements in plant biotechnology have been directed at enhancing agronomic traits including pest resistance, herbicide tolerance, and disease resistance (e.g., viruses and fungi). More recently, significant progress has been made in improving the nutritional value of many cereal grains.

Regulatory Considerations

Historically, the regulatory approval of genetically engineered products has varied widely from country to country and from product to product. The safety and regulatory approval process for therapeutic proteins has generally proceeded without any major impediments. Basically, approval processes for what are usually considered to be "biologics" have long been established. Alternate sources of insulin for the treatment of diabetes mellitus and of growth hormone for treating children with growth deficiencies have not been controversial and posed no significant risk. Still other more novel therapeutic proteins have undergone more rigorous testing to ensure safety and efficacy prior to any regulatory approval. Agribiotechnology involving improvements in food quality, however, has not been universally accepted by the public nor, in every instance, by regulatory agencies. Generally, the somewhat slower progress in obtaining regulatory approval of genetically modified foods has not been an issue of human safety, but of potential environmental impact on wildlife and nontarget species.

Many countries have well-established regulatory committees or groups for biotechnology-derived drugs or biopharmaceuticals (Table 3). In the United States, the Food and Drug Administration (FDA), through the Center for Biologic Evaluation and Research (CBER), has oversight responsibilities for ensuring safety and efficacy of biopharmaceuticals.

Regulatory responsibility for biotechnology-derived foods may involve several U.S. agencies including the Department of Agiculture (USDA), the Environmental Protection Agency (EPA), and the FDA. The FDA again has oversight on food safety as it relates to new plant varieties (e.g., dairy products, food processing aids, etc.). The USDA regulates poultry products and meats and has responsibility for the oversight of field tests for genetically modified crop plants. The EPA, through it regulatory responsibility for pesticides, oversees aspects of genetically altered plants with traits such as herbicide tolerance, pest resistance, and potential effects on nontarget wildlife species.

Table 3
Regulatory Aspects of Biotechnology

Country(s)	Regulatory committee/agency
Europe (EU)	Committee or Proprietary Medicinal Products (CPMP)
	Safety Working Party (SWP)
	Biotechnology Quality Working Party (BQWP)
	Operations Working Party (OWP)
	Council Directive 87.22 EEC
	Concertation Procedure
	Notes for Guidance (e.g., preclinical safety) (1987)
Japan (Ministry of Healthy and Welfare)	Pharmaceutical Affairs Bureau (PAB)
	Central Pharmaceutical Affairs Council (CPAC)
	Committee on Drugs
	Committee on Antibiotic Drugs
	Committee on Blood Products
	National Institute of Hygienic Sciences
	Notification no. 243 (1984)
	Notification no. 10 (1988)
United States (FDA)	Center for Drug Evaluation and Research (CDER)
	Center for Biologic Evaluation and Research (CBER)
	Public Health Services Act
	Food, Drug and Cosmetic Act
	Federal Register: biotechnology notice (1984)
	Federal Register: regulation of biotechnology
	Product (class) oriented "Points to Consider" (PTC)

Sources: Bass *et al.* (1992); Thomas (1995).

With the globalization of commerce, many agencies worldwide have become involved in the regulation of genetically modified organisms, which may involve both plant and animal biotechnologies. Several organizations dealing with GMOs have been identified (Table 4). Many of these organizations have no regulatory authority, yet have proposed various guidelines for the use of GMOs. Continuing efforts have been devoted to harmonization and standardization of safety protocols for both biotherapeutic agents and agribiotechnology products.

Table 4

Selected Organizations Involved in Plant and Animal Biotechnology

AcNFP	Advisory Committee on Novel Foods and Processes (U.K.)
BQWP	Biotechnology Quality Work Party (EU)
CBD	Convention on Biological Diversity (multinational)
CBER	Center for Biologic Evaluation and Research (U.S.A.)
CDER	Center for Drug Evaluation Research (U.S.A.)
CPMP	Committee for Proprietary Medicinal Products (EU)
GIBiP	Green Industry Biotechnology Platform (multinational)
ICH	International Conference on Harmonization (multinational)
NIBSC	National Institute for Biological Standards and Control (U.K.)
OECD	Organisation for Economic Cooperative and Development
UNCED	U.N. Conference on Environment and Development
UNEP	U.N. Environmental Program
UNIDO	U.N. Industrial Development Organisation

There are several advantages and disadvantages to regulatory guidelines (Table 5). Importantly, the advantages outweigh the disadvantages. Harmonization or adoption of consistent review protocols is imperative and will lead to better overall assessment of safety and efficacy of new biotechnology-derived products. It is important that the review be science based, and that it continue to evolve and keep abreast of the latest breakthroughs in technologies and food processing.

Table 5

Regulatory Guidelines

Advantages	Disadvantages
Formalization of a consensus scientific opinion	Nonuniformity in study designs
Promotion of consistency of review	Uncertainties encourage retreat to "check-the-box" approach
Improvement of quality of studies performed	Disincentive to industry to be creative
Assistance to industry	Failure to appropriately incorporate new and evolving technologies

Source: Henck *et al.* (1996).

BIOTHERAPEUTICS

INTRODUCTION

The development of recombinant DNA technologies has enabled the pharmaceutical-related industries to manufacture large quantities of macromolecules in relatively pure form. This has led to the production of such therapeutic proteins as human insulin, human growth hormone (hGH), recombinant plasminogen activator (rTPA), and human erythropoietin (EPO) (Thomas, 1998). While hormone substitutes (e.g., insulin, hGH) represented an early entry in biopharmaceuticals, many other classes of therapeutic agents have been produced or are otherwise undergoing development or clinical trials. Since most of the early biopharmaceuticals were represented by simple hormone replacement therapies (e.g., insulin), they did not undergo the same extensive preclinical safety and efficacy studies such as those that would be required for interferon(s) (Vial and Descotes, 1994) and antisense drugs. Several biopharmaceutical products have been approved for a variety of clinical indications (Table 6). Several hGH preparations have received regulatory approval. New clinical indications have been approved for the interferons and for other immunomodulating agents. Deficiencies or disorders of blood factor(s) have benefited from these biotechnological advances, as have medical specialties requiring blood clot dissolution proteins. There have been an increasing number of regulatory approvals for monoclonal antibodies. It is noteworthy that the regulatory approval of rhGH was expedited by the discovery that some cadaveric hGH was possibly contaminated with latent neural viruses responsible for Creutzfeldt–Jacob disease.

In the new few years, a particularly exciting area that uses single-nucleotide polymorphism (SNPs) will yield information about promising drug targets for a host of diseases. Gene-based medicine, known also as functional genomics, may hold the key to understanding and treating various diseases (e.g., Alzheimer's disease, diabetes mellitus).

The preclinical safety assessment of new biotherapeutic agents for human medical therapies has highlighted the issue of immunogenicity between species (Cavagnaro, 1997; Thomas, 1999). Macromolecules derived through biotechnology are expected to challenge the immune system of the laboratory animal used for toxicological testing protocols. Despite some of these unique and early concerns, guidelines for the preclinical testing of high molecular weight biopharmaceuticals has proceeded, and has generally been adopted by regulatory agencies. The challenge to the design of toxicological protocols represented by macromolecules has led also to challenges in the formulation and delivery of biopharmaceuticals (Panchagnula and Thomas, 2000). Macromolecules also represent a challenge in noninvasive drug delivery

Table 6

Selected Biopharmaceutics

Generic name	Trade name	Classification	Indication(s)
Alteplase (rTPA)	Activase	Recombinant protein	Acute myocardial infarction Pulmonary embolism
Tenectaphase	TNKase	Recombinant protein	Acute myocardial infarction
Epoetin alfa (EPO)	Epogen Procrit	Recombinant protein	Anemia of chronic renal failure
			Anemia secondary to zidovudine treatment in HIV-infected patients
Antihemophilic factor	Recombinate, Humate-P,Refacto	Recombinant protein	Hemophilia A
Factor VIII	KoGENAteFS, Helix-ate FS	Recombinant protein	Hemophilia A
Factor VIIIc	Monoclate-P	Recombinant protein	Hemophilia A
Factor IX	L-Nine SC	Recombinant protein	Hemophilia B
TNF	Embrel	Recombinant protien	Active rheumatoid arthritis
Filgrastim (G-CSF)	Neupogen	Recombinant protein	Neutropenia following chemotherapy
Sargramostim	Leukine	Recombinant protein	Myeloid reconstitution after bone marrow transplantation
Interferon-α-2a	Roferon A	Recombinant protein	Hairy cell leukemia AIDS-related Kaposi's sarcoma
Interferon-α-2b	Intron A	Recombinant protein	Hairy cell leukemia AIDS-related Kaposi's sarcoma, hepatitis
	non-A, non-B/C		Condylomata acuminata
Interferon-α	Wellferon	Recombinant protein	Chronic hepatitis C
Interferon-α-n3	Alferon N	Human Protein	Genital warts
Interferon-β-1a	Avonex	Recombinant protein	Multiple sclerosis
Interferon-β-1b	Betaseron	Recombinant protein	Multiple sclerosis
Interferon-γ-1b	Actimmune	Recombinant protein	Chronic granulomatous disease
I-IGF-1	Myotropin	Recombinant protein	Amyotrophic lateral sclerosis (ALS)
Aldesleukin (IL-2)	Proleukin	Recombinant protein	Kidney cancer
Dornase-α	Pulmozyme	Recombinant protein	Cystic fibrosis

(*continues*)

Table 6 (*continued*)

Generic name	Trade name	Classification	Indication(s)
Human insulin	Humulin	Recombinant protein	Diabetes mellitus
Somatrem (rMehGH)	Protropin	Recombinant protein	Growth hormone deficiency
Somatropin (rhGH)	Humatrope Saizen Genotropin Nutrotropin	Recombinant protein	Growth hormone deficiency
FSH	Puregon	Recombinant protein	To induce ovulation
Somatotropin (rbGH)[a]	Posilac	Recombinant protein	To increase bovine lactation
Pegaspargase (PEG-1-asparaginase)	Oncaspar	Poly (ethylene glycol) modified protein	Acute lymphocytic leukemia
Hemophilus B conjugate vaccine	HibTITER	Recombinant protein	Prophylaxis against *Hemophilus influenzae*
Hepatitis B vaccine	Engeridx-B Recombivax HB	Recombinant protein	Prophylaxis against hepatitis B infection
Satumomab	OncoScint CR103	Monoclonal antibody	Colorectal cancer imaging
Satumomab	OncoScint OV103	Monoclonal antibody	Ovarian cancer imaging
Muromonab CD3	Orthoceone (OKT3)	Monoclonal antibody	Acute allograft rejection in renal transplants
Alemtuzumab	Compath	Monoclonal antibody	Chronic lymphocytic leukemia
	Herceptin	Monoclonal antibody	Breast cancer
	Rituxan	Monoclonal antibody	Non-Hodgkin's lymphoma

[a]Veterinary product.

systems, but the use of supercritical fluid technology may produce dry powder formulations suitable for inhalation or needle-free administrations (Tservistus *et al.*, 2001).

The New Biologies

Progress in the scientific disciplines of genetics, immunology, and molecular biology has provided a strong impetus for advances in biotechnology, and particularly in biopharmaceuticals (Tomlinson, 1992). Indeed, biotechnology-derived products could not have progressed to their current state without the coincidental and independently developed technology of monoclonal antibody production. Certainly, recombinant DNA technologies and monoclonal antibody production are two areas that are inseparably important to the development of new biopharmaceuticals. Monoclonal antibodies (MAb) are essential in the identification, extraction, and purification of most macromolecules. The term "molecular biology" has come to signify the biochemical study of nucleic acids, advanced by the discovery of a series of enzymes (e.g., restriction enzymes or endonucleases) that allows for specific manipulation of RNA and DNA.

There are several techniques or approaches used in molecular biology (Table 7) (Ausubel *et al.*, 1995). These technologies are advancing at a very rapid pace. The ability to manipulate genetic material of a cell to be able to modify gene expression by means of a variety of new techniques has facilitated the discovery of new biopharmaceuticals. Using the appropriate mammalian expression system for the development of biopharmaceuticals is very important (Werner *et al.*, 1998). New advances in the development of highly sensitive and highly specific analytical methods (e.g., HPLC MS/MS) and *caco-2* cell techniques will facilitate the entry of new biopharmaceuticals into the market (Panchagnula and Thomas, 2000).

Table 7
Some Molecular Cell Biology Tools

Recombinant DNA techniques
Site-directed mutagenesis
Single-strand conformational polymorphism (SSCP)
Ligated gene fusions
Hybridoma technology for production of antibodies
Carbohydrate engineering
Novel instrumental techniques
Polymerase chain reaction (PCR) and reverse transcription PCR (RT-PCR)
Subtractive hybridization differential display
Immunoblotting
Confocal laser microscopy
Fluorescence-activated cell sorting (FACS)
Scanning tunneling electron microscopy
Pulsed field electrophoresis
Northern blotting and solution hybridization
In situ hybridization
Southern blotting: gene deletions

Sources: Demain (1991); Davis (1996).

Therapeutic Proteins

Developments in recombinant DNA technologies have focused upon mammalian macromolecules, often peptides, for discovering therapeutically active entities (Kelley, 1996). Mammalian biotechnology has advanced more rapidly than agribiotechnology, although the latter field promises to be very fast moving. Early successes witnessed the use of prokaryotic cell systems (e.g., *E. coli*) capable of producing rather complex mammalian proteins. Subsequently, and with more advances in techniques employed in molecular biology (Table 8), prokaryotic cell systems were found to be able to secrete even more complex proteins and glycoproteins. Improved methods of molecular characterization made it possible to create more complex and more highly purified therapeutic entities. Mammalian organisms possess a plethora of proteins that modulate intricate physiological systems, and through biotechnology some of these will be discovered for their therapeutic usefulness.

Table 8

Advances in the Characterization of Biological Products

Method	Examples
In vitro bioassays	Cell culture techniques for hormones and cytokines
Gene analysis	PCR analysis (viral contaminations of blood products)
Immunological tests	Enzyme-linked immunosorbent assay (ELISA)
	Epitope mapping by panels of monoclonal antibodies
	Immunoblotting of gels (Western blots)
	Testing for process-related impurities
Advanced chromatographic techniques	Size exclusion and reversed-phase high performance liquid chromatography (HPLC)
	Capillary zone electrophoresis
Analysis of three-dimensional structure	Circular dichroism Nuclear magnetic resonance spectrometry
Protein analysis by mass spectrometry	Fast atom bombardment mass spectrometry (FAB-MS) (antigen binding) Electrospray mass spectrometry (ES-MS)
Analysis of glycosylation isoforms	Detailed structure–function studies required

Source: Modified from Tomlinson (1992).

Modern molecular biology, particularly through the use of the polymerase chain reaction (PCR), has become exceedingly important in the discovery and production of recombinant proteins. Acting as a molecular "photocopier," PCR has been employed extensively to amplify DNA sequences. Subsequently, the design of new biopharmaceuticals has proceeded with quantities of purified material sufficient to permit investigators to undertake both preclinical and clinical studies. Nowadays, robot gene sequencers (e.g., ABI Prism 3700) and high-speed computers have greatly facilitated the discoveries associated with the Human Genome Project.

The majority of new biopharmaceuticals usually must be administered parenterally. Proteins or large peptides are vulnerable to gastric degradation and experience a loss of biological activity. There is considerable interest and need to develop drug delivery systems for these macromolecules (cf. Tservistas *et al.*, 2001). Selective drug targeting is a characteristic that is highly desirable. Progress has been made with some biotherapeutic agents by modifying their biological dispersion characteristics (Table 9). Such approaches have included site-directed mutagenesis, hybrid site-specific proteins, and supercritical fluid technologies. In addition, significant progress has been made toward the development of site-specific drug delivery systems for protein and peptide biopharmaceuticals (Pettit and Gombotz, 1998).

Table 9

Approaches for Modifying the Biological Dispersion of Therapeutic Proteins

Approach	Therapeutic protein
Site-directed mutagenesis	α_1-Antitrypsin
Hybrid site-specific proteins	Immunoglobulin and toxin
Linked synthetically	Fragments
Fused-gene products	Growth factors
Protectants: poly(ethylene glycols)	Interleukins
Administration	
Frequency and rate	Growth hormone (effectiveness altered)
Route	Insulin (site of injection)
Staging	Combinations of interferon and tumor necrosis factor

Source: Demain (1991).

Safety Evaluation

General Considerations

Assessing the "safety" of drugs produced by biotechnology resembles the assessment of conventional new chemical entities (NCE), but with certain major differences in testing protocols (Zbinden, 1990; Dayan, 1995; Cavagnaro, 1997). Several safety considerations must be incorporated into the preclinical testing protocol (Table 10) (Tomlinson, 1992). The quality of each biotechnology-derived product necessitates careful control because of concerns about immunogenic proteins or peptides, endotoxins released by the harvesting of prokaryotic cell systems that secrete the product, and other chemical contaminants that may emanate from processing procedures. There are at least three areas of concern regarding biotechnology-derived products: toxicology issues pertaining to differences in pharmacodynamic properties, "intrinsic" toxicity (i.e., adverse effects due to the molecule itself and not directly related to pharmacological actions), and "biological" toxicity, or responses resulting from the activation of a physiological mechanism (e.g., antigen–antibody reaction) (Zbinden, 1990).

Several safety issues form the basis for toxicity testing. There are safety issues pertaining to the intrinsic actions of the product itself, and there are safety issues pertaining to the cellular systems or processes that lead to the production or secretion of the macromolecule (Thomas and Thomas, 1993; Dayan, 1995; Thomas, 1995; Cavagnaro, 1997). Product-related impurities or variants might include mutants, aggregated forms, and aberrant glycosylation. Process-related impurities might include pyrogens, host-cell-derived proteins, viruses, and possibly prions (Jeffcoate, 1992). Table 11 summarizes

Table 10

Some Safety Considerations of Biotechnology-Derived Products

Aggregated forms
DNA and potential oncogenicity
Genetic variants and mutants
Glycosylation patterns
Host-cell-derived proteins
Process-related impurities
Product-related impurities or variants
Pyrogens
Viruses (human, simian, murine, bovine)

Source: Modified from Tomlinson (1992).

Table 11

Selected Guidelines for Evaluating Biopharmaceutics

Toxicological issues arising from the producing system
Prokaryotic production system (recombinant DNA): Correct gene and promoter, stable expression, contaminating toxins, etc.
Eukaryotic Production System (recombinant DNA): Correct gene and promoter; stable expression, presence of antigens, other bioactive peptides, etc.
Other: chemical modifications, infection-free animal vectors, etc.
Toxicological issues arising from the production process
De- and renaturation of protein(s)
Presence of other bioactive molecules
Presence of chemical residues
Microbial contamination (e.g., endotoxins)
Toxicological issues arising from biopharmaceutics
Pharmacodynamics
Pharmacological and toxicological actions
Immunological
Other (e.g., live or attenuated vaccine)

Source: Modified from Dayan (1995).

some of the more important aspects of evaluating the safety of a potential biopharmaceutical. It is of fundamental importance to have reliable cell systems, expressing the correct gene and ensuring its proper insertion along with any promoter. The secretion or presence of other bioactive moieties must always be considered and needs to be addressed through proper extraction and purification processes. Protein or peptide contaminants can produce both immunological and nonimmunological actions that are unwanted. Clinical or conventional toxicological protocols may not be relevant in attempting to characterize the phrmacodynamics of a macromolecule. Nonimmunological testing depends on the product's potential therapeutic use, and at a minimum testing should include biodistribution, developmental and reproductive, and mutagenicity/oncogenicity assessments. Protocols that define the kinetics and safety of infused recombinant proteins include bolus delivery, continuous infusion, and other physiological parameters (Yaksh *et al.*, 1997). It is very important that a panel for detecting possible immune alterations be an integral component of the toxicology testing protocol (Table 12). The first tier consists of a screening panel that evaluates immunopathology, cell-mediated immunity, and humoral immunity. Tier 1 should also identify agents that may elicit an immune response. Agents testing positive in tier I

Table 12
Panel for Detecting Immune Alterations

Tier I
Hematology (e.g. leukocyte counts)
Weights (body, spleen, thymus, kidney, liver)
Cellularity (spleen, bone marrow)
Histology of lymphoid organ
IgM antibody plaque-forming cells (PFCs)
Lymphocyte blastogenesis
T-cell mitogens (phytohemagylutinin, concanavalin A)
T-cell mitogens (mixed leukocyte response MLR)
β-cell mitogens (lipopolysaccharide, LPS)
Natural killer (NK) cell activity
Tier 2
Quantitation of splenic B and T cells
Lymphocytes (surface markers)
Enumeration of IgG antibody PFC response
Cytotoxic T lymphocytes (CTL)
Cytolysis or delayed hypersensitivity response (DHR)
Host resistance
Syngeneic tumor cells
PYB6 sarcoma (tumor incidence)
B16F10 melanoma (lung burden)
Bacterial models
Listeria monocytogenes (morbidity)
Streptococcus species (morbidity)
Viral models: influenza (morbidity)
Parasite models: *Plasmodium yoelii* (parasitemia)

Source: Condensed from Luster *et al.* (1992).

must undergo further evaluation. Tier 2 seeks to evaluate the presence of any immunopathological effects.

Antisense Drugs

The use of antisense oligonucleotide is probably the most mature component in the development of macromolecular drugs (cf. Crooke, 1992; Juliano *et al.*, 2001). Base pairing between relatively short oligonucleotides (12–18 residues) and complementary sequences in mRNA provides highly selective recognition that can affect gene expression.

In antisense technology, the primary sequencing information available with the completion of the sequencing of the human genome, can be used to design short oligonucleotides to hybridize to a specific mRNA (Bennett *et al.*, 1998). Once bound to the targeted mRNA in a cell, the antisense oligonucleotide prevents expression of the protein product encoded by the targeted RNA. Antisense oligonucleotides can interrupt gene expression. Antisense molecules are highly selective, and when inserted into a cell can bind particular mRNA molecules and prevent them from being translated into proteins.

Antisense drug therapy may affect pathophysiological processes at the genetic level (Harvey, 1998). Such therapies involve using modified oligonucleotides that bind specifically to target sequences of RNA to affect translational arrest or to sections of DNA to block transcription. It is noteworthy that there are naturally occurring antisense RNAs (Dolnick, 1997). Naturally occurring antisense RNAs may act as mediators of alterations in gene expression.

Certain technical issues need to be overcome before antisense drug development can be fully exploited. Such issues include the size of the molecule needed to ensure specificity, the frequent need to enhance the agent's chemical stability, and the problems associated with controlling uptake and tissue distribution. Unmodified oligonucleotides are negatively charged and do not readily enter cells. However, modification of the internucleoside phosphate linkages to uncharged methylphosphonates enables the oligonucleotides to gain access to a cell's interior (cf. Persaud and Jones, 1994).

An oligonucleotide may inhibit target gene expression by several mechanisms (cf. Gibson, 1994; Bennett *et al.*, 1998; Toulme, 2001). Oligonucleotides can induce degradation of the target DNA by the cellular enzyme RNase H. Other mechanisms proposed to explain how protein synthesis might be inhibited include the steric blocking of ribosome assembly and splicing, and capping – all events involved in mRNA production. The maturation of mRNA or its translation into protein has also been suggested.

The first generation of oligonucleotides has met with limited success primarily because natural DNA is degraded rapidily by serum and cellular nucleases. Chemical modfications, including oligonucleotides that are resistant to nuclease degradation, have had some limited therapeutic effectiveness. The most commonly used modified oligonucleotides are the phosphorothioate oligonucleotides in which one of the nonbridging oxygen atoms in the phosphate backbone is replaced with sulfur (cf. Bennett *et al.*, 1998).

Modified oligonucleotides have half-lives that range from 30 to 60 minutes following intravenous administration. Oral bioavailability is low, due principally to nucleases present in the gastrointestinal tract. Biodistribution reveals relatively high concentrations in the liver and kidney.

The pharmacological implications for antisense therapy may encompass virology, oncology, and the cardiovascular system. Pulmonary, immuno-, and neuropharmacology are also potential targets. Cancer-related targets such as Bcl-2, Raf-1, the PKC isoenzymes, and BCR-Abl have attracted considerable attention. Members of the Bcl-2 family are important regulators of apoptosis and represent attractive targets for the development of anticancer drugs. Targets involved in inflammation (e.g., ICAM-1, intercellular adhesion molecule 1) and viral diseases are also being developed (Juliano *et al.*, 2001). Perhaps the most emphasis has been placed on the development of antiviral therapy, cancer therapy, and anti-inflammatory therapy. Several antisense oligonucleotides are undergoing clincial trials or have completed them. Forevirsen (ISIS 2922), an agent indicated for cytomegalovirus (CMV) retinitis, is commercially available. Vitravene is an antisense agent used for the topical treatment of CMV retinitis in AIDS patients.

The antisense phosphorothioate nucleotides exhibit four distinct types of toxicity (Bennett *et al.*, 1998):

1. Sequence-specific toxicity due to exaggerated pharmacological actions
2. Sequence-specific toxicity due to serendipitous hybridization to nontarget RNA
3. Sequence-specific toxicity due to non-antisense effects
4. Sequence-independent toxicity; non-antisense effects

It appears that the major toxicological concerns associated with antisense drugs are related to their interaction with targets that are not nucleic acids. Expectedly, species differences play a role in the toxicity of the antisense drugs. In rodents, immune cell activation appears to be a dose-limiting toxicity. In primates, acute effects on the complement system and anticoagulation effects have been observed. Transient fluctuations in blood pressure occurring during preclinical toxicology assessment may be due to the activation of the complement system.

New technologies are emerging that will allow the effective intracellular delivery of large molecules, including antisense oligonucleotides. Despite some early failures, antisense technology will undoubtedly play an important role in the development of new biotherapeutics.

Gene Products

Gene therapy has been defined as the transfer of exogenous genes to somatic cells of a patient in an effort to correct an inherited or acquired gene defect or to introduce a new function or characteristic. Gene therapy products can be considered biotherapeutic agents. Human gene therapy products include naked DNA and viral as well as nonviral vectors containing nucleic acids (cf. Verdier and Descotes, 1999). Some disappointing preliminary clinical results nothwithstanding, gene therapy has become an established concept in medicine. The future success of gene therapy depends on further development of the basic sciences in the discovery of new disease-related genes as well as on the identification of suitable animal models (Barzon *et al.*, 2000). It has been estimated that nearly 400 gene therapy clinical trials are under way.

By insertion of plasmid DNA into target cells, it may be possible to rectify genetic disorders, and to produce *in situ* therapeutic agents in the form of peptides and proteins or antigens that will stimulate the immune system (cf. Davis, 1997). Advances in gene therapy may also one day allow for controlling pain (Yang and Wu, 2001). For example, a therapeutic gene might code for an antinociceptive receptor required for neurotransmitter synthesis. Thus, gene therapy could even be used to block the production of pronociceptive molecules. Other attractive targets for gene therapy include the epithelial surfaces of the lungs and gastrointestinal tract, endothelial cells lining the blood vessels, muscles, myoblasts, and skin fibroblasts. Three principal means of gene-delivery systems are: viral vectors, nonviral vectors (e.g., particles, polymers), and direct injection (e.g., "gene guns").

Generally, nonviral vector systems have been favored, but they too possess certain disadvantages. For nonviral vectors to be effective, it is essential that the negatively charged plasmid DNA be condensed into a nanoparticulate structure. The use of cationic lipids and cationic polymers in which the interaction between the selected cationic materials and the anionic DNA results in a condensed/compacted structure. Such complexes provide increased stability of the genetic material as well as improved uptake into target cells (cf. Davis, 1997).

The preclinical safety evaluation of gene therapy products entails a number of toxicological issues (cf. Harris and Dayan, 1998; Verdier and Descotes, 1999). While some of these issues are common to all biological products, others are specific to gene therapy products. Table 13 lists specific preclinical safety issues for gene therapy products. Of course all test materials must be in compliance with Good Manufacturing Practices (GMP). Both *in vivo* and *in vitro* assays for adventitious viral contaminants must be undertaken. Human cells employed in the production of adenovirus vectors may be assayed for tumorigenicity in nude mice. Establishing the absence of replication-competent viruses is essential. Modification of the gene expression construct may potentially alter the product's biodistribution and/or gene expression, thus necessitating additional toxicological testing.

Table 13
Preclinical Safety Issues for Gene Therapy

Preclinical Safety Issues for Gene Therapy
Characteristics of gene therapy product
Quality assurance of gene therapy product
Safety evaluation of the vector
Toxicity of expressed protein(s)
Selection of animal species
Dose selection
Selection of route of administration

Source: Verdier and Descotes (1999).

No general test is available to predict the virulence or pathogenicity of recombinant viral vectors (cf. Smith *et al.*, 1996). Some viral vectors may induce toxic shock and possess other specific cytotoxic properties.

In considering the toxicity of the expressed protein(s), the amount of protein equivalent to the expected total gene expression is compared against the dose of protein used in the safety evaluation (cf. Verdier and Descotes, 1999) The selection of the animal species is critical, and generally nonhuman primates provide the most relevant model. Dose selection is often based on preliminary dose-ranging studies in animals, but the proposed dose used in human clinical trials should be taken into consideration. Likewise, the selection of the route of administration should be the same in both the preclinical and clinical studies with the intended therapeutic use being consistent.

Monoclonal Antibodies (Mab)

The advent of monoclonal antibody technology has revealed that the immunoglobulins exhibit a wide range of biological activities. They are very specific. Antigen-combining sites of antibody molecules have great potential for developing bioactive peptides. Earlier, technical problems arose in Mab production because of host anti-antibody immune responses. Subsequently, "humanized" antibodies have been developed by inserting the rodent hypervariable regions into human framework regions. In addition, monoclonal antibodies have been obtained by selecting antibodies from recombinatorial phage display libraries (cf. Dougall *et al.*, 1994).

Monoclonal antibodies have been used for both diagnostic and therapeutic indications, and a number of these agents have been clinically approved (see Table 14). Monoclonal antibodies have been used diagnostically in colorectal cancer imaging and ovarian cancer imaging. Antibodies (e.g., hMN-14) are under development for the treatment of ovarian, breast, and medullary

Table 14

Applications of Monoclonal Antibodies

Disease	Humanized antibody	Antigen
Coronary artery disease	C7E3Fab	Platelet glycoprotein 11b/111a receptor
Mycosis fungoides	Anti-CD4	CD4
Non-Hodgkin's B-cell lymphoma	CAMPATH-1H	CAMPATH-1; lymphoid/monocytes
Systemic vasculitis	CAMPATH-1H	CAMPATH-1
Refractory rheumatoid arthritis	CAMPATH-1H	CAMPATH-1
Generalized pustular psoriasis	Anti-CD4	CD4
Severe psoriasis	Anti-CD4	CD4
Cardiac transplant	Anti-Tac-H	CD4
Metastatic colorectal C/A	IgG1 antibody 17-1A	Glycoprotein antigen
Rheumatoid arthritis	CD4 (CM-T412)	CD4
Septic shock	HA-1A	Lipid A region, Gram negative

Source: Modified and condensed from Vaswani and Hamilton (1998).

thyroid cancers. Still other antibody-targeted chemotherapeutic agents represent a new class of anticancer therapies (e.g., CMA-676-Mylotarg). This recombinant humanized antibody is linked with a cytotoxic antitumor antibiotic (e.g., calicheamicin) and may be indicated in the treatment of acute myelogenous leukemia (AML).

Regulatory Considerations

There are some difficulties in conceiving and applying guidelines for the safety evaluation of biotechnology-derived products (Claude, 1992; Cohen-Haguenauer, 1996; Cavagnaro, 1997). Safety is highly dependent on the particular industrial process and its quality control. Pharmaceutical companies and regulatory agencies responsible for public health must establish relevant and meaningful guidelines for the preclinical and clinical evaluation of new biotherapeutic agents. When possible, there should be efforts to compare the new biopharmaceutical with a natural biologic entity. Study designs often use a case-by-case approach (Cavagnaro, 1997). Regulatory guidelines must be mindful of the importance of selecting the appropriate animal model.

Good Manufacturing Practices for biopharmaceuticals are an integral part of the safety and regulatory process (Jeffcoate, 1992; Federici, 1994).

Table 15
Aspects of Regulatory Procedures for Manufacturing

Source materials
Genetically engineered microorganisms
Transformed mammalian cell lines
Hybridoma technology
Manufacture
Clear production strategy and in-process controls
Validation of virus inactivation and removal
Purification of final product
Fraction and chromatographic procedures
Affinity purification (e.g., monoclonal antibodies)
Pasteurization
Lyophilization
End product quality
Accurate and precise methods
Rigorous specifications

Source: Modified from Jeffcoate (1992).

Other manufacturing steps that convert the new biologies into biopharmaceuticals require the exercise of due diligence throughout the entire process (Table 15).

Biopharmaceuticals are generally classified as "biologicals" by regulatory agencies. This classification would ordinarily include any virus, therapeutic serum, toxin, antitoxin, or analogous product applicable to the prevention, treatment or cure of diseases or injuries to man. The quality control process of biopharmaceuticals engenders essentially the same measures as those applied to the analysis of conventional or low molecular weight pharmaceutical products or drugs.

AGRIBIOTECHNOLOGY

INTRODUCTION

Genetically Modified Foods and World Population

Plants are one of many novel hosts that can be used not only in the production of more nutritious foods but also in the production of biopharmaceuticals. Since the early discoveries of Mendel, geneticists have been interested in the prospects of directed genetic change. Genetic modification of crop plants, undertaken to improve many of their qualities or traits, has led to a proliferation of recombinant products. Biotechnology has been used for centuries, but its newest form has enabled scientists to transfer traits between different plant species. Modern-day agribiotechnology techniques are more precise and more rapid. Agribiotechnology has many potential benefits, including increased crop yields and livestock productivity, enhanced micronutrient composition, and improved pest control through the development of herbicide-resistant crops. This technology may also improve food processing and even provide diagnostic tools for detecting plant pathogens. Still another important dimension of transgenic plants is the development and production of edible vaccines and other biotherapeutic agents. Genetic engineering of plants is not unlike the types of modification that have been used by earlier generations of plant breeders, but the techniques are more complex and diverse.

In the 1950s, there was a doubling of the world's food supply. Three decades later, the productivity of U.S. farmers exceeded the population growth curve, leading to an increase in per-capita grain output of approximately 40%. In the United States this resulted in significant agricultural surpluses. Simultaneously, the world's food supply continued to shrink and stockpiles of grain have been diminishing since the mid-1980s. These supplies continue to dwindle, even as the demand for food worldwide nearly tripled from the 1950s to the 1990s. It is noteworthy that organic farming methods (i.e., farming without the use of GM crops or vegetables), can provide less than 2% of the U.S. food supplies. The world's population is about 6 billion and growing at a rate of approximately 90 million/year. As the amount of tillable land remains unchanged, the need for agribiotechnology will assist in increasing worldwide food production.

There are several agronomic traits that can be enhanced through agribiotechnology (Stark *et al.*, 1993) (Table 16). Transgenically introduced traits that have received considerable attention include herbicide tolerance (e.g., glyphosate), virus resistance, and insect resistance. Other traits that are under active investigation or development include resistance to fungal and bacterial infections, and tolerance to different stresses such as extremes of salt concentrations, heavy metals, temperature, nitrogen levels, and phytohormones.

Representative agribiotechnology-derived products introduced as having altered traits include both food and vaccines (Table 17). Canola can be modified by using an acyl carrier protein thioesterase gene. A glufosinate-tolerant canola can be made by using a phosphenothricin acetyltransferase gene obtained from *Streptomyces viridochromogenes*. A glyphosate-tolerant cotton can be made by using the enolpyruvylshikimate-3-phosphate gene

Table 16

Advantages for Crop and Food Production Using Biotechnology

Insect resistance
Bacterial and fungal resistance
Viral resistance
Herbicide resistance
Stress tolerance
Extended shelf life
Nutrient modification

Table 17

Selected Biotechnology Products

Product	Altered trait	Source of gene(s)
Canola	High lauric acid	Turnip, oilseed, rape, etc.
Cotton	Herbicide resistance	Bacteria, virus
Potato	Resistance to beetles	Bacteria
Soybean	Herbicide resistance	Bacteria, virus
Squash	Viral resistance	Virus
Tomato	Delayed ripening	Tomato, bacteria, virus
Vaccinia virus	Vaccine	Rabies virus

Source: Paoletti and Pimentel (1996).

from *Agrobacterium* sp. strain CP4. Likewise, canola and soybeans can be rendered glyphosate tolerant. Insect-protected potatoes and corn can be made using the *cryIIIA* gene from *Bacillus thuringiensis*. Squash can be rendered virus resistant by using cota protein genes from the watermelon mosaic viruses or zucchini yellow mosaic virus. Viral-resistant GM papayas (e.g., grown in Hawaii) and viral-resistant sweet potatoes (under development in Kenya) have flourished locally, but regulatory and cultural issues in other countries have precluded their usefulness and distribution. A tomato with unconventional ripening characteristics can be achieved using several different gene modifying approaches. The insert of several genes into a plant, often referred to as "gene stacking" can lead to several new species with enhanced levels of various micronutrients. For example, rice can be nutritionally enhanced with both increased levels of iron and with β-carotene (a precursor to vitamin A). Genes from the daffodil, (whose yellow flowers contain high levels of endogenous β-carotene, can be transfected into rice, giving rise not only to a "golden" color but more importantly producing a crop plant with enhanced vitamin A It should be noted that there are several areas involved in food biotechnology, not just the alteration of plant traits and other forms of resistance. Actually, food biotechnology constitutes several areas:

- Processing aids (e.g., enzymes) and additives produced by fermentation with GM microorganisms
- GM microbial starter cultures for the production of fermented foods (dairy products, cereals, vegetables, etc.)
- GM crops for fresh and processed food
- GM animals (e.g., husbandry of lean or less fatty animals)

All these areas of food technology are important advancements, but the public focus has been on GM crop alterations that relate to human safety and perceptions of environmental risk.

Field Trials and Field Releases

Over the past few years, there has been a tremendous increase in the number of different crop species that have been field tested. Well over 50 different crops have been tested in field experiments in over a dozen countries since the first trials with tobacco in the late 1980s (Dale, 1995). The United States leads in the number of field trials, although China purportedly has more acreage devoted to transgenic crops. Commercial planting of GM crops in China began in about 1992 (Brown, 2001). Several transgenic plant varieties have been approved for commercial use. It is estimated that over 109 million acres of farmland worldwide are planted with GM crops (Brown, 2001). The plant trait that seems to be most often modified is herbicide tolerance (Table 18). Insect tolerance and viral resistance are two other commonly modified traits, and crops with these traits have undergone field releases. The number of field trials of transgenic crops has increased tremendously over the past decade. However, different methods of categorizing and counting procedure have made it difficult to obtain precise comparisons from worldwide inventories (Giddings, 1996). Transgenic plant field trials in the United States are subject to regulatory oversight by the Animal and Plant Health Inspection Service (APHIS) of the USDA and by the EPA. Depending upon the commercial end use of the product, the FDA may also be involved in regulatory oversight.

Field releases of genetically engineered plants have called attention to safety concerns about the environment and nontarget species (e.g., butterflies, bees). Hence, principles or guidelines have been promulgated to ensure adequate environmental monitoring of field experiments. Some ecologists have

Table 18

Plants Modified by Transformation

Modified characteristic	Field releases (1986–1993)
Insect resistance	127
Viral resistance	164
Bacterial resistance	11
Fungal resistance	29
Herbicide tolerance	340
Modified quality	184
Resistance to marker genes	93
Multiple traits	26
Unspecified	31

Source: Green Industry Biotechnology Platform database, see also, Dale (1995).

expressed concern regarding the possible environmental risk and GM crops (cf. Paoletti and Pimentel, 1996). It has been asserted that the release of GM organisms might adversely impact both tropical and temperate zone biodiversity. It has been speculated that GM plants might generate so-called superweeds and also affect nontarget species, but measuring this risk has proved to be difficult (Marvier, 2001). So while public attention has been focused on potential risks to human health, there has also been concern about possible environmental consequences. Ironically, it has been estimated that between 75 and 100% of naturally occurring agricultural crops already contain some degree of natural host plant resistance – a resistant trait acquired by classical or non-GM breeding (Oldfield, 1984).

Substantial Equivalence

The most practical method of the establishing of a GM foods (or components) is to determine whether it is "substantially equivalent" to a conventional food product. Such equivalency takes into account biochemical and chemical composition, nutritional value comparisons, and several other factors (e.g., antinutritional components). To demonstrate "substantial equivalence," several criteria need to be considered:

- The composition and characteristics of the conventional food to which GM food is being compared
- A knowledge of any new components (i.e., introduced gene) either expressed or otherwise present in the GM food.

It is also important to understand what influence various food-processing techniques might exert on the new GM food. During the entire process of ensuring similar composition and characteristics, it is necessary to establish the food's safety (e.g., allergenicity). The establishment of the concept of "substantial equivalence" between a conventional food and a GM food provided the basis for regulatory approval in the United States and many, but not all, other countries.

Altering Genetic Traits

Herbicide Tolerance

A herbicide is a chemical agent that kills plants or significantly interrupts their growth and development. Every year, U.S. growers use an estimated 971 million pounds of pesticides, mostly to kill insects, weeds, and fungi (Brown, 2001). These chemicals interfere with metabolic processes that are essential for plant growth and vitality. Their phytotoxicity is due to adverse effects on normal enzyme activity. Crops and weeds are naturally resistant to many herbicides. This resistance can be due to metabolic detoxification, prevention of the herbicide from reaching its site of action, and resistance at its site of action.

Glyphosate-based weed control products are among the most widely used broad-spectrum herbicides in the world (Giesy *et al.*, 2000). The herbicidal properties of glyphosate were discovered over three decades ago. Glyphosate-based herbicides are used in agriculture, in industry, in ornamental gardening, and in residential weed management. In agriculture, they are used in applications involving genetically modified plant varieties selected for their capacity to withstand glyphosate treatment (e.g., Round Up Ready).

The herbicide glyphosate (e.g., Round Up) is a nonselective, broad-spectrum herbicide that ordinarily cannot be used on crops without severe plant injury. Once glyphosate enters the plant, it inhibits the enzyme 5-enolpyruvylshikimate-3-phosphate synthase (EPSPS), thereby preventing the plant from producing aromatic amino acids essential for protein synthesis. EPSPS is the only physiological target of glyphosate.

Glyphosate-resistant (i.e., glyphosate-tolerant) crops employ a target site modification wherein a glyphosate-resistant EPSPS is introduced into the plant by genetic engineering. The plant is unaffected upon exposure to glyphosate because of the continued action of the glyphosate-resistant EPSPS enzyme provides for the plant's needs for amino acid synthesis. An adjacent weed, for example, would not be protected from the glyphosate, hence would die because of the inhibition of amino acid synthesis.

There are a number of herbicide-tolerant crops now available commercially (e.g., glufosinate/bialaphos, bromoxynil, etc.), but glyphosate was one of the first to be marketed. It has also undergone extensive safety testing and therefore is a representative prototype. The development of glyphosate-resistant (i.e., glyphosate-tolerant) crops has withstood two decades of safety testing and field trials.

The mechanism of action of glyphosate is through its binding and blockade of EPSPS, an important enzyme necessary in the synthesis of aromatic amino acids both in crops and in weeds. Glyphosate's inhibition of EPSPS prevents a plant from producing aromatic amino acids that are essential for protein synthesis. EPSPS can be found in all plants, in bacteria, and in fungi. EPSPS is not found in animals, and therefore aromatic amino acids must be present in the diet.

Herbicide-tolerant crops are more environmentally-friendly because they lead to a much lower use of chemical herbicides. Furthermore, and in the case of glyphosate, it is readily degraded to metabolites that are nontoxic and nonresidual.

Glyphosate in the environment tends to bind tightly to soil and to particulate matter. It is essentially not available to plants and other soil organisms (Giesy *et al.*, 2000).

Bacillus thuringiensis Crops

The gene for the *Bt* toxin (*Bacillus thuringiensis*) was discovered over a hundred years ago. This soil bacterium has been used extensively in so-called organic farming. Since 1991, the *Bt* gene has been inserted into more than 50 plant crops (cf. Paoletti and Pimentel, 1996). There have been some concerns about the development of resistance to *Bt* toxin both in field and laboratory tests (Stone *et al.*, 1991; Lambert and Peferoen, 1992; Tabashnik *et al.*, 1992). Despite some of these early concerns, and now with the benefit of another 5 to 10 years of evaluation, *Bt* crops have been approved by all the appropriate U.S. regulatory agencies and other agencies throughout the world. Again, it should be emphasized that *Bt* crops can be treated with lower amounts of chemical pesticide spray than are required for conventional crop.

Insect resistance can be provided by a gene from the soil bacterium *Bt*. This gene (e.g., *crylAb*) directs cells to synthesize a crystalline protein that is toxic to selected insects, particularly caterpillars and beetles that destroy food crops. The toxin gene in various strains of *Bt* can affect different insects; hence a seed can be tailor-made to provide the best protection.

So-called *Bt*-protected plants provide a safe and highly effective method of insect control (Betz *et al.*, 2000). Corn, cotton, and potatoes protected by *Bt* were introduced into U.S. agriculture over five years ago. Not only do they produce higher yields to the farmer, but they also reduce levels of fungal toxins (e.g., fumonisin in corn). Non-*Bt* corn and other species do not take up the toxin released into the soil in root exudates of *Bt* corn. The persistence of the toxin in soil for up to 180 days after its release indicates that the pesticide remains bound to surface-active particles that protect the toxin from biodegradation (Saxena *et al.*, 1999).

Extensive safety evaluation tests, including acute, subacute, and chronic exposures, have been undertaken with *Bt*-protected plants. All have been fully approved by U.S. regulatory agencies including the FDA, EPA, and the USDA. These *Bt*-protected crops have been determined to be substantially equivalent in both food and animal feeds derived from non-GM or conventional foods.

Safety Evaluation

Safety issues concerning GM plants have involved both foods for human consumption and animal feeds. Toxicological studies have been aimed at ensuring safety and at establishing compositional equivalency. Substantial equivalence establishes important safety criteria for GM foods and animal feeds.

Safety issues are not solely the concern for human and domestic animals, but also for wildlife species and with respect to possible adverse effects on the environment and its ecosystems. The issue of nontarget species (e.g., monarch butterflies) has been raised by environmentalists. Will innocent nontarget species be affected by insecticides that are incorporated into GM crops? The risk appears to be very small, yet continued surveillance and additional research are necessary. Will "superweeds" arise as genes inserted into crops migrate or jump to non-GM plants? Since pollen from some GM plants might fertilize weedy relatives, non-GM crops should not be grown nearby. Rather, appropriate crop planting distances (e.g, "refuges") should be maintained as prescribed by regulatory agencies. Thus far, no scientific studies have found evidence of GM crops causing "superweeds." Finally, in terms of safety considerations, will insects evolve a tolerance to GM crops leading to an immunity to herbicides sprayed over fields of herbicide-tolerant GM plants? To date, this has been only a hypothetical concern.

Allergenicity and Digestible Proteins

To assess allergy-inducing potential, it is necessary to examine the chemical composition of each novel protein produced by the altered GM plant against a known allergen. It is estimated that 500 or more proteins are allergens. Several food allergen sequences can be retrieved from public domain databases for proteins and nucleic acids (cf. Metcalfe *et al.*, 1996).

It has been estimated that there are thousands of different proteins in the body. Only a small percentage of these are capable of causing food allergies. Because genetically engineered crop plants introduce new proteins, there is always the possibility that the protein encoded into this new genetic material might produce an allergic response. Many foods, not genetically modified, including milk, eggs, tree nuts, shellfish, and certain legumes (e.g., peanuts, soybeans) are commonly allergenic. Allergies and intolerances to foods differ significantly across countries (Woods *et al.*, 2001). Hence, a food crop (e.g., legume) that is genetically modified is apt to be just as allergenic in a sensitive person as in the sensitive person consuming the same legume from a non-genetically modified version of the same crop.

There is no evidence that recombinant proteins in newly developed foods are more allergenic than naturally occurring proteins. In fact, evidence suggests that the vast majority of these proteins are safe to consume (Lehrer *et al.*, 1996). The stability (or lack of it) of food allergens to gastric digestion can determine potential immunogenicity (Astwood *et al.*, 1996). Moreover, often overlooked are the temperature extremes that might be used in the processing of the final food product and the effect such exposure to heat and cold may have on allergenicity.

A decision tree for evaluating recombinant food proteins for the immunogenicity can be very useful (Lehrer *et al.*, 1996). Recombinant proteins from known allergens can be readily tested by using a tier of *in vitro* immunochemical assays. There are many biochemical and immunotoxicity tests that can be used to evaluate allergenicity. Unfortunately, recombinant proteins from unknown allergenic sources are not easily recognized, nor are they necessarily anticipated. Hence, it may be difficult to assay recombinant proteins from sources with undetermined allergenic activity. Importantly, there is no evidence that recombinant proteins are more allergenic than naturally occurring proteins. In fact, the vast majority of proteins present in new food products are safe for consumption. Nevertheless, it is important that new food products, regardless of source, be prudently developed and tested for any potential immunogenicity.

The ability of food allergens to cross the mucosal membanes of the gastrointestinal tract is generally a prerequisite for its allergenicity. High temperatures associated with food processing and the acidity of the stomach do not necessarily result in digestion of allergens. A protein that is stable under such environmental conditions (i.e., proteolysis) has an increased probablility of reaching the intestinal mucosa intact. Many allergens exhibit stability despite such challenges (cf, Metcalfe *et al.*, 1996).

It is possible to compare the relative stability of proteins engineered into plants with a number of commonly known allergenic food proteins.

There are several major classes of food allergens including egg white allergens, milk allergens, soybean allergens, peanut allergens, and mustard allergens.

The toxicological evaluation of many of these allergens can be carried out by using similated gastric and digestive models (e.g. pepsin resistant) of mammalian digestion as decribed in the U.S. Pharmacopeia. It is possible to compare the relative stability of proteins engineered into plants (e.g., *B.t.* insecticidal protein, CP4 EPSPS synthase, etc.) with a number of commonly known allergenic food proteins. The *in vitro* digestibility of *Bt* Cry proteins in similated gastric fluid ranges from 30 to 60 seconds (Betz *et al.*, 2000). Such *in vitro* models can be used to evalulate the digestibility plant and animal proteins as well as certain food additives. Not surprisingly, many of the known food allergens are stable to digestion in these similated gastrointestinal digestive models. The allergen (or a protein fragment) may be stable for at least 2 minutes; major allergens are typically stable for as long as an hour. On the other hand, common food proteins with no history of allergenicity undergo rapid degradation in these simulated gastric and intestinal models. Thus, it is possible to add this *in vitro* battery of safety testing to the older tools for evaluating new GM foods or animal feeds.

Feeding Studies

Preclinical feeding studies are an integral part of the overall safety assessment of GM foods and/or animal feeds. Animal feeding studies, whether using laboratory animals or domestic animals, provide insight into the nutritional status of the new GM product. Such research also allows for the monitoring of growth rates of the test animals. Daily dietary regimens using several different intake levels and durations (e.g., 4–10 weeks depending on the species) aid in establishing normal growth patterns and in performing nutritional balance studies. From such studies, it may be possible to set any safety margins that might prove necessary. Often, several different species can be used to determine nutritional parameters, and these might include laboratory rats and domestic animals commonly found in the human food chain. Feeding studies involving invertebrates (e.g., catfish) may also be undertaken not only for assessing another food chain component but also for any potential ecological effect(s). The chicken is an excellent animal feeding model because it can be easily studied through its life span. It also undergoes a several fold increase in body weight in just over 6 weeks and provides additional information about nutrient value of the new GM food or feed. The chicken is a very sensitive model for assessing nutritional equivalency because it must ingest significant amounts of protein to accomodate such a rapid body weight increase during this short time period.

Non target Species

The overall safety assessment of GM food and the safety of animal feeds that are derived from GM plants must include on evaluation of the effects of such crops on nontarget species (e.g., monarch butterflies, swallowtail butterfly larvae, honeybees, etc.). Field studies that include any potential harmful effect on nontarget species should be conducted as a routine evaluation of a GM food or feed. Such studies, if conducted in a laboratory setting, must closely approximate those conditions found in the natural setting. The environmental risks that GM crops might pose on nontarget species are not completely known, but experiments to date have not revealed any adverse effects. Further, the cultivation of GM food crops (e.g., *Bt*-protected plants) may result in fewer adverse impacts on nontarget organisms than would be incurred from the use of chemical pesticides. Glyphosate formulations do not affect honeybees, nor do they affect nontarget terrestrial plants (Giesy *et al.*, 2000). Glyphosate does not bioaccumulate in fish or other animals.

There must be continued vigilance to determine any possible ecological consequences that might arise to harm nontarget species of wildlife. The weight of evidence indicates no unreasonable adverse effects of *Bt* Cry proteins expressed into plants to nontarget wildlife or beneficial invertebrates (e.g. earthworms, soil microbials) or flora.

Nutritional Improvement

The nutritional health and well-being of humans is dependent on plant foods that are either directly or indirectly consumed by certain domestic animals. Crop foods provide most essential vitamins and minerals, along with a host of other nutrition-related phytochemicals. Oftentimes, micronutrients are found only in low concentrations in staple food crops. Essential micronutrients in the human diet consist of about 17 minerals and 13 vitamins. Unfortunately, the agricultural systems in many developing or underdeveloped countries in the world are unable to provide proper or adequate nutrition. GM foods, particularly those that increase yield and productivity, may allay malnutrition. It has been estimated that about 800 million people, mostly children, are malnourished to some degree.

Modern agriculture, particularly in the last few decades, has witnessed advances in crop breeding that have led to increases in productivity and yields. While efforts have been focused on attaining caloric needs, genetically engineered crops now afford the opportunity to enhance the micronutrient composition of crop foods (cf. Comaci, 1993). Through nutritional genomics it is now possible to increase plant micronutrients, thereby leading to an improvement in human health (DellaPenna, 1999). By inserting foreign genes into various crop foods, it is possible to achieve nutritional enhancement of certain micronutrients. Initial efforts to attain GM foods with traits that endowed them with herbicidal resistance ordinarily involved the insertion of one or two foreign genes. Multiple gene insertions, sometimes referred to as the "stacking" of gene, may involve over a half-dozen gene insertions and more than one metabolic pathway in the same plant.

An early entry into nutritional genomics via multiple gene insertions was so-called golden rice. Millions of people depend on rice as a staple food. Through nutritional genomics it is possible to enhance rice with increased levels of micronutrients, namely, β-carotene and iron. "Golden mustard oil" with high levels of β-carotene also can be made from GM plants. Four encoded enzymes that provide rice with the ability to synthesize β-carotene and three other encoded enzymes allow the kerneals to accumulate additional iron. Both vitamin A deficiency and iron deficiency, particularly in children, can potentially be treated by ingesting this novel form of GM rice. Efforts are under way to even further increase the levels of these micronutrients in this GM rice as well as other food crops. By enhancing the levels of carotenoids and the glucosinolates in GM foods, it may be possible to reduce the risk of certain forms of cancer. Likewise, the enhancement of phytoestrogens as well as selected phytophenolic compounds might reduce the risk of other diseases, including cardiovascular diseases, osteoporosis, and certain hormone-related cancers. Aqua culture of GM salmon is under development and may prove to offer an important alternate source of protein. Commercial "fish farms" growing GM salmon that have been transfected with a growth hormone from other fish species (e.g., ocean pout fish, which grows four times faster than salmon) have been produced and are to be reviewed for regulatory approval. Upon insertion into fish eggs, the growth hormone gene will be present in every cell in the body of every fish born from a treated clutch of eggs. Use of this integrated gene can create a breeding stock of a new, faster growing variety.

Agribiotechnology and Medicines

Clearly, biotherapeutics, which involves the genetic manipulation of mammalian cells, is focused on the discovery of new medicines and other therapeutic products for the treatment of various diseases. Early successes in developing alternate sources of often scarce or animal-origin hormones through recombinant technologies led to the production of synthetic human insulin and human growth hormone. These were indeed scientific breakthroughs enjoyed by the then new biotechnologies. Progress in developing new medicinal therapies through agribiotechnology has not been as rapid nor as successful, yet there are many potential indications (Giddings *et al.*, 2000) (Table 19). Many of the expression systems used in the production of biopharmaceuticals have employed *Agrobacterium*-mediated transformations (AMT). Also, the production from the potato or rice plant of so-called nutraceuticals may be useful in the eventual treatment of deficiencies in provitamin A and certain amino acids (Giddings *et al.*, 2000). Improved crop traits have produced healthier food such as margarines free of trans isomers and improved protein and vitamins levels.

One very promising area of agribiotechnology is the use of GM crop foods as a vector for the delivery of edible vaccines (Richter 1997; Richter and Kipp, 1999) (Table 20). When plants genetically engineered to produce bacterial or viral proteins are consumed as food, they can trigger an immune response (Arntzen, 1998; Arakawa *et al*, 1998). Oral immunizations produced from transgenic crops could provide a low-cost approach to protecting populations against several different diseases. Cholera vaccines, using the potato as a

Table 19

Production of Biopharmaceuticals in Transgenic Plants

Potential application/indication	Plant	Protein
Anticoagulants		
Protein C pathway	Tobacco	Human protein C (serum)
Indirect thrombin inhibitors	Tobacco, oilseed, Ethiopian mustard	Human hirudin variant 2
Recombinant hormones/proteins		
Neutropenia	Tobacco	Human granulocyte-colony-stimulating factor
Anemia	Tobacco	Human erythropoietin
Antihyperanalgesic (opiate)	Thale cress, oilseed	Human enkephalins
Wound repair	Tobacco	Human epidermal growth factor
Hepatitis C and B	Rice, turnip	Human interferon alfa
Liver cirrhosis	Potato, tobacco	Human serum albumin
Blood substitute	Tobacco	Human hemoglobin
Collagen	Tobacco	Human collagen
Protein/peptide inhibitors		
Cyctic fibrosis	Tobacco	Human α_1-antitrypsin
Trypsin inhibitor	Maize	Human aprotinin
Hypertension	Tobacco, tomato	Acetylcholinesterase inhibitors
HIV	*Nicotiana bethamiana*	α-Trichosanthin (TMV-Ul)
Recombinant enzymes		
Gaucher's disease	Tobacco	Glucocerebrosidases

Source: Modified and condensed from Giddings *et al.* (2000).

vehicle, could provide an immunity to this disease. Likewise, a GM potato could be used as a vehicle for developing vaccines against enterogenic *E. coli* (ETEC) infections. Transgenic bananas, since they can be eaten raw, might provide the necessary vehicle for a hepatitis B vaccine that would be destroyed by the heat of cooking. In an effort to control dosage, the transgenic banana could be homogenized into a measured baby food to ensure the effectiveness of the vaccine. Thus pharmaceutical foodstuffs containing oral immunization from a transgenic plant may provide the basis for new immunomodulating drugs and vaccines.

SUMMARY

Agribiotechnology, while slower than biotechnology to gain public acceptance, continues to witness major scientific advances. Such advances will be

Table 20

Production of Vaccines from Plants

Potential Application/indication	Plant	Protein
Hepatitis B	Tobacco	Recombinant Hepatitis B surface Antigen
	Tobacco	Marine hepatitis epitope
Dental caries	Tobacco	*Streptococcus mutans*
Autoimmune diabetes	Potato	*Vibrio cholerae* toxin B
	Potato	Glutamic acid decarboxylase
Cholera and *E. coli* diarrhea	Tobacco, potato	*E. coli* enterotoxin LT-B
Oral vaccine (cholera)	Potato	*V. cholerae* toxin CtoxA/Ctox-B
Mucosal vaccines	Cowpea	D2 peptide (fibronectin-B) *Staphylococcus aureus*
Diarrhea (Norwalk virus)	Tobacco, potato	Coat protein (Norwalk virus)
Rabies	Tobacco, spinach	Rabies virus glycoprotein
HIV	Tobacco, blackeyed pea	HIV epitope (gp 120)
	Cowpea	HIV epitope(gp 41)
Rhinovirus	Blackeyed pea	Human rhinovirus epitope (HR14)
Foot and mouth disease	Blackeyed pea	Foot & Mouth virus epitope (VPl)
Malaria	Tobacco	Malarial B-cell epitope
Influenza	Tobacco	Hemagglutinin
Cancer	Tobacco	c-Myc

Source: Modified and condensed from Giddings *et al.*, (2000).

driven, in part, by burgeoning world population and by the need to provide adequate food supplies and also foods with enhanced nutritional value. The scope of products derived from agribiotechnology will increase significantly in the areas of nutritional genomics and plant-based medical products.

REFERENCES

Arakawa, T., Chong, D. K., and Lanridge, W. H. R. (1998). Efficacy of a food-based oral cholera toxin B subunit vaccine. *Nat. Biotechnol.*, **16**, 292–297.

Arntzen, C. J. (1998). Pharmaceutical foodstuffs – Oral immunization with transgenic plants. *Nat. Med.*, **4**, 502–503

Astwood, J. D., Leach, J. N., Fuchs, R. L. (1996). Stability of food allergens to digestion *in vitro*. *Nat. Biotechnol.*, **14**, 1269–1273.

Ausubel, F. M., Brent, R., and Kingston, R., eds. (1995). *Current Protocols in Molecular Biology*. Wiley, New York.

Barzon, L., Bonaguro, R., Palu, G., and Boscaro, M. (2000). New perspectives for gene therapy in endocrinology. *Eur. J. Endocrinol.*, **143**, 447–466.

Bass, R., Kleeberg, U., Schrode, R. H. and Scheibner, E. (1992). Current guidelines for the preclinical safety assessment of therapeutic proteins. *Toxicol. Lett.*, **64/65**, 339–347.

Bennett, C. R., Dean, N. M., and Monia, B. R. (1998). Antisense oligonucleotide therapeutics, in *Advances Drug Discovery Techniques*, Harvey, A.L., edit. pp. 173–204. Wiley, New York.

Betz, F. S., Hammond, B. G., and Fuchs, R. L. (2000). Safety and advantages of *Bacillius thuringiensis*-protected plants to control insect pests. *Regul. Toxicol. Pharmacol.*, **32**, 156–173.

Brown, K. (2001). Seeds of concern. *Sci. Am.*, **284**, 52–57.

Cavagnaro, J. A. (1997). Considerations in the pre-clinical safety evaluation of biotechnology-derived products, in *Comprehensive Toxicology*, Vol. 2, I. G. Sipes, *et al.*, eds., pp. 291–298, Pergamon Press, Oxford, U. K.

Claude, J.-R. (1992). Difficulties in conceiving and applying guidelines for the safety evaluation of biotechnologically-produced drugs: Some examples. *Toxicol. Lett.*, **64/65**, 349–355.

Cohen-Haguenauer, O. (1996). Safety and regulation at the leading edge of biomedical biotechnology. *Curr. Opin. in Biotechnol.*, **7**, 265–272.

Comaci, L. (1993). Impact of plant genetic engineering on foods and nutrition. *Annu. Rev. Nutr.*, **13**, 191–215.

Crooke, S. T. (1992). Therapeutic applications of oligonucleotides. *Annu. Rev. Pharmacol. Toxicol.*, **32**, 329–376.

Dale, P. J. (1995). R & D regulation and field trialling of transgenic crops. *Trends Biotechnol.*, **13**, 398–403.

Davis, J. R. E. (1996). Molecular biology techniques in endocrinology. *Clin. Mol. Endocrinol.*, **45**, 125–133.

Davis, S. S. (1997). Biomedical applications of nanotechnology – Implications for drug targeting and gene therapy. *Trends Biotechnol.*, **15**, 217–224.

Dayan, A. D. (1995). Safety evaluation of biological and biotechnology-derived medicines. *Toxicology*, **105**, 59–68.

Della Penna, D. (1999). Nutritional genomics: Manipulating plant micronutritents to improve human health. *Science*, **285**, 375–379.

Demain, A. L. (1991). An overview of biotechnology. *Occup. Med. State Art Rev.*, **6**, 157–168.

Dolnick, B. J. (1997). Naturally occurring antisense RNA. *Pharmacol. Ther.*, **75**, 179–184.
Dougall, W. C., Peterson, N. C., and Greene, M. I. (1994). Antibody-structure-based design of pharmacological agents. *Trends Biotechnol.*, **12**, 372–379.
Federici, M. M. (1994). The quality control of biotechnology products. *Biologicals*, **22**, 151–159.
Gibson, I. (1994). Antisense DNA and RNA strategies: New approaches to therapy. *J. R. Coll. Physicians (London)*, **28**, 507–511.
Giddings, L. V. (1996). Transgenic plants on trial in the USA. *Curr. Opin. Biotechnol.*, **7**, 275–280.
Giddings, L. V., (2001). Transgenic plants as factories for biopharmaceuticals. *Nature*, **18**: 1151–1155
Giesy, J. P., Dobson, S., and Solomon, K. R. (2000). Ecotoxicological risk assessment for Roundup herbicide. *Rev. Environ. Contam Toxicol.*, **167**, 35–120.
Goodnow, C. C. (1992). Transgenic mice and analysis of β-cell tolerance. *Annl. Rev. Immunol.*, **10**, 489–518.
Grosveld, F., and Kollias G., eds. (1992). *Transgenic Animals*. Academic Press, San Diego, CA.
Harris, P., and Dayan, A. (1998). Designing a safety evaluation programme for gene therapy products: Recommendations and the way forward. *Hum. Exp. Toxicol.*, **17**, 63–83.
Harvey, A. L. (1998). Sense, nonsense and antisense: Different approaches to drug discovery, in *Advances in Drug Discovery Techniques*, A. L. Harvey, ed., pp. 1–11. Wiley, New York.
Henck, J. W., *et al.* (1996). Reproductive toxicity testing of therapeutic biotechnology agents. *Teratology*, **53**, 185–195.
Jeffcoate, S. L. (1992). New biotechnologies: Challenges for the regulatory authorities. *J. Pharm. Pharmacol.* **44**, 191–194.
Juliano, R. L., Astriab-Fisher, A. and Falke, D. (2001). Macromolecular therapeutics. *Mol. Interventions*, **1**, 40–53.
Kelley, W. S. (1996). Therapeutic peptides: The devil is in the details. *Biotechnology*, **14**, 28–31.
Lambert, B., and Peferoen, M. (1992). Insecticidal promise of *Bacillus thuringiensis*: Facts and mysteries about a successful biopesticide. *BioScience*, **42**, 112–122.
Lehrer, S. B., Horner, W. E., and Reese, G. (1996). Why are some proteins allergenic? Implications for biotechnology. *Crit. Rev. Food Sci. Nutr.*, **36**, 553–564.
Liberman, D. F., *et al.* (1991). Risk assessment of biological hazards. *Occup. Med. State Art Rev.*, **6**, 285–299.
Luster, M. I., *et al.* (1992). Risk assessment in immunotoxicology; sensitivity and predictability of immune tests. *Fundam. Appl. Toxicol.*, **18**, 200–210.
Marvier, M. (2001). Ecology of transgenic crops. *Am. Sci.*, **89**, 160–167.
Metcalfe, D. D., Astwood, J. D., Towsend, R., Sampson, H. A., Taylor, S. L., and Fuchs, R. L. (1996). Assessment of allergenic potential of foods derived from genically engineered crop plants. *Crit. Rev. Food Sci. Nutr.*, **36**, S165–S186.
Oldfield, M. L. (1984). *The Value of Conserving Genetic Resources*. Washington, DC, U.S. Department of the Interior, National Park Service.
Panchagnula, R., and Thomas, N. S. (2000). Biopharmaceutics and pharmacokinetics in drug research. *Int. J. Pharm.* **201**, 131-150.
Paoletti, M. G., and Pimentel, D. (1996). Genetic engineering in agriculture and the environment. *Bioscience*, **46**, 665–673.
Persaud, S. J., and Jones, P. M. (1994). Antisense oligonucleotide inhibition of gene expression: Application to endocrine systems. *J. Mol. Endocrinol.*, **12**, 127–130.
Pettit, D. K., and Gombotz, W. R. (1998). The development of site-specific drug-delivery systems for protein and peptide biopharmaceuticals. *Trends Biotechnol.*, **16**, 343–349.
Richter, L. (1997). Transgenic plants for oral vaccines against diarrhoeal disease. *Proc. R. Coll. Physicians (Edinb.)*, **27**, 16–21.

Richter, L. and Kipp, P. B. (1999). Transgenic plants as edible vaccines. *Curr. Top. Microbiol. Immunol.*, **240**, 159–176.
Saxena, D., Flores, S. and Stotzky, G. (1999), Insecticidal toxin in root exudotes from Bt corn. *Nature*, **402**: 480-491.
Smith, K. T., Shepard, A. J., Boyd, J. E., and Lees, G. M. (1996). Gene delivery systems for use in gene therapy: An overview of quality assurance and safety issues. *Gene Ther.*, **3**, 190–200.
Stark, D. M., Barry, G. F., and Kishore, G. M. (1993). Impact of plant biotechnology on food and food ingredient production, in *Science for the Food Industry of the 21st Century*, Yalpani, M., Edit. Chap. 8, ATL Press. Mt. Prospect, IL.
Stone, T., Sims, S. R., MacIntosh, S. C., Fuchs, R. L., and Marrone, P. G. (1991). Insect resistance to *Bacillus thuringiensis*, in. *Biotechnology of Biological Control of Pests and Vector*, K. Maramorosch, ed., pp. 53–68. CRC Press, Boca Raton, FL.
Tabashnik, B. A., *et al.* (1992). Inheritance of resistance to *Bacillus thuringiensis* in diamondback moth (*Lepidoptera plutellidae*). *J. Econ. Entomol.*, **85**, 1046–1055.
Thomas, J. A. (1995). Recent developments and perspectives of biotechnology-derived products. *Toxicology*, **105**, 7–22.
Thomas, J. A. (1997). Use of animal models in biomedical research. in *Comprehensive Toxicology*, Vol. 2, I. G. Sipes, *et al.*, eds., pp. 227–238, Pergamon Press, Oxford, U.K.
Thomas, J. A. (1998). Biotechnology: Therapeutic and nutritional products, in *Biotechnology and Safety Assessment*, 2nd ed., J. A. Thomas, ed., pp. 283–300. Taylor & Francis, London.
Thomas, J. A. (1999). Biotechnology: Safety evaluation, in *General and Applied Toxicology*, 2nd ed., B. Ballantyne *et al.*, eds., pp. 1963–1975. Macmillan, New York.
Thomas, J. A., and Thomas, M. J. (1993). New biologics: Their development, safety, and efficacy, in, *Biotechnology and Safety Assessment*, 1st ed., J. A. Thomas and L. A. Myers, eds, pp. 1–22. Raven Press, New York.
Tomlinson, E. (1992). Impact of the new biologies on the medical and pharmaceutical sciences. *J. Pharm. Pharmacol.*, **44**, 147–159.
Toulme, J. J. (2001). New candidates for true antisense. *Nat. Biotechnol.*, **19**, 17–18.
Tservista, M., Levy, M. S., Lo-Yim, M. Y. A., O'Kennedy, R. D., York, P., Humphrey, G. O., and Hoare, M. (2001). The formation of plasmid DNA loaded pharmaceutical powders using supercritical fluid technology. *Biotechnol. Bioeng.*, **72**, 12–18.
Vaswani, S. K., and Hamilton, R. G. (1998). Humanized antibodies as potential therapeutic drugs. *Ann. Allergy Asthma Immunol.*, **81**, 105–119.
Verdier, F., and Descotes, J. (1999). Preclinical safety evaluation of human gene therapy products. *Toxicol. Sci.*, **47**, 9–15.
Vial, T., and Descotes, J. (1994). Clinical toxicity of interferons. *Drug Saf.*, **10**, 115–150.
Werner, R. G., Noe, W., and Schluter, M. (1998). Appropriate mammalian expression systems for biopharmaceuticals. *Arzneim.-Forsch./Drug Res.*, **48**, 870–880.
Woods, R. K., Abramson, M., Bailey, M. and Walters, E. H. (2001). International prevalences of reported food allergies and intolerances. Comparisons arising from European Community Respiratory Health Survey (ECRHS) 1991–1994. *Eur. J. Clin. Nutr.*, **55**, 298–304.
Yaksh, T. L., Rathbun, M. L., Dragani, J. C., Malkmus, S., Bourdeau, A. R., Richter, P., Powell, H., Myers, R. R., and LeBel, C. P. (1997). Kinetic and safety studies on intrathecally infused recombinant methionyl human brain-derived neurotropic factor in dogs. *Fundam. Appl. Toxicol.*, **38**, 89–100.
Yang, J., and Wu, C. L. (2001). Gene therapy for pain. *Am. Sci.*, **89**, 126–135.
Zbinden, G. (1990). Safety evaluation of biotechnology products. *Drug Saf.*, **5**, 58–64.

Chapter 13

The Potential of Plant Biotechnology for Developing Countries

Jennifer A. Thomson,
Department of Molecular and Cell Biology,
University of Cape Town, Cape Town,
South Africa

Developing countries need genetically modified crops as part of their fight for sustainable agriculture and to feed their people. It has been calculated that Africa will have a grain shortfall of approximately 90 million tons by the year 2025. New technologies will be required to cope with this problem, and the use of transgenic crops is part of this. One of the greatest challenges for sub-Saharan Africa is the improvement of the nutrient status of its soil. Since Africa's crop production per unit cultivated is the lowest in the agricultural world, planting of genetically modified crops can greatly assist in bringing about this improvement. The Green Revolution has done much to improve agricultural productivity, but for the poor to take advantage of its benefits they need access not only to seed but also to fertilizers and irrigation. Genetically modified crops could increase their independence from both these expensive aids. The most important crops for sub-Saharan Africa are

Biotechnology and Safety Assessment, 3rd edition

maize, cassava, bananas, and sweet potatoes. Problems being addressed by modern biotechnology include viruses, fungal and insect pests, cyanide in cassava, weevils, and the weed Striga, as well as one of the future scourges of Africa, drought. Another potentially valuable aspect of plant biotechnology is the development of edible vaccines in transgenic plants. This would eliminate the need for expensive hypodermic syringes and also would decrease the dependence on cold storage of tropical residents. Finally it should be noted that of the 11% growth in transgenic crop area from 1999 to 2000, 84% occurred in developing countries. Thus, the area of transgenic crops in developing countries grew by 51% compared with a 2% growth in industrial countries. Future directions include crops high in vitamin A, the use of phytosterols to decrease cholesterol levels, and phytoremediation of toxic compounds by genetically modified plants.

INTRODUCTION

Hunger is commonplace in developing countries. Of the 600 million people who live in sub-Saharan Africa, nearly 200 million are chronically undernourished and some 40 million children are severely underweight. Over 50 million people, mostly children, suffer from vitamin A deficiency, and 65% of women of childbearing age are anemic (Vitamin Information Centre, 1999). Africans are becoming poorer and hungrier. Decades of war, weak governments, one-party states, and widespread corruption are part of the problem.

These human-made tragedies obscure other sets of problems encountered by farmers and their families in Africa. These problems, which scientists could help to address, include the failure of technologies developed to boost agriculture during the Green Revolution to benefit Africa by more than a modest extent. Few of the technologies focused on Africa's staple crops or growing conditions.

There is a solution to this, according to Gordon Conway, president of the Rockefeller Foundation. In his a book *The Doubly Green Revolution* (1997), he states that three-quarters of a billion people who live in a world where food is plentiful have almost no access to this food. If we add up the world's food production and divide it equally among the world's population, each man, woman and child would receive a daily average of over 2700 calories, an amount sufficient to prevent hunger and probably enough to allow everyone to lead a healthy life.

In reality, however, people living in sub-Saharan Africa eke out an existence, while Europeans and North Americans enjoy an average of 3500 calories per day. Calculations that by the year 2025 there will be a grain shortage of about 90 million tons in sub-Saharan Africa (Dyson, 1999) are based on the average cereal yield during the period 1989 to 1991, which comes to 1.165 tons per hectare (tons/ha). Dyson projects an average yield of 1.536 tons/har by the year 2025, but even this will leave the subcontinent with a massive shortage. What is the solution?

In Conway's eyes the solution lies in the "Doubly Green Revolution." What does this mean? The Green Revolution began in the 1940s in Mexico. It was aimed at improving yields of basic food crops such as maize and wheat. Plant breeders used indigenous varieties suited to Mexican conditions and achieved remarkable success in producing high-yielding lines. In 1948 farmers planted 1400 tons of improved maize seed, and the country did not need to import maize for the first time since 1910. By the 1960s maize yields averaged over 1000 kg/har. Similar increases were achieved in other countries in Latin America and Asia.

However, the impact in Africa was far less impressive. Yield improvement was not solely attributable to the new varieties. Their potential could be realized only if the crops were supplemented with fertilizers and optimal supplies of water, both of which are in short supply for most African farmers. Therefore the Green Revolution was least successful in sub-Saharan Africa, where cereal yields have changed little over the past 40 years and cereal production per capita has steadily declined.

If this is the Green Revolution, what is the Doubly Green Revolution? Conway argues that the only way to improve crop production in the twenty-first century is to combine conservation of the environment with productivity. Scientists and farmers must forge partnerships to design better crops specifically suited to a variety of environments. Alternatives must be found to fertilizers and pesticides. Ways must be found to improve soil quality and water utilization. Conway argues that the answer lies in genetic engineering to improve crops. This technology has the potential to create new plant varieties that not only deliver higher yields but also provide solutions to biotic and abiotic challenges. These challenges include viruses, fungi, bacteria, and insects, as well as desiccation and salt stresses. An added advantage of genetic engineering is that it could be a valuable tool for production in poor as well as rich soils. It could increase productivity and help to achieve agricultural stability and sustainability in developing countries.

AGRICULTURE IN AFRICA

The soil in Africa is probably the poorest in the world in terms of quality. Reasons for this include the high rates of erosion, the level of general acidity, and the poor nutrient status of agricultural lands. Therefore one of the greatest challenges for the continent is to improve the nutrient status of

Africa's arable lands. Some soils are naturally richer in nutrients than others and can be exploited for a while. Eventually, however, they lose nutrients, which must then be replaced. African farmers cannot, on the whole, afford synthetic nutrients. They therefore have to rely on growing more organic matter in the soil and plowing this back to return nutrients to the soil. Productivity is usually low, and this reduces the amount of organic matter available to be returned to the soil after harvesting. It is therefore essential that Africans increase the amount of organic matter grown in the soil.

In Africa, crop production is the lowest in the agricultural world. Florence Wambugu, former Director of the AfriCenter of the International Service for the Acquisition of Agri-Biotech Applications (ISAAA) in Nairobi, Kenya, cites the example of sweet potato, a staple crop in many African countries. The yield in Africa is 6 tons/hr, compared with the global average of 14 tons/hr (Wambugu, 1999). China produces three times the African average. Wambugu predicts that African production could double if viral diseases could be controlled. She also states that at present Africa imports at least 25% of its grain, but this could be decreased dramatically if genetically modified crops were introduced into the region.

Africans are among the poorest people in the world, and poverty destroys environments. Anything done to alleviate poverty will serve the environment and human health. The Green Revolution did much to improve agricultural productivity and alleviate poverty, but as pointed out earlier, for the poor to take advantage of this they need not only access to seed but to fertilizers and irrigation, as well. If genetically modified crops could be introduced, to reduce dependency on fertilizers and irrigation technology, farmers would need only to purchase seed.

Critics who claim that Africans cannot benefit from plant biotechnology often state that subsistence farmers will be forced to buy genetically modified seed. Yet African farmers have for years benefited from buying seed, including hybrid seed, obtained from local and multinational companies. Genetically engineered, or transgenic, seed is simply an added-value improvement on these hybrids. When Africans are asked their opinions on the advantages of planting genetically modified crops for sustainable agriculture, the overwhelming response is positive.

AFRICAN CROPS AND THEIR PROBLEMS

Soil acidity and drought stress, combined with diseases and pests, constitute Africa's major crop problems. In most African soils the pH values are between 3.5 and 4.5, and the continent is chronically short of water. In addition, the crops are plagued with a variety of plant viruses, fungi, bacteria, and insects.

Cassava is known to most western societies as a source of tapioca, but it is the staple food of many Africans. The leaves and starchy roots, when powdered, fried, or fermented, constitute the world's third-largest source of calories, after rice and maize. Plant breeders in East Africa have succeeded in increasing the size and number of edible roots, but these improvements have been offset by diseases. In some years cassava mosaic virus has almost wiped out the entire crop in some African countries. Scientists in the United States (Missouri) are collaborating with colleagues in Africa to develop cassava genetically modified to be resistant to this virus.

Bananas and their close relatives plantains are regarded by the western world as a snack and a dessert. However, in western and central Africa they provide more than a quarter of food calories. Indeed, the UN Food and Agriculture Organisation ranks bananas as one of the world's fourth most important food crop. Scientists in Kenya have succeeded in using modern tissue culture techniques to increase the yields of bananas.

Sweet potatoes are a subsistence crop for many people in eastern and southern Africa. Scientists from the United States, South Africa, Kenya, and Uganda have improved the protein content of sweet potatoes by a factor of 4, from the traditional 3% to 12% (Qaim, 1999). This could have a significant effect on the lives of many people in Africa. In addition, however, sweet potatoes are sensitive to viruses and weevils. In a joint project, Monsanto, the Kenya Agricultural Research Institute, ISAAA, and the International Potato Centre are using genetic engineering to address these problems.

Maize streak virus is endemic in Africa causing major losses to commercial and subsistence maize farmers alike. In another collaboration between the public and private sectors, research in South Africa is yielding very promising results on transgenic crops resistant to this virus.

Striga is a parasitic weed that is an enormous pest in Africa. It is a particular scourge because it gravitates toward weak plants such as maize, rice, and sorghum. It can be contained with a modern herbicide, but the crops must first be genetically modified to tolerate the herbicide.

VACCINES FOR AFRICA

One of the reasons for the success of the program to eradicate smallpox was that it was launched before the onset of HIV/AIDS, when hypodermic needles could be shared. Needles and cold storage are two of the most expensive aspects of a vaccination program. Scientists are trying to circumvent these problems by producing oral vaccines in edible plants. If effective, such vaccines could have an enormous impact on developing countries, with hundreds of rural children lining up at clinics for a banana that could immunize them against life-threatening diseases such as dysentery and diarrhea. The World Health Organisation estimates that over 12 million children under the age of 5 die each year from infectious diseases.

Orally administered vaccines are most appropriate for protection against pathogens that enter the body via mucosal surfaces such as the digestive tract, respiratory system, or urogenital tracts. In South Africa scientists are developing oral and other vaccines that act against human papillomavirus, one of the most important causative agents of cervical cancer in Africa. They are also working on a vaccine for the African-specific variety of HIV. Their approach is not an edible virus but the use of transgenic tobacco to produce the vaccines. Tobacco is an extremely hardy crop, well suited to growing even in marginal soils. The production of vaccines in this "pharming" system also has the advantages of being cheap, and the vaccines would escape possible contamination by animal viruses, which could potentially infect humans. The aim of these research programs is to produce vaccines for African diseases that cost cents rather than dollars.

GENETICALLY MODIFIED CROPS IN SOUTH AFRICA

It might be of interest to view the status of genetically modified crops in South Africa. Figure 1 shows the number of applications for permits to undertake trials or commercial releases between the years 1990 and 2000. The

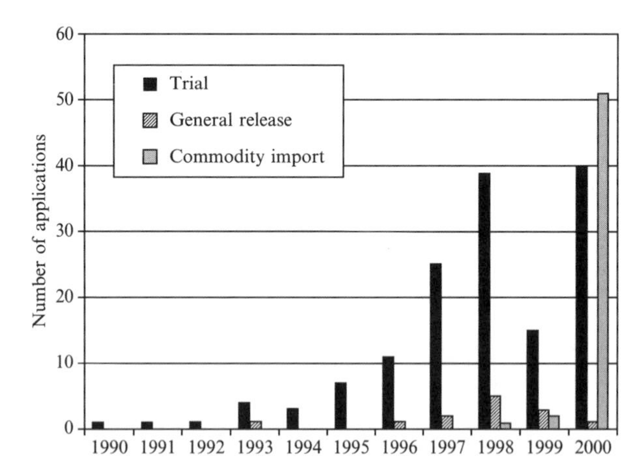

Figure 1 Number of applications for trial releases, (black bars), general releases (hatched bars), and commodity imports (gray bars) in South Africa from 1990 to 2000 (Personal Communication, M. Koch, Innovation Biotechnology)

South African Genetic Experimentation Committee (SAGENE) handled these applications prior to the implementation of the Genetically Modified Organisms (GMO) Act toward the end of 1999. The decrease in the numbers of trials and commercial releases during the first half of 1999 probably occurred because SAGENE had by then ceased to function effectively. There was something of a hiatus between the time when SAGENE wounddown and the GMO Act came into action. Figure 2 gives an indication of the crops and organisms that were considered for permits during this period. Most of them were for cotton and maize, followed by soybean and various microorganisms.

Small-scale farmers in the KwaZulu Natal Province of South Africa have enthusiastically taken to growing insect-resistant cotton expressing the *Bacillus thuringiensis* (*Bt*) insecticidal protein as they have begun to see the financial benefits. In 1997 only four farmers took part in field trials. The next year, 75 farmers planted 200 ha of *Bt* cotton. In 1999 this number increased to 410 and 789 ha respectively. During the 2000 growing season 644 farmers planted 1250 ha. This accounts for approximately 50% of the total area planted to cotton in that region.

GENETICALLY MODIFIED CROPS IN DEVELOPING COUNTRIES

According to a report published in 2000, the world's four principal crops at that time were soybean, maize, cotton, and canola (James, 2000). In 2000 15%

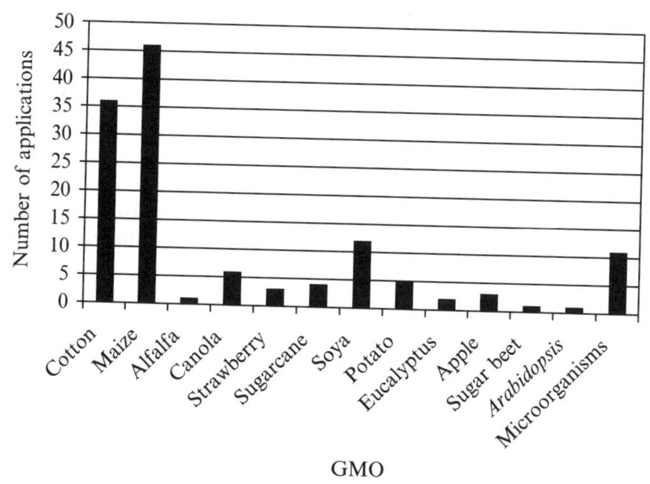

Figure 2 Number of applications for permits of different crops and organisms considered in South Africa from 1990 to 2000 (Personal Communication, M. Koch, Innovation Biotechnology)

of these crops were transgenic varieties. This represents 44.2 million hectares, which is equivalent to almost twice the total land area of the United Kingdom. The increase in area of transgenic crops between 1999 and 2000 was 11%, equivalent to 4.3 million hectares. Thirteen countries have contributed to this increase: eight are industrialized nations and five are from the developing world.

In China alone, within a period of 2 years, more than 1.5 million small-scale farmers were growing an average of 0.15 ha of insect-resistant *Bt* cotton. The area planted to *Bt* cotton was estimated to have increased by approximately 0.2 million hectares in 2000 to 0.5 million hectares. Argentina leads the list of developing countries planting genetically modified crops, with a total of 10 million hectares in 2000. This represented an increase of 49% over the plantings in 1999. There was significant increase in the growth of transgenic soybean and corn and a modest increase in cotton. The countries growing transgenic crops in 2000 included two eastern European countries, Romania, which grew soybean and potatoes, and Bulgaria, growing herbicide-tolerant maize. Uruguay reported the commercialization of transgenic crops for the first time in 2000, growing a small area (3000 ha) of herbicide-tolerant soybean.

Of the four countries that grew 99% of global transgenic crops, the United States accounted for 68%, Argentina 23%, Canada 7%, and China 1%. Nine countries together grew the other 1% with South Africa and Australia being the only countries in that group growing more than 100,000 ha of transgenic crops.

Figure 3 shows the relative areas of transgenic crops in industrial and developing countries from 1996 to 2000 (James, 2000). It shows that although the substantial share of up to 85% of global transgenic crops was grown in the developed world, the proportion of genetically modified crops grown in the developing world has increased consistently over this period. It increased from 14% in 1997 to 16% in 1998 to 18% in 1999 and 24% in 2000. Table 1, which compares the global area of transgenic crops grown during 1999 and 2000 in developed and developing countries, indicates that in 2000 approximately one-quarter of the global transgenic crop area of 44.2 million hectares was grown in developing countries. Moreover, growth continued to be strong between 1999 and 2000.

Table 1 shows that transgenic crop areas have increased from 39.9 million hectares in 1999 to 44.2 million hectares in 2000. This constitutes a global increase of 4.3 million hectares in 2000, which is equivalent to 11% growth over 1999. Of this 4.3 million hectares, 3.6 million hectares, equivalent to 84%, was grown in developing countries. This compares with only 16%, equivalent to 0.7 million hectares, in developed countries. Thus the area of genetically modified crops in developing countries grew by 51% from 7.1 million hectares

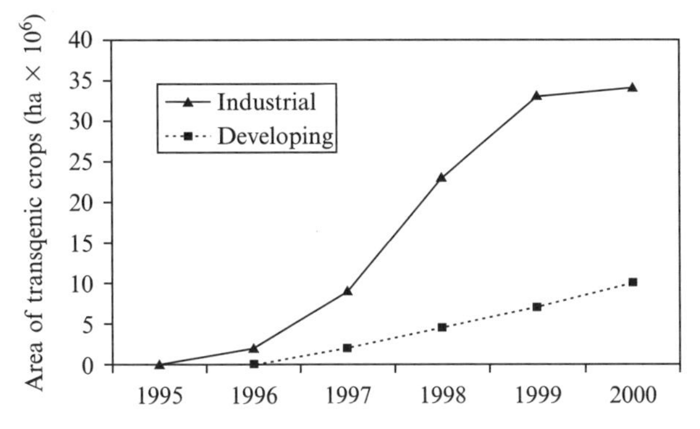

Figure 3 Global area of transgenic crops in industrial (triangles) and developing (squares) countries from 1996 to 2000. (From James, 2000)

Table 1

Global Area of Transgenic Crops (ha x 10^6) in 1999 and 2000: Industrial and Developing Countries

	1999		2000			
	Crop area	% of total available land	Crop area	% of total available land	Global increase (%)	Gain over 1999 (%)
Industrial countries	32.8	82	33.5	76	+0.7	+2
Developing countries	7.1	18	10.7	24	+3.6	+51
Total	39.9	100	44.2	100	+4.3	+11

Source: James (2000)

in 1999 to 10.7 million hectares in 2000. This compares with a 2% growth in industrial countries, where the area planted with transgenic crops increased from 32.8 million hectares in 1999 to 33.5 million hectares in 2000.

THE ANTI-GMO LOBBY AND DEVELOPING COUNTRIES

How might people in Africa and other developing regions respond to the current outcry in Europe against genetically modified crops and foods? First, it could be pointed out that developing area do not have enough food to feed their populations, and agricultural productivity is far less than in developed countries. Moreover, people in the developing world are particularly dependent on agriculture and, unlike countries in Europe, the distinction between farmers and consumers is often blurred.

Although African scientists such as Florence Wambugu (1999) applaud the use of biotechnology to improve crop and food production in developing countries, there are journalists who disagree. They contend that America finds a ready market for genetically modified foods in the form of food aid. They imply a conspiracy between the U.S. government and the UN World Food Programme (WFP) to dump unsafe crops and food as emergency aid for the world's starving and displaced. In fact the WFP accepts only food donations that comply with the safety standards of the donor country. Of greater concern should be the decline of international support for emergency food aid from about 10 million tons in 1994 to about 7.7 million tons in 1998.

One country in Africa that has come out very strongly in favor of biotechnology is Nigeria. President Obasanjo has promised more than U.S. $200 million per year for the next three years to support this growing industry.

FUTURE DIRECTIONS

Apart from the important crop traits addressing biotic and abiotic stresses mentioned earlier, what other traits would be important for developing countries? Golden rice for people living in Asia immediately springs to mind. This rice provides consumers with both iron and the precursor to vitamin A. The grains of the transgenic rice are a light golden-yellow because some of the genes came from the daffodil.

Vitamin A deficiency is a global problem. According to the World Health Organisation (WHO), 250 million children are at risk annually, and vitamin A deficiency is reponsible for significant illness and death in about 10 million people. Since vitamin A affects the absorption and use of amino acids, a deficiency results in impaired vision, reduced immune function, and protein malnutrition. A diet relying heavily on conventional rice can exacerbate iron deficiency. This shortage of a primary micronutrient afflicts up to 3.7 billion people, particularly women, leaving them weakened by anemia and susceptible to complications during childbirth (Gura, 1999).

The best sources of vitamin A are the carotenes, particularly β-carotene, found in many fruits and vegetables. The body converts carotenes into vitamin A, and it is generally accepted that it is safer to consume carotenes than vitamin A itself. Fruits and vegetables with high carotene contents, such as mangoes, spinach, carrots, and pumpkins, are not routinely available at affordable prices to poor people in Africa and Asia, particularly in the urban areas. Fortification of crops that can be grown by small-scale farmers on these continents could do much to alleviate vitamin A deficiencies. Africa needs "Golden maize."

Cardiovascular diseases, linked to high levels of dietary cholesterol, are becoming ever more prevalent in developing countries as their inhabitants move to the cities and become exposed to high-cholesterol diets. We have known for some time that plant sterols, phytosterols, can reduce cholesterol in humans by 10 to 15% by interfering with cholesterol absorption in the gastrointestinal tract. Plant sterols are not currently available in adequate quantities in the foods we eat, and scientists are actively engaged in increasing the phytosterol content of several grains (Kishore and Shewmaker, 1999).

Metals such as aluminum, the earth's most abundant metal, can be enemies of plant growth. Normally aluminum is locked up in mineral compounds and in this form is not dangerous to plants. However, in many parts of Africa, soils are acidic, and because of their low pH, aluminum ions that can poison plant roots are liberated. Stunted growth and poor harvests result. Plant breeders have coped with this problem by crossing metal-sensitive plant varieties with the few species that can thrive in soil containing aluminum ions. Unfortunately, such varieties are few, and classical breeding is slow. Recently, however, several research groups have identified metal-resistance genes. Some allow plants to thrive in soils containing four times the level of aluminum that stunts the growth of normal plants (Moffat, 1999a). If such plants were to become widely available, it would indeed be good news for farmers in Africa.

Plants can also be used in environmental bioremediation. When grown in soil contaminated with metals, such plants absorb the metals and can be harvested and deposited in landfill sites. It has been estimated that the cost of using plants to clean up polluted soils could be less than one-tenth the cost of making the soil into concrete or digging it up and transporting it to a hazardous waste landfill site. Plants are already being used to clean up mercury, a lethal waste product found at various industrial sites in both the developed and developing worlds. A gene has been introduced into a number of plant species, including canola, tobacco, and poplar. This gene allows the plants to grow on mercury-laden media and release the metal into the air. While some might disapprove of plants emitting trails of mercury vapor, scientists argue that compared with the existing concentration of mercury in the air, plants growing on contaminated sites would merely add trace amounts (Moffat, 1999b).

CONCLUSION

It is clear that the developing world needs genetically modified crops, and it is also clear that a number of countries are planting more and more such plants. It is also clear that many more would use plant biotechnology if they had the regulatory, biosafety, and scientific expertise required. It would be a tragedy if public opinion in Europe and elsewhere were to cause the poor people of the world to be deprived of this technology.

REFERENCES

Conway, G. (1997). *The Doubly Green Revolution: Food for All in the Twenty-first Century*. Penguin, Ithaca, NY.

Dyson, T. (1999). World food trends and prospects for 2025. *Proc. Natl. Acad. Sci. USA*, **96**, 5929–5933.

Gura, T. (1999). New genes boost rice nutrients. *Science*, **285**, 994–995.

James, C. (2000). Global review of commercialized transgenic crops: 2000. International Service for the Acquisition of Agri-Biotech Applications Brief 17. ISAAA, Ithaca, NY.

Kishore, G. G., and Shewmaker, C. (1999). Gene discovery and product development for grain quality traits. *Science*, **285**, 372–375.

Moffat, A. S. (1999a). Engineering plants to cope with metals. *Science*, **285**, 369–370.

Moffat, A. S. (1999b). Crop engineering goes South. *Science* **285**, 370–371.

Qaim, M. (1999). The economic effects of genetically modified orphan commodities: Projections for sweet potato in Kenya. International Service for the Acquisition of Agri-Biotech Applications Brief 13. ISAAA, Ithaca, NY.

Vitamin Information Centre. (1999). Helping Africa feed itself. *Science*, **289**, 1685.

Wambugu, F. (1999). Why Africa needs agricultural biotech. *Nature*, **400**, 15–16.

Chapter 14

Preclinical Safety Evaluation of Vaccines

François Verdier
Aventis Pasteur
Marcy L'Étoile
France

The safety evaluation of vaccines is complex because composition varies from attenuated microorganisms to recombinant proteins and gene transfer products. Vaccines act through a multistage mechanism in which the immunizing agent by itself is like a prodrug, antibodies and activated lymphocytes being the actual effectors. Therefore, several potential toxicities must be considered: direct toxicity of the test article (also designated as intrinsic toxicity), toxicity linked to the pharmacodynamic activity of the vaccine, activation of preexisting disorders, toxicity of contaminants and impurities, and other adverse reactions due to interaction between the various components. In addition, the context of the prophylactic use of vaccines requires that every effort be made to ensure safe use.

Quality control strongly contributes to the safety assessment of vaccines. However, the repeated-dose study, mimicking the human immunization schedule and performed in the most relevant animal species, is pivotal. Immunological and safety pharmacology parameters should be adapted to the specific properties of vaccines and added to this type of study.

Vaccines intended for pregnant women and other women of childbearing age also require embryo–fetal and postnatal studies with an adapted design to

Biotechnology and Safety Assessment, 3rd edition

obtain appropriate fetal and maternal exposure during gestation with continuation into the postnatal period. Tests exist to detect hypersensibility or autoimmune reactions, but further validation is necessary.

In addition to this tailor-made approach, any adjuvant or active component added to the vaccine formulation must itself be assessed via studies routinely performed for new drugs. From this review, vaccine toxicology would appear to be a separate discipline on its own whose predictive power will be increased by the development of new methods.

INTRODUCTION

The preclinical safety assessment of vaccines merits a full chapter even though vaccines represent a minor share of the global pharmaceutical market (about 1.5%) for they are a fast-growing business with a fivefold increase over the past decade and even more rapid growth projected for the coming years (Gréco, 2002). Vaccines also act more and more through innovative mechanisms with peculiar safety issues. Several aspects of the use of vaccines need to be mentioned if one is to clearly understand the level of safety requirement.

Vaccines still represent a major weapon against many infectious diseases and more recently also promising methods of prevention or treatment for disorders of other types such as cancer and autoimmune disease. Their composition has significantly evolved from the whole-microorganism approach (attenuated or inactivated) to the use of crude fractions or purified parts of these organisms (polysaccharides and toxoids) to protein–polysaccharide conjugates and now to recombinant proteins or polypeptides (Table 1). Gene transfer products (naked DNA and vector-based vaccines), currently in prelicensing phases, also represent a new approach in vaccinology (Bonnet *et al.*, 2000); they are already being successfully used as veterinary vaccines for rabies or distemper.

Table 1

Vaccine Categories Developed Today

Vaccine categories
Whole organisms (attenuated or inactivated)
Crude fractions or purified parts (polysaccharides, toxoids)
Conjugates
Recombinant proteins or polypeptides
Gene transfer products (naked DNA, vector-based vaccines)
Combinations of various types

Up until recently, the unjustified statement that all vaccines are safe for all people was still supported by some in this field. Nowadays, few if any wise scientists would make such an assertion. This caution comes from a better knowledge of the immune system and also from improving communication about all adverse events declared to be associated with vaccination even without a causal relationship (Table 2).

Vaccines undoubtedly fall within the scope of medicinal products (as defined by the amended Council Directive 65/65/EEC,) even if they are considered as a separate category as illustrated by the dedicated Office of Vaccines Research and Review in the Center for Biologic Evaluation and Research (CBER) of the U.S. Food and Drug Administration (U.S. FDA). Thus strict nonclinical and clinical safety evaluations must be completed before licensing. One of the major differences between vaccines and other pharmaceuticals is with respect to distribution: vaccines are given to large numbers of healthy people, and predominantly to healthy children. Thus significant emphasis must be placed on their safety. In developing countries, vaccination remains the main strategy in the fight against contagious diseases and particularly childhood epidemics, which still kill numerous children. The situation is markedly different in developed countries, where the global beneficial effect is not appreciated at all and the risk/benefit ratio is looked at on an individual level. The public expects zero risk from vaccination even though it is as unrealistic here as it is for any other medical intervention.

With this description of the current vaccine climate in mind, the toxicologist's responsibility for the nonclinical safety evaluation of candidate vaccines is crucial.

An appropriately designed and thereby relevant safety assessment will also provide a scientific argument against potential unfounded concerns raised during license assessment or by lay individuals after licensing. Several safety concerns associated with vaccines have been the subject of recent meetings and papers; the primary question being whether a causal association exists or merely

Table 2

Questionable Associations (unproven causal relationships) between Vaccines and Diseases

Mumpst measles–rubella vaccine and autism (Wakefield *et al.*, 1998)
Influenza vaccine and Guillain–Barré syndrome (Lasky *et al.*, 1998)
Combination vaccine and autoimmune diabetes (Karvonen *et al.*, 1999)
Lyme disease and autoimmune arthritis (Gross *et al.*, 1998)
Hepatitis B vaccine and demyelinating diseases (Ascherio *et al.*, 2001)
Aluminum hydroxide and macrophagic myofaciitis (WHO, 1999)

temporal or ecological associations (coincidental timing or environmental factors respectively) have been observed (McPhillips and Marcuse, 2001).

The main challenge in establishing a predictive safety assessment is that vaccines act through a highly complex, multistage mechanism in which the vaccine by itself is not the final triggering component. Vaccine-produced antibodies or activated T cells are the actual effectors. In addition the immune response can be enhanced by the use of adjuvant. Considering this multilevel interaction between the organism and the vaccine, five distinct categories of potential toxicological effect can be identified, and appropriate investigations need to be designed accordingly.

1. *Intrinsic toxicity of the test* article [i.e., caused directly by the vaccine component(s) per se]. This direct toxicity is the major concern for chemical entities. With vaccines, given the small quantity of antigen in a human dose and the limited number of administrations, this type of toxic effect should be very rare. However, such direct toxicity will be evident in the acute studies, which are generally included as quality control release tests, and also in the general toxicity studies, described later. The intrinsic toxicity associated with adjuvants, excipients, or preservatives in the final formulation must be evaluated and can require specific studies.

2. *Toxicity linked to the pharmacodynamic activity of the candidate vaccine or its components* (e.g., cross-reactivity between autoantigens and vaccine-produced antibodies). As indicated earlier, during the vaccination process the active component is not the antigen itself but the effectors of the immune response (T and B cells and antibodies), which are ready to mount a defense against any intrusion. Any unwanted targeted action of these active components can result in a toxic effect. The detection of such toxic effects is more complex than for direct toxicity and requires careful selection of a relevant animal species.

3. *The "biological toxicity", as designated by Prof. Zbinden* (i.e., the adverse response that is related to the activation of preexisting biological processes and pathways, such as the exacerbation of existing autoimmune disorders). This kind of toxicity cannot be excluded but is extremely difficult to predict. The lack of validated disease models is one of the problems encountered.

4. *Toxicity of contaminants and impurities* (a category that falls within the scope of both quality control and toxicology). Absence of contaminants is an essential criterion for biological and biotechnology-derived products, where a major objective is to avoid the risk of introduction of adventitious agents, and in particular, contamination with prion proteins, by removing all substances of animal origin. The potential risk associated with host cell contaminants such as protein from monkey or human cell lines in the final vaccine preparation is also a subject of importance, but the toxicological consequences remain obscure.

5. *Evaluation of potential adverse reactions due to interaction between components.* Particularly relevant for combined vaccines are potential adverse reactions due to interaction between the various vaccine components.

The species specificity of the immune function is frequently mentioned as being a major obstacle to safety assessment of vaccines in animals before proceeding to trials in man. Differences as well as homologies have also been identified between laboratory animal and human immune systems. Thus, it is the responsibility of vaccine manufacturers to investigate all potential non-clinical methods to ensure that use in humans is supported by the maximum safety data.

REGULATORY FRAMEWORK

Two categories of document exist: those that are applicable to all pharmaceuticals including biotechnology-derived products, and those specifically designed for vaccines. From the first category only the S6 guideline of the International Conference on Harmonization (ICH, 1997A) is mentioned in this chapter. The major interest of this document is that it clearly explains the flexible strategy recommended for biotechnology-derived pharmaceuticals including vaccines. The scope of this document includes recombinant protein vaccines, but conventional bacterial and viral vaccines or DNA vaccines are not covered.

Probably the most comprehensive set of specific guidelines is Note for Guidance on Preclinical Pharmacological and Toxicological Testing of Vaccines, promulgated by the Committee for Proprietory Medicinal Products (CPMP) of the European Medicines Evaluation Agency (EMEA, 1997). Although several criticisms were made when the document was released, it is the only guide solely dedicated to the preclinical safety evaluation of vaccines. Its formal use is restricted to Europe, but it lists all the items that should be addressed and allows some flexibility in the selection of investigations. There is no such a guideline in the United States, and the Guidance for Industry for the Evaluation of Combination Vaccines for Preventable Diseases: Production, Testing and Clinical studies (U.S. FDA, 1997) includes only a short section dealing with nonclinical studies. In 2001 the World Health Organisation released a document presenting the WHO guidelines for regulatory expectations on the clinical evaluation of vaccines. This document (WHO, 2001) includes a chapter dealing with the toxicology studies for vaccines and mentions that "a more detailed document which deals with the preclinical and laboratory evaluation of vaccines is under development". Monitoring WHO publications for the future release of this notice is recommended.

Interestingly, the FDA has released a draft guidance on reproductive toxicology for vaccines intended for women of childbearing age and those that may be used to immunize pregnant women (U.S. FDA, 2000). The recommendations have been adapted from document ICH S5a in order to include the specific properties of vaccines.

For vaccines considered to be in the gene therapy or gene transfer medicinal product category (e.g., pox vector and naked DNA), the following specific guidelines are of use:

- FDA, CBER, December 1996: Points to Consider on Plasmid DNA Vaccines for Preventive Infectious Disease Indications
- FDA, CBER, March 1998: Guidance for Human Somatic Cell Therapy and Gene Therapy
- EMEA, CPMP, April 2001: Note for Guidance on the Quality, Preclinical and Clinical Aspects of Gene Transfer Medicinal Products

In addition, the nonclinical evaluation of vaccines was the topic of an EMEA workshop, and the corresponding literature is available on request (EMEA, 2001).

Guidelines and position papers are regularly published or amended by drug agencies, particularly for new therapeutic approaches such as gene transfer products. The list of documents given is susceptible to modification, and one should check the latest version on the drug agency website (www.fda.gov). Several vaccine safety information sources are given in Table 3.

OUTLINE OF THE PROPOSED STUDIES

Quality Assessment

As for all biologicals, the quality of the product is an essential part of the safety assessment. This quality assessment starts with the raw materials and also applies to the different batches produced. Recently, the risk of prion contamination seriously impacted the biotechnology industry. New cases of bovine spongiform encephalopathy (BSE) and the related new variant of Creutzfeldt–Jacob disease (CJD) led to strict rules concerning materials of animal origin. Not only should vaccine manufacturers use BSE-free countries for their materials of animal origin, but they should document the full process from the origin of bacterial or viral seed to their current products. Any undefined step with potential contamination by products of unknown animal origin

Table 3

Vaccine Safety Information Sources

Institution	Website
Non Profit Groups and Universities	
• The Institute for Vaccine Safety	www.vaccinesafety.edu/index.html
• Vaccine Education Center at the Children's Hospital of Philadelphia	vaccine.chop.edu
• The Vaccine Page: vaccine news and database	vaccines.org
Government Organizations	
• European Medicines Evaluation Agency	www.eudra.org
• International Conference on Harmonization	ifpma.org/ich1.html
• U.S. Centers for Disease Control and Prevention	www.cdc.gov
• U.S. Food and Drug Administration (FDA)	www.fda.gov/cber
• U.S. National Immunization Program and Vaccine Safety	www.cdc.gov/nip/vacsafe
• U.S. National Institute of Allergy and Infectious Diseases	www.niaid.nih.gov
• World Health Organisation's pages on vaccines and immunization	www.who.int/vaccines

Source: Adapted from McPhillips and Marcuse (2001).

will require rederivation of the seed by means of methods such as ribonucleic acid (RN) transfection. The lack of other adventitious agents, and particularly viruses, is an essential safety requirement for all biotechnological products. *In vitro* and *in vivo* tests are not described in this chapter. Details about essential phases of viral safety can be found in the quality guidelines (ICH, 1997a).

In addition to viral safety tests, any new cell banks should be studied for their tumorigenicity in chemically or genetically immunosuppressed rodent models.

Several vaccines still use attenuated or inactivated microorganisms. Specific toxicity tests are performed as part of the quality control test battery to verify the attenuation or the inactivation of the microorganism. For instance, rabies vaccine is tested by the intracerebral route in mice, and oral polio vaccine is still tested by intracerebral route in monkeys, although alternative methods are currently being evaluated. Specific toxicity tests are also used to control the inactivation of bacterial toxins (e.g., diphtheria and tetanus toxins tested by the intraperitoneal route in the guinea pig and *Bordetella pertussis* toxin tested by the same route in mice). In complement to this specific toxicity test, abnormal toxicity (also called general safety) tests are also performed at various steps of the vaccine production (final product, and, in some cases, bulk and primary seed) by intraperitoneal injection of the product to mice and guinea pigs. Mortality and body weight gain are recorded during these tests.

To complete this section on safety evaluation as part as the quality assessment of vaccines, pyrogenicity tests should be carried out in the rabbit model, particularly for polysaccharide vaccines.

Toxicity Studies

Concerning the toxicological studies in their strictest sense, animal models remain the best option for mimicking the human situation. Despite many efforts to develop *in silico* and *in vitro* models, such replacement models remain of limited use because they are not able to reproduce the complexity of the human immune system.

Thus, a critical challenge is the identification of the "relevant" animal model. Among the criteria used are the following (in order of importance): development of an immune response similar to the expected human response after vaccination, demonstration of a similar binding profile for the induced antibodies (Finne *et al.*, 1987; Kantor *et al.*, 1992), susceptibility to the targeted pathogen and, perhaps the most difficult to fulfil the ability to develop exacerbation reactions (Kakuk *et al.*, 1993). These criteria imply the need to be able to generate and interpret immunogenicity data in the species used for toxicological studies. Rats and monkeys remain the most frequently used species. Recently the use of newborn or aged rodents has been considered for the evaluation of vaccines intended for neonates or elderly people, since the immune responses of these populations present some peculiarities (Siegriest, 2001).

The Single-Dose Toxicity Study

The evaluation of the potential intrinsic (direct) toxicity of the test article is the main outcome of the single-dose toxicity study; therefore the relevance of species selection is less critical than for the subsequent animal studies and rodents are used. The study should be performed in at least one animal species (EMEA, 1997), but two species are recommended in the ICH documents. This is normally done as part of the quality control test battery. If the single-dose study is the pivotal study supporting a clinical trial, the design should be modified to include more parameters (i.e., clinical pathology and microscopic examination of a limited list of tissues).

The Repeated-Dose Toxicity Study

The repeated-dose study or studies are considered to be pivotal and should be performed in full compliance with Good Laboratory Practice (GLP). Several criteria have to be justified in the design of this type of study. The fundamental selection of the animal model was mentioned earlier. The route of administration and the dosing regimen should mimic the proposed human schedule, although the time between injections can be reduced for practical reasons. This is probably the easiest way to fulfill the "be relevant" principle. Dose levels are also a subject of discussion, since vaccines and their potential adverse effects do not always follow a linear relationship. When two dose levels are tested, the lower should correspond to the "pharmacological" dose (the dose known to induce a significant immune response in the selected species) and the higher should be equal to one human dose where dose volumes permit this. Parameters monitored include clinical signs, ophthalmology, clinical pathology (hematology and serum clinical chemistry), and, at necropsy, organ weights and results of gross and histopathological examination. In the interests of reducing animal use and study costs, the assessment of local tolerance during either a single-or repeated-dose toxicity study can be made when the route of administration used is the same as for humans (EMEA, 2000).

To take into account, the prolonged action of vaccines through the immune response, animals can be split into two subgroups: the first is sacrificed immediately after the first administration and the second 15 days later, as close as possible to the immune response peak.

During the *in vivo* phase of repeated-dose studies, additional parameters of safety pharmacology such as body temperature and blood pressure also may be considered. Their inclusion is recommended for cost and ethical reasons as long as they do not interfere with the other parameters. The traditional safety pharmacology test battery does not seem relevant for vaccines. Efforts to develop vaccine-relevant safety pharmacology parameters, such as interference with existing immune disease, are encouraged. Indeed, in all pivotal studies, a basic screening of the immune system should be performed (e.g., histopathology of the lymphoid tissues and blood cell counts). Additional immunological indicators (e.g., cytokine production and immune complex assays) can be added on a case-by-case basis depending on the mechanisms involved and the expected effects.

Embryo–Fetal and Postnatal Toxicology

Besides general toxicity evaluation, creative solutions are required for other specific studies. For instance, vaccination of women of childbearing age (e.g., with rabies vaccine) and even pregnant women (e.g., with influenza vaccine) is common (Munoz and Englund, 2000; Paradiso, 2002), and this necessitates an evaluation of embryo–fetal and perinatal toxicity before licensing. Not only is successful gestation dependent on the status of the immune system (Raghupathy, 1997), but antibodies from pregnant animals can cross the placenta (Ghetie and Ward, 1997) and have harmful effects on the fetus.

The classical division of reproductive toxicology studies into three segments as used for new chemical entities is not wholly applicable. A treatment design (Figure 1) including immunization(s) before mating is recommended for vaccine embryo–fetal toxicity studies to give antibody exposure during the entire embryo–fetal period. Post–parturition immunization is suggested for females not submitted to cesarean delivery for fetus examination and their pups will be necropsied at the end of the lactation period to study the effects of antibodies in the newborn. A preliminary study to define and justify the laboratory animal species chosen should demonstrate appropriate antibody transfer with exposure of the embryo to vaccine-induced IgG. This is recommended because immunoglobulin transfer during gestation can render the animal model more susceptible to developmental effects following prenatal exposure to the vaccine immunoglobulin.

As part of the study in the main species, fetal examination, developmental monitoring of the litters, and pup necropsy should be included.

Immunotoxicology

Potential adverse effects related to interaction with the immune system may be detected during repeated-dose toxicology studies as already described. In addition, hypersensitivity and autoimmune reactions are the most commonly perceived adverse effects associated with vaccination (Shoenfeld and Aron-Maor, 2000; Kumagai *et al.*, 2000). These adverse reactions are likely to be rare, and so the use of animal models to predict them is a real challenge.

Systemic anaphylactic models and biologicals, including vaccines, have always been a subject of controversy. Systemic anaphylactic models have

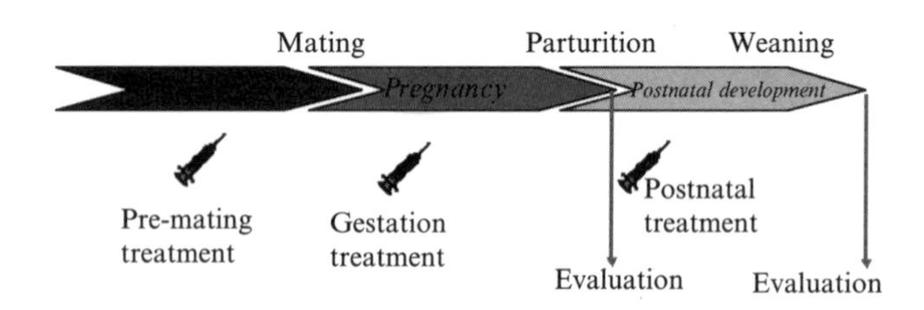

Figure 1 Treatment design for embryo–fetal and postnatal toxicology study.

been required by the Japanese guidelines for many years, but rarely are considered relevant in the ICH guideline for biotechnology products (ICH, 1997b). A proposed protocol in the guinea pig allows for the detection of high molecular weight drugs with this type of toxicity. The test consists of two or three subcutaneous administrations followed by an intravenous challenge approximately 2 weeks later (Verdier *et al.*, 1994). False positive results can however be obtained when human proteins are used (Brutzkus *et al.*, 1997). To enhance the frequency or severity of the adverse effects being sought, these models should incorporate an artificial element such as an adjuvant (e.g., aluminum hydroxide, cholera toxin) or should use only laboratory animals with a responsive genetic background (Verdier *et al.*, 1994). A protocol using intranasal sensitization in the guinea pig and in the mouse is being validated for testing new vaccines or their components (excluding proteins from mammalian origin), intended to be given by the intranasal route in humans (e.g., nasal flu vaccine).The local lymph node assay was promoted as an alternative solution to the guinea pig model to detect both immediate and delayed-type hypersensibility reactions. However validation data were obtained mainly from chemicals and not from pharmaceuticals. One can regret the lack of research work in this field, which would put us in a better position to decide whether to study hypersensibility reactions on new vaccine components.

To focus on autoimmunity, there are two aspects to be considered: the induction *de novo* of autoimmune disease and the exacerbation of existing autoimmune disorders.

In the first category, autoimmune reactions induced by molecular mimicry occur when the host and the microbial/vaccine determinants are largely similar and are present in such a way that there is a break in immune tolerance. Several presentations from a recent international symposium on autoimmunity induced by infection or immunization described this mechanism as a link between infections and autoimmune diseases (Bach, 2001; Regner and Lambert, 2001). This molecular mimicry was also suspected with some new vaccines, such as the recently commercialized Lyme vaccine, where the selected antigen, namely, the outer surface protein Osp A from *Borrelia burgdorferi*, presents sequence homology with human lymphocyte function associated antigen-1 (LFA-1) and is suspected of triggering treatment-resistant Lyme arthritis (Hemmer *et al.*, 1999; Steere *et al.*, 2001). Risk of molecular mimicry was also considered for a polysaccharidic vaccine against meningitis. The group B meningococcal capsular polysaccharide used for this vaccine is a polymer of scialic acid, which mimics the polysialylated form of neural cell adhesion molecules present particularly on mammalian embryo cells (Finne *et al.*, 1987).

Advances in computer software such as LifeSeq from Incyte and the availability of the human genome sequence allow rapid comparison between the protein sequence alignment of a vaccine antigen and a host protein.

However results obtained by using these tools present some limitations as predictive methods. For instance, for B-cell epitopes, sequence comparison will look for primary structure mimicry, but similar antigenic surfaces (conformational mimicry) are technically very difficult or even impossible to detect. Moreover, T-cell epitope mimicry requires not only an appropriate presentation of the identified common peptide by antigen-presenting cells but also the presence of autoreactive T cells that can recognize the mimicking peptide–HLA complexes.

From these hypothesises, a recommended strategy would be to avoid any vaccine antigen presenting a mimicry with a host antigen involved in an autoimmune disease. Applying this rule, *Helicobacter pylori* catalase was excluded from the screening of vaccine antigens because first it showed sequence homology with human catalase and second human catalase is reported to be an autoantigen in inflammatory bowel disease (Roozendaal *et al.*, 1998). In any case, surrogate markers can be included in the protocol of toxicology and clinical studies to detect any sign of pathogenic autoimmune response.

The risk of autoimmunity is also taken into consideration for cancer vaccines because tumor-associated antigens (TAA) are also generally expressed in normal human tissue. This risk is limited by using TAA/self antigens expressed in immunoprivileged sites such as the testis (e.g., MAGE antigen for melanoma cancer vaccine) or fetal or embryonic antigens that are also expressed in cancer cells (e.g., carcinoembryonic antigen for colorectal cancer vaccine) (Gilboa, 2001). This antigen selection does not exclude the choice of appropriate animal models and markers to look for potential toxic effects linked to an autoimmune reaction during preclinical studies.

The second identified category of autoimmune adverse effects is the exacerbation of a preexisting pathological condition as reported for influenza infection and multiple sclerosis (De Keyser *et al.*, 1998). Animal models for autoimmune diseases (e.g., MRL/Mpj-lpr, NZB/NZW, SJL, and NOD mice, BB rats) have been proposed (Taneja and David, 2001), but we do not yet have sufficient validated data. Moreover, observations of biological markers for autoimmune or hypersensitivity reactions are not necessarily linked to pathogenic consequences. For instance, and this is an important point, the presence of autoantibodies does not necessarily indicate the induction of autoimmune disease (Verdier *et al.*, 1997).

Evaluation of Adjuvants

To ensure that all safety issues related to vaccines are covered, the evaluation of new adjuvants or any other components of the vaccine formulation should also be considered, and dedicated studies are usually required. Novel adjuvants are classified as active compounds, and they are evaluated as new chemical or biological entities independently from the vaccine formulation. Recommended studies are listed in Table 4. These studies include genotoxicity assays and repeated dose and single-dose studies. Specific attention is given to pharmacokinetic studies as illustrated by the work done for *E. coli* and cholera toxins administered by the intranasal route (Van Ginkel *et al.*, 2000). Specific studies are added, taking into consideration the mechanism of action of the selected adjuvant. For instance, an adjuvant derived from lipopolysaccharide (LPS) should have been modified to avoid LPS toxic activity. Preclinical tests such as endotoxin shock in the mouse should be performed to confirm the inactivation of intrinsic properties.

The hard task of removing mercury-containing preservatives, which are potentially neurotoxic based on long term exposure to organic mercury, emphasizes the need to guarantee the safety of extraneous components added to a vaccine formulation. Experimental data should be collected if the scientific literature does not cover all safety aspects.

CONCLUSION

The most predictive nonclinical assessment will rely on the selection of the most relevant models and on careful thought to justify the strategy chosen.

Table 4

Nonexhaustive List of Toxicology Studies Applicable to Vaccine and Adjuvant Candidates

On the adjuvant alone	On the vaccine + adjuvant
Acute toxicity study in rodents (IV and human route of administration)	Repeated dose (mimicking design) in the relevant species (before phase I) by the human route of administration
Repeated dose (14 or 28 days) in the rat (daily administrations) (route to be defined)	Pharmacokinetics of the adjuvant or any other extraneous components using the vaccine formulation
Pharmacokinetic (distribution) of the adjuvant alone	Repeated dose (mimicking design) in a second species (optional, on a case-by-case basis)
Genotoxicity tests (at least two tests: Ames and mouse lymphoma)	Reproduction toxicology study in rodents or rabbits (before phase II/III)
Potential to trigger immediate hypersensitivity reaction.	
Specific test related to its mechanism of action (e.g., endotoxin shock for LPS-derived product)	

It is impossible to calculate the real predictive value of nonclinical test data by the retrospective analysis of human data, since the detection of even a remote toxicological risk in nonclinical models generally leads to cancellation of the clinical development of the product. However, we do know from experience, such as that with the oral rotavirus vaccine associated with intussusception in infants, that nonclinical studies will not pick up all adverse effects.

It is our belief that there is a clear need for the development of a scientific discipline of vaccine toxicology to provide a panel of validated methods applicable to vaccine candidates that will allow the industry to work proactively in identifying any antigens and adjuvants associated with risk early in a candidate's evaluation.

REFERENCES

Ascherio, A., Zhang, S. M., Hernan, M. A., Olek, M. J., Coplan, P. M., Brodovicz, K., and Walker, A. M. (2001). Hepatitis B vaccination and the risk of multiple sclerosis. *N. Engl. J. Med.*, **344**, 327–332.

Bach, J. F. (2001). Proceedings of the International Symposium: Autoimmunity induced by infection or immunization. *Int. J. Autoimmun.* **16**, 173–371.

Bonnet, S., Tartaglia, J., Verdier, F., Kourilsky, P., Lindberg, A., Klein, M., and Moingeon, P. (2000). Recombinant viruses as a tool for therapeutic vaccination against human cancer. *Immunol. Lett.*, **74**, 11–25.

Brutzkus, B., Coquet, B., Danve, B., and Descotes, J. (1997). Systemic anaphylaxis in guinea-pigs: Intra-laboratory validation study. *Fundam. Appl. Toxicol.*, **36**, (suppl.), S192.

De Keyser, J., Zwanikken, C., Boon, M. (1998). Effects of influenza vaccination and influenza illness on exacerbations in multiple sclerosis. *J. Neurol.Sci*, **159**, 51–53.

EMEA (1997). Note for Guidance on Preclinical Pharmacological and Toxicological Testing of Vaccines. CPMP/SWP/465/95. European Medicines Evaluation Agency, London.

EMEA (2000). Note for Guidance on Non clinical Local Tolerance Testing of Medicinal Products. CPMP/SWP/2145/00. European Medicines Evaluation Agency, London.

EMEA (2001). Note for Guidance on the Quality, Preclinical and Clinical Aspects of Gene Transfer Medicinal Products. CPMP/BWP/3088/99. European Medicines Evaluation Agency, London.

Finne, J., Bitter-Suermann, D., Goridis, C., and Finne, U. (1987). An IgG monoclonal antibody to group B meningococci cross-reacts with developmentally regulated polysialic acid units of glycoproteins in neural and extraneural tissues. *J. Immunol.*, **138**, 4402–4407.

Ghetie, V., and Ward, E. S. (1997). FcRn: The MHC class I-related receptor that is more than an IgG transporter. *Immunol. Today*, **18**, 592–598.

Gilboa, E. (2001). The risk of autoimmunity associated with tumor immunotherapy. *Nat. Immunol.*, **2**, 789–792.

Gréco, M. (2002). The future of vaccines: An industrial perspective. *Vaccine*, **20**, S101–S103.

Gross, D. M., Forsthuber, T., Tary-Lehmann, M., Etling, C., Ito, K,. Nagy, Z. A., Field, J. A., Steere, A. C., and Huber, B. T. (1998). Identification of LFA-1 as a candidate autoantigen in treatment-resistant Lyme arthritis. *Science*, **281**, 703–706.

Hemmer, B., Gran, B., Zhao, Y., Marques, A., Pascal, J., Tzou, A., Kondo, T., Cortese, I., Bielekova, B., Straus, S. E., McFarland, H. F., Houghten, R., Simon, R., Pinilla, C., and

Martin, R. (1999). Identification of candidate T-cell epitopes and molecular mimics in chronic Lyme disease. *Nat. Med.*, **5**, 1375–1382.

ICH (1997a). Note for Guidance on Quality of Biotechnological Products: Viral Safety Evaluation of Biotechnology Products Derived from Cell Lines of Human or Animal Origin. CPMP/ICH/295/95, European Medicines Evaluation Agency, International Conference on warmonization, April.

ICH (1997b). Preclinical Safety Evaluation of Biotechnology-Derived Pharmaceuticals. CPMP/ ICH/302/95, European Medicines Evaluation Agency, International Conference on warmonization, September, Step 4 on July 16.

Kakuk, T. J., Soike, K., Brideau, R. J., Zaya, R. M., Cole, S. L., Zhang, J. Y., Roberts, E. D., Wells, P. A., and Wathen, M. W. (1993). A human respiratory syncytial virus (RSV) primate model of enhanced pulmonary pathology induced with a formalin-inactivated RSV vaccine but not a recombinant FG subunit vaccine. *J. Infect. Dis.*, **167**, 553–561.

Kantor, J., Irvine, K., Abrams, S., Snoy, P., Olsen, R., Greiner, J ., Kaufman, H., Eggensperger, E., and Schlom, J. (1992). Immunogenicity and safety of a recombinant vaccinia virus vaccine expressing the carcinoembryonic antigen gene in a nonhuman primate. *Cancer Res.*, **52**, 6917–6925.

Karvonen, M., Cepaitis, Z., and Tuomilehto, J. (1999) Association between type 1 diabetes and *Haemophilus influenzae* type B vaccination: Birth cohort study. *Br. Med. J.*, **318**, 1169–1172.

Kumagai, T., Ozaki, T., Kamada, M., Igarashi, C., Yuri, K., Furukawa, H., Wagatuma, K., Chiba, S., Sato, M., Kojima, H., Saito, A., Okui, T., and Yano, S. (2000). Gelatin-containing diphtheria–tetanus–pertussis (DTP) vaccine causes sensitization to gelatin in the recipients. *Vaccine*, **18**, 1555–1561.

Lasky, T., Terracciano, G. J., Magder, L., Koski, C. L., Ballesteros, M., Nash, D., *et al.* (1998). Guillain–Barré syndrome and the 1992–93 and 1993–94 influenza vaccines. *N. Engl. J. Med.*, **339**, 1797–1802.

McPhillips, H., and Marcuse, E. K. (2001). Vaccine safety. *Curr Probl. Pediatr.*, **31**, 95–121.

Munoz, F. M., and Englund, J. A. (2000). A step ahead: Infant protection through maternal immunization. *Pediatr. Clin. North Am.*, **47**, 449–463.

Paradiso, P. R. (2002). Maternal immunization: The influence of liability issues on vaccine development. *Vaccine*, **20**, S73–S74.

Raghupathy, R. (1997). Th1-type immunity is compatible with successful pregnancy. *Immunol. Today*, **18**, 478–482.

Regner, M., and Lambert, P. H. (2001) Autoimmunity through infection or immunization. *Nat. Immunol.*, **2**, 185–188.

Roozendaal, C., Zhao, M. H., Horst, G., Lockwood, C. M., Kleibeuker, J. H., Limburg, P. C., Nelis, G. F., and Kallenberg, C. G. M. (1998). Catalase and alpha-enolase: Two novel granulocyte autoantigens in inflammatory bowel disease (IBD). *Clin. Exp. Immunol.*, **112**, 10–16.

Shoenfeld, Y., and Aron-Maor, A. (2000). Vaccination and autoimmunity—"Vaccinosis": A dangerous liaison? *J Autoimmun.*, **14**, 1–10.

Siegrist, C. A. (2001). Neonatal and early life vaccinology. *Vaccine*, **19**, 3331–3346.

Steere, A. C., Gross, D., Meyer, A. L., and Huber, B. T. (2001). Autoimmune mechanisms in antibiotic treatment – resistant Lyme arthritis. *J. Autoimmun.*, **16**, 263–268.

Taneja, V., and David, C. S. (2001). Lessons from animal models for human autoimmune diseases. *Nat. Immunol.*, **2**, 781–784.

U.S. FDA, CBER (1996). Addendum to the points to consider in human somatic cell and gene therapy (1991). Docket No. 96N-0400.

U.S. FDA, CBER (1997). Guidance for Industry for the Evaluation of Combination Vaccines for Preventable Diseases: Production, Testing and Clinical Studies. Docket No. 97N-0029.

U.S. FDA, CBER (1998). Guidance for Human somatic Cell Therapy and Gene Therapy.

U.S. FDA, CBER (2000). Guidance for industry. Considerations for Reproductive Toxicity Studies for Preventive Vaccines for Infectious Disease Indications.

Van Ginkel, F. W., Jackson, R. J., Yuki, Y., and McGhee, J. R. (2000). Cutting edge: The mucosal adjuvant cholera toxin redirects vaccine proteins into olfactory tissues. *J. Immunol.*, **165**, 4778–4782.

Verdier, F., Chazal, I., and Descotes, J. (1994). Anaphylaxis models in the guinea-pig. *Toxicology*, **93**, 55–61.

Verdier, F., Patriarca, C., and Descotes, J. (1997). Autoantibodies in conventional toxicity testing. *Toxicology*, **119**, 51–58.

Wakefield, A. J., Murch, S. H., Anthony, A., Linnell, J., Casson, D. M., Malik, M., *et al.* (1998). Ileal–lymphoid–nodular hyperplasia, non-specific colitis, and pervasive developmental disorder in children. *Lancet*, **351**, 637–641.

WHO. (1999). Weekly epidemiological record 41, pp. 337–340. World Health Organisation, Geneva.

WHO. (2001). WHO Guidelines on Clinical Evaluation of Vaccines: Regulatory expectations. Quality Assurance and Safety of Biologicals (QSB), Draft version. World Health Organisation, Geneva.

Chapter 15

Gene Flow from Transgenic Plants

Mike Wilkinson
Department of Agricultural Botany, Plant Sciences Laboratories
Reading, United Kingdom

Transformation technology offers unprecedented opportunities for the improvement of crops. At the same time, there is concern that the use of these plants may lead to unwanted change to the broader environment. Gene flow from genetically modified (GM) crops is perhaps the most apparent route by which such change could be mediated and has been the subject of intense study. There are essentially four sorts of recipient destination into which a transgene might move by gene flow: fields of non-GM cultivars, feral populations of the crop, conspecific weeds or wild populations, and populations of a related species. The process of gene flow can itself be broken down into several components to form an interdependent pathway of intermediate stages. Examples are used to illustrate our current state of knowledge of three key stages in the phenomenon: the formation of a F_1 *hybrid, gene stabilization by introgression, and transgene spread to other populations. The likelihood of completing the pathway so that a transgene becomes widespread and leads to some form of environmental change is dependent on a complex array of factors. These include characteristics of the crop and the transgene, the geographical location and the size, proximity, and the biology of the recipient.*

In attempting to quantify the risks posed by a particular transgene–crop–location combination, a progressive stepwise approach is therefore advocated.

INTRODUCTION

Improvements to biotechnology and the application of genomics research have led to proliferation in the number of crops for which viable transformation protocols are available and also in the number of transgene constructs that can be inserted into them. In turn, this has led to a global increase in the number of commercial GM crops and cultivars and the area of land used for their cultivation, and in the complexity of the transgene constructs they contain (James and Krattiger, 1999; Phipps and Beever, 2000). At the same time, concern over the possibility of environmental or agricultural consequences arising from the cultivation of such crops has also grown (e.g., Raybould and Gray, 1993; Ellstrand *et al.*, 1999; Gregory *et al.*, 2001).

The recognition of the stages in the pathway to unwanted environmental change and their assemblage into a structured order represents the beginning of the process leading to quantification of the risks posed by a GM cultivar. The far more challenging task lies in quantifying the likelihood of each stage being completed. It is at this point that quantitative information is needed. The data required for this purpose can be broadly divided into four categories:

1. Generic information that is likely to hold true for all GM lines of a particular crop
2. Generic information relating to a particular class of transgenes in the target crop
3. Information that is specific to a certain transgene or construct
4. Information that has relevance only to the transgene–plant genotype combination under consideration

Regulators clearly need to consider information from all categories. However, the initial goal of current research into GM risk assessment needs to center on generating information from the first three categories, with the highest priority being placed on data sets for quantifying the exposure to stages that are basal in the network leading to environmental change.

In this chapter, I will confine myself to the first section of the pathway in which gene flow from GM crops leads to unwanted transgene spread. In this respect, the most basal stage for all networks is the possible formation of hybrids between GM crops and non-GM recipients. The formation of hybrids between GM and organic crops could be viewed as problematic in itself, although in many ways its true significance lies in enabling the realization of more significant changes later in the network. For this reason, it is important that the probability of initial hybrid formation is evaluated early in the risk assessment process. Indeed, much of the literature on gene flow from GM crops has concentrated on this aspect.

INITIAL HYBRID FORMATION

The first stage in quantifying initial hybrid formation for any crop lies in the identification of possible recipients and the context in which hybridization may occur. There are essentially four categories of recipient population into which transgenes from a GM crop can move.

1. GM crop to non-GM crop
2. GM crop to feral/naturalized crop (i.e., escapes from agriculture)
3. GM crop to conspecific wild or weedy relative
4. GM crop to a wild or weedy relative of a different species

These scenarios each have distinctive characteristics that influence the likelihood of hybrid formation and so have different exposure characteristics.

Gene Flow from GM Crop to Non-GM Crop

Gene movement from a GM cultivar and into a non-GM equivalent can occur by any one of three main routes: pollen-mediated gene flow between fields; seed-mediated gene flow by contamination during propagation, transportation, or sowing; or gene movement through an intermediate host where volunteers act as the source for gene flow.

Pollen-mediated gene flow between fields is heavily influenced by farm practice (nature of any crop rotation and field size) and by the pollen dispersal characteristics of the crop. It is possible that the action of a minority of transgenes may affect the likelihood or rate of this kind of gene movement. These would include transgenes that confer male sterility (Rosellini *et al.*, 2001), those incorporating "terminator technology" (e.g., Oliver *et al.*, 1999), and those that are sited on the chloroplast genome (Daniell *et al.*, 1998). For the vast majority of GM lines, however, the risk profile for crop-to-crop gene flow is largely a function of the crop itself and so remains constant for the species. This consistency has been utilized historically by breeders and growers to establish "safe" isolation distances at which cross-pollination between plots does not rise above unacceptable levels to ensure that the genetic purity of stocks is maintained. It soon became apparent, however, that tolerance of transgene movement by hybridization is much lower than that required for maintaining genetic purity. For this reason, many studies attempted to establish appropriate isolation distances for GM containment by quantifying gene movement in the agricultural setting. Early works were based on direct observation of pollen movement and/or on the direct detection of hybrids from small donor plots of GM plants. It was later realized that plot size has a significant bearing on the dispersal of pollen, particularly that mediated by wind. For instance, McCartney and Lacey (1991) reported rapid decline of *Brassica* napus pollen concentration (C) from small plots could be described by the function $C = e^{-r/\alpha}$, where α is the characteristic dispersal distance. The authors estimated that for oilseed rape, α lies between 2.3 and 8.8 m. In later work, Timmons *et al.* (1996) discovered that commercial fields of oilseed rape behave quite differently and have much greater dispersal capacity, with α estimated with 95% confidence to lie between 128 and 172 m. Low levels of viable oilseed rape pollen (leading to seed formation) were detected up to 2.5 km from the nearest source field. Giddings *et al.* (1997a,b) demonstrated similar capacity for low-frequency, long-range pollination in the wind-pollinated genus *Lolium* based on extrapolations modeled from plot experiments. Likewise, Rognli *et al.* (2000) used isozyme markers to study conspecific gene flow from a small circular donor plot of *Festuca pratensis* plants to concentric rings of acceptor plants. Decline in gene flow with distance was initially rapid up to 75 m but fell much more slowly thereafter. Critically, when acceptor plants were positioned in pairs, markedly fewer hybrids were detected than in individual acceptor plants. This effect, attributed to competition from the local (acceptor) pollen, demonstrates that the size of the recipient plot is also important in determining the rate of hybrid formation. In the context of gene flow between fields, both donor and recipient blocks contain very large numbers of plants. Thus, competition from local acceptor pollen is likely to be intense in most wind-pollinated crops.

There are surprisingly few studies that provide direct observational data of gene flow between commercial fields. Timmons *et al.* (1999) used vernalization requirement, RAPD and ISSR-PCR to monitor gene flow between neighboring fields of an autumn-sown cultivar (cv. Falcon) and a spring-sown cultivar (cv. Comet). Intercultivar hybrids were detected at a frequency of 0.4% at 2 m in the recipient field and down to 0.04% at 100 m. The authors pointed out that these figures represent an approximate four fold underestimate because the flowering period two fields coincided for only 2 out of 8 weeks of flower production. Nevertheless, these figures are far lower than hybrid frequencies observed at similar distances into small populations of recipients (Timmons *et al.*, 1996). It can be inferred from this that low-frequency hybridization occurs between neighboring oilseed rape fields (0.2–1.6%) but probably only at negligible levels between fields that are separated by intervening fields. Interestingly, in a 3-year U.K. ground survey of Angus and northeastern Fife, Timmons *et al.* (1995) found that some 20% of autumn-sown oilseed rape lay within 100 m of another field of the same crop. This region is not exceptional in the United Kingdom for the amount of oilseed rape grown. It follows that low-level genetic exchange will be relatively widespread at least between fields of this crop. The significance of such levels of transgene movement is to some extent a subjective judgment but would also depend on the nature of the construct being considered.

One concern is that transgenes conferring tolerance to different herbicides may accumulate in volunteers of the crop and thereby exacerbate an existing control problem. The cultivation of GM crops across North America and elsewhere has led to the first anecdotal reports of genetic exchange of this sort between commercial fields of GM crops. For instance, Leahy (1999) noted that a grower of oilseed rape in Canada had observed oilseed rape volunteers growing after the application of glyphosate in a field where no glyphosate-resistant cultivars had been planted. It was inferred that the plants probably arose as a result of gene flow from a nearby field. Such informal observations require rigorous verification before they can be considered for risk assessment purposes, but the case does serve to illustrate the nature of the problem.

Crop to Feral/Naturalized Crop

Several works have focused on the possibility that transgenic crops may themselves become invasive of natural environments (e.g., Hails *et al.*, 1997; Crawley *et al.* 2001; Stump and Westra, 2000). Either transgenic plants could invade new habitats *de novo* or the transgenes could move into existing feral populations by gene flow and thence increase the invasiveness of the recipient. The former is almost impossible to predict on a generic level except to predict that recruitment into the new habitat would need to be mediated by seed or vegetative propagules. Williamson (1994) extrapolated from experience of invasive plants, and the state of feral populations in the United Kingdom to suggest that about 10% of transgenic crops will become established outside agriculture and 1% will become pests. However, the validity of a comparison between the impact of an introduced alien species, where much of the genome is divergent from the existing flora, and a new GM cultivar, which differs from existing cultivars only in the possession of one or a few genes is, at best, questionable. The importance that the introduced transgene(s) will have in affecting the invasiveness of a crop will clearly depend heavily on the phenotype conferred. It is impossible to foresee such properties without a priori knowledge of the function of the gene and a detailed knowledge of the ecology of the recipient feral plants. It is nevertheless difficult to imagine that any but a tiny fraction of the crop–transgene combinations will cause feral escapes to behave sufficiently differently from the nontransgenic crop plants or existing ferals to occupy new niches.

There are numerous examples of feral populations of a crop already existing outside agriculture. To some extent, the ability of a transgene to influence the persistence or invasiveness of such populations depends on the current status of these populations. For many crops, it can be reasoned that pollen-mediated gene flow from crop to feral populations is little importance because it is evident that such populations are both ephemeral and rare (e.g., wheat or oats in the United Kingdom). For these plants to become invasive, the transgene would need to confer a significant selective advantage. The persistence of feral populations of some other species is less clear. For instance, Crawley and Brown (1995) made casual observations over 2 years on the location and size of feral oilseed rape populations growing along the M25 motorway (London, U.K.). Upon noting considerable turnover in site occupancy, with only 22% of sites being occupied over both years, they inferred a strong tendency toward site extinction in 3 years. The authors also monitored three feral populations deemed to have arisen through the movement of topsoil. In this instance, all three populations became extinct after 3 years, largely because of competition by perennial grasses. It was deduced that in the absence of disturbance, oilseed rape typically becomes locally extinct after 2 to 4 years. Nevertheless, this premise is apparently contrary to the findings of Pessel *et al.* (2001), who reported relict plants in roadside populations with genetic fingerprints, based on isozymes and chemical content, identical to those of a cultivar not grown in the area for at least 8 years. While this result may indeed be indicative of occasional persistence in feral oilseed rape populations, it is difficult to discriminate between the effect of seeds germinating from the original spill and those returned by plants that occupied the site in subsequent seasons. There is certainly a large body of evidence from observational data (Lutman, 1993), seed burial experiments (Crawley *et al.*, 1993), and the appearance of volunteers (Talbot, 1993) to show that seed of oilseed rape can remain viable in the soil for many years. Wilkinson *et al.* (1995) sampled between 10 and 22 soil cores from six feral populations of oilseed rape both before and after pod dehiscence. They found an approximate 40-fold increase in seed content after pod dehiscence across all populations, with soil from four populations showing a marked increase, while two remained unchanged. Thus, it is possible for some feral oilseed rape populations to make substantial returns of seed to the soil such that the seed bank is increased. It is also clear from observational studies, however, that the vast majority of feral populations of oilseed rape decline to apparent extinction within a short time frame. Whether a small number of populations are able to persist for longer remains open to question. Should this be the case, the conditions that favor persistence merit investigation. The information generated may also provide clues to the categories of transgene most likely to extend population longevity.

The mode of transgene recruitment into feral populations can be a significant factor influencing the likelihood of spread by controlling the size of the founder population. In the case of oilseed rape, new feral populations generally arise by seed spillage or through the movement of agricultural soil (Wilkinson *et al.*, 1995). This suggests that transgene recruitment into feral populations is most probable by fresh spillage of seed (to create new populations or replenish the seed bank of extant populations) rather than by pollen-mediated gene flow from a field containing a GM cultivar. Such spillage events will often give rise to populations that are fixed for the possession of a transgene. In contrast, recruitment by pollen-mediated gene flow will result in sporadic transgenic individuals in an otherwise non-GM population.

There are a relatively small number of cultivated species in which conspecific feral populations are both stable and abundant in natural or seminatural habitats. The forage and amenity grasses perhaps provide the most obvious examples of plants belonging to this group. *Lolium perenne* for instance, is widespread throughout the United Kingdom. Seed, vegetative and pollen-mediated gene flow between cultivated and wild populations of this species has occurred on such a large scale that it is now impossible to distinguish between native populations and escapes from cultivation (e.g., Hubbard, 1968). In these examples, the frequency of gene flow between crop and recipient would be common enough (Warren *et al.*, 1998) to become an almost an irrelevant part of the risk assessment process. Emphasis here must be placed on characterizing downstream consequences of hybrid formation.

Crop-to-Conspecific Wild Relatives

The absence of interspecific breeding barriers means that the incidence of hybrid formation between GM crops and conspecific weedy or wild relatives is largely dependent upon the ability of the crop to deliver pollen to the recipient. In turn, this is heavily influenced by the degree of sympatry shared by the crop and its recipient, in terms of both their geographical and ecological distribution, and on the strength of mechanisms favoring inbreeding over outcrossing. For instance, Beebe *et al.* (1997, 2001) found evidence that spontaneous hybrids occur between wild and cultivated forms of *Phaseolus vulgaris* across Central and South America in places where both forms grow in close proximity to each other. In common with many crops, the wild form of *Phaseolus vulgaris* frequently occurs as a weed or in the margins of agriculture. This property greatly increases the likelihood of gene flow from the crop. On the other hand, and also in common with most important crops, the geographic range of the *P. vulgaris* is far wider than that of its wild conspecific progenitor. This means that initial hybrid formation is not possible for most of the land area over which the crop is grown (i.e., outside a region stretching from Mexico to northern Argentina). This is also true for notable other crops with conspecific wild relatives, including barley (*Hordeum vulgare*), maize (*Zea mays*), cotton (*Gossypium hirsutum*), sorghum (*Sorghum bicolor*), sunflower (*Helianthus annuus*), and cocoa (*Theobroma cacoa*). The extent of hybridization in the case of geographical sympatry can to some extent be compensated for by ecological specialization of the wild relative. In the case of the cabbages, kale, broccoli, and cauliflower (*B. oleracea*), for example, the wild progenitor is restricted to marine cliff faces over much of its range and so is only rarely locally sympatric with the cultivated crop (Mitchell and Richards, 1979).

The product for which the crop is grown may also have bearing on the likelihood of genetic exchange with a conspecific relative. Crops grown for their vegetative parts may be harvested prior to flowering and so present reduced opportunity for hybrid formation. Hybrid formation is nevertheless possible in such plants where occasional individuals within the crop spontaneously flower. In this way, rare flowering plants in cultivated fields of sugar beet (*Beta vulgaris*) have facilitated hybrid formation with wild populations of sea beet (*B. vulgaris* subsp. *maritima*) (Santoni and Bervillé, 1992; Boudry *et al.*, 1993).

Interspecific Hybridization from GM Crop to Wild Recipients of a Different Species

Most modern crops have close wild relatives belonging to different species but with which hybrids can be formed. Ellstrand *et al.* (1999) reviewed evidence relating to interspecific hybrid formation in the 13 most economically important crops (14 species). The authors found reasonably strong genetic evidence for interspecific hybrid formation in eight species, some genetic evidence suggestive of interspecific hybrids in a further three species, and the presence of morphological intermediates (putative hybrids) for two species. Thus, there was only one species (groundnut, *Arachis hypogea*) for which there was no evidence of any spontaneous interspecific hybrids. The absence of hybrids in this crop was attributed to the fact that all *Arachis* species are geocarpic and so presumably almost entirely self-pollinated, thereby largely preventing opportunities for spontaneous cross-pollination. Significantly, experimental crosses have demonstrated that fertile interspecific hybrids can be formed between groundnut and *A. monticola*. There are numerous interspecific breeding barriers that can be variously effective in restricting the likelihood or ability of crops to form hybrids with their wild relatives. In attempting to anticipate the probability and frequency of hybrid formation, it is important first to consider the geographical area the assessment process aims to address. The review of Ellstrand *et al.* (1999) clearly demonstrates that almost all crops are capable of yielding interspecific or intergeneric hybrids when considered in a global context.

Conversely, much of the cultivated range of many crops contains no cross-compatible wild relatives. Raybould and Gray (1993) reviewed the evidence relating to the probability of gene flow from 30 crops grown in the United Kingdom. In seven instances (grape, *Vitis vinifera*; sunflower, *Helianthus annuus*; cucumber, *Cucumis sativus*; pea, *Pisum sativum*; runner bean, *Phaseolus vulgaris*; French bean, *P. coccineus*; and maize, *Zea mays*), there were no wild or weedy species sufficiently closely related to the crop to provide any possibility for hybrid formation. In contrast, there were three crops (apple, *Malus pumila*; plum, *Prunus domestica*; and poplar, *Populus* spp.) having one or more congenerics with an established history of cross-fertilization and a further seven crops with conspecific wild relatives. These were all considered as exhibiting a high likelihood of gene flow. The remaining 13 crops all had wild relatives in the United Kingdom that have the potential to form hybrids. In six cases where natural hybrids were not reported (cabbage, *B. oleracea*; oilseed rape, *B. napus*; wheat, *Triticum aestivum*; barley, *Hordeum vulgare*; strawberry, *Fragaria x ananassa*; and lettuce, *Lactuca sativa*), relevant crosses made under controlled conditions had succeeded in producing hybrids.

In the light of subsequent evidence from field studies, at least two of these cross combinations have since been shown to form viable F_1 interspecific hybrids in wild populations. Wilkinson *et al.* (2000) found a natural hybrid between *B. napus* and wild *B. rapa*, and Guadagnolo *et al.* (2001) provided molecular evidence suggesting that introgression has occurred between cultivated *Triticum aestivum* and natural populations of *Hordeum marinum*. The last example also throws open the question of hybrid formation between *Hordeum marinum* and the closer cultivated relative, barley *(H. vulgare*).

Uncertainty of this kind is inevitable for species combinations with strong isolation barriers. Where hybrid formation is possible, then the strength of the barrier is highly influential in determining the frequency of gene flow between the species. The sheer scale of the agricultural environment means that for many species combinations, the number of potential cross-pollination events between donor and recipient is vast in any one year. Thus, rare hybrids may form with regularity across the area of contact of the two species, even where hybridization is difficult or experimentally impossible by hand pollination. The scarcity of such occurrences itself creates a serious problem of how to detect rare F_1 hybrids in the natural environment given that they will appear stochastically among otherwise pure stands of the recipient species. There is a clear need, therefore, to develop new systems for the detection of rare hybrids. To achieve this, cognizance must be taken of the complexity of the dynamic mosaic of the agricultural environment brought about by crop rotation and variability in farm management practices. These factors may themselves influence the likelihood of hybrid formation and survival.

The number of hybrid plants generated in any single year and in any one geographic area has variable importance in influencing the probability of spread and ultimately, of causing ecological change. At one extreme, where interspecific hybrids form readily, fresh recruitment of transgenes into the recipient species could be frequent enough to ensure the constant presence of transgenic individuals in exposed populations, regardless of even moderate fitness costs. At the other, where F_1 hybrids appear only at very low frequencies, gene flow will be uncertain during the commercial lifetime of a particular GM cultivar, and both persistence and spread of the transgene will be subject to several other factors. These include hybrid fertility and fitness, coincidence in the flowering of hybrid and recipient, chromosome number of both species, transgene integration site, the degree of outcrossing in the recipient, genetic drift, pollinator behavior, F_1 seed and seedling viability, and last, the size of any selective advantage conferred by the transgene.

At either extreme of hybrid frequency, then, the fitness conferred by any transgene may not be the key factor affecting the likelihood of stable introgression and spread. The forage grass *Lolium perenne*, for example, is wind pollinated and hybridizes freely with many of the broad-leaf fescues, of which two are common and grow in close proximity (*Festuca pratense* and *F. arundinacea*). Giddings (2000) modeled pollen dispersal from large fields of *L. perenne* and concluded that pollen output was so high that under certain circumstances, small recipient conspecific populations are likely to be overwhelmed. The ease of hybridization between *Lolium perenne* and *L. multiflorum*, and *Festuca pratense* and *F. arundinacea* (Stace 1975), and their frequent sympatry (Perring and Walters, 1976), suggest that hybrids would be so frequent that even detrimental transgenes are likely to appear somewhere for as long as the GM cultivars are grown. Conversely, a single hybrid between *Brassica napus* (oilseed rape) and *Sinapis arvensis* (wild mustard) has been produced under controlled conditions with extreme difficulty, using ovule culture (Bing *et al.*, 1995). The hybrid plant was highly sterile and capable of yielding only two F_2 offspring that failed to backcross to *S. arvensis*. The authors later tried and failed to obtain spontaneous hybrids. They concluded that "crosses between the cultivated and weedy species are practically impossible under field conditions in Saskatchewan, and that the escape of transgenes from transgenic cultivars of *B. napus, B. rapa* and *B. juncea* into *S. arvensis* is basically zero in this region" (Bing *et al.*, 1995).

In a similar study, Lefol *et al.* (1996) screened 2.9 million offspring from *S. arvensis* plants for the presence of hybrids with a herbicide-tolerant GM cultivar of *B. napus*. No hybrids were detected. The 50,000 flowers of a male sterile oilseed rape cultivar screened for reciprocal cross yielded just six hybrids. Artificial hybrids between the two species failed to set seed when grown in the presence of *Sinapis*. The authors surmised that a flower of either species has a probability of less than 10^{-10} of yielding an intergeneric hybrid. Frequencies as low as this could be regarded as negligible over the commercial life span of aGM cultivar where either the recipient is rare or else is rarely in contact with the donor crop (such as wild *Solanum acaule* and the cultivated potato, *S. tuberosum*).

When the recipient is a common weed like *Sinapis arvensis*, however, the number of flowers exposed to cross-pollination by the donor crop is several orders of magnitude greater than can be assessed experimentally. It is therefore possible that isolated hybrids may arise sporadically over the entire range of sympatry. Where hybrids do appear, persistence of the transgene to establishment within the population will be rendered unsure by stochastic events regardless of any selective advantage conferred, although its subsequent spread will be heavily reliant on selective advantage. Thus, the more often such hybrids appear, the greater the likelihood that a transgene that confers an advantage will establish and spread. This means that cognizance must be taken of exposure even where the frequency of hybrid formation appears extremely low. It can also be deduced that the founder population will almost certainly be a single hybrid and so unlikely to appear twice in the same area. Transgene detection at several nearby sites is consequently likely to be indicative of establishment and spread, thereby providing indirect evidence of some level of selective advantage.

Quantifying Hybrid Formation in the Natural Environment

For recipients where hybrid formation is expected to be modest or low, compilation of a geographically explicit probability profile for the distribution of hybrids is important for identifying areas in which hybridization is likely to be most or least frequent. Such information allows targeted efforts for monitoring and crucially is the starting point for modeling the spread of transgenes between populations of recipients following initial hybrid formation. For species where high frequencies of hybrids are expected, information of this kind is less valuable, although it may identify potential trial sites where hybrid formation is less frequent or improbable. There are essentially three prerequisite data sets that need to be completed before a geographically explicit model predicting hybrid frequency can be assembled:

1. Overlap of geographic and regional distributions of donor and recipient
2. Local sympatry (i.e., proximity and size of donor and recipient populations in overlap area)
3. Hybridization frequency under conditions identified in item 2

The distribution of cultivation for all crops of any significance is well known and freely available (e.g., Simmonds and Smartt, 1999). Figures indicating the regional abundance of a crop within a particular country, however, are variable both in availability and in detail. Likewise, the distributions of wild recipients is rarely known to a precision exceeding 10 km^2 (e.g., Perring and Walters, 1976). There is usually no information on the relative abundance of a recipient species with occupied 10-km squares. Thus, while it is often possible to state in broad terms where wild species and crop will co-occur, it is difficult to be sure where the two will coincide most commonly on a local scale.

Wilkinson *et al.* (2000) attempted to address this problem by estimating the number of locally sympatric sites between oilseed rape (*B. napus*) and its wild progenitors, *B. rapa* and *B. oleracea*. Both wild recipients have fairly strictly defined habitat preferences outside the agricultural environment. *Brassica oleracea* is restricted to a limited number of well described maritime sea cliff localities (Mitchell and Richards, 1979), whereas wild *B. rapa* grows along the banks of rivers. The authors used remote sensing to identify all potential donor fields of oilseed rape in a 15,000-km^2 region of southern England in 1998 and subsequently surveyed by foot all fields sited adjacent to a river (for *B. rapa*) and those next to cliffs reported to contain *B. oleracea*. In this way, two populations of *B. rapa* and one of *B. oleracea* were identified as growing adjacent to fields of oilseed rape. These populations were screened the following year for hybrids recruited into the sympatric populations. One was found in a *B. rapa* population, but none were found among the *B. oleracea* population. The authors were cautious about extrapolating their results over a wider area, although in principle, the strategy could be used to cover a much larger area for these species. Difficulty arises when the recipient has a broad ecological range or is a common agricultural weed. Under these circumstances, a polling approach is probably the most appropriate means of estimating co-occurrence.

The manner in which the two species coexist with each other also has a bearing on the probability of genetic exchange. For example, *Brassica rapa* growing as an agricultural weed (probably a persistent agricultural volunteer from a previous turnip or *rapa* rape crop) is subject to immense pollen load from the surrounding oilseed rape plants, is unable to self-pollinate (self-incompatible), and may be some distance from the nearest conspecific pollen source. In consequence, Jørgensen *et al.* (1996) reported hybrid frequencies of up to 93% on such plants. In contrast, wild *B. rapa* plants growing in natural populations on riverbanks next to oilseed rape are isolated from the crop by 1 to 10 m, are surrounded by conspecific individuals, and occupy habitats that support their natural pollinators. These factors greatly reduce the proportion of interspecific hybrids to between 0.4 and 1.5% for populations 1 and 5 m from the source field, respectively (Scott and Wilkinson, 1998).

In instances where initial hybrid formation is a rare event in the natural context, the introduction of cultural or genetic measures to further restrict hybridization might conceivably reduce frequency over the lifetime of a GM cultivar to negligible levels. The imposition of isolation distances is perhaps the simplest of the precautions that can be used for this purpose. There are many studies that have quantified gene dispersal from plots of a range of crops including wheat, cotton, maize, and potato and into recipient plots of the same species. Studies of this kind have utility in establishing isolation distances to minimize gene flow between experimental plots for breeding and multiplication purposes but are of less value for establishing 'safe distances' from wild relatives to avoid interspecific hybrid formation or for preventing gene flow from commercial fields of the crop. One difficulty for wind-pollinated species arises from the fact that size and shape of the donor and recipient plots heavily influence the airborne pollen load and so the extent of hybrid formation. Likewise, for insect-pollinated species, the behavior of pollinators can be strongly influenced by donor and recipient plot sizes. This means that pollen emitted from agricultural fields will vary greatly according to field shape and size. The size of the recipient population is likely to be even more variable than that of the donor. This too will influence gene flow rates.

Rognli *et al.* (2000) studied gene flow between artificial populations of *Festuca pratensis* Huds and reported that the ability of donor pollen to fertilize receptor plants was heavily dependent upon the density and size of the recipient plot. Pairs of plants consistently yielded fewer hybrids than isolated receptor plants. Added to this, there can be a genotype component affecting the degree of genetic transfer between plots of conspecific plants. Hucl (1996), for instance, examined variability in the outcrossing rates of 10 wheat cultivars and found rates varied between 0.3% (cv. CDC Makwa) and 6.05% (cv. Oslo). The author concluded that the isolation distances used at the time (3–10 m) were insufficient to ensure seed purity. In the case of crop-to-wild genetic exchanges, the recipient wild population is likely to be far more genetically variable than the cultivar, and this will almost invariably be unquantified. Thus, the presence of unquantified breeding barriers that may change between recipient–donor population combinations, the large variability in recipient population size, and the uneven overlap of flowering time collectively mean that "safe isolation distances" are less valuable as a concept for control of interspecific gene flow.

Targeted manipulation of the transgene contruct within the GM cultivar to minimize the risk of hybrid formation offers an alternative approach. Daniell *et al.* (1998) integrated the transgene for enolpyruvyl-3-shikimate phosphate synthase (EPSPS) into the chloroplast genome of tobacco and suggested GM crops of this type (transplastomic crops) would effectively prevent pollen-mediated transgene escape for species where chloroplasts are obligately maternally inherited. This proposition stimulated considerable debate (e.g., Daniell and Varma, 1998; Chamberlain and Stewart, 1999), centered largely on the validity of the apparent assumption that chloroplasts are always maternally inherited in crops. The mode of chloroplast inheritance is known for the majority of cultivated species but can be influenced by both genetic and environmental factors (Stewart and Prakash, 1998). Scott and Wilkinson (1999) were the first to study chloroplast inheritance in 47 natural interspecific hybrids between *B. napus* and *B. rapa*. The discovery that all had the maternal haplotype was taken to suggest that in this case at least, chloroplast-borne transgene movement from the agricultural environment would be reliant upon seed dispersal from the crop and into the wild population. Such an eventuality obviously varies according to crop and context. In the case of *B. napus* to *B. rapa*, chloroplast capture is relatively likely for populations of weeds growing among the crop where introgression is possible, with the hybrids and subsequent generations acting as the recurrent female parents. This is less true for natural populations, however, which usually are isolated from the crop by several meters and so would require seed dispersal from the initial hybrid over at least this distance (depending on the position of the hybrid in the field).

A genetic solution to the problem of interspecific gene flow has already been generated in GM material of commercial worth. GM cultivars produced by several companies contain complex constructs that require the application of a chemical stimulant that, through the action of an inducible promoter, prevents expression of a lethal toxin from coinciding with anther development (e.g., Bridges *et al.*, 2001) or seed germination (e.g., Oliver *et al.*, 1999). This allows for seed multiplication but means that hybrid seed generated from the crop would be inviable or the resultant plants would be male sterile. The technology (popularly termed "terminator technology") was almost certainly originally developed primarily as a means of protecting the intellectual property rights of the companies that owned GM technology by preventing the use of farmer-saved seed (Masood, 1998). However, the technology also offers the prospect of eliminating or severely restricting the frequency of interspecific hybrid formation for many crop–wild relative combinations. Thus far, the author is aware of no empirical evidence comparing interspecific hybrid frequency in GM lines carrying this technology with non-GM equivalents. To some extent, however, the relevance of such data has been diminished by the strong reaction against the technology by farmers, particularly from developing countries, who have a strong tradition of using farmer-saved seed (Masood, 1998). It is open to question whether the use of such an approach only for the prevention or dramatic reduction of initial hybrid formation would be viable in a legislative or commercial sense.

TRANSGENE INTROGRESSION

Initial formation of hybrids between GM crops and the recipient populations marks the start of a reticulate network of pathways that could lead to detrimental or beneficial environmental change. The complexity of the possible routes leading to change is to a large extent dependent on the crop, the geographical location, and the construct concerned. When clonal spread of the hybrid is not possible, the first stage in this process is the stablilization of the transgene into the recipient population. This will be achieved by introgression of the transgene into the genetic background of the recipient's genome by recurrent backcrossing to the recipient. The likelihood of a transgene becoming stabilized in this way will vary hugely between different donor–recipient combinations. Clearly, when crop and wild recipient are conspecific, are genetically similar, and share a common chromosome number (e.g., *Theobroma cacoa*), meiotic pairing between homologues from the two parental genomes should be full, both in the F_1 hybrid and in subsequent generations. Introgression of transgenes should be highly probable in such cases, regardless of integration site, and relatively few generations should have to pass before transgenic individuals begin to share the phenotype of the recipient except for those traits affected by the transgene(s).

As the donor and recipient genomes diverge, however, there is a strong possibility that pairing and meiotic exchange between homeologous chromosomes may be limited. Certainly, it has been long recognized by wheat breeders that the restriction of homeologous pairing in wheat–alien hybrids represents a significant barrier to the transfer of genes into the crop by introgression (Riley *et al.*, 1959). While breeders have used genetic manipulations and irradiation to overcome these problems for crop introgression, these barriers remain a significant factor in limiting introgression in the reverse direction to facilitate transgene stabilization into a wild relative.

Problems associated with limited homeologous pairing leading to loss of the transgene during introgression are made more acute when donor and recipient species differ in chromosome number. For instance, Mikkelsen *et al.* (1996) produced partly fertile triploid F_1 hybrids (AAC, $2n = 3x = 29$)

between *Brassica napus* (AACC, $2n = 4x = 38$) and its diploid progenitor *B. rapa* (AA, $2n = 2x = 20$) and demonstrated that stable introgression is possible by selfing or backcrossing to the diploid. Two generations of backcrossing was sufficient to restore the diploid complement. A later study also showed that although backcross and F_2 plants tend to be less fit than either parent, some individuals show no loss of vigor (Hauser *et al.*, 1998). Thus, reduced fitness is not an insurmountable barrier to stable introgression. Furthermore, the apparent transmission of markers from the C genome (Mikkelsen *et al.*, 1996) indicates that transgene introgression could occur from either genome, although this is not evidence in itself that transmission is equally likely from the A and C genomes.

In a recent theoretical study, Tomiuk *et al.* (2000) attempted to predict the likelihood of C-genome transmission on the basis of a model assuming that homeologous and homologous pairing occurred at equal frequency. The authors were able to conclude that should this basic assumption prove to be valid, C-genome transmission is as likely to be observed as are A-genome chromosomes. Detailed cytological data on the relative frequencies of AC versus AA pairing in hybrids are lacking. If preferential pairing between homologues does occur in the hybrids, it is also likely that some homeologues within the genome will show a greater tendency to pair than do others. Transgenes located on C chromosomes that exhibit the least tendency to pair with their homeologues would clearly be less likely to be transmitted into later generations than would those with a high propensity to pair and exchange genetic material.

There are also reasons for the tendency of some parts of the genome to display bias toward greater or reduced transmission rates on a much finer scale. Should the transgene be integrated into a site close to a gene that is beneficial in the agricultural environment but detrimental in the habitat of the recipient species, then linkage drag may inhibit its passage into subsequent generations. This scenario is most probable where the recipient has an ecological preference that is distinct from the ruderal agricultural environment (e.g. *Daucus carota* subsp. *sativus* (carrot) and the maritime *D. carota* subsp. *gummifer*) or where the crop and recipient have divergent genetic backgrounds (e.g., *Brassica napus* and *Sinapis arvensis*). Here, genes important for survival in the agricultural context may well carry a cost in the wild habitat. In such instances, the likelihood of stable transmission may depend significantly on the proximity of a transgene to a locus that imparts a penalty in the recipient environment and on the size of that penalty. An extension of this line of reasoning is that experiments based on field trials or greenhouse crosses may have limited utility in predicting the pattern of transmission into natural populations, although they will have relevance to weed populations, that share essentially the same habitat as the crop.

There are innate difficulties, however, in studying transmission in natural populations. For most crop–wild relative combinations, F_1 hybrids will be relatively rare in any one population. The scarcity of hybrids in turn makes real-time monitoring of introgression difficult. While screening techniques can be used to identify initial hybrids on a fairly large scale (Davenport *et al.*, 2000; Wilkinson *et al.*, 2001), the subsequent monitoring of introgression would require detailed paternity analysis of the subsequent generations. An alternative approach is to seek evidence of historical introgression. This is certainly the most plausible means of demonstrating introgression where hybrids are rare. Guadagnuolo (2001) found numerous putatively wheat-specific DNA markers in one individual of *Hordeum marinum*. However, there are also difficulties in the interpretation of this finding. It is problematic for any worker to be certain that markers detected in the recipient population truly originated from the crop and does not represent natural variability present in the recipient species. Theoretically, this ambiguity could be avoided through the use of closely linked markers that contain rare alleles specific to a formerly common cultivar and screening for evidence of linkage disequilibrium. This requires a reasonable knowledge of the recipient genome and ready access to mapped markers. It may therefore not be practicable for all cases. A possible alternative strategy may be to manipulate hybrid frequency in recipient populations and then monitor the efficacy of introgression in subsequent generations. In this way, transmission rates from different parts of the genome could be compared directly.

TRANSGENE SPREAD

Once a gene has stabilized in the recipient genetic background, the spread of the transgene to other populations of the same species or to other species is reliant upon a wide range of factors. Perhaps the most amenable of these to modeling is transgene migration to other populations through pollen, vegetative, or seed movement between populations. Again, several approaches can be used. Raybould *et al.* (1998) used estimates based on F statistics to calculate the relationship between distance and gene migration in *Beta maritima* populations. Approaches relying on linked markers have also been used to study genetic migration between populations of plants and animals (e.g., Wilson and Balding, 1998; Beaumont, 1999). Such models are valuable but insufficient in themselves to predict the behavior of transgenes that enhance fitness. In these cases, it is vital that other factors be accommodated in the model, especially the stage(s) in the life history at which selection is being exerted, the nature of selection and the distribution and size of the recipient populations.

CONCLUSION

The processes by which a transgene could move from the agricultural environment and ultimately lead to ecological change are numerous and complex. Given the diversity of crop–transgene–locality combinations likely to arise from advances in the genomics and postgenomics era, estimating all eventualities for all submissions is clearly impractical. Instead, a structured approach is advocated in which each stage of the process is examined in a progressive manner. Early data sets should aim to determine whether completion of a stage in the pathway to environmental change is possible for a particular crop–recipient combination. The first stage is therefore demonstration that hybrids between a crop and a potential recipient population can be formed under laboratory conditions. Work should then progressively aim to determine the probability of the hybridization occurring under natural conditions. Research should then progress to show that hybrids form under field conditions and can be detected in the wild. Finally, it is important to attempt to quantify the frequency of natural hybrids in a spatially explicit manner.

Likewise, for transgene stabilization, early studies center on demonstrating that stabilization can occur. These should be followed by experimental trials attempting to quantify transmission and to seek evidence over whether site of integration affects the probability of transmission. Finally, efforts need to focus on modeling the relationship between integration site on the host genome and the frequency of stable transmission in the natural environment under natural selection pressures. Study of subsequent spread is an even more complex undertaking and will require understanding of the distribution, size, and density of recipient populations, together with the relationship between distance and gene migration and knowledge of the life history of the recipient species. Nevertheless, commonality in the patterns of inheritance shown by transgenes and genes of the host genome means that studies to predict transgene spread can be based on the existing or historical movement of marker genes between non-GM recipient populations.

REFERENCES

Beaumont, M. A. (1999). Detecting population expansion and decline using microsatellites. *Genetics*, **153**, 2013–2029.

Beebe S., Rengifo J., Gaitan E., Duque M. C., and Tohme J. (2001). Diversity and origin of Andean landraces of common bean. *Crop Sci.*, **41**, 854–862.

Beebe S., Toro O., Gonzalez A. V., Chacon M. I., and Debouck D. G. (1997). Wild-weed-crop complexes of common bean (*Phaseolus vulgaris* L., Fabaceae) in the Andes of Peru and Colombia, and their implications for conservation and breeding. *Genet. Resour. Crop Eval.*, **44**, 73–91.

Bing, D. J., Downey, R. K., and Rakow, G. F. W (1995). An evaluation of the potential of intergeneric gene transfer between *Brassica napus* and *Sinapis arvensis*. *Plant Breed.*, **114**, 481–484.

Boudry, P., Mörchen, M., Saumitou-Laprade, P., Vernet, P., and Van Dijk, H. (1993). The origin and evolution of wild beets: Consequences for the breeding and release of herbicide-resistant transgenic sugar beets. *Theor. Appl. Genet.* **87**, 471–478.

Bridges, I. G., Bright, S. W. J., Greenland, A. J., and Schuch, W. W. (2001). Plant gene construct encoding a protein capable of disrupting the biogenesis of viable pollen. U.S. Patent **6**, 172, 279.

Chamberlain, D., and Stewart, C. N. (1999). Transgene escape and transplastomics. *Nat. Biotechnol.*, **17**, 330–331.

Crawley, M. J., and Brown, S. L. (1995). Seed limitation and the dynamics of feral oilseed rape on the M25 motorway. Proc. R. *Soc. London B, Biol.*, **259**, 49–54.

Crawley, M. J., Hails, R. S., Rees, M., Kohn, D., and Buxton, J. (1993). Ecology of transgenic oilseed rape in natural habitats. *Nature*, **363**, 620–623.

Crawley, M. J., Brown, S. L., Hails, R. S., Kohn, D. D., and Rees, M. (2001). Biotechnology—Transgenic crops in natural habitats. *Nature*, **409**, 682–683

Daniell H., and Varma, S. (1998). Chloroplast-transgenic plants: Panacea no! Gene containment yes! *Nat. Biotechnol.*, **16**, 602–602.

Daniell, H., Datta, R., Varma, S., Gray, S., and Lee, S.-B. (1998). Containment of herbicide resistance through genetic engineering of the chloroplast genome. *Nat. Biotechnol.*, **16**, 345–348.

Davenport, I. J., Wilkinson, M. J., Mason, D. C., Jones, A. E., Allainguillaume, J., Butler, H. T., and Raybould A. F. (2000). Quantifying gene movement from oilseed rape to its wild relatives using remote sensing. *Int. J. Remote Sensing*, **21** 3567–3573.

Ellstrand N. C., Prentice H. C., and Hancock J. F. (1999). Gene flow and introgression from domesticated plants into their wild relatives. *Annu. Rev. Ecol. Syst.*, **30**, 539–563.

Giddings, G. (2000). Modelling the spread of pollen from *Lolium perenne*. The implications for the release of wind-pollinated transgenics. *Theor Appl. Genet.*, **100**, 971–974.

Giddings G. D., Hamilton N. R. S., and Hayward M. D. (1997a). The release of genetically modified grasses. 1. Pollen dispersal to traps in *Lolium perenne. Theor. Appl. Genet.*, **94**, 1000–1006.

Giddings G. D., Hamilton N. R. S., and Hayward M. D. (1997b). The release of genetically modified grasses 2. The influence of wind direction on pollen dispersal. *Theor. Appl. Genet.*, **94**, 1007–1014.

Gregory, P., von Grebmer, K., and Ehart, O. (2001). Risk assessment data for GM crops. *Science*, **292**, 638.

Guadagnuolo, R., Savova-Bianchi, D., Keller-Senften, J., and Felber, F. (2001). Search for evidence of introgression of wheat (*Triticum aestivum* L.) traits into sea barley (*Hordeum marinum* s. str. Huds.) and bearded wheatgrass (*Elymus caninus* L.) in central and northern Europe, using isozymes, RAPD and microsatellite markers. *Theor Appl. Genet.*, **103**, 191–196.

Hails R. S., Rees M., Kohn D. D., and Crawley M. J. (1997). Burial and seed survival in *Brassica napus* subsp. *oleifera* and *Sinapis arvensis* including a comparison of transgenic and non-transgenic lines of the crop. *Proc. R. Soc. London. B., Biol.*, **264**, 1–7.

Hauser T. P., Jørgensen R. B., and Ostergard H. (1998). Fitness of backcross and F-2 hybrids between weedy *Brassica rapa* and oilseed rape (*B. napus*). *Heredity*, **81**, 436–443.

Hubbard, C. E. (1968). *Grasses: A guide to Their Structure, Identification, Uses and Distribution in the British Isles*. Penguin Books, London, pp. 150–151.

Hucl, P. (1996). Out-crossing rates for 10 Canadian spring wheat cultivars. *Can. J. Plant Sci.*, **76**, 423–427.

James, C., and Krattiger, A., (1999). The role of the private sector, in *Biotechnology for Developing-Country Agriculture: Problems and Opportunities*, (G. J., Persley, ed. 2020 Vision Focus 2, Brief 4 of 10. International Food Policy Research Institute, Washington, DC.

Jørgensen, R. B., Andersen, B., Landbo, L., and Mikkelsen, T. R. (1996). Spontaneous hybridization between oilseed rape (*Brassica napus*) and weedy relatives. *Acta Hortic.* **407**, 89–91.

Leahy (1999). First GM tolerance transfer is feared in Canadian OSR. *Farmers Wkly.*, p. 52, January 15.

Lefol, E., Danielou, V., and Darmency, H. (1996). Predicting hybridization between transgenic oilseed rape and wild mustard. *Field Crops Res.*, **45**, 153–161.

Lutman P. J. W. (1993). The occurrence and persistence of volunteer oilseed rape (*Brassica napus*). *Aspects Appl. Biol.*, **35**, 29–43.

Masood, E. (1998). Monsanto set to back down over "terminator" gene? *Nature*, **396**, 503–503.

McCartney, H. A. and Lacey, M. E. (1991). Wind dispersal of pollen from crops of oilseed rape (*Brassica napus* L.). *J. Aerosol. Sci*, **22**, 467–477.

Mikkelsen T. R., Jensen J., and Jørgensen R. B. (1996). Inheritance of oilseed rape (*Brassica napus*) RAPD markers in a backcross progeny with *Brassica campestris*. *Theor. Appl. Genet.*, **92**, 492–497.

Mitchell N. D., and Richards, A. J. (1979) Biological Flora of the British Isles No. 145, *Brassica oleracea* ssp. *oleracea*. *J. Ecol.*, **67**, 1087–1096.

Oliver, M. J., Quisenberry, J. E., Trolinder, N. L. G., and Keim, D. L. (1999). Control of plant gene expression. U.S. Patent **5**, 925, 808.

Perring, F. H., and Walters, S. M. (1976). *Atlas of the British Flora*. Botanical Society of the British Isles, London.

Pessel, F. D., Lecomte J., Emeriau V., Krouti M., Messean A., and Gouyon P. H. (2001). Persistence of oilseed rape (*Brassica napus* L.) outside of cultivated fields. *Theor. Appl. Genet.*, **102**, 841–846.

Phipps, R. H., and Beever, D. E. (2000). New technology: Issues relating to the use of genetically modified crops. *J. Anim. Feed Sci.*, **9**, 543–561.

Raybould A. F., and Gray, A. J. (1993). Genetically modified crops and hybridization with wild relatives: A U.K. perspective. *J. Appl. Ecol.*, **30**, 199–219.

Raybould, A. F., Mogg, R. J., Aldam, C., Gliddon, C. J., Thorpe, R. S., and Clarke, R. T. (1998). The genetic structure of sea beet (*Beta vulgaris* ssp. *maritima*) populations. III. Detection of isolation by distance at microsatellite loci. *Heredity*, **80**, 127–132.

Riley, R., Chapman, V., and Kimber, G. (1959). Genetic control of chromosome pairing in intergeneric hybrids with wheat. *Nature*, **183**, 1244–1246.

Rognli, O. A., Nilsson, N. O., and Nurminiemi, M. (2000). Effects of distance and pollen competition on gene flow in the wind-pollinated grass *Festuca pratensis* Huds. *Heredity*, **85**, 550–560.

Rosellini, D., Pezzotti, M., and Veronesi, F. (2001). Characterization of transgenic male sterility in alfalfa. *Euphytica*, **118**, 313–319.

Santoni, S., and Bervillé, A. (1992). Evidence for gene exchanges between sugar beet (*Beta vulgaris* L.) and wild beets: Consequences for transgenic sugar beets. *Plant Mol. Biol.*, **20**, 560–578.

Scott, S. E., and Wilkinson, M. (1998). Transgene risk is low. *Nature*, **393**, 320.

Simmonds, N. W., and Smartt, J. (1999). *Principles of Crop Improvement*. Blackwell Science, Oxford, U.K.

Stace, C. A. (1975). *Hybridization and the British Flora*. Academic Press, London and New York.

Stewart, Jr., C. N., and Prakash, C. S. (1998). Chloroplast-transgenics are not a gene flow panacea. *Nat. Biotechnol*, **16**, 401.

Stump, W. L., and Westra, P. (2000) The seedbank dynamics of feral rye (*Secale cereale*). *Weed Technol.*, **14**, 7–14.

Talbot, M. N. (1993). The occurrence of volunteers as weeds of arable crops in Great Britain. *Aspects Appl. Biol.*, **35**, 231–235.

Timmons, A. M. O'Brien, E. T. Charters, Y. M. Dubbels, S. J., and Wilkinson, M. J. (1995). Assessing the risks of wind pollination from fields of genetically modified *Brassica napus* ssp. *oleifera*. *Euphytica*, **85**, 417–423.

Timmons A. M., Charters Y. M., Crawford J. W., *et al.* (1996). Risks from transgenic crops. *Nature*, **380**, 487.

Timmons, A. M., Dubbels, S. J., Wilson, N. J., and Wilkinson, M. J. (1999) Assessing the risks of wind pollination from fields of genetically modified *Brassica napus* ssp. *oleifera*. Department of the Environment Transport and the Regions (DETR). Environmental impact of genetically modified crops, Research Report 10, Genetically Modified Organisms, pp. 10–18.

Tomiuk, J., Hauser, T. P., and Bagger-Jørgensen, R. (2000). A- or C-chromosomes, Does it matter for the transfer of transgenes from *Brassica napus*? *Theor. Appl. Genet.*, **100**, 750–754.

Warren, J. M., Raybould, A. F., Ball, T., Gray, A. J., and Hayward, M. D. (1998) Genetic structure in the perennial grasses *Lolium perenne* and *Agrostis curtisii*. *Heredity*, **81**, 556–562.

Wilkinson, M. J., Timmmons, A. M., Charters, Y. M., Dubbels, S., Robertson, A., Wilson, N., Scott, S., O'Brien, E., and Lawson, H. M. (1995). Problems of risk assessment with genetically modified oilseed rape. *Brighton Crop Protection Conference 1995—Weeds*, **3**, 1035–1044.

Wilkinson, M. J., Davenport, I. J., Charters, Y. M., Jones, A. E., Allainguillaume, J., Butler, H. T., Mason, D. C., and Raybould, A. F. (2000). A direct regional scale estimate of transgene movement from genetically modified oilseed rape to its wild progenitors. *Mol. Ecol.*, **9**, 983–991.

Williamson, M. (1994). Community response to transgenic plant release—Predictions from British experience of invasive plants and feral crop plants. *Mol. Ecol.*, **3**, 75–79.

Wilson, I. J., and Balding, D. J. (1998). Genealogical inference from microsatellite data. *Genetics*, **150**, 499–510.

Chapter 16

Safety Assessment of Insect-Protected Cotton

Kathryn A. Hamilton, Richard E. Goodman, and Roy L. Fuchs
Monsanto Company
St. Louis, MO

Insect-protected, Bollgard cotton, developed by Monsanto and field tested since 1992, produces an insect control protein (Cry1Ac) derived from the naturally occurring soil bacterium Bacillus thuringiensis *subsp.* kurstaki (*Btk*). *Production of the Cry1Ac protein in the cotton plant provides effective season-long protection against key lepidopteran insect pests, including tobacco budworm, pink bollworm, and cotton bollworm. Microbial formulations of* Bacillus thuringiensis *that contain the Cry1Ac insecticidal protein have been registered in numerous countries worldwide and have been safely used for control of lepidopteran insect pests for more than 40 years. The Cry1Ac protein produced in Bollgard cotton is nearly identical in structure and activity to the Cry1Ac protein found in nature and in commercial* Btk *microbial formulations.* Bacillus thuringiensis *and* Btk *microbial formulations are specific to the target insect pests and do not have deleterious effects to nontarget organisms such as beneficial insects, birds, fish, and mammals, including humans.*

Biotechnology and Safety Assessment, 3rd edition

The chapter provides information on the methods used to develop Bollgard cotton event 531 and a summary of the food, feed, and environmental safety studies that support the safety of Bollgard cotton. In addition to the molecular characterization, the following safety studies were conducted: safety of the produced proteins, food/feed composition, and environmental safety. On the basis of this evaluation, Bollgard cotton and its processed fractions were found to be comparable to conventionally bred cotton, taking into consideration the natural variation seen among cotton varieties, with the exception of the expression of the Cry1Ac and NPTII proteins. The Cry1Ac and NPTII proteins were shown to be safe for human and animal consumption and to the environment.

INTRODUCTION

Cotton is the leading plant fiber crop produced in the world and the most important in the United States. Lepidopteran insects are the main pest problem on most of the cotton acres produced in the United States and many parts of the world. The incorporation of the *Bt* protein into cotton reduces, and in some cases eliminates, the need to spray chemical insecticides to control the major caterpillar pests. Bollgard cotton has value not only as a replacement for insecticide applications for specific pests, but also as a pest management tool that can provide benefits above and beyond the reduction in insecticide costs (Falck-Zepeda *et al.*, 1998, 2000; Wier *et al.*, 1998; Gianessi and Carpenter, 1999; Xia *et al.*, 1999, Klotz-Ingram *et al.*, 1999; Traxler and Falck-Zepeda, 1999; Betz *et al.*, 2000; Economic Research Service/USDA, 2000; Fernandez-Cornejo and McBride, 2000; Carpenter and Gianessi, 2001; Edge *et al.*, 2001; Perlak *et al.*, 2001). These additional benefits include reduced risk to grower health, improved environment for beneficial insects and wildlife, and a more stable economic outlook and productivity for the cotton industry.

Bollgard cotton was produced using *Agrobacterium tumefaciens*–mediated transfer of the *cry1Ac* gene, which encodes for insecticidal activity against lepidopteran insect pests, into the genome of Coker 312 cotton. The Coker 312 cultivar was chosen because of its favorable response to the tissue culture system used in the process to produce transgenic plants. Although the Coker 312 is no longer widely grown, it is still a commercially acceptable cultivar. The Bollgard trait has since been transferred to other commercial cotton varieties by means of traditional breeding techniques.

The primary benefits of Bollgard cotton are reduced insecticide use, improved control of target insect pests, improved yield, reduced production costs, improved profitability, reduced farming risk, and improved opportunity to grow cotton, resulting in improved economics for the cotton growers.

Planting of Bollgard cotton since 1996 in the United States has resulted in reductions in insecticide use (insecticidal active ingredients cut back by 2.7 million pounds) and in insecticide applications (decreased by 15 million). U.S. cotton growers planting Bollgard cotton showed a 260-million-pound increase in cotton production per year, which resulted in an estimated increase in net income in 1999 of $99 million. There also are a number of secondary benefits associated with the reduction in insecticide use, which include enhanced populations of beneficial insect and wildlife populations, reduced potential runoff of insecticides, and improved safety for farm workers, whose potential exposure to insecticides is reduced.

The genetically improved Bollgard cotton product was produced by means of *Agrobacterium tumefaciens*–mediated transfer of the *cry1Ac* gene into the genome of a conventional cotton variety, Coker 312, using a binary plasmid vector. The *npt*II gene, which encodes a selectable marker enzyme, neomycin phosphotransferase II (NPTII), was also present on the plasmid to facilitate selection of insect-protected plants. The NPTII protein served no other purpose and has no pesticidal properties. The plasmid also contained the antibiotic resistance *aad* gene, which encodes the bacterial selectable marker enzyme 3″(9)-*O*-aminoglycoside adenylyltransferase (AAD). This gene, which confers resistance to the antibiotics spectinomycin and streptomycin, facilitated the selection of bacteria containing the plasmid in the initial steps of transforming the cotton tissue. The *aad* gene is under the control of a bacterial promoter, and the encoded protein is not detected in Bollgard cotton plant tissue.

In assessing the nutritional and compositional equivalence of Bollgard cotton to conventional cotton varieties, more than 2500 separate analyses were performed on 67 components of the cottonseed and oil. Protein, fat, moisture, calories, minerals, amino acid, cyclopropenoid fatty acid, and gossypol levels were among those analyzed. The results of these analyses clearly demonstrate that, other than the production of the Cry1Ac and NPTII proteins, Bollgard cotton is compositionally equivalent to and as safe as conventional cotton varieties currently available.

The *A. tumefaciens* transformation system is well understood and has been utilized for many years in the genetic modification of many dicotyledonous plants. The plasmid vector was modified to prevent the transformation system from transmitting crown gall disease. This transformation system stably inserts genes from the plasmid vector into the chromosome of the plant cell. Molecular characterization showed that two T-DNA (transferred DNA) inserts were integrated into the cotton genome to produce Bollgard event 531. The *cry1Ac* gene segregated in a manner consistent with the presence of a single active copy of the coding sequence and was stably transferred to numerous other commercial cotton varieties by traditional breeding techniques.

The plasmid vector in the *A. tumefaciens* used to produce Bollgard cotton event 531 contains the fully sequenced *cry1Ac, npt*II and *aad* genes (Figure 1). The *cry1Ac* gene was derived from the common soil microbe *Bacillus thuringiensis* subsp *kurstaki (Btk)* and encodes an insecticidal protein, Cry1Ac. This *cry1Ac* gene cassette contains an e-35S promoter and a 7S 3′ transcriptional termination sequence. The *npt*II gene encodes a selectable marker enzyme, neomycin phosphotransferase II (NPTII), which was used to identify cotton cells containing the Cry1Ac protein. The *npt*II gene is driven by a cauliflower mosaic virus 35S promoter and is followed by a nopaline synthase (*nos*) 3′ region that directs polyadenylation of the mRNA. The NPTII protein served no other purpose and has no pesticidal properties. The *aad* gene encodes the bacterial selectable marker enzyme 3″(9)-*O*-aminoglycoside adenylyltransferase (AAD), which allowed for the selection of the *Agrobacteria* on media containing spectinomycin or streptomycin. The *aad* gene is under the control of a bacterial promoter, and therefore the encoded protein is not expressed in plants derived from Bollgard cotton event 531.

The gene donor organisms include cauliflower mosaic virus (e-35S promoter), *A. tumefaciens* (nopaline synthase terminator), *E. coli* (*npt*II and *aad*), and soybean (*Glycine max*; 7S 3′ termination signal). Cauliflower mosaic virus and *A. tumefaciens* are phytopathogens. *E. coli* is a Gram-negative, nonpathogenic bacterium used for DNA cloning and vector construction.

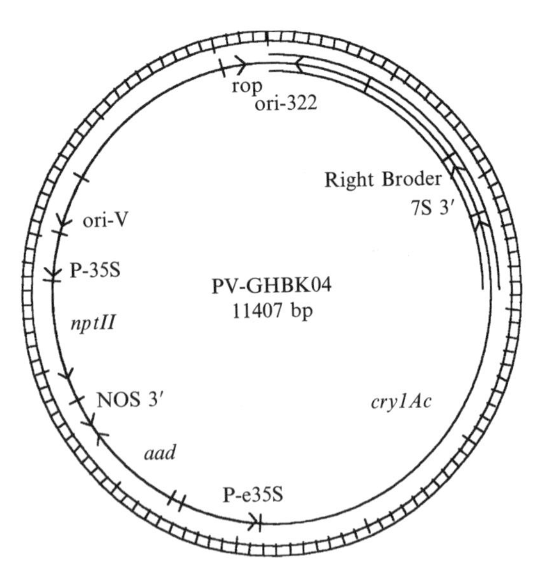

Figure 1 Plasmid map of PV-GHBK04.

The characteristics of the cauliflower mosaic virus and *A. tumefaciens* donor organisms do not warrant any analytical or toxicological tests, as only the specific sequenced genes encoding enzymes were transferred to the host organism.

MOLECULAR CHARACTERIZATION OF BOLLGARD COTTON

Molecular characterization of Bollgard cotton event 531 demonstrated there are two T-DNA inserts (Figure 2). The primary functional insert contains single copies of the full-length *cry1Ac* gene, the *npt*II gene, and the *aad* antibiotic resistance gene. This T-DNA insert also contains an 892-bp portion of the 3′ end of the *cry1Ac* gene fused to the 7S 3′ transcriptional termination sequence. This segment of DNA is at the 5′ end of the insert, is contiguous and in the reverse orientation with the full-length *cry1Ac* gene cassette, and does not contain a promoter. A transcript detected by means of the reverse transcription polymerase chain reaction (RT-PCR) corresponds to this 3′ segment of the *cry1Ac* gene and the adjacent genomic DNA. In the unlikely event that this RNA were translated, the resulting peptide would be highly homologous to the corresponding portion of the C-terminus of the Cry1Ac protein. The protein, if produced, would have been a constituent in all safety studies conducted with either the Cry1Ac protein or with Bollgard cotton plants or cottonseed, since both these inserts were shown to be present in the materials used for these studies as well as in the materials used for the extensive regulatory field trials conducted with Bollgard cotton.

The second T-DNA insert contains 242 bp of a portion of the 7S 3′ polyadenylation sequence from the terminus of the *cry1Ac* gene. No RNA transcript was detected by RT-PCR that corresponds to or would have been

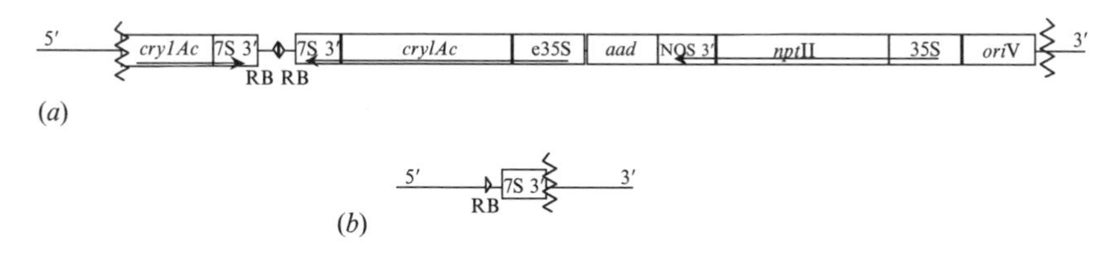

Figure 2 Schematic representation of two predicted DNA inserts in Bollgard cotton event 531: (*a*) The primary functional insert and (*b*) the 242-bp portion of the 7S 3’ genetic element. Vertical serrated lines indicate junctions between truncated genetic elements and the plant genomic DNA. Arrows within the genetic elements imply the direction of the genetic element. Triangles represent the border region of the T-DNA and are shown for the orientation purposes.

transcribed from the 242-bp 7S 3′ T-DNA insert, hence, as expected, no peptide is produced.

The selfed data from the crosses to other commercial cotton varieties demonstrates the stability of the transfer of the functional insert from generation to generation. The *cry1Ac* gene in Bollgard cotton has been demonstrated to be stably integrated into the chromosome on the bases of molecular analyses, data on phenotypic expression, and inheritance patterns. The results of these studies are summarized as follows:

- Southern blot analyses of numerous generations of Bollgard cotton performed over 8 years provided an identical Southern blot pattern, indicating stability of the functional *cry1Ac* gene insert.
- Enzyme-linked immunosorbent assays (ELISA) of seed obtained from multisite trials over 8 years showed similar levels of the Cry1Ac and NPTII proteins.
- Production of the Cry1Ac protein has been confirmed by immunodetection and/or efficacy data under different environmental conditions and in numerous Bollgard cotton varieties.
- Mendelian inheritance of the Bollgard trait is observed after self-pollination or backcrossing with other cotton varieties.
- The insecticidal efficacy has been maintained during the development of this product and since its marketing in 1996; total U.S. production came to over 17 million acres planted.
- The seed quality of Bollgard cotton has been maintained after transfer of the *cry1Ac* gene into different genetic backgrounds.

Based on this information, there is no evidence or likelihood of instability either genetically or with respect to efficacy. These data confirm that the Bollgard trait is stably integrated in the cotton genome.

CRY1AC AND NPTII PROTEIN LEVELS IN BOLLGARD COTTON PLANTS

Cry1Ac and NPTII proteins are produced at low levels in the various tissues of the Bollgard cotton plant. Data generated from 1992 samples are presented in Tables 1 and 2. Cry1Ac and NPTII proteins were detected in event 531 and were not detected, as expected, in the Coker 312 parental line. The mean levels of the Cry1Ac protein in 1992 were 1.56 and 0.86 μg/g fresh weight in leaf and raw cottonseed, respectively. The mean levels of the NPTII protein in 1992 were 3.15 and 2.45 μg/g fresh weight, respectively, for leaf and raw cottonseed. In 8 years of field testing of samples from numerous sites, the mean level of the Cry1Ac protein in raw cottonseed ranged from

Table 1
Levels of Cry1Ac, NPTII, and AAD Proteins in Cotton Leaf Tissue in 1992

Analyte	Protein ($\mu g/g$ tissue fresh weight)[a]	
	Coker 312	Line 531
Cry1Ac		
Mean	ND	1.56
Range	NA	1.18–1.94
SE	NA	0.15
NPTII		
Mean	ND	3.15
Range	NA	2.46–3.84
SE	NA	0.270
AAD		
Mean	ND	ND
Range	NA	NA
SE	NA	NA

[a]Mean expression level across all field test locations.
$N = 36$; six samples from each of six sites.
ND, nondetectable; NA, not applicable.

approximately 1 to 9 μg/g fresh weight. In raw cottonseed, the mean level of the NPTII protein ranged from 2.0 to 15 μg/g fresh weight over the same sites and years of field testing. Cry1Ac and NPTII proteins are present at low levels in whole plants collected just prior to defoliation. In field tests from the 1992 season, mature Bollgard plants contained an estimated 0.08 μg Cry1Ac protein/g and 3.3 μg NPTII protein/g fresh weight of a mature, whole plant (≈ 10 μg of Cry1Ac protein per plant). Two additional years of data from the 1998 and 1999 field seasons showed Cry1Ac protein levels in whole plant ranging from under 0.07 μg/g (limit of detection of the assay) to 0.19 μg/g fresh weight. The Cry1Ac protein levels remained sufficiently high for effective control of the targeted insect pests throughout the season.

The Cry1Ac protein was not detected in nectar collected from Bollgard cotton by means of an assay with a limit of detection of 1.6 ng/g fresh weight of the nectar. The Cry1Ac protein is present in pollen at levels just above the limit of detection of the assay used to evaluate the Cry1Ac protein concentrations: 11.5 ng/g fresh weight of the pollen.

Table 2
Levels of Cry1Ac, NPTII, and AAD Protein Expression in Cottonseed Tissue in 1992

	Protein ($\mu g/g$ tissue fresh weight)[a]	
Analyte	Coker 312	Line 531
Cry1Ac		
Mean	ND	0.86
Range	NA	0.40–1.32
SE	NA	0.18
NPTII		
Mean	ND	2.45
Range	NA	1.97–2.93
SE	NA	0.19
AAD		
Mean	ND	ND
Range	NA	NA
SE	NA	NA

[a]Mean expression level across all field test locations.
$N = 36$, six samples from each of six sites.
ND, nondetectable; NA = not applicable.

After processing, the amounts of Cry1Ac protein were reduced to nondetectable levels in the major cottonseed processed products: refined oil, linter brown stock, and cottonseed meal. The Cry1Ac protein was not detected by ELISA or insect bioassay in the processed cottonseed meal. The total protein content of refined cottonseed oil was found to be below the limit of detection of the assay (1.3 ppm). The presence of the Cry1Ac protein in linters from Bollgard cottonseed was measured by means of Western blot analysis and bioactivity. The Cry1Ac protein was detected at 0.1 μg/g weight of raw linters. After the linters had been processed to linter brownstock or to more highly purified cotton linters, the Cry1Ac protein was not detected at a limit of detection of 0.08 μg/g weight.

The AAD protein was not detected in the leaf or seed tissue from Bollgard cotton at the limit of detection of 0.008 and 0.005 μg/g fresh weight for leaf and seed, respectively. This result was expected, since the *aad* gene is driven by a bacterial promoter and expression in the cotton plant had not been anticipated.

SAFETY ASSESSMENT OF THE CRY1AC AND NPTII PROTEINS IN BOLLGARD COTTON

Safety assessments of the Cry1Ac and NPTII proteins expressed in Bollgard cotton event 531 include demonstrating the lack of similarity to known allergens and toxins and the long history of safe consumption of comparable proteins in microbial formulations, rapid digestion in simulated gastric and intestinal fluids, mode of action/specificity of the Cry1Ac protein, and lack of acute oral toxicity in mice.

MODE OF ACTION AND SPECIFICITY OF THE CRY1AC PROTEIN

The Cry1Ac protein is produced as an insoluble crystal in the *B. thuringiensis* microbe. The crystal protein is composed of the protoxin form of the protein. Insecticidal activity of the Cry1Ac protein requires that the protein be ingested. In the insect gut, a high-pH environment, the protein is solubilized and then proteolytically cleaved to the active core of the protein, which is resistant to further degradation by the insect gut proteases. The core protein binds to specific receptors on the midgut of lepidopteran insects, inserts into the membrane, and forms ion-specific pores (English and Slatin, 1992). These events disrupt the digestive processes and cause the death of the insect. The digestive tract tissues of nontarget insects, mammals, birds, and fish do not contain receptors that bind the Cry1Ac protein. Therefore the Cry1Ac protein cannot disrupt digestion and is, therefore, nontoxic to species other than lepidopteran insects (Hofmann *et al.*, 1988; Betz *et al.*, 2000).

CHARACTERIZATION AND HISTORY OF SAFE CONSUMPTION OF THE CRY1AC AND NPTII PROTEINS

There is a history of safe use of Cry1Ac protein in microbial *Bt*-based products (U.S. EPA, 1988; IPCS, 2000). Both EPA and WHO (through IPCS, the International Programme on Chemical Safety) have recognized the potential for dietary exposure to Cry proteins from use of microbial sprays on food crops: "The use patterns for *B. thuringiensis* may result in dietary exposure with possible residues of the bacterial spores on raw agricultural commodities. However, in the absence of any toxicological concerns, risk from the consumption of treated commodities is not expected for both the general population and infants and children" (U.S. EPA, 1998) and "*B.t.* has not been reported to cause adverse effects on human health when present in drinking-water or food" (IPCS, 2000).

The amino acid sequence of the Cry1Ac protein expressed in Bollgard cotton has been predicted based on nucleotide sequence of the coding sequence. The Cry1Ac protein produced in Bollgard cotton is more than 99.4% identical to the protein produced by the *B. thuringiensis* subsp. *kurstaki* (*Btk*) bacterial strain. Strains of *B. thuringiensis* have been used safely as commercial microbial pesticides for over 40 years (Lüthy *et al.*, 1982, Baum *et al.*, 1999). The naturally occurring Cry proteins produced in *Btk* have been shown to have no deleterious effects to fish, avian species, mammals, and other nontarget organisms (U.S. EPA, 1988; Betz *et al.*, 2000). The safety of the Cry proteins to nontarget species is attributed to their highly specific mode of action and rapid digestibility.

The NPTII protein expressed in Bollgard cotton is chemically and functionally similar to the naturally occurring NPTII protein, which is found in a number of naturally occurring bacteria (Fuchs *et al.*, 1993; U.S. FDA, 1994).

Digestion of CRY1AC and NPTII Proteins in Simulated Gastric and Intestinal Fluids

In addition to the lack of receptors for the Cry1Ac protein, the absence of toxic effects in humans and other mammals is further supported by the rapid degradation of the protein in an *in vitro* gastric digestion study. The rate of degradation of the Cry1Ac protein was evaluated separately in simulated gastric (pepsin, pH 1.2) and intestinal (pancreatin, pH 7.5) fluids, the simulated gastric and intestinal fluids were constituted based on recommended levels in U.S. Pharmacopeia (1995). The degradation of the Cry1Ac protein was assessed by Western blot analysis and insect bioactivity. The study showed that the Cry1Ac protein and peptides degrade in approximately 30 seconds upon exposure to gastric fluid (Betz *et al.*, 2000). The acid conditions of the stomach denature the native conformation of the Cry1Ac protein, facilitating its rapid degradation. In intestinal fluid, the Cry1Ac protein was converted to the protease-stable form and remained intact and bioactive for at least 21 hours. This result was expected because protease-resistant core proteins of *Bt* insecticidal proteins are known to be resistant to further trypsin digestion. *In vivo*, the Cry1Ac protein would be exposed to gastric conditions prior to entering the intestinal lumen. The low pH and pepsin in the stomach would be expected either to fully digest the protein or to render it susceptible to intestinal digestion.

The NPTII protein was shown to degrade rapidly under simulated mammalian digestive conditions. The degradation of the NPTII protein in digestion fluids was assessed over time by Western blot analysis. The enzymatic activity of the NPTII protein was shown to be destroyed after a 2-minute incubation in simulated gastric fluid and a 15-minute incubation in simulated intestinal fluid (Fuchs *et al.*, 1993).

Lack of Acute Oral Toxicity of Cry1Ac and NPTII Proteins in Mice

Few proteins are toxic when ingested. When proteins are toxic, they are known to act by acute mechanisms and at low dose levels (Sjoblad *et al.*, 1992). Results of a mammalian acute oral toxicity study support the specificity and safety of the Cry1Ac protein. There was no evidence of toxicity even at extremely high dose levels (4200 mg/kg body weight) when the Cry1Ac protein was administered orally to mice (Betz *et al.*, 2000). Therefore, the Cry1Ac protein is not considered to be toxic, except to target insect pests. Also, the Cry1Ac protein produced in the cotton plant is not expected to present a risk of dermal or inhalation toxicity. First, the expression level of the Cry1Ac protein in cotton is low, and the protein is found internally within the cell walls of the plant tissues, with little or no potential for dermal or inhalation exposure. Second, proteins that are nontoxic by the oral route are not expected to be toxic by the dermal or pulmonary route. Similarly, the NPTII protein caused no deleterious effects in mice when administered by gavage at dosages up to 5000 mg/kg body weight (Fuchs *et al.*, 1993).

Lack of Sequence Similarity of CRY1AC and NPTII Proteins to Known Protein Toxins

One method for the assessment of potential toxic effects of proteins introduced into plants is to compare the amino acid sequences of the protein and known toxic proteins. Homologous proteins derived from a common ancestor have similar amino acid sequences, are structurally similar, and share common function. Therefore, it is undesirable to introduce a DNA that encodes for a protein that is homologous to any toxin. Homology is determined by using published criteria to find the degree of amino acid similarity between proteins (Doolittle *et al.*, 1990). The Cry1Ac protein does not show meaningful amino acid sequence similarity when compared with known protein toxins (except other Cry proteins) present in the Protein Identification Resource (PIR), European Molecular Biology Laboratory (EMBL), Swis-sProt, and GenBank protein databases. The NPTII protein does not show meaningful amino acid sequence similarity in comparison to known protein toxins present in these protein databases.

Assessment of Exposure of Humans to CRY1AC and NPTII Proteins from Bollgard Cotton

Cottonseed oil and processed cotton linters are the only cotton products used for human food (NCPA, 1989). Analysis of refined cottonseed oil derived from both the parental Coker 312 control line and Bollgard cotton event 531 confirmed that there is no detectable protein in cottonseed oil at a limit of detection for the assay of 1.3 ppm total protein. This is consistent with other reports that conclude the absence of protein in cottonseed oil (Jones and King, 1993). Analysis of processed linters also confirmed there was no detectable protein (Sims *et al.*, 1996). Therefore, significant human consumption of the Cry1Ac and NPTII proteins and any degradation products of these proteins present in Bollgard cotton varieties is extremely unlikely. Furthermore, direct food challenge of individuals allergic to proteins contained in the meal derived from oilseed crops (e.g., soybean, peanut, sunflower) with the oil from these respective crops has established that refined oil does not elicit an allergenic response (Taylor *et al.*, 1981; Bush *et al.*, 1985; Halsey *et al.*, 1986). This is consistent with the lack of detectable protein in the oil (Tattrie and Yaguchi, 1973). This information provides a strong basis for concluding that Bollgard cottonseed oil poses no significant allergenic concerns, based solely on lack of significant exposure.

Lack of Sequence Similarity of CRY1AC and NPTII Proteins to Known Allergens

Although there are no single predictive bioassays available to assess the allergenic potential of proteins in humans (U.S. FDA, 1992), the physicochemical and human exposure profile of the protein provides a basis for assessing potential allergenicity by comparing it against known protein allergens. Thus, important considerations contributing to the allergenicity of proteins ingested orally include exposure and an assessment of the factors that contribute to exposure, such as stability to digestion, prevalence in the food, and consumption pattern (amount) of the specific food (Metcalfe *et al.*, 1996; Kimber *et al.*, 1999).

A key parameter contributing to the systemic allergenicity of certain food proteins is stability to gastrointestinal digestion, especially stability to acid proteases like pepsin, found in the stomach (FAO, 1995; Astwood *et al.*, 1996; Astwood and Fuchs, 1996; Fuchs and Astwood, 1996; Kimber *et al.*, 1999). Important food allergens tend to be stable to peptic digestion and the acidic conditions of the stomach if they are to reach the intestinal mucosa where an immune response can be initiated. As noted earlier, the *in vitro* assessment of the digestibility of Cry1Ac and NPTII proteins showed that these proteins and their degradation products are readily digested.

Another significant factor contributing to the allergenicity of certain food proteins is their high concentrations in foods (Taylor *et al.*, 1987; Taylor, 1992; Fuchs and Astwood, 1996). Most allergens are present as major protein components in the specific food, representing from 2 to 3% up to 80% of total protein (Fuchs and Astwood, 1996). In contrast, the Cry1Ac and NPTII proteins are present at low levels in Bollgard cotton plants and are not detectable in the components of cotton that are used for food.

It is also important to establish that the protein does not represent a previously described allergen and does not share potentially cross-reactive amino acid sequence segments or structure with a known allergen. An efficient way to determine whether the added protein is an allergen or is likely to contain cross-reactive structures is to compare the amino acid sequence with that of all known allergens. A database of protein sequences associated with allergy and celiac disease has been assembled from publicly available genetic databases (GenBank, EMBL, PIR, and SwissProt). The amino acid sequences of the Cry1Ac and NPTII proteins were compared with these sequences, and neither showed any meaningful amino acid sequence similarity to the known allergens (Astwood *et al.*, 1996).

In addition, the NPTII protein has been approved by the U.S. Food and Drug Administration as a processing aid food additive for tomato, cotton, and canola (U.S. FDA, 1994) and was exempted from the requirement of a tolerance as an inert ingredient by the U.S. Environmental Protection Agency (U.S. EPA, 1994). These approvals included an assessment of potential allergenic effects for the NPTII protein, and both agencies concluded that there were no significant concerns.

In summary, these data and analyses support the conclusion that Cry1Ac and NPTII proteins are not detectable in cotton products used for human food, do not pose a significant allergenic risk, are not derived from allergenic sources, do not possess immunologically relevant sequence similarity with known allergens, and do not possess the following characteristics of known protein allergens as described in Taylor (1992) and Taylor *et al.* (1987).

Characteristic	Allergens	Cry1Ac	NPTII
Stable to digestion	Yes	No	No
Stable to processing	Yes	No	No
Similarity to known allergens	Yes	No	No
Prevalent protein in food	Yes	No	No

The conclusion regarding the safety of Cry1Ac and NPTII in human foods is supported by the lack of any reports of sensitization to the commercial microbial formulations and the lack of allergic concerns with the Cry proteins (McClintock *et al.*, 1995).

COMPOSITIONAL ANALYSIS AND NUTRITIONAL ASSESSMENT OF BOLLGARD COTTON

Compositional analyses were conducted to assess the levels of key nutrients and anti-nutrients in cotton to learn whether there were any changes in these components in comparison to conventional cotton varieties, including the parental control from which Bollgard was generated. These data, which demonstrate that Bollgard cotton is compositionally equivalent to conventional cotton, are summarized in Tables 3 to 8. The components important for food and feed uses were assessed. These analyses included the following:

- *Proximate analysis*: protein, fat, ash, water, carbohydrate, calories (Table 3).
- *Fatty acid profile*: total lipid content, percentage of individual fatty acids (Table 4).
- *Amino acid composition*: percentage of individual amino acids (Table 5).
- *Levels of three antinutrients*: gossypol, and cyclopropenoid fatty acids and aflatoxin (Tables 6 and 7)

Results of these analyses (Tables 3–6) demonstrate that seed from Bollgard cotton is compositionally equivalent to, and as nutritious as, seed from the parental cotton variety and other commercial cotton varieties (Berberich *et al.*, 1996).

In addition to seed, the refined oil from Bollgard cotton was shown to be equivalent to products produced from the control cotton cultivar (Table 7). The refined oil was evaluated for fatty acid profile (including cyclopropenoid fatty acids), free and total gossypol content, and tocopherol levels. The fatty acid profile was typical of commercial cottonseed oil. Free gossypol was reduced to undetectable levels after processing, and the tocopherol levels were comparable to levels found in commercial cottonseed oil.

Toasted cottonseed meal was evaluated for free gossypol and total gossypol content. After toasting, free gossypol was reduced to levels acceptable for feed use in both the Bollgard and conventional cotton varieties. The total gossypol content was within the acceptable range to allow the meal to be used as a protein supplement for feed (Table 8). Linters, the shorter fibers associated with the seed after ginning, are composed primarily of cellulose and are highly processed for both chemical and nonchemical uses. Yields of linters

Table 3

Proximate Analysis of Cottonseed from Bollgard Cotton Event 531 and Coker 312 (1993 Field Trials)

	C312		Bollgard 531		Literature range/mean value	Ref.
Characteristic[a]	Mean[b]	(Range)[c]	Mean[d]	(Range)[c]		
Protein, %	27.00	(23.3–28.4)	27.56	(22.8–31.0)	18.8–22.9	Turner *et al.*, (1976)
					23.5–29.5	Cherry *et al.* (1978a)
					12–32	Kohel, *et al.* 1985.
Fat, %	22.96	(19.6–25.1)	23.23	(22.2–25.8)	23.2–25.7	Cherry *et al.* (1978b).
					21.4–26.8	Cherry *et al.* (1978a)
Ash, %	4.63	(4.3–5.0)	4.53	(3.9–4.7)	4.1–4.9	Cherry *et al.* (1978b)
					3.8	Belyea *et al.* (1989)
Carbohydrate, %	45.40	(42.8–47.6)	44.68	(42.0–46.7)		
Calories/100 g	496.32	(479–508)	498.11	(495–511)		
Moisture, %	12.36	(9.6–15.9)	13.43	(11.2–14.7)	5.4–10.1	Cherry *et al.* (1978a)

[a]Protein, fat, ash, carbohydrate, and calories reported as percentage of dry sample weight.
[b]Value reported is least-squares mean of five samples, one from each field site where bulk seed from cultivar C312 was collected.
[c]Range denotes the lowest and highest individual values across sites for each cultivar or event.
[d]Value reported is least squares mean of four samples, one from each field site where bulk seed from Bollgard cotton event 531 was collected. No statistically significant differences from the Coker 312 control.
Source: Berberich *et al.* (1996).

from Bollgard cotton were comparable to those of the control and to ranges reported for other cotton varieties (Table 9). Therefore, insertion of the DNA including the *cry1Ac* coding sequence in the cotton genome did not alter the processing characteristics of the cottonseed.

In summary, a detailed compositional analysis has been performed on the Bollgard cotton event 531 (Berberich *et al.*, 1996). Bollgard event 531 does not differ in composition from the parent or commercial cultivars regarding nutrients such as total protein, fat, fiber, ash, and amino acids, demonstrating that there is no significant quantitative or qualitative difference between Bollgard cotton event 531 and the cotton cultivar from which it was derived, or values for other reported varieties, with regard to these components. Therefore Bollgard cottonseed is substantially equivalent in nutritional value to other cotton varieties currently available.

Table 4
Lipid and Fatty Acid Composition of Cottonseed from Bollgard Event 531 and Coker 312 (1993 Field Trials)

Component	C312[a,b]		Bollgard 531[a,b]	
	Mean	Range[c]	Mean	Range
Lipid	33.5	30.9–35.5	33.5	30.8–35.9
Myristic acid (14:0)	0.94	0.67–1.07	0.88	0.75–0.98
Pentadecanoic acid (15:0)	0.40	0.32–0.60	0.62	0.32–0.90
Palmitic acid (16:0)	26.5	24.8–27.8	26.3	25.1–27.2
Palmitoleic acid (16:1)	0.64	0.48–0.71	0.61	0.54–0.64
Margaric acid (17:0)	0.16	0.13–0.20	0.18	0.14–0.27
Stearic acid (18:0)	2.63	2.32–3.26	2.90	2.71–3.26
Oleic acid (18:1)	15.3	14.8–16.0	16.8	14.8–19.1
Linoleic acid (18:2)	47.8	46.4–49.9	45.6	41.6–49.0
Linolenic acid (18:3)	0.20	0.13–0.29	0.14	0.13–0.18
Arachidic acid (20:0)	0.29	0.26–0.31	0.28	0.22–0.33
Behenic acid (22:0)	0.15	0.12–0.17	0.14	0.13–0.15
Malvalic acid (C-17)	0.37	0.22–0.45	0.38	0.23–0.47
Sterculic acid (C-18)	0.59	0.48–0.70	0.62	0.54–0.69
Dihydrosterculic acid (C-19)	0.36	0.29–0.50	0.49	0.24–0.84

[a]Value of lipid is percentage of dry sample weight. Value of fatty acid is percentage of total lipid.
[b]Values presented are least squares mean and ranges of five samples for C312 and four samples for Bollgard event 531.
[c]Range denotes the lowest and highest individual value across sites for each variety. No statistically significant differences from the Coker 312 control
Source: Berberich *et al.* (1996).

In addition to the compositional studies, the nutritional wholesomeness of seed from Bollgard cotton was demonstrated by feeding rats and dairy cows diets that contained raw cottonseed from both the Bollgard cotton and control cotton cultivars. At completion of the rat study, there were no significant differences in weight gain or feed intake between rats consuming Bollgard cotton and the control cotton diet (Table 10). Results of the dairy cow study showed that the cottonseed from Bollgard cotton performed comparable to the control cottonseed (Table 11). There were no significant differences in milk yield, milk composition, and body condition score of the cows (Castillo *et al.*, 2001). The results from these studies led to the conclusion that the newly expressed proteins and the cottonseed containing these

Table 5

Amino Acid Composition of Cottonseed[a] from Bollgard Cotton Event 531 and Coker 312 (1993 Field Trials)

Amino acid	Literature Max[b]	Literature Min[b]	C312[c]	Bollgard 531[d]
Aspartic acid	9.5	8.8	9.72	9.49
Threonine	3.2	2.8	3.40	3.42
Serine	4.4	3.9	4.62	4.67
Glutamic acid	22.4	20.5	19.56	18.21*
Proline	4.0	3.1	4.22	4.03
Glycine	4.5	3.8	4.32	4.18
Alanine	4.2	3.6	4.12	4.03
Cysteine	3.4	2.3	1.60	1.68
Valine	4.7	4.3	4.50	4.09*
Methionine	1.8	1.3	1.48	1.94*
Isoleucine	3.4	3.0	3.26	3.02*
Leucine	6.1	5.5	5.98	5.93
Tyrosine	3.3	2.8	2.92	3.08*
Phenylalanine	5.6	5.0	5.32	5.28
Lysine	4.1	3.9	4.50	4.73*
Histidine	2.8	2.6	2.72	2.94*
Arginine	12.3	10.9	11.20	11.68
Tryptophan	1.4	1.0	1.04	1.00

[a]Amino acids reported as milligrams per kilogram of dry weight of protein in the cottonseed.
[b]Lawhon, *et al.* (1977): only one relevant reference could be cited, resulting in a narrow range for comparison that probably is not representative of variation in levels of amino acids between different cotton varieties.
[c]Value reported is least-squares mean of five samples, one from each field site where bulk seed from cultivar C312 was collected.
[d]Value reported is least-squares mean of four samples, one from each field site where bulk seed from Bollgard event 531 was collected.
*Significantly different from the control cultivar C312, at the 5% level (paired *t* test). Additional years' data have shown that these differences are not consistent across years and in different genetic backgrounds, indicating that differences observed in 1993 are not attributable to the genetic trait.
Source: Berberich *et al.* (1996).

proteins pose no safety concerns. Furthermore, Bollgard cotton has been planted on a total of more than 17 million acres commercially since 1996 with no reports of differences in animal feed performance.

Table 6

Aflatoxin and Gossypol Levels Determined in Cottonseed from Bollgard Cotton Event 531 and Coker 312 (1993 Field Trials)

Variety	Total gossypol (%)[a]		Aflatoxin
	Mean	Range	
C312	1.16[b]	(0.97–1.43)[b]	ND[c]
Bollgard 531	1.10	(0.86–1.29)	ND

[a]Expressed as percent dry weight of seed: literature range is 0.39–1.7% (Berardi and Goldblatt, 1980).
[b]Values reported for seed samples are the least-squares mean (from statistical analyses); ranges represent the lowest and highest values among six samples per variety, one sample per site where bulk seed samples collected.
[c]Not detected at a limit of detection of 1 ppb.
Source: Berberich *et al.* (1996).

Table 7

Fatty Acid[a] Profile of Refined Oil from Bollgard Cotton Event 531 and Coker 312 (1993 Field Trials)

Fatty acid	Literature range	Refined oil (% of total fatty acids)	
		Coker 312	Bollgard 531
Myristic (14:0)	(0.5–2.5)[b] (0.68–1.16)[c]	0.98	0.77
Palmitic (16:0)	(17–29)[b] (21.63–26.18)[c]	25.42	25.08
Palmitoleic (16:1)	(0.5–1.5)[b] (0.56–0.82)[c]	0.64	0.58
Stearic (18:0)	(1.0–4.0)[b] (2.27–2.88)[c]	2.53	2.67
Oleic (18:1)	(13–44)[b] (15.17–19.94)[c]	14.92	15.89
Linoleic (18:2)	(33–58)[b] (49.07–57.64)[c]	50.27	50.88
Linolenic (18:3)	(0.1–2.1),[b](0.23)[d]	0.16	0.17
Arachidic (20:0)	(<0.5),[b] (0.41)[d]	0.21	0.30
Behenic (22:0)	(<0.5)[b]	0.12	0.13
Malvalic (C-17)	(0.22–1.44)[e]	0.36	0.46

(*continues*)

Table 7 (*continued*)

Sterculic (C-18)	(0.08–0.56)[e]	0.48	0.43
Dihydrosterculic (C-19)	NA	0.22	0.16
Total gossypol[f]	0.01% (1 ppm)[c]	0.09	ND
Free gossypol[f]	0.01% (1 ppm)[c]	ND	ND
α-Tocopherol[g]	136–660[h]	638	568

[a]Reported as a percent of total lipids.
[b]Ranges adopted by the FAO/WHO Codex Alimentarius Committee on Fats and Oils (Jones and King, 1993).
[c]Cherry and Leffler (1984).
[d]Cherry (1983).
[e]Jones and King (1993); values reported for crude cottonseed oil.
[f]Free and total gossypol are reported as percent weight.
[g]α tocopherol reported as milligrams per kilogram of oil.
[h]Rossel (1991); Dicks (1965).
ND, not detected; NA, not available.
Source: Berberich *et al*., 1996.

Table 8

Total and Free Gossypol Levels Determined in Raw Cottonseed Meal and in Toasted Meal from Bollgard Cotton Event 531 and Coker 312[a]

	Gossypol (%)	
Meal	Total	Free
Raw		
C312	1.06	0.667
Bollgard 531	1.05	0.687
Toasted		
C312	1.11	0.011
Bollgard 531	0.87	0.008

[a]Values were obtained from analysis of one composite sample comprising of seed from all field sites where bulk cottonseed was collected in the 1993 field test.
Source: Berberich *et al.* (1996).

Table 9

Yield of Linters Fractions from Processing Cottonseed

	Yield		
	lbs	%	Across Cultivars (%)
C312	5.3	11.8	9.9–12.4,[a] 8.4[b]
Bollgard 531	7.5	14.7	

[a]Cherry and Leffler (1984).
[b]NCPA (1989).

Table 10

Summary of Rat Weight Gain in One-Month Rat Feeding Study with Bollgard Cottonseed Meal[a]

	Body weight (g)			
	Male		Female	
	Pretest	28 days	Pretest	28 days
C312				
Mean	174.5	321.9	140.0	191.9
SD	10.29	18.87	6.51	10.31
Sample size	10	10	10	10
Bollgard 531				
Mean	174.8	323.8	140.9	189.3
SD	10.38	14.39	8.15	10.31
Sample size	10	10	10	10

[a]Dietary concentration of approximately 10%.

Table 11

Effect of Bollgard Cottonseeds (DP50B) on Daily Dry Matter Intake (DMI), Milk Yield, Milk Composition, and Body Condition Score of Dairy Cows.[a]

Parameter	DP50	DP50B
DMI, kg/day	23.4	23.8
Cottonseed, kg/day	2.25	2.29
Milk yield, kg/day	26.9	26.7
Milk composition		
Fat, %	3.59	3.60
Protein, %	3.16	3.14
Lactose, %	4.97	5.01
Non-fat solids, %	8.84	8.92
N-Urea mg/100ml	18.77	19.49
Body condition score (scale of 1 to 5)	2.30	2.30

[a]Analysis based on 4×4 latin square design with four treatments. No significant differences ($p < 0.05$) were observed between treatments.
Source: Modified from Castillo *et al.* (2001).

HORIZONTAL GENE TRANSFER AND THE ASSESSMENT OF MARKER GENES

Horizontal gene transfer is defined as the transfer of DNA from one species to another. With respect to crop plants that are developed through biotechnology, a number of assessments have been performed to evaluate the possibility that antibiotic resistance marker genes used to facilitate the selection of the transformed plants might be transferred to bacteria either in the field or in animals that have consumed the crop. The reason for the assessment is that some species of bacteria found in soil, in the rumen, or in the intestine can receive DNA from other organisms through three mechanisms of transfer (Morrison, 1996; Davison, 1999). However, only one mechanism, transformation, is relevant to the possible transfer of DNA from plants to bacteria and subsequent expression of the encoded protein product. The other two mechanisms, conjugation (exchange of plasmid DNA between compatible bacteria) and transduction (viral transfer of DNA into bacteria) are specific to restricted forms of transfer and are not relevant to the potential transfer of DNA from plants (Thomson, 2000). In general, bacterial species differ markedly in their ability to accept DNA from the environment, and the frequency of transformation even under ideal circumstances is very low. The DNA that was transferred into cotton to produce Bollgard cotton was incorporated into the genomic DNA of the plant and represents a small fraction of cotton genome. The probability that a bacterium would take up the marker genes from the transformation is the same as from any other randomly chosen piece of DNA from the plant.

Horizontal Gene Transfer in the Field

The factors affecting possible "horizontal" gene transfer between genetically modified plants expressing antibiotic resistance marker genes and microorganisms in the environment have been extensively studied (Prins and Zadoks, 1994; Schlüter *et al.*, 1995; Nielsen *et al.*, 1998; Smalla *et. al.*, 2000). To date, there is no experimental evidence that any antibiotic resistance marker gene from a plant has transformed a bacterium either under laboratory conditions or in the field (Schlüter *et al.*, 1995; Broer *et al.*, 1996; Nielsen *et al.*, 1997). Most bacteria in natural environments are not competent to accept DNA. Even under laboratory conditions, studies specifically designed to detect the transfer of functional marker genes from plants into bacteria have failed to demonstrate such an occurrence.

Horizontal Gene Transfer from Food and Feed Products

In addition to investigations in the field environment, several studies have addressed the potential for the horizontal transfer of antibiotic selectable marker genes from transgenic plants to microflora in the gut of humans, ruminants, or other animals. The probability of this event occurring is virtually zero (Prins and Zadoks, 1994; Schlüter *et al.*, 1995; Nielsen *et al.*, 1998, Beever and Kempe, 2000).

If a marker gene were to be transferred, an important question would be whether there is any added risk regarding the abundance of antibiotic-resistant bacteria. Recently, Smalla *et al.* (2000) published a thorough review of the potential hazard associated with horizontal gene transfer of an antibiotic-resistant marker from a plant to a microorganism and concluded that "it is unlikely that antibiotic resistance genes used as markers in transgenic crops will contribute significantly to the spread of antibiotic resistance in bacterial populations." As such, the risk associated with an antibiotic resistant marker in a modified crop is considered to be minimal.

Bollgard cotton contains two antibiotic marker genes, *aad* and *npt*II. The *aad* gene was isolated from transposon Tn7, commonly found in Gram-negative bacteria (Shaw *et al.*, 1993). If somehow acquired by a gut bacterium, the *aad* gene would have no selective advantage in the absence of spectinomycin or streptomycin. The AAD protein is ubiquitous in nature and therefore is consumed as part of our natural diet. Even if the AAD protein were present in the gut, it could not compromise the therapeutic efficacy of these antibiotics, since to function, the AAD enzyme needs specific cofactors at appropriate concentrations that are not found in the gut. Databases of protein sequences were screened with the AAD amino acid sequence; no similarities to known toxins or allergens were revealed (Kärenlampi, 1996). The *npt*II gene was isolated from transposon Tn5, which is found in a number of Gram-negative bacteria, including strains that naturally colonize the human gut (Kärenlampi, 1996). Additionally, if the *npt*II gene was transferred to a microbe, it would not be expressed unless it was integrated into a region containing a bacterial promoter as the *npt*II gene is regulated by a plant promoter.

The origin of replication for plasmid maintenance at high copy number in *E. coli*, ori-322, contained on the plasmid PV-GHBK04 that was used for transformation, was not transferred into the cotton plant genome. Therefore the antibiotic resistance genes in Bollgard cotton cannot be mobilized by excision of the marker gene with other inserts to create a functional plasmid. To replicate and be passed on through reproduction, the DNA would have to be integrated into the recipient's genome or plasmid.

The question of the transfer of antibiotic resistance marker genes was discussed in detail by scientific experts in the European Union in relation to an application to market an insect-protected maize under Directive 90/220 and at a seminar organized by the Biomolecular Engineering Commission and the Genetic Engineering Commission. The European Commission requested the opinion of three scientific committees, which focused in particular on the risks of transfer of the *bla* gene, which confers ampicillin resistance to bacteria in the gastrointestinal tract of humans and animals. The scientific committees concluded that "(a) the possibility of transfer of a functional *bla*-gene construct" is virtually zero, and (b) "If the virtually impossible event occurred, it would have no clinical significance."[1] A similar conclusion was reached by Salyers (1998).

ENVIRONMENTAL ASSESSMENT

COTTON

Cotton is of the genus *Gossypium*, of the tribe Gossypieae, and of the family Malvaceae. Four species of cotton are of agronomic importance worldwide: the two diploid Asiatic species, *G. arboreum* and *G. herbaceum*, and the two allotetraploid New World species, *G. barbadense* and *G. hirsutum*. Although the diploid species remain important in restricted areas of India, Asia, and Africa, the two New World species account for approximately 98% of world cotton fiber production. Wild species of *Gossypium* typically occur in arid parts of the tropics and subtropics. Wild populations *of G. hirsutum* are relatively rare and tend to be widely dispersed.

OUTCROSSING POTENTIAL

Cotton is predominantly a self-pollinating crop, but it can be cross-pollinated by certain insects. However, outcrossing of the *cry1Ac* gene from Bollgard cotton to other *Gossypium* species or to others of the Malvaceae family is extremely unlikely for the following reasons (Percival *et al.*, 1999):

- Cultivated cotton is an allotetraploid and is incompatible with cultivated or wild diploid cotton species; therefore, it cannot cross and produce fertile offspring.
- Although outcrossing to wild or feral allotetraploid *Gossypium* species can occur, commercial cotton production generally does not occur in the geographical locations in which the wild relatives grow. For example, out-

[1] For the complete text, see the EV Website: http://europa.eu.int/comm/food/fs/sc/oldcomm6/out01_en.html

crossing to *G. tomentosum* in Hawaii is possible, but no commercial cotton is grown in Hawaii.

- There are no identified noncotton plants that are sexually compatible with cultivated cotton.

If the *cry1Ac* gene were to be transferred to a wild population of a tetraploid cotton species, and if this was considered undesirable, the size of the plants, their perennial growth habit, their restricted habitat, and their low natural fecundity would make them easy to control. Crossing of the insect protection trait into other cultivated cotton genotypes is possible, should the plants be in close proximity; however, studies have shown that this occurs at a very low frequency and is not considered to be a concern because it is unlikely to cause any adverse impact to the environment (Green and Jones, 1953; Mehetre, 1992).

Agronomic Performance

Field test data concerning yields and visual observations of agronomic properties including susceptibility to diseases and insects indicate that the agronomic performance of Bollgard 531 cotton does not differ from that of nonmodified varieties. Any plant pest risk to other plants and the environment that might be posed by Bollgard cotton is no different from risks associated with conventional cotton varieties. This is demonstrated in the subsections that follow.

Weediness Potential

Bollgard cotton does not have any weediness characteristics different from those of other conventional cotton varieties. Cotton is not considered to have weediness characteristics, such as seed dormancy, soil persistence, germination under diverse environmental conditions, rapid vegetative growth, a short life cycle, or high seed output and dispersal. Bollgard cotton exhibits no agronomic or morphological traits that differ from those of controls which would confer a competitive advantage over other species in the ecosystem in which it is grown. Also, there is little probability that any *Gossypium* species crossing with Bollgard cotton could become more weedy. All wild and feral relatives of cotton are tropical, woody, perennial shrubs other than a few herbaceous shrubs (Percival *et al.*, 1999). In most instances, the distribution of these species is determined by soil and climatic conditions. As perennials, the plants are not particularly programmed to produce seed each year. Based on these mechanistic arguments and field experience, there is no indication that insertion of the *cry1Ac* gene into the cotton genome would have any effect on the weediness traits of the cotton plant.

Lack of Effect on Nontarget Organisms

There is extensive information about microbial preparations of *Bacillus thuringiensis* subsp. *kurstaki* containing Cry proteins, including the Cry1Ac protein, to demonstrate that these proteins are nontoxic to nontarget organisms (U.S. EPA, 1988; Betz *et al.*, 2000). The literature has established that the Cry proteins are extremely selective for the lepidopteran insects, bind specifically to receptors on the midgut of lepidopteran insects, and have no deleterious effect on beneficial/nontarget insects.

To confirm and expand on results obtained for the microbial products that contain the same Cry1Ac protein as Bollgard cotton, the potential impact of the Cry1Ac protein on nontarget organisms was assessed on several representative organisms. The nontarget insect species included larvae and adult honey bee (*Apis mellifera* L.), a beneficial insect pollinator; green lacewing larvae (*Chrysopa carnea*), a beneficial predaceous insect commonly found on cotton and other cultivated crops; parasitic Hymenoptera (*Nasonia vitripennis*), a beneficial parasite of the housefly; the ladybird beetle (*Hippodamia convergens*), a beneficial predaceous insect that feeds on aphids and other plant bugs commonly found on stems and foliage of weeds and cultivated plants; and collembolan (*Folsomia candida* and *Xenylla grisea*) nontarget soil organisms (Betz *et al.*, 2000).

There were no deleterious effects on the growth and development of the insects and test organisms (Betz *et al.*, 2000). No effects were observed when *Folsomia candida* and *Oppia nitens* (Acari: Orbatidae) were fed transgenic cotton leaf material containing the Cry1Ac protein (Yu *et al.*, 1997; Betz *et al.*, 2000).

An assessment of potential impacts on birds present in Bollgard fields was conducted. Bobwhite quail chicks were fed a diet containing 10% raw cottonseed meal from Bollgard cotton and control cotton. This feeding level of cottonseed approximates consumption of 400 seeds/kg body weight per bird. There was no difference in the feed consumption or weight gain for chicks eating the diet with Bollgard cottonseed meal and those feeding on the control cottonseed meal (Betz *et al.*, 2000).

The purified Cry1Ac protein was administered orally, by gavage, to male and female mice at 500, 1000, and 4200 mg/kg body weight. The growth and feed consumption of the mice was unaffected by the Cry1Ac protein (Betz *et al.*, 2000). The levels of protein exposed to the mice represented a safety factor of more than 50,000 times the amount that a cow would consume when eating raw cottonseed.

Environmental Fate of CRY1AC Protein

The U.S. Department of Agriculture has conducted environmental assessments of Cry proteins and has issued findings of no significant impact (FONSI) for the Cry1Ac protein (USDA, 1995). Cry protein crystals have been found to degrade readily in the field as a result of solar radiation and temperature (Palm *et al.*, 1993, 1994, 1996).

The environmental fate of purified Cry proteins has been extensively studied. The published literature has demonstrated that Cry protein adsorption to soil is rapid and complete within 30 minutes (Venkateswerlu and Stotzky, 1992). Numerous other studies of the biodegradation and binding of Cry proteins in soil have been conducted, including Tapp *et al.*, (1994), Tapp and Stotzky (1995, 1998), Koskella and Stotzky (1997), and Crecchio and Stotzky (1998). These studies demonstrate that isolated Cry proteins could bind to clay particles and humic acids in artificial soil mixes.

The Cry1Ac protein levels were measured in whole mature plants obtained from field tests at the end of the 1992 and 1993 seasons. Those data were used to estimate the amount of Cry1Ac protein that would enter the environment after harvest when the plants are plowed into the soil. For the two years evaluated (1992 and 1993), the load per area of soil was estimated to be 1.44 and 0.6 g of Cry1Ac protein, respectively. Based on these values, an *in vitro* soil degradation study was conducted in which insecticidal activity was used to measure degradation of the protein. This study showed that the Cry1Ac protein was rapidly degraded in the soil in both the purified form of the protein and as part of the cotton plant tissue. The half-life of the Cry1Ac protein in plant tissue was calculated to be 41 days, which is comparable to the degradation rates reported for *Bt* microbial formulations (Betz *et al.*, 2000). The half-life for the purified protein was less than 20 days. These values are similar to the degradation rates observed by Palm *et al.*, (1993, 1994, 1996) for transgenic plants producing Cry proteins.

Summary

The use of Bollgard cotton has reduced the number and cost of insecticide applications needed for control of cotton bollworm, tobacco budworm, and pink bollworm, and a number of secondary benefits are offered as well. The introduced Cry1Ac protein is comparable to Cry proteins that have been safely used for over 40 years. Detailed food, feed, and environmental safety assessments confirm the safety of this product. These analyses included detailed molecular characterization of the introduced DNA, safety assessments of the expressed Cry1Ac and NPTII proteins, compositional analysis of cottonseed, oil, and meal, and environmental impact assessment of the Cry1Ac protein and Bollgard cotton plants. These studies demonstrate the Cry1Ac protein poses minimal risk to nontarget organisms, including humans, animals, and beneficial insects. Based on the available data and experience collected to date the risks to the environment posed by Bollgard cotton are comparable to or fewer than those posed by traditional cotton treated with commercially approved insecticides. Rather, the reduction of insecticide applications as a result of using Bollgard affords significant environmental benefits. Additionally, Bollgard cotton plants, cottonseed, cottonseed oil, and fiber were shown to be equivalent to conventional cotton varieties, hence as safe as the latter materials.

ACKNOWLEDGMENTS

The authors thank the following contributors: Jeannine G. Augustin, J. Austin Burns, David C. Kolwyck, Bibiana E. Ledesma, Mark W. Naylor, Thomas Nickson, Joel E. Ream, Andrew J. Reed, Steven Reiser, and James Surber.

REFERENCES

Astwood, J. D., and Fuchs, R. L. (1996). Food allergens are stable to digestion in a simple model of the gastrointestinal tract. *J. Allergy Clin. Immunol.*, **97**, 241.

Astwood, J. D., Leach, J. N., and Fuchs, R. L. (1996). Stability of food allergens to digestion *in vitro*. *Nature Biotechnol.*, **14**, 1269–1273.

Baum, J. A., Johnson, T. B., and Carlton, B. C. (1999). *Bacillus thuringiensis* natural and recombinant bioinsecticide products. in *Methods in Biotechnology*. Vol. 5, *Biopesticides: Use and Delivery*, F. R. Hall and J. J. Mean, eds., pp 189–209. Humana Press, Totowa, N. J.

Beever, D. E., and Kemp, C. F. (2000). Safety issues associated with the DNA in animal feed derived from genetically modified crops. A review of scientific and regulatory procedures. *Nutri. Abstr. Rev. Ser. B: Livest. Feeds* Feeding, **17**, 175–182.

Belyea, R. L., Steevens, B. J. Restrepo, R. J., and Clubb, A. P. (1989). Variation in composition of by-product feeds. *J. Dairy Sci.*, **72**, 2339–2345.

Berardi, L. C., and Goldblatt, L. A. (1980). Gossypol. in *Toxic Constituents of Foodstuffs*, 2nd ed., I. E. Liener, ed., pp. 184–237. Academic Press, New York.

Berberich, S. A., Ream, J. E., Jackson, T. L., Wood, R., Stipanovic, R., Harvey, P., Patzer, S., and Fuchs, R. L. (1996). The composition of insect-protected cottonseed is equivalent to that of conventional cottonseed. *J. Agric. Food Chem.*, **44**, 365–371.

Betz, F. S., Hammond, B. G., and Fuchs, R. L. (2000). Safety and advantages of *Bacillus thuringiensis*-protected plants to control insect pests. *Regul. Toxico. Pharmacol.*, **32**, 156–173.

Broer, I., Dröge-Laser, W., and Gerke, M. (1996). Examination of the putative horizontal gene transfer from transgenic plants to agrobacteria, in *Transgenic Organisms and Biosafety*. E. R. Schmidt, and T. Hankeln, eds., pp. 67–70. Springer-Verlag, Berlin, Heidelberg, and New York.

Bush, R. K., Taylor, S. L., Nordlee, J. A., and Busse, W. W. (1985). Soybean oil is not allergenic to soybean-sensitive individuals. *J. Allergy Clin. Immunol.*, **76**(2), 242–245.

Carpenter, J. E., and Gianessi, L. P. (2001). *Agricultural Biotechnology: Updated Benefit Estimates*. National Center for Food and Agricultural Policy, Washington, DC.

Castillo, A. R., Gallardo, M. R., Maciel, M., Giordano, J. M., Conti, G. A., Gaggiotti, M. C., Quaino, O., Gianni, C., and Hartnell, G. F. (2001). Effect of feeding dairy cows with either Bollgard, BollgardII, Roundup Ready or control cottonseeds on feed intake, milk yield and milk composition. *J. Dairy Sci.*, **84**,(suppl. 1), abstr. 1712.

Cherry, J. P. (1983). Cottonseed oil. *J. Am. Oil Chem. Soc.*, **60**(2), 360–367.

Cherry, J. P., and Leffler, H. R. (1984). Seed. in *Cotton*, R. J. Kohel, and C. F. Lewis, eds. Agronomy Series, No. 24, pp. 511–569. American Society of Agronomy, Crop Science Society of America, Soil Science Society of America, Madison, WI.

Cherry, J. P., Simmons, J. G., and Kohel, R. J. (1978a). Potential for improving cottonseed quality by genetic and agronomic practices, in *Nutritional Improvement of Food and Feed Proteins, M. Friedman, ed.*, pp 343–364. Plenum Press: New York.

Cherry, J. P., Simmons, J. G., and Kohel, R. J. (1978b). Cottonseed composition of national variety test cultivars grown at different Texas locations. in *Proceedings of the Beltwide Cotton Production Research Conference, Dallas, TX*, J. M. Brown, ed., pp. 47–50. National Cotton Council: Memphis, TN.

Crecchio, C. and Stotzky, G. (1998). Insecticidal activity and biodegradation of the toxin from *Bacillus thuringiensis* subsp. *kurstaki* bound to humic acids from soil. *Soil Biol. Biochem.*, **30**, 463–470.

Davison, J. (1999). Genetic exchange between bacteria in the environment. *Plasmid*, **42**, 73–91.

Dicks, M. W. (1965). Vitamin E Content of Foods and Feeds for Human and Animal Consumption. Bullet in 435, University of Wyoming Agricultural Experiment Station.

Doolittle, R. F., Feng, D. F., Anderson, K. L., and Alberro, M. R. (1990). A naturally occurring horizontal gene transfer from a eukaryote to a prokaryote. *J. Mol. Evol.*, **31**, 383–388.

Economic Research Service/USDA. (2000). Genetically engineered crops: Has adoption reduced pesticide use? *Agric. Outlook*, August, pp. 13–17.

Edge, J. M., Benedict, J. H., Carroll, J. P., and Reding, H. K. (2001). Bollgard cotton: An assessment of global economic, environmental, and social benefits. *J. Cotton Sci.*, **5**, 1–8.

English, L., and Slatin, S. L. (1992). Mode of action of delta-endotoxin from *Bacillus thuringiensis*: A comparison with other bacterial toxins. *Insect Biochem. Mol. Biol.*, **22**(1), 1–7.

FAO (1995). Report of the FAO Technical Consultation on Food Allergies, Rome, November 13–14, 1995. Food and Agriculture Organization, Rome.

Falck-Zepeda J. B., Traxler, G., and Nelson, R. G. (1998). Rent creation and distribution from biotechnology innovations: The case of Bt cotton and herbicide-tolerant soybeans in (1997). *Agribusiness*, **16**, 1–25. ISAAA Briefs No. 14. International Service for Agri-Biotech Appliations, Ithaca, NY.

Falck-Zepeda, J. B., Traxler, G., and Nelson, R. G. (2000). Surplus distribution from the introduction of a biotechnology innovation. *Am. J. Agric Econ.*, **82**, 360–369.

Fernandez-Cornejo, J., and McBride, W. D. (2000). Genetically engineered crops for pest management in U. S. agriculture: Farm level effects. Agricultural Economic Report 786, Economic Research Service/U. S. Department of Agriculture.

Fuchs, R. L., and Astwood, J. D. (1996). Allergenicity assessment of foods derived from genetically modified plants. *Food Technol.*, **50**, 83–88.

Fuchs, R. L., Ream, J. E., Hammond, B. G., Naylor, M. W., Leimgruber, R. M., and Berberich, S. A. (1993). Safety assessment of the neomycin phosphotransferase II (NPTII) protein. *Bio/Technology*, **11**, 1543–1547.

Gianessi, L. P., and Carpenter, J. E. (1999). Agricultural Biotechnology: Insect Control Benefits. National Center for Food and Agricultural Policy, Washington, DC. http://www.bio.org/food&ag/ncfap.htm

Green, J. M., and Jones, M. D. (1953). Isolation of cotton for seed increase. *Agron. J.*, **45**, 366–368.

Halsey, A. B., Martin, M. E., Ruff, M. E., Jacobs, F. O., and Jacobs, R. L. (1986). Sunflower oil is not allergenic to sunflower seed-sensitive patients. *J. Allergy Clin. Immunol*, **78**(3), 408–410.

Hofmann, C., Lüthy, P., Hutter, R., and Pliska. V. (1988). Binding of the delta endotoxin from *Bacillus thuringiensis* to brush-border membrane vesicles of the cabbage butterfly (*Pieris brassicae*). *Eur. J. Biochem.*, **173**, 85–91.

IPCS (2000). *Bacillius thuringiensis*. Environmental Health Criteria of the International Programme on Chemical Safety, No. 217.
http://www.who.int/pcs/docs/ehc_217.html

Jones, L. A., and King, C. C., eds. (1993). *Cottonseed Oil*. National Cottonseed Products Associations and the Cotton Foundation, Memphis, TN.

Kärenlampi, S. (1996). Health effects of marker genes in genetically engineered food plants. *TemaNord*, 530.

Kimber, I., Kerkvliet, N. I., Taylor, S. L., Astwood, J. D., Sarlo, K., and Dearman, R. J. (1999). Toxicology of protein allergenicity: Prediction and characterization. *Toxicol.* Sci. **48**, 157–162.

Klotz-Ingram, C., Jans, S., Fernandez-Cornejo, J., and McBride, W. (1999). Farm-level production effects related to the adoption of genetically modified cotton for pest management. *AgBioForum*, **2**(2), 73–84.

Kohel, R. J., Glueck, J., and Rooney, L. W. (1985). Comparison of cotton germplasm collections for seed-protein content. *Crop Sci.*, **25**, 961–963.

Koskella, J., and Stotzky., G. (1997). Microbial utilization of free and clay-bound insecticidal toxins from *Bacillus thuringiensis* and their retention of insecticidal activity after incubation with microbes. *Appl. Environ. Microbiol.*, **63**(9), 3561–3568.

Lawhon, J. T., Cater, C. M., and Mattil, K. F. (1977). Evaluation of the food use potential of sixteen varieties of cottonseed. *J. Am. Oil Chem. Soc.*, **54**, 75–80

Lüthy, P., Cordier, J. L., and Fischer, H. M. (1982). *Bacillus thuringiensis* as a bacterial insecticide: Basic considerations and applications, in *Microbial and Viral Pesticides*, E. Kurstak, ed., pp 35–74. Dekker, New York.

McClintock, J. T., Schaffer, C. R., and Sjoblad, R. D. (1995). A comparative review of the mammalian toxicity of *Bacillus thuringiensis*–based pesticides. *Pestic. Sci.*, **45**, 95–105.

Mehetre, S. S. (1992). Natural crossing in cotton (*Gossypium* sp): Its significance in maintaining variety purity and production of hybrid seed using male sterile lines. *J. Cotton Res. Dev.*, **6**(2), 73–97.

Metcalfe, D. D., Astwood, J. D., Townsend, R., Sampson, H. A. Taylor, S. L., and Fuchs, R. L. (1996). Assessment of the allergenic potential of foods derived from genetically engineered crop plants. *Criti. Revi. Food Sci. Nutri.*, **36**(suppl.), S165–S186.

Morrison, M. (1996). Do ruminal bacteria exchange genetic material? *J. Dairy Sci.*, **79**, 1476–1486.

NCPA. (1989). *Cottonseed and Its Products*, 9th ed. National Cottonseed Products Association Memphis, TN.

Nielsen, K. M., Gebhard, F., Smalla, K., Bones, A. M., and van Elsas J. D. (1997). Evaluation of possible horizontal gene transfer from trangenic plants to the soil bacterium *Acinetobacter calcoaceticus* Bd413. *Theor. Appl. Geneti.*, **95**(5/6), 815–821.

Nielsen, K. M., Bones, A. M., Smalla, K., and van Elsas, J. D. (1998). Horizontal gene transfer from transgenic plants to terrestrial bacteria – A rare event? *FEMS Microbiol. Revi.*, **22**, 79–103.

Palm, C. J., Seidler, R. J., Donegan, K. K., and Harris. D. (1993). Transgenic plant pesticides: Fate and persistence in soil. *Plant Physiol. Suppl.*, **102**, 166.

Palm, C. J., Donegan, K. K., Harris, D., and Seidler, R. J. (1994). Quantification in soil of *Bacillus thuringiensis* var. *kurstaki* delta-endotoxin from transgenic plants. *Mol. Ecol.*, **3**, 145–151.

Palm, C. J., Schaller, D. L., Donegan, K. K., and Seidler, R. J. (1996). Persistence in soil of transgenic plant produced *Bacillus thuringiensis* var. *kurstaki* delta-endotoxin. *Can. J. Microbiol.*, **42**, 1258–1262.

Percival, A. E., Wendel, J. F., and Stewart, J. M. (1999). Taxonomy and germplasm resources, in *Cotton: Origin, History, Technology, and Production.*, W. C. Smith, (ed.), pp 33–63. Wiley, New York.

Perlak, F. J., Oppenhuizen, M., Gustafson, K., Voth, R., Sivasupramaniam, S., Herring, D., Carey, B., Ihrig R. A., and Roberts, J. K. (2001). Development and commercial use of Bollgard cotton in the USA – Early promises versus today's reality. *Plant J.*, **27**, 489–501.

Prins, T. W., and Zadoks, J. C. (1994). Horizontal gene transfer in plants, a biohazard? Outcome of a literature review. *Euphytica* **76**, 133–138.

Rossell, J. B. (1991). Vegetable oils and fats, in *Analysis of Oilseeds, Fats, and Fatty Foods.* J. B. Rossell, and J. L. R. Pritchard eds., pp 261–327. Elsevier Applied Science Publisher, New York.

Salyers, A. (1998). Genetically engineered foods: safety issues associated with antibiotic resistance genes.
http://www.healthsci.tufts.edu/apua/salyersreport.htm

Schlüter, K., Fütterer, J., and Potrykus. I. (1995). "Horizontal" gene transfer from a transgenic potato line to a bacterial pathogen (*Erwinia chrysanthemi*) occurs – if at all-at an extremely low frequency. *Bio/Technology* **13**, 1094–1098.

Shaw, K. J., Rather, P. N., Hare, R. S., and Miller, G. H. (1993). Molecular genetics of aminoglycoside resistance genes and familial relationships of the aminoglycoside modifying enzymes. *Microbiol. Rev.*, **57**, 138–163.

Sims, S. R., Berberich, S. A., Nida, D. L., Segalini, L. L., Leach, J. N., Ebert, C. C., and Fuchs, R. L. (1996). Analysis of expressed proteins in fiber fractions from insect-protected and glyphosate-tolerant cotton varieties. *Crop Sci.*, **36**(5), 1212–1216.

Sjoblad, R. D., McClintock J. T., and Engler. R. (1992). Toxicological considerations for protein components of biological pesticide products. *Regul. Tox-col. Pharmacol.*, **15**, 3–9.

Smalla, K., Borin, S., Heuer, H., Gebhard, F., van Elsas, J. D., and Nielsen. K. (2000). Horizontal transfer of antibiotic resistance genes from transgenic plants to bacteria, in *Proceedings of the sixth International Symposium on the Biosafety of GMOs*, C. Fairbairn, G. Scoles, and A. McHughen, eds., pp. 146–154. University Extension Press, University of Saskatchewan, Saskatoon.

Tapp, H., and Stotzky, G. (1995). Insecticidal activity of the toxins from *Bacillus thuriengiensis* subsp. *kurstaki* and *tenebrionis* adsorbed and bound on pure and soil clays. *Appl. Environ. Microbiol.*, **61**(5), 1786–1790.

Tapp, H., and Stotzky, G. (1998). Persistence of the insecticidal toxin from *Bacillus thuriengiensis* subsp. *kurstaki* in soil. *Soil Biol. Biochem.*, **30**, 471–476.

Tapp, H., L. Calamai, L., and Stotzky. G. (1994). Adsorption and binding of the insecticidal proteins from *Bacillus thuringiensis* subsp. *kurstaki* and subsp. *tenebrionis* on clay minerals. *Soil Biol. Biochem.* **26**, 663–679.

Tattrie, N. H., and Yaguchi. M. (1973). Protein content of various processed edible oils. *J. Inst. Can. Sci. Technol. Aliment.*, **6**(4), 289–290.

Taylor, S. L. (1992). Chemistry and detection of food allergens. *Food Technol.*, **46**, 146–152.

Taylor, S. L., Busse, W. W. Sachs, M. I. Parker, J. L., and Yunginger, J. W. (1981). Peanut oil is not allergenic to peanut-sensitive individuals. *J. Allergy Clin. Immunol.*, **68**(5), 372–375.

Taylor, S. L., Lemanske, Jr., R. F. Bush, R. K., and Busse, W. W. (1987). Food allergens: Structure and immunologic properties. *Ann. Allergy*, **59**(5), 93–99.

Thomson, J. (2000). Gene Transfer: Mechanisms and Food Safety Risks. Topic 11, Biotech 00/13. Joint FAO/WHO Expert Consultation on Foods Derived from biotechnology. May 29–June 2, (2000). World Health Organisation, Geneva.

Traxler, G., and Falck-Zepeda. J. (1999). The distribution of benefits from the introduction of transgenic cotton varieties. *AgBioForum*, **2**(2), 94–98.

Turner, J. H., Ramey, Jr., H. H., and Worley S., Jr. (1976). Influence of environment on seed quality of four cotton cultivars. *Crop Sci.*, **16**, 407–409.

USDA. (1995). Availability of determination of non-regulated status for genetically engineered cotton. *Fed. Regis.*, **60**(134), 36096–36097.

U.S. EPA. (1988). Guidance for the Re-registration of Pesticide Products Containing *Bacillus thuringiensis* as the Active Ingredient. NTIS Publication 89–164198. National Technical Information Service, Springfield, VA.

U.S. EPA. (1994). Neomycin phosphotransferase II; tolerance exemption. *Fed. Regist.*, **59**(187), 49351–49353.

U.S. EPA. (1998). R. E. D. Facts: *Bacillus thuringiensis*. EPA 738-F-98-001.

U.S. FDA. (1992). Statement of policy: Foods derived from new plant varieties. *Fed. Regist.*, **57**(104), 22984–23005.

U.S. FDA. (1994). Secondary direct food additives permitted in food for human consumption; food additives permitted in feed and drinking water of animals; aminoglycoside 3-phosphotransferase II. *Fed. Regist.*, **59**(98), 26700–26711.

U.S. Pharmacopeia. (1995). United States Pharmacopeia, Vol. 23. U. S. Pharmacopeial Convention, Rockville, MD.

Venkateswerlu, G., and Stotzky. G. (1992). Binding of the protoxin and toxin proteins of *Bacillus thuringiensis* subsp. *kurstaki* on clay minerals. *Curr. Microbiol.*, **25**, 225–233.

Wier, A. T., Mullins, J. W., and Mills, J. M. (1998). Bollgard cotton – Update and economic comparisons including new varieties. *Proceedings of the Beltwide Cotton Conference*, Vol. 2, 1039–1040. National Cotton Council. Memphis, TN.

Xia, J. Y., Cui, J. J., Ma, L. H., Dong S. L., and Cui, X. F. (1999). The role of transgenic *Bt* cotton in integrated insect pest management. *Acra* Gossypii Sin., **11**, 57–64.

Yu, L., Berry R. E., and Croft, B. A. (1997). Effects of *Bacillus thuringiensis* toxins in transgenic cotton and potato on *Folsomia candida* (Collembola: Isotomidae) and *Oppia nitens* (Acari: Orbatidae). *J. Econ. Entomol.*, **90**(1), 113–118.

Subject Index

B

H

I

O

S

T